T0310654

STATISTICAL METHODS FOR CLIMATE SCIENTISTS

This book provides a comprehensive introduction to the most commonly used statistical methods relevant in atmospheric, oceanic, and climate sciences. Each method is described step-by-step using plain language, and illustrated with concrete examples, with relevant statistical and scientific concepts explained as needed. Particular attention is paid to nuances and pitfalls, with sufficient detail to enable the reader to write relevant code. Topics covered include hypothesis testing, time series analysis, linear regression, data assimilation, extreme value analysis, Principal Component Analysis, Canonical Correlation Analysis, Predictable Component Analysis, and Covariance Discriminant Analysis. The specific statistical challenges that arise in climate applications are also discussed, including model selection problems associated with Canonical Correlation Analysis, Predictable Component Analysis, and Covariance Discriminant Analysis. Requiring no previous background in statistics, this is a highly accessible textbook and reference for students and early career researchers in the climate sciences.

TIMOTHY M. DELSOLE is Professor in the Department of Atmospheric, Oceanic and Earth Sciences, and Senior Scientist at the Center for Oceanic Atmospheric, and Land Studies, at George Mason University, Virginia. He has published more than 100 peer-reviewed papers in climate science and served as co-editor-in-chief of the *Journal of Climate*.

MICHAEL K. TIPPETT is an associate professor at Columbia University. His research includes forecasting El Niño and relating extreme weather (tornadoes and hurricanes) with climate, now and in the future. He analyzes data from computer models and weather observations to find patterns that improve understanding, facilitate prediction, and help manage risk.

Includes both the mathematics and the intuition needed for climate data analysis.

–Professor Dennis L Hartmann, *University of Washington*

STATISTICAL METHODS FOR CLIMATE SCIENTISTS

TIMOTHY M. DELSOLE

George Mason University

MICHAEL K. TIPPETT

Columbia University

CAMBRIDGE
UNIVERSITY PRESS

CAMBRIDGE
UNIVERSITY PRESS

University Printing House, Cambridge CB2 8BS, United Kingdom

One Liberty Plaza, 20th Floor, New York, NY 10006, USA

477 Williamstown Road, Port Melbourne, VIC 3207, Australia

314-321, 3rd Floor, Plot 3, Splendor Forum, Jasola District Centre, New Delhi – 110025, India

103 Penang Road, #05–06/07, Visioncrest Commercial, Singapore 238467

Cambridge University Press is part of the University of Cambridge.

It furthers the University's mission by disseminating knowledge in the pursuit of education, learning, and research at the highest international levels of excellence.

www.cambridge.org
Information on this title: www.cambridge.org/9781108472418
DOI: 10.1017/9781108659055

© Cambridge University Press 2022

This publication is in copyright. Subject to statutory exception
and to the provisions of relevant collective licensing agreements,
no reproduction of any part may take place without the written
permission of Cambridge University Press.

First published 2022

A catalogue record for this publication is available from the British Library.

Library of Congress Cataloging-in-Publication Data
Names: DelSole, Timothy M., author.
Title: Statistical methods for climate scientists / Timothy M. DelSole and Michael K. Tippett.
Description: New York : Cambridge University Press, 2021. | Includes
bibliographical references and index.
Identifiers: LCCN 2021024712 (print) | LCCN 2021024713 (ebook) |
ISBN 9781108472418 (hardback) | ISBN 9781108659055 (epub)
Subjects: LCSH: Climatology–Statistical methods. | Atmospheric
science–Statistical methods. | Marine sciences–Statistical methods. |
BISAC: SCIENCE / Earth Sciences / Meteorology & Climatology
Classification: LCC QC866 .D38 2021 (print) | LCC QC866 (ebook) |
DDC 551.601/5118–dc23
LC record available at https://lccn.loc.gov/2021024712
LC ebook record available at https://lccn.loc.gov/2021024713

ISBN 978-1-108-47241-8 Hardback

Additional resources for this publication at www.cambridge.org/9781108472418.

Cambridge University Press has no responsibility for the persistence or accuracy of
URLs for external or third-party internet websites referred to in this publication
and does not guarantee that any content on such websites is, or will remain,
accurate or appropriate.

Contents

Preface

This book provides an introduction to the most commonly used statistical methods in atmospheric, oceanic, and climate sciences. The material in this book assumes no background in statistical methods and can be understood by students with only a semester of calculus and physics. Also, no advanced knowledge about atmospheric, oceanic, and climate sciences is presumed. Most chapters are self-contained and explain relevant statistical and scientific concepts as needed. A familiarity with calculus is presumed, but the student need not solve calculus problems to perform the statistical analyses covered in this book.

The need for this book became clear several years ago when one of us joined a journal club to read "classic" papers in climate science. Specifically, students in the club had difficulty understanding certain papers because these papers contained unfamiliar statistical concepts, such as empirical orthogonal functions (EOFs), significance tests, and power spectra. It became clear that our PhD curriculum was not adequately preparing students to be "literate" in climate science. To rectify this situation, we decided that students should take a statistics class. However, at that time, there did not exist a single self-contained course that covered all the topics that we considered to be essential for success in climate science. Therefore, we designed a single course that covered these topics (which eventually expanded into a two-semester course). This book is based on this course and embodies over a decade of experience in teaching this material.

This book covers six key statistical methods that are essential to understanding modern climate research: (1) hypothesis testing; (2) time series models and power spectra; (3) linear regression; (4) Principal Component Analysis (PCA), and related multivariate decomposition methods such as Canonical Correlation Analysis (CCA) and Predictable Component Analysis, (5) data assimilation; and (6) extreme value analysis. Chapter 1 reviews basic probabilistic concepts that are used throughout the book. Chapter 2 discusses hypothesis testing. Although the likelihood ratio provides a general framework for hypothesis testing, beginners often find this framework daunting. Accordingly, Chapter 2 explains hypothesis testing based on heuristic arguments for Gaussian distributions, which most students find intuitive. The framework discussed in Chapter 2 provides the foundation for hypothesis testing that is used in the rest of the book. The related concept of confidence intervals, as well as bootstrap methods and distribution-free tests, is discussed in Chapters 3 and 4. Fundamental concepts in time series analysis, especially stochastic processes and power spectra, are discussed in Chapters 5 and 6, respectively. Certain topics that typically are included in statistical texts are omitted because they are seldom used in climate science; for instance, moving average models are not discussed in detail because they are used much less often in climate science than autoregressive models.

The second half of this book covers multivariate methods. We have striven to convey our hard-learned experience about these methods collected over many years. Basic concepts in linear algebra and multivariate distributions are outlined in Chapter 7. Linear regression is discussed in Chapters 8 and 9. Pitfalls in linear regression are discussed in detail, especially model selection (Chapter 10) and screening (Chapter 11). These concepts are critical for proper usage and interpretation of statistical methods, especially in statistical prediction, but are not easy to find in introductory texts. Principal Component Analysis is the most commonly used multivariate method in climate science, hence our discussion in Chapter 12 is very detailed. Subsequent chapters discuss field significance (Chapter 13), Multivariate Linear Regression (Chapter 14), Canonical Correlation Analysis (Chapter 15), Covariance Discriminant Analysis (Chapter 16), Analysis of Variance (Chapter 17), and Predictable Component Analysis (Chapter 18). An introduction to extreme value theory is provided in Chapter 19. Data assimilation and ensemble square root filters are discussed in Chapters 20 and 21 with the goal of introducing essential ideas and common practical problems that we believe every user of data assimilation products should be aware of.

This book is designed for either a one-semester or a two-semester course. Considerable effort has been made to select and arrange the material in a logical order that facilitates teaching and learning. We have used this book to teach a one-semester course covering Chapters 1–13 at approximately one chapter per week. For more advanced students, a second-semester course is offered covering Chapters 14–21. The homework sets are available at the Cambridge University Press website associated with this book.

The multivariate part of this book is distinguished from previous books in an important way. Typical climate data sets are much bigger in the spatial dimension than in the time dimension. This creates major difficulties for applying such multivariate techniques as Canonical Correlation Analysis, Predictable Component Analysis, and Covariance Discriminant Analysis to climate data, although these difficulties are rarely discussed in standard statistics texts. In the climate literature, the standard approach to this problem is to apply these techniques to a few principal components of the data, so that the time dimension is much bigger than the state dimension. The most outstanding barrier in this approach is choosing the number of principal components. Unfortunately, no standard criterion for selecting the number of principal components exists for these multivariate techniques. This gap was sorely felt each time this material was taught and motivated us to conduct our own independent research into this problem. This research culminated in the discovery of a criterion that was consistent with standard information criteria and could be applied to all of the problems discussed in this book. For regression models and CCA, this criterion is called Mutual Information Criterion (MIC) and is introduced in Chapter 14 (for full details, see DelSole and Tippett, 2021a). After formulating this criterion, we discovered that it was consistent with many of the criteria derived by Fujikoshi et al. (2010) based on likelihood ratio methods, which supports the soundness of MIC. However, MIC is considerably easier to derive and apply. We believe that MIC will be of wide interest to statisticians and to scientists in other fields who use these multivariate methods.

The development of this book is somewhat unique. Initially, we followed our own personal experience by giving formal lectures on each chapter. Inspired by recent educational research, we began using a "flipped classroom" format, in which students read each chapter and sent questions and comments electronically *before* coming to class. The class itself was devoted to going over the questions/comments from students. We explicitly asked students to tell us where the text failed to help their understanding. To invite feedback, we told students that we needed their help in writing this book, because over the ten years that we have been teaching this topic, we have become accustomed to the concepts and could no longer see what is wrong with the text. The resulting response in the first year was more feedback than we had obtained in all the previous years combined. This approach not only revolutionized the way we teach this material but gave us concrete feedback about where precisely the text could be improved. With each subsequent year, we experimented with new material and, if it did not work, tried different ways. This textbook is the outcome of this process over many years, and we feel that it introduces statistical concepts much more clearly and in a more accessible manner than most other texts.

Each chapter begins with a brief description of a statistical method and a concrete problem to which it can be applied. This format allows a student to quickly ascertain if the statistical method is the one that is needed. Each problem was chosen after careful thought based on intrinsic interest, importance in real climate applications, and instructional value.

Each statistical method is discussed in enough detail to allow readers to write their own code to implement the method (except in one case, namely extreme value theory, for which there exists easy-to-use software in R). The reason for giving this level of detail is to ensure that the material is complete, self-contained, and covers the nuances and points of confusion that arise in practice. Indeed, we, as active researchers, often feel that we do not adequately understand a statistical method unless we have written computer code to perform that method. Our experience is that students gain fundamental and long-lasting confidence by coding each method themselves. This sentiment was expressed in an end-of-year course evaluation, in which one of our students wrote, "Before this course, I had used someone else's program to compute an EOF, but I didn't really understand it. Having to write my own program really helped me understand this method."

The methods covered in this book share a common theme: to quantify and exploit dependencies between X and Y. Different methods arise because each method is tailored to a particular probability distribution or data format. Specifically, the methods depend on whether X and Y are scalar or vector, whether the values are categorical or continuous, whether the distributions are Gaussian or not, and whether one variable is held fixed for multiple realizations of the other. The most general method for quantifying X-Y dependencies for multivariate Gaussian distributions is Canonical Correlation Analysis. Special cases include univariate regression (scalar Y), field significance (scalar X), or correlation (scalar X and scalar Y). In climate studies, multiple realizations of Y for fixed X characterize *ensemble* data sets. The most general method for quantifying X-Y dependencies in ensemble data sets is Predictable Component Analysis (or equivalently, Multivariate Analysis of Variance). Special cases include Analysis of Variance (scalar Y), and the t-test (scalar X and scalar Y). Many of these techniques have non-Gaussian versions. Linear regression provides a framework for exploiting dependencies to predict one variable from the other. Autoregressive models and power spectra quantify dependencies across time. Data assimilation provides a framework for exploiting dependencies to infer Y given X while incorporating "prior knowledge" about Y. The techniques for the different cases, and the chapter in which they are discussed, are summarized in Table 0.1.

Table 0.1. *Summary of methods for quantifying dependencies between X and Y.*

Y	X	Statistic or Procedure	Chapter
Vector	Vector	Canonical Correlation Analysis	15
Scalar	Vector	Multiple regression	9
Vector	Scalar	Field significance	13
Scalar	Scalar	Scalar regression or correlation	1
Ensemble and vector	Categorical	Predictable Component Analysis	18
Ensemble and scalar	Categorical	Analysis of Variance	17
Ensemble and scalar	Two categories	t-test	2

It is a pleasure to acknowledge helpful comments from colleagues who graciously gave up some of their busy time to read selected chapters in this book, including Jeffrey Anderson, Grant Branstator, Ian Jolliffe, and Jagadish Shukla. We thank our (former) students whose feedback was invaluable to finding the best pedagogical approach to this material, especially Paul Buchman, Xia Feng, Rachael Gaal, Olivia Gozdz, Liwei Jia, Keri Kodama, Emerson LaJoie, Douglas Nedza, Abhishekh Srivastava, Xiaoqin Yan, and M. Tugrul Yilmaz. We are indebted to Anthony Barnston, Grant Branstator, Ping Chang, Ben Kirtman, Andy Majda, Tapio Schneider, Jagadish Shukla, and David Straus for discussions over many years that have shaped the material presented in this book. We thank Vera Akum for assistance in acquiring the permissions for the quotes that open each chapter. Special thanks to Tony Barnston for suggesting the example used in Chapter 9. Any errors or inaccuracies in this book rest solely with the authors. We will be grateful to readers who notify us of errors or suggestions for improvement of this book.

1

Basic Concepts in Probability and Statistics

> Probability theory is nothing more than common sense reduced to calculation.
>
> *Pierre Simon Laplace*

This chapter reviews some essential concepts of probability and statistics, including the following:

- line plots, histograms, scatter plots
- mean, median, quantiles, variance
- random variables
- probability density function
- expectation of a random variable
- covariance and correlation
- independence
- the normal distribution (also known as the Gaussian distribution)
- the chi-squared distribution.

These concepts provide the foundation for the statistical methods discussed in the rest of this book.

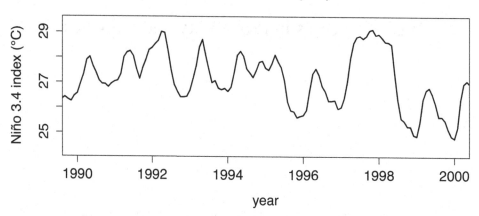

Figure 1.1 A time series of the monthly Niño 3.4 index over the period 1990–2000.

1.1 Graphical Description of Data

Scientific knowledge is based on observations. However, a mere list of observational facts rarely advances science. Instead, the data need to be organized in ways that help the scientist interpret the data in a scientific interpret the data in a scientific framework and formulate new hypotheses that can be checked in independent data or experiments. To illustrate ways of describing the main characteristics of a data set, consider a specific observable quantity: the area-average sea surface temperature in the equatorial Pacific in the region $170°W - 120°W$ and $5°S - 5°N$. This quantity is called the Niño 3.4 index and is an indicator of seasonal climate variations. The monthly average value of this index over a period of 50 or more years is readily available from various data portals. What are some ways of describing such a data set?

Data taken sequentially in time are known as *time series*. A natural way to visualize time series is to plot them as a function of time. A *time series plot* of Niño 3.4 is shown in Figure 1.1. The figure reveals that peaks and valleys occur at nearly periodic intervals, reflecting the annual cycle for this region. The figure also reveals that the time series is "smooth" – the value at one time is close to the value at neighboring times. Such time series are said to be *serially correlated* or *autocorrelated* and will be studied in Chapter 5. Another feature is that the minimum values generally decreased from 1993 to 2000, suggesting a possible long-term change. Methods for quantifying long-term changes in time series will be discussed in Chapters 8 and 9. Note how much has been learned simply by plotting the time series.

Another way to visualize data is by a *histogram*.

Definition 1.1 (Histogram) *A histogram is a plot obtained by partitioning the range of data into intervals, often equal-sized, called bins, and then plotting a*

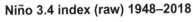

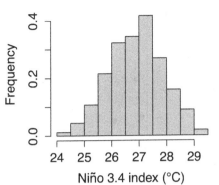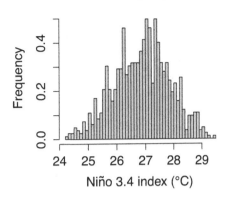

Figure 1.2 Histograms of the monthly mean Niño 3.4 index over the period 1948–2017. The two histograms show the same data, but the left histogram uses a wider bin size than the right.

rectangle over each bin such that the area of each rectangle equals the empirical frequency with which samples fall into the bin. The total area of the rectangles equals one. (Sometimes, histograms may be defined such that the total area of the rectangles equals the total number of samples, in which case the area of each rectangle equals the number of samples that fall into that bin.)

Histograms of the Niño 3.4 index for different bin sizes are shown in Figure 1.2. The figure shows that this index varied between 24°C and 29.5°C over the period 1948–2017. Also, values around 27° occur more frequently than values around 25° or 29°. However, the shape of the histogram is sensitive to bin size (e.g., compare Figures 1.2a and b); hence, the conclusions one draws from a histogram can be sensitive to bin size. There exist guidelines for choosing the bin size, e.g., Sturges' rule and the Freedman–Diaconis rule, but we will not discuss these. They often are implemented automatically in standard statistical software.

The *scatterplot* provides a way to visualize the relation between *two* variables. If X and Y are two time series over the same time steps, then each point on the scatterplot shows the point $(X(t), Y(t))$ for each value of t. Some examples of scatterplots are illustrated in Figure 1.3. Scatterplots can reveal distinctive relations between X and Y. For instance, Figure 1.3a shows a tendency for large values of X to occur at the same time as large values of Y. Such a tendency can be used to *predict* one variable based on knowledge of the other. For instance, if X were known to be at the upper extreme value, then it is very likely that Y also will be at its upper extreme. Figure 1.3b shows a similar tendency, except that the relation is weaker, and therefore a prediction of one variable based on the other would have more uncertainty. Figure 1.3c does not immediately reveal a relation between the two variables. Figure 1.3d shows that X and Y tend to be *negatively* related to each other,

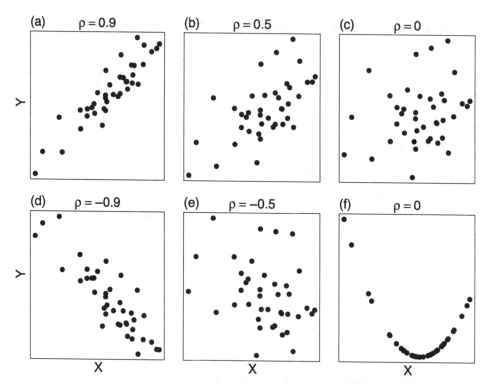

Figure 1.3 Scatterplots of X versus Y for various types of relation. The correlation coefficient ρ, given in the title of each panel, measures the degree of linear relation between X and Y. The data were generated using the model discussed in Example 1.7, except for data in the bottom right panel, which was generated by the model $Y = X^2$, where X is drawn from a standardized Gaussian.

when one goes up, the other goes down. Methods for quantifying these relations are discussed in Section 1.7.

1.2 Measures of Central Value: Mean, Median, and Mode

Visual plots are informative, but ultimately data must be described *quantitatively*. A basic descriptor of a set of numbers is their *central value*. For instance, the central value could be identified with the *most frequent* value, called the *mode*. The mode could be estimated by the location of the peak of a histogram, although this definition would depend on bin size. Also, for the Niño 3.4 time series, each value occurs *only once*, so there is no "most frequent value." Other measures of central value are the *mean* and *median*. When these quantities are computed from data, the qualifier *sample* is used to emphasize its dependence on data.

Definition 1.2 (Sample Mean) *The sample mean (or average) of N numbers $X_1, \ldots, X_N$ is denoted $\hat{\mu}_X$ and equals the sum of the numbers divided by N*

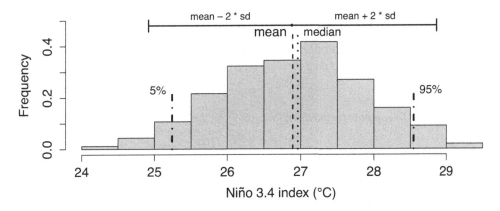

Figure 1.4 Histogram of the monthly mean (raw) Niño 3.4 index over the period 1948–2017, as in Figure 1.2, but with measures of central value and dispersion superimposed. The mean and median are indicated by dashed and dotted vertical lines, respectively. The dash-dotted lines indicate the 5th and 95th percentiles. The horizontal "error bar" at the top indicates the mean plus or minus two standard deviations. The empirical mode is between 27°C and 27.5°C.

$$\hat{\mu}_X = \frac{X_1 + X_2 + \cdots + X_N}{N} = \frac{1}{N} \sum_{n=1}^{N} X_n. \tag{1.1}$$

The mean of the Niño 3.4 index is indicated in Figure 1.4 by the dashed vertical line. The mean is always bounded by the largest and smallest elements.

Another measure of central value is the median.

Definition 1.3 (Sample Median) *The sample median of N numbers $X_1, \ldots, X_N$ is the middle value when the data are arranged from smallest to largest. If N is odd, the median is the unique middle value. If N is even, then two middle values exist and the median is defined to be their average.*

The median effectively divides the data into two equal halves: 50% of the data lie above the median, and 50% of the data lie below the median. The median of the Niño 3.4 index is shown by the dotted vertical line in Figure 1.4 and is close to the mean. In general, the mean and median are equal for symmetrically distributed data, but differ for asymmetrical distributions, as the following two examples illustrate.

Example 1.1 (The Sample Median and Mean for N Odd) *Question: What is the mean and median of the following data?*

$$2 \quad 8 \quad 5 \quad 9 \quad 3. \tag{1.2}$$

Answer: To compute the median, first order the data:

$$2 \quad 3 \quad 5 \quad 8 \quad 9. \tag{1.3}$$

The middle value is 5, hence the median is 5. The mean is

$$\frac{2+3+5+8+9}{5} = 5.4. \tag{1.4}$$

Example 1.2 (The Sample Median and Mean for N Even) *Question: What is the mean and median of the following data?*

$$2 \quad 8 \quad 5 \quad 9 \quad 3 \quad 100. \tag{1.5}$$

Answer: To compute the median, first order the data:

$$2 \quad 3 \quad 5 \quad 8 \quad 9 \quad 100. \tag{1.6}$$

The two middle values are 5 and 8, hence the median is their average, namely 6.5. In contrast, the mean is 21.2, which differs considerably from the median (contrary to example 1.1). Note that if the value of 100 were changed to some higher value X, the median would remain at 6.5 regardless of the value X, but the mean would increase with X. This example shows that the mean is sensitive to extreme values in a data set, whereas the median is not.

1.3 Measures of Variation: Percentile Ranges and Variance

Two data sets can have similar central values but differ by how they *vary* about the central value. Two common measures of variation are *quantile ranges* and *variance*. Sample quantiles are points that divide the sample into equal parts. Common quantiles have special names. For instance, *terciles* divide the sample into three equal parts; *quartiles* divide a sample into four equal parts. One of the most common quantiles is the *percentile*.

Definition 1.4 (Sample Percentiles) *A (sample) percentile is indicated by a number p, such that after the data are ordered from smallest to largest, at least $p \cdot 100\%$ of the data are at or below this value, and at least $100(1-p)\%$ are at or above this value. The resulting value is said to be the $100p$-th percentile (e.g., the 90th percentile corresponds to $p = 0.9$).*

The median is a special case of a percentile: It is the 50th percentile (i.e., $p = 0.5$). The above definition states merely that *at least $p \cdot 100\%$* of the data lies below the $100p$'th percentile, hence the sample percentile is not unique. There are several definitions of sample quantiles; for instance, Hyndman and Fan (1996) discuss nine different algorithms for computing sample quantiles. The differences between these sample quantiles have no practical importance for large N and will not be of concern in this book. Mathematical software packages such as Matlab, R, and Python have built-in functions for computing quantiles.

The *percentile range* is the interval between two specified percentile points. For instance, the 5–95% range includes all values between the 5th and 95th precentiles. This percentile range is a measure of variation in the sense that it specifies an interval in which a random number from the population will fall 90% of the time. The 5th and 95th percentiles of the Niño 3.4 index are indicated in Figure 1.4 by the two dash-dot lines.

Another measure of variation is the variance.

> **Definition 1.5** (Sample Variance) *The sample variance of N numbers* $X_1, \ldots, X_N$ *is denoted* $\hat{\sigma}_X^2$ *and defined as*

$$\hat{\sigma}_X^2 = \frac{1}{N-1} \sum_{n=1}^{N} (X_n - \hat{\mu}_X)^2, \tag{1.7}$$

> *where* $\hat{\mu}_X$ *is the sample mean of the data, defined in (1.1).*

The reader ought to be curious why the sum in (1.7) is divided by $N-1$, whereas the sum for the mean (1.1) was divided by N. The reason for this will be discussed in Section 1.10 (e.g., see discussion after Theorem 1.4). Based on its similarity to the definition of the mean, the variance is approximately the average squared difference from the sample mean.

> **Definition 1.6** (Standard Deviation) *The standard deviation is the (positive) square root of the variance:*

$$\hat{\sigma}_X = \sqrt{\hat{\sigma}_X^2}. \tag{1.8}$$

> *The standard deviation has the same units as X.*

Among the different measures listed above, the ones that will be used most often in this book are the *mean* for central tendency, and the *variance* for variation. The main reason for this is that the mean and variance are algebraic combinations of the data (i.e., they involve summations and powers of the data); hence, they are easier to deal with theoretically compared to mode, median, and percentiles (which require ranking the data). Using the mean and variance, a standard description of variability is the mean value plus and minus one or two standard deviations. For the Niño 3.4 index shown in Figure 1.4, the mean plus or minus two standard deviations is indicated by the error bar at the top of the figure.

Selected Properties of the Sample Variance

If $\hat{\sigma}_X^2$ is the sample variance of $X_1, \ldots, X_N$ and k is a constant, then

- variance of k times each X_n: $\hat{\sigma}_{(kX)}^2 = k^2 \hat{\sigma}_X^2$.
- variance of k plus each X_n: $\hat{\sigma}_{(X+k)}^2 = \hat{\sigma}_X^2$.

An identity that is occasionally useful is

$$\hat{\sigma}_X^2 = \left(\hat{\mu}_{(X^2)} - \hat{\mu}_X^2 \right) \frac{N}{N-1}. \tag{1.9}$$

Numerically, computation of sample variance based on (1.7) requires *two* passes of the data: one to compute the mean, and a second to compute deviations from the mean. With (1.9), the sample variance can be computed from one pass of the data, but requires tracking two quantities, namely the means of X and X^2. The sample variance is nonnegative, but in practice (1.9) can be (slightly) negative owing to numerical precision error.

1.4 Population versus a Sample

An observation is defined as the outcome of an experiment performed on nature. We will conceive of a theoretical collection of all possible outcomes, and then interpret an observation as a random draw from this theoretical collection. The theoretical collection of all possible observations is called the *population*, while a random draw from this collection is called a *sample* or *realization*. The goal of statistics is to make inferences or decisions about a population based on information derived from a sample.

In nature, population properties are never known with complete certainty. Knowledge of population properties is tantamount to knowledge of the "inner machinery" of the system. Except in idealized settings, we never know the inner workings of the system on which we experiment, and therefore we can never be sure about the population properties. Rather, we can only *infer* population properties based on the outcome of experiments. We might attempt to approximate the population probability of an event by measuring the relative frequency with which the event occurs in a large number of independent samples, but this approach meets fundamental difficulties with defining "large," "approximate," and "independent." These and other subtle problems can be avoided by *defining* probability in axiomatic terms, much like geometry is developed strictly from a set of axioms and rules of logic. This is the approach mathematicians have adopted. For the problem considered in this book, this axiomatic abstraction is not required. Therefore, we briefly review basic concepts in probability theory that are needed to get started. Most text books on statistics and probability cover these concepts in detail and can be consulted for further information.

1.5 Elements of Probability Theory

What is the probability of tossing a fair coin and getting heads? A typical 10-year-old child knows that the probability is 50%. However, that same 10-year-old child can become confused by an experiment where 6 out of 10 tosses are heads, since 6/10 is not 50%. The child eventually learns that "50% probability" refers to the idea that in a *long sequence* of coin tosses the relative frequency of heads *approaches* 50%. However, the relative frequency of heads in a small number of experiments can differ considerably from 50%. Asserting that heads occurs with 50% probability is tantamount to asserting knowledge of the "inner machinery" of nature. We refer to the "50% probability" as a *population property*, to distinguish it from the results of a particular experiment, e.g., "6 out of 10 tosses," which is a *sample property*. Much confusion can be avoided by clearly distinguishing population and sample properties. In particular, it is a mistake to equate the relative frequency with which an event occurs in an experiment with the probability of the event in the population.

A *random variable* is a function that assigns a real number to each outcome of an experiment. If the outcome is numerical, such as the temperature reading from a thermometer, then the random variable often is the number itself. If the outcome is not numerical, then the role of the function is to assign a real number to each outcome. For example, the outcome of a coin toss is heads or tails, i.e., not a number, but a function may assign 1 to heads and 0 to tails, thereby producing a random variable whose only two values are 0 and 1. This is an example of a *discrete* random variable, whose possible values can be counted. In contrast, a random variable is said to be *continuous* if its values can be any of the infinitely many values in one or more line intervals.

Sometimes a random variable needs to be distinguished from the value that it takes on. The standard notation is to denote a random variable by an uppercase letter, i.e. X, and denote the specific value of a random draw from the population by a lowercase letter, i.e. x. We will adopt this notation in this chapter. However, this notation will be adhered to only lightly, since later we will use uppercase letters to denote matrices and lowercase letters to denote vectors, a distinction that is more important in multivariate analysis.

If a variable is discrete, then it has a countable number of possible realizations $X_1, X_2, \ldots$. The corresponding probabilities are denoted $p_1, p_2, \ldots$ and called the *probability mass function*. If a random variable is continuous, then we consider a class of variables X such that the probability of $\{x_1 \leq X \leq x_2\}$, for all values of $x_1 \leq x_2$, can be expressed as

$$P(x_1 \leq X \leq x_2) = \int_{x_1}^{x_2} p_X(x)dx, \qquad (1.10)$$

where $p_X(x)$ is a nonnegative function called the *density function*. By this definition, the probability of X falling between x_1 and x_2 corresponds to the *area under the density function*. This area is illustrated in Figure 1.5a for a particular distribution. If an experiment always yields some real value of X, then that probability is 100% and it follows that

$$\int_{-\infty}^{\infty} p_X(x)dx = 1. \qquad (1.11)$$

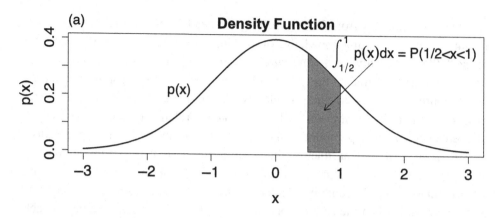

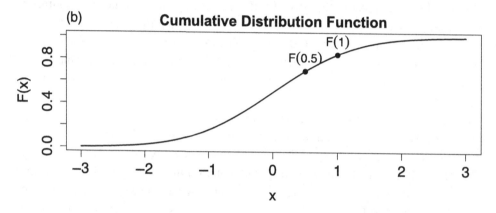

Figure 1.5 Schematic showing (a) a probability density function for X and the fact that the probability that X lies between 1/2 and 1 is given by the area under the density function $p(x)$, and (b) the corresponding cumulative distribution function $F(x)$ and the values at $x = 0.5$ and $x = 1$, the difference of which equals the area of the shaded region in (a).

The histogram provides an estimate of the density function, provided the histogram is expressed in terms of relative frequencies. Another function is

$$F(x) = P(X \leq x) = \int_{-\infty}^{x} p_X(u)du, \tag{1.12}$$

which is called the *cumulative distribution function* and illustrated in Figure 1.5b. The probability that X lies between x_1 and x_2 can be expressed equivalently as

$$P(x_1 \leq X \leq x_2) = F(x_2) - F(x_1). \tag{1.13}$$

The above properties do not uniquely specify the density function $p_X(x)$, as there is more than one $p_X(x)$ that gives the same left-hand side of (1.10) (e.g., two density functions could differ at isolated points and still yield the same probability of the same event). A more precise definition of the density function requires *measure theory*, which is beyond the scope of this book. Such subtleties play no role in the problems discussed in this book. Suffice it to say that distributions considered in this book are *absolutely continuous*. A property of this class is that the probability of the event $\{X = x_1\}$ vanishes:

$$P(X = x_1) = \int_{x_1}^{x_1} p_X(x)dx = 0. \tag{1.14}$$

Although continuous random variables are defined with integrals, you need not explicitly evaluate integrals to do statistics – all integrals needed in this book can be obtained from web pages, statistical software packages, or tables in the back of most statistical texts.

1.6 Expectation

Just as a sample can be characterized by its mean and variance, so too can the population.

Definition 1.7 (Expectation) *If X is a continuous random variable with probability density p(x), then the expectation of the function f(X) is*

$$\mathbb{E}_X[f(X)] = \int_{-\infty}^{\infty} f(x)p_X(x)dx. \tag{1.15}$$

The special case f(x) = x gives

$$\mu_X = \mathbb{E}_X[X] = \int_{-\infty}^{\infty} xp_X(x)dx, \tag{1.16}$$

which is called the "population mean," or the "mean," of X.

If the random variable is *discrete* and takes on discrete values $X_1, X_2, \ldots, X_N$ with probabilities $p_1, p_2, \ldots, p_N$, then the expectation is defined as

$$\mathbb{E}_X[X] = \sum_{n=1}^{N} X_n p_n. \tag{1.17}$$

This expression is merely the discrete version of (1.16).

The term "mean" has been used to characterize both (1.1) and (1.16), yet the expressions look different. To understand why the term is appropriate, consider samples of a discrete random variable in which values $X_1, \ldots, X_K$ occur $N_1, \ldots, N_K$

times, respectively. Then, the sum in (1.1) involves N_1 terms equal to x_1, N_2 terms equal to x_2, and so on; hence, the sample mean is

$$\hat{\mu}_X = \frac{x_1 N_1 + x_2 N_2 + \ldots + x_K N_K}{N_1 + N_2 + \ldots + N_K} = \sum_{n=1}^{N_K} x_n f_n, \qquad (1.18)$$

where $f_n = N_n/(N_1 + N_2 + \cdots + N_K)$ is the *relative frequency* of X_n. Comparing (1.18) and (1.17) reveals the connection: The sample mean and the expectation are both expressible as a sum of values weighted by the respective frequency or probability. Just to be clear, an important difference between these expressions is that the sample mean in (1.18) is a random variable while (1.15)–(1.17) define fixed properties of the population.

Notation
We use *carets* ˆ to distinguish sample quantities from population quantities; for example,

$$\hat{\mu}_x = \frac{1}{N} \sum_{n=1}^{N} x_n, \quad \hat{\mu}_X = \frac{1}{N} \sum_{n=1}^{N} X_n \quad \text{and} \quad \mu_X = \mathbb{E}_X[X]. \qquad (1.19)$$

The sample mean $\hat{\mu}_x$ is a *specific numerical value* obtained from a given sample, $\hat{\mu}_X$ is a *random variable* because it is a sum of random variables, and μ_X is a *fixed population quantity*. Because $\hat{\mu}_X$ is a random variable, it can be described by a probability distribution with its own expectation. This fact can lead to potentially confusing terminology, such as "the mean of the mean." In such cases, we say "expectation of the sample mean."

Selected Properties of Expectation (these should be obvious)
If k_1 and k_2 are constants and X and Y are random variables, then

- $\mathbb{E}_X[k_1] = k_1$
- $\mathbb{E}_X[k_1 X] = k_1 \mathbb{E}_X[X]$
- $\mathbb{E}_X[k_1 X + k_2 Y] = k_1 \mathbb{E}_X[X] + k_2 \mathbb{E}_X[Y]$

> **Definition 1.8** (Variance) *The variance of the random variable X is defined as*
>
> $$\text{var}[X] = \mathbb{E}_X[(X - \mathbb{E}_X[X])^2]. \qquad (1.20)$$
>
> *The variance of X often is denoted by σ_X^2. The standard deviation σ_X is the positive square root of the variance.*

Interpretation
Variance is a measure of dispersion or scatter of a random variable about its mean. Small variance indicates that the variables tend to be concentrated near the mean.

Selected Properties of Variance

If k is a constant and X is a random variable, then

- $\text{var}[k] = 0$
- $\text{var}[X] = \mathbb{E}_X[X^2] - (\mathbb{E}_X[X])^2$
- $\text{var}[kX] = k^2 \, \text{var}[X]$
- $\text{var}[X + k] = \text{var}[X]$

Example 1.3 (Standardized Random Variables) *Question: Suppose X has population mean μ_X and variance σ_X^2. A random variable of the form*

$$Z = \frac{X - \mu_X}{\sigma_X} \tag{1.21}$$

is called a standardized *random variable. Show that $\mathbb{E}[Z] = 0$ and $\text{var}[Z] = 1$. This result is very useful and will be used many times in this book.*

Answer:

$$\mathbb{E}[Z] = \mathbb{E}\left[\frac{X - \mu_X}{\sigma_X}\right] = \frac{\mathbb{E}[X] - \mu_X}{\sigma_X} = \frac{\mu_X}{\sigma_X} - \frac{\mu_X}{\sigma_X} = 0.$$

and

$$\text{var}[Z] = \mathbb{E}\left[(Z - \mathbb{E}[Z])^2\right] = \mathbb{E}\left[Z^2\right] = \mathbb{E}\left[\left(\frac{X - \mu_X}{\sigma_X}\right)^2\right]$$

$$= \frac{\mathbb{E}\left[(X - \mu_X)^2\right]}{\sigma_X^2} = \frac{\sigma_X^2}{\sigma_X^2} = 1.$$

1.7 More Than One Random Variable

The concept of a probability distribution can be generalized to more than one variable. Instead of considering a single event, such as $X \leq x_2$, we consider a collection of events, called a *joint event*. The joint event is said to occur if and only if every event in the collection occurs. Joint events can be interpreted as propositions about events connected by the word "and." For example, a joint event could be a heads on the first toss *and* a heads on the second toss of a coin. The probability of the joint event $\{x_1 \leq X \leq x_2\}$ *and* $\{y_1 \leq Y \leq y_2\}$ is defined through a *joint probability density p_{XY}* as

$$P(x_1 \leq X \leq x_2, y_1 \leq Y \leq y_2) = \int_{x_1}^{x_2} \int_{y_1}^{y_2} p_{XY}(x, y)dxdy. \tag{1.22}$$

In general, a comma separating two events is shorthand for "and." The probability of the single event $\{x_1 \leq X \leq x_2\}$ can be computed from the joint density $p(x, y)$ by integrating over all outcomes of Y:

$$P(x_1 \leq X \leq x_2, -\infty \leq Y \leq \infty) = \int_{x_1}^{x_2} \int_{-\infty}^{\infty} p_{XY}(x, y) dx dy. \tag{1.23}$$

However, this probability already appears in (1.10). Since this is true for all x_1 and x_2, it follows that

$$p_X(x) = \int_{-\infty}^{\infty} p_{XY}(x, y) dy. \tag{1.24}$$

To emphasize that only a single variable is considered, the density $p_X(x)$ often is called the *unconditional* or *marginal* probability density of X.

Notation
Technically, the distributions of X and Y should be denoted by separate functions, say $p_X(x)$ and $p_Y(y)$. Moreover, the arguments can be arbitrary, say $p_X(w)$ and $p_Y(z)$. For conciseness, the subscripts are dropped *with the understanding that the argument specifies the specific function in question*. For example, $p(x)$ denotes the density function $p_X(x)$. Similarly, $p(x, y)$ denotes the joint density $p_{XY}(x, y)$. The expectation follows a similar convention: *The expectation is taken with respect to the joint distribution of all random variables that appear in the argument*. For instance, $\mathbb{E}[f(X)]$ denotes the expectation of $f(x)$ with respect to $p_X(x)$, and $\mathbb{E}[g(X, Y)]$ denotes the expectation with respect to the *joint* distribution of X and Y; that is,

$$\mathbb{E}[g(X, Y)] = \int_{-\infty}^{\infty} \int_{-\infty}^{\infty} g(x, y) p_{XY}(x, y) dx dy. \tag{1.25}$$

An important measure of the *joint* relation between random variables is the covariance.

Definition 1.9 (Covariance) *Let X and Y be two random variables with respective means μ_X and μ_Y and joint density p(x,y). Then, the covariance between X and Y is defined as*

$$\text{cov}[X, Y] = \int_{-\infty}^{\infty} \int_{-\infty}^{\infty} (x - \mu_X)(y - \mu_Y) p(x, y) dx dy, \tag{1.26}$$

or equivalently,

$$\text{cov}[X, Y] = \mathbb{E}[(X - \mathbb{E}[X])(Y - \mathbb{E}[Y])], \tag{1.27}$$

where the outermost expectation denotes the expectation over the joint distribution p(x, y).

Variance is a special case of covariance: $\text{var}[X] = \text{cov}[X, X]$. The *sign* of the covariance indicates how two variables are linearly related. For instance, if X and Y tend to go up and down together, as in Figure 1.3a, then the covariance is positive. Conversely, if X tends to increase when Y decreases, as illustrated in Figure 1.3d, then the covariance is negative.

Selected Properties of the Covariance
Let X, Y, Z be three arbitrary random variables, and let k_1, k_2 be constants. Then:

- $\text{cov}[X, Y] = \mathbb{E}[XY] - \mathbb{E}[X]\mathbb{E}[Y]$.
- $|\text{cov}[X, Y]| \leq \sigma_X \sigma_Y$.
- $\text{cov}[X, k_1] = 0$.
- $\text{cov}[k_1 X + k_2 Y, Z] = k_1 \text{cov}[X, Z] + k_2 \text{cov}[Y, Z]$.

> **Example 1.4** (Variance of a Sum)
> **Question:** *What is the variance of $X + Y$, where X and Y are random variables?*
> **Answer:**
>
> $$\begin{aligned}
> \text{var}[X + Y] &= \mathbb{E}[(X + Y - \mathbb{E}[X + Y])^2] \quad \text{(definition of variance (1.20))} \\
> &= \mathbb{E}[((X - \mathbb{E}[X]) + (Y - \mathbb{E}[Y]))^2] \\
> &= \mathbb{E}[(X - \mathbb{E}[X])^2] + \mathbb{E}[(Y - \mathbb{E}[Y])^2] + 2\mathbb{E}[(X - \mathbb{E}[X])(Y - \mathbb{E}[Y])] \\
> &= \text{var}[X] + \text{var}[Y] + 2\text{cov}[X, Y] \quad\quad\quad\quad\quad\quad\quad\quad (1.28)
> \end{aligned}$$
>
> *The variance of a sum generally differs from the sum of the individual variances.*

The covariance of X and Y depends on the (arbitrary) units in which the variables are measured. However, if the two variables are standardized, then the resulting covariance is independent of measurement units, and has other attractive properties, as discussed next.

> **Definition 1.10** (Correlation Coefficient) *The correlation coefficient between X and Y is*
>
> $$\rho_{XY} = \frac{\text{cov}[X, Y]}{\sqrt{\text{var}[X]\,\text{var}[Y]}}. \quad\quad (1.29)$$
>
> *ρ_{XY} also is called Pearson's product-moment correlation coefficient, to distinguish it from other measures, such as the rank correlation (see Chapter 4). The correlation coefficient can be written equivalently as the expectation of the product of standardized variables:*
>
> $$\rho_{XY} = \mathbb{E}\left[\left(\frac{X - \mu_X}{\sigma_X}\right)\left(\frac{Y - \mu_Y}{\sigma_Y}\right)\right]. \quad\quad (1.30)$$
>
> *It can be shown that the correlation coefficient satisfies $-1 \leq \rho_{XY} \leq 1$.*

Selected Properties of the Correlation Coefficient
- The correlation coefficient is exactly 1 if and only if $Y = \beta X + k$ for $\beta > 0$.
- The correlation coefficient is exactly -1 if and only if $Y = \beta X + k$ for $\beta < 0$.
- The correlation coefficient is symmetric with respect to X and Y; that is, $\rho_{XY} = \rho_{YX}$.

- The correlation coefficient does not change under the transformations $X \rightarrow aX + b$ and $Y \rightarrow cY + d$, for any a, b, c, d where $ac > 0$. The above transformations are called *affine transformations*. If $ac > 0$, then the transformations are *invertible*. The correlation coefficient is said to be invariant to invertible affine transformations of the two variables.

The above properties suggest that ρ_{XY} is a measure of the degree of *linear* relation between X and Y: values at ± 1 indicate that X and Y are exactly linearly related, while independent variables (which are defined in the next section) have zero correlation. Examples of scatter plots and the associated correlation are shown in Figure 1.3.

1.8 Independence

A fundamental concept in statistical analysis is *independence*:

> **Definition 1.11** (Independence) *The variables X and Y are independent if and only if*
>
> $$p(x, y) = p(x)p(y), \tag{1.31}$$
>
> *for all x and y.*

If two variables are independent, then any functions of them are also independent. If two random variables are not independent, then they are *dependent*. Dependence between two random variables can be quantified by its conditional distribution.

> **Definition 1.12** (Conditional Distribution) *The conditional distribution of Y given that $X = x$ has occurred is denoted $p(y|x)$ and defined by*
>
> $$p(y|x) = \frac{p(x, y)}{p(x)} \quad \text{provided } p(x) \neq 0. \tag{1.32}$$

The symbol "$p(y|x)$" is read "probability density of Y given that the random variable X equals x," or more concisely, "the probability of Y given x." The conditional distribution tells how the probability of an event depends on the occurrence of another event.

> **Example 1.5** *Question: What is $p(y|x)$ if X and Y are independent?*
> *Answer:*
>
> $$p(y|x) = \frac{p(x, y)}{p(x)} = \frac{p(x)p(y)}{p(x)} = p(y). \tag{1.33}$$
>
> *This result ought to be intuitive: If X and Y are independent, then knowing X tells us nothing about Y that was not already known from the unconditional distribution $p(y)$.*

Important property
If X and Y are independent, then $\text{cov}[X, Y] = 0$. This fact can be shown as follows:

$$
\begin{aligned}
\text{cov}[X, Y] &= \int_{-\infty}^{\infty}\int_{-\infty}^{\infty}(x - \mu_X)(y - \mu_Y)p(x, y)dxdy \\
&= \int_{-\infty}^{\infty}\int_{-\infty}^{\infty}(x - \mu_X)(y - \mu_Y)p(x)p(y)dxdy \\
&= \left(\int_{-\infty}^{\infty}(x - \mu_X)p(x)dx\right)\left(\int_{-\infty}^{\infty}(y - \mu_Y)p(y)dy\right) \\
&= \mathbb{E}[X - \mu_X]\mathbb{E}[Y - \mu_Y] \\
&= 0.
\end{aligned}
\tag{1.34}
$$

It follows as a corollary that if X and Y are independent, then $\rho_{XY} = 0$. This fact is one of the most important facts in statistics!

While covariance vanishes if two variables are independent, the converse of this statement is not true: the covariance can vanish even if the variables are dependent. Figure 1.3f shows a counter example (see also Section 1.11). The fact that independence implies vanishing of the covariance is a valuable property that is exploited repeatedly in statistics.

Example 1.6 (Variance of a Sum of Independent Variables) **Question:** *What is the variance of $X + Y$, where X and Y are independent?* **Answer:** *According to example (1.4)*

$$
\text{var}[X + Y] = \text{var}[X] + \text{var}[Y] + 2\,\text{cov}[XY] \tag{1.35}
$$
$$
= \text{var}[X] + \text{var}[Y], \tag{1.36}
$$

where we have used $\text{cov}[X, Y] = 0$ for independent X and Y. This result shows that if variables are independent, the variance of the sum equals the sum of the variances. This result will be used repeatedly throughout this book.

Example 1.7 (A Model for Generating X and Y with Prescribed Correlation) **Question:** *Suppose X and Z are independent random variables each with zero mean and unit variance. Define*

$$
Y = \rho X + Z\sqrt{1 - \rho^2}. \tag{1.37}
$$

Show that the correlation between Y and X is ρ. This model is extremely useful in numerical simulations. In particular, it provides a way to generate random numbers (X, Y) from a population with a prescribed correlation ρ. (In fact, this is how Figure 1.3 was generated.)

Answer: *Because X and Z are independent, the variance of their sum equals the sum of their variances. Therefore,*

$$
\text{var}[Y] = var[\rho X + \sqrt{1 - \rho^2}Z] = \rho^2\,\text{var}[X] + (1 - \rho^2)\,\text{var}[Z] = 1. \tag{1.38}
$$

The last equality follows from the fact that $\text{var}[X] = \text{var}[Z] = 1$. *The covariance is*

$$\text{cov}[X, Y] = \text{cov}[X, \rho X + \sqrt{1 - \rho^2} Z] = \rho \, \text{var}[X] + \sqrt{1 - \rho^2} \, \text{cov}[X, Z] = \rho, \tag{1.39}$$

where we have used selected properties of the covariance, $\text{cov}[X, X] = \text{var}[X]$, *and* $\text{cov}[X, Z] = 0$ *because X and Z are independent. Consolidating these results yields*

$$\text{cor}[X, Y] = \frac{\text{cov}[X, Y]}{\sqrt{\text{var}[X] \, \text{var}[Y]}} = \rho. \tag{1.40}$$

1.9 Estimating Population Quantities from Samples

In practice, the population is unknown. Its characteristics are inferred using samples drawn from that population. However, a finite sample cannot fully specify a continuous density function. Accordingly, we will be satisfied with characterizing a population with a few numbers, known as *population parameters*. For instance, the mean μ_X and variance σ_X^2 are population parameters that characterize the probability distribution of X. Our goal is to estimate population parameters using samples drawn from that population. A function of random variables that is used to estimate a population parameter is known as an *estimator*.

Example 1.8 *Question: Suppose* $X_1, X_2, \ldots, X_N$ *are drawn from a distribution with the same population mean. What is the expectation of the sample mean?* **Answer:**

$$\mathbb{E}[\hat{\mu}_X] = \mathbb{E}\left[\frac{1}{N} \sum_{i=1}^{N} X_i\right] = \frac{1}{N} \sum_{i=1}^{N} \mathbb{E}[X_i] = \frac{1}{N} \mathbb{E}[X] N = \mathbb{E}[X]. \tag{1.41}$$

This example shows that the expectation of the sample mean equals the population mean. As a result, the sample mean is a useful *estimator* of the population mean. The expectation $\mathbb{E}[X] = \mu$ is a population parameter, while the sample mean $\hat{\mu}$ is an estimator of μ.

Definition 1.13 (Unbiased Estimator) *If the expectation of an estimator equals the corresponding population parameter, then it is called an unbiased estimator. Otherwise it is called a biased estimator.*

Example 1.9 *The sample mean is an unbiased estimator of* μ *because* $\mathbb{E}[\hat{\mu}] = \mu$ *(see Example 1.8).*

Example 1.10 *The sample variance (1.7) is an unbiased estimator; that is,* $\mathbb{E}[\hat{\sigma}_X^2] = \sigma_X^2$. *Had the sample variance been defined by dividing the sum by N instead of N − 1, as in*

$$\hat{\sigma}_B^2 = \frac{1}{N} \sum_{n=1}^{N} (x_n - \hat{\mu}_X), \tag{1.42}$$

then the resulting estimator $\hat{\sigma}_B^2$ would have been biased, in the sense that $\mathbb{E}[\hat{\sigma}_B^2] \neq \sigma_X^2$. In fact, it can be shown that $\mathbb{E}[\hat{\sigma}_B^2] = \sigma_X^2(N-1)/N$.

Example 1.11 (Variance of the Sample Mean: Independent Variables) *Let X_1, $X_2, \ldots, X_N$ be independent random variables drawn from the same distribution with population mean μ_X and variance σ_X^2. What is the variance of the sample mean $\hat{\mu}_X$?*

$$
\begin{aligned}
\text{var}[\hat{\mu}_X] &= \mathbb{E}[(\hat{\mu}_X - \mu_X)^2] = \mathbb{E}[(\hat{\mu}_X - \mu_X)(\hat{\mu}_X - \mu_X)] \\
&= \mathbb{E}\left[\left(\frac{1}{N} \sum_{i=1}^{N} (x_i - \mu_X) \right) \left(\frac{1}{N} \sum_{j=1}^{N} (x_j - \mu_X) \right) \right] \\
&= \mathbb{E}\left[\frac{1}{N^2} \sum_{i=1}^{N} \sum_{j=1}^{N} (x_i - \mu_X)(x_j - \mu_X) \right] \\
&= \frac{1}{N^2} \sum_{i=1}^{N} \sum_{j=1}^{N} \text{cov}[X_i, X_j]
\end{aligned}
\tag{1.43}
$$

The first line follows by definition of variance. The second line follows by definition of sample mean, and the fact that μ_X is constant. Importantly, the two summations in the second line have different indices. The reason for this is that the sum should be computed first then squared, which is equivalent to computing the sum two separate times then multiplying them together. The third line follows by algebra. The last line follows from the definition of covariance. Since the variables are independent and identically distributed,

$$
\text{cov}[X_i, X_j] = \begin{cases} 0 & \text{if } i \neq j \\ \sigma_X^2 & \text{if } i = j \end{cases}.
\tag{1.44}
$$

Thus, the double sum in (1.43) vanishes whenever $i \neq j$, and it equals σ_X^2 whenever $i = j$. It follows that (1.43) can be simplified to

$$
\text{var}[\hat{\mu}_X] = \frac{1}{N^2} \sum_{i=1}^{N} \sigma_X^2 = \frac{\sigma_X^2}{N}.
\tag{1.45}
$$

The standard deviation $\sigma_X/\sqrt{N}$ is known as the standard error of the mean.

This result is important! It states that as the number of independent samples grows, the variability of their arithmetic mean shrinks. In the limit of large N, the variance of the sample mean shrinks toward zero, implying that the sample mean *converges* (in some mathematical sense) to the population mean; that is, $\hat{\mu}_X \to \mu_X$ as $N \to \infty$. Intuitively, samples from a population are scattered randomly about the mean: sometimes the value is above the mean and sometimes it is below the mean. When the arithmetic average is computed, random fluctuations above and below the mean tend to cancel, yielding a number that is closer to the mean (on average) than the individual random variables.

This result embodies a fundamental principle in statistics: arithmetic averages have less variability than the variables being averaged. One way or another, every statistical method involves some type of averaging to reduce random variability.

Acronym

Samples that are drawn independently from the same population are said to be *independent and identically distributed*. This property is often abbreviated as *iid*.

1.10 Normal Distribution and Associated Theorems

The most important distribution in statistics is the Normal Distribution.

Definition 1.14 (Normal Distribution) *The probability density of a normally distributed random variable X is*

$$p(x) = \frac{1}{\sqrt{2\pi}} \frac{1}{\sigma} e^{-\frac{(x-\mu)^2}{2\sigma^2}}, \tag{1.46}$$

where $\sigma > 0$. For this distribution, $\mathbb{E}[X] = \mu$ and $\mathrm{var}[X] = \sigma^2$. The normal distribution is also known as the Gaussian distribution.

This distribution is illustrated in Figure 1.6. It has a characteristic "bell shape," is symmetric about μ, and decays monotonically to zero away from μ. We often need to know the probability that x is between two numbers. A standard quantity is $z_{\alpha/2}$, defined as

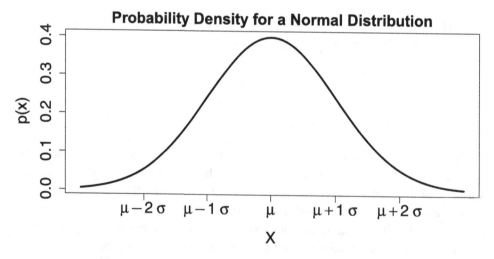

Figure 1.6 The normal probability density function p(x).

Table 1.1. *Values of* $(\alpha, z_{\alpha/2})$ *that satisfy* $P(-z_{\alpha/2} < Z < z_{\alpha/2}) = 1 - \alpha$ *for a standardized normal distribution. The value of* $1 - \alpha$ *is the fractional area under a standardized normal distribution contained in the interval* $(-z_{\alpha/2}, z_{\alpha/2})$.

$z_{\alpha/2}$	3.00	2.58	2.00	1.96	1.645	1.00
α	0.27%	1%	4.55%	5%	10%	31.70%
$1 - \alpha$	99.73%	99%	95.45%	95%	90%	68.27%

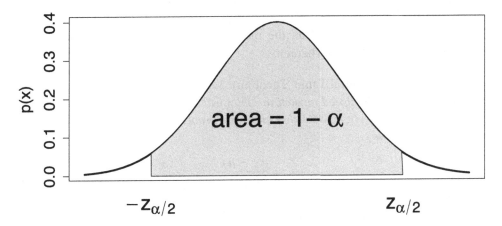

Figure 1.7 Illustration of the definition of $z_{\alpha/2}$ for a standardized normal distribution (i.e., a normal distribution with zero mean and unit variance).

$$P\left(-z_{\alpha/2} \le Z < z_{\alpha/2}\right) = 1 - \alpha, \tag{1.47}$$

or equivalently, as

$$P(\mu - \sigma z_{\alpha/2} < X < \mu + \sigma z_{\alpha/2}) = 1 - \alpha, \tag{1.48}$$

where $Z = (X - \mu)/\sigma$ (i.e., a standardized Gaussian). The meaning of $z_{\alpha/2}$ is illustrated in Figure 1.7, and values of z_α for some common choices of α are tabulated in Table 1.1.

Notation
The statement that X is normally distributed with mean μ and variance σ^2 is denoted as $X \sim \mathcal{N}(\mu, \sigma_X^2)$. The symbol $\sim$ means "is distributed as." A *standardized normal distribution* is a normal distribution with zero mean and unit variance, $\mathcal{N}(0, 1)$.

> **Example 1.12** *If* $X \sim \mathcal{N}(\mu, \sigma)$, *what is the probability that* X *is within two standard deviations of its mean?* **Answer:** *The question asks for the probability that* $\{\mu - 2\sigma \le X < \mu + 2\sigma\}$, *or equivalently, in terms of the standardized variable* $Z = (X - \mu)/\sigma$, *the probability that* $\{-2 \le Z < 2\}$. *According to Table 1.1, the probability is 95.45%.*

The normal distribution can be generalized to multiple random variables. The associated joint density function is expressed most simply in terms of vectors and matrices, hence its expression is postponed until Chapter 7. An important fact is that *any linear combination of joint normally distributed random variables is normally distributed*. In the case of two variables, the joint normal distribution is known as the *bivariate normal distribution*. The bivariate normal distribution is described completely by five population parameters: the means and variances of the two variables, and the *correlation* between the two variables.

Why is the normal distribution the most important distribution in statistics? Because of the Central Limit Theorem:

Theorem 1.1 (Central Limit Theorem) *Suppose $X_1, \ldots, X_N$ are independent and identically distributed random variables with mean μ_X and finite variance σ_X^2, but the distribution from which they are drawn is not necessarily Gaussian. Then the random variable*

$$Z = \frac{\hat{\mu}_X - \mu_X}{\left(\frac{\sigma_X}{\sqrt{N}}\right)} \tag{1.49}$$

approaches (in some mathematical sense) the normal distribution $\mathcal{N}(0, 1)$ as $N \to \infty$. Equivalently, as $N \to \infty$

$$\hat{\mu}_X \sim \mathcal{N}\left(\mu_X, \frac{\sigma_X^2}{N}\right). \tag{1.50}$$

In essence, the Central Limit Theorem states that the sum of *iid* variables tends to have a normal distribution, even if the original variables do not have a normal distribution.

To illustrate the Central Limit Theorem, consider a discrete random variable X that takes on only two values, -1 or 1, with equal probability. Thus, the probability mass function of X is $P(X = -1) = P(X = 1) = 1/2$, and is zero otherwise. A histogram of samples from this distribution is shown in the far left panel of Figure 1.8. This histogram is very unlike a normal distribution. The mean and variance of this distribution are derived from (1.17):

$$\mathbb{E}[X] = \frac{1}{2}(-1) + \frac{1}{2}(1) = 0 \tag{1.51}$$

$$\text{var}[X] = \frac{1}{2}(-1)^2 + \frac{1}{2}(1)^2 = 1. \tag{1.52}$$

Now suppose the arithmetic mean of $N = 10$ random samples of X is computed. Computing this arithmetic average repeatedly for different samples yields the histogram in the middle. The histogram now has a clear Gaussian shape. Although the

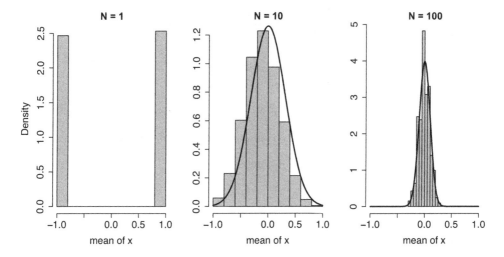

Figure 1.8 An illustration of the Central Limit Theorem using randomly generated $\pm$1s. The random number is either $+1$ or -1 with equal probability. A histogram of a large number of samples of this random variable is shown in the far left panel – it is characterized by two peaks at $\pm$1, which looks very different from a Gaussian distribution (e.g., it has two peaks rather than one). The mean of the population distribution is zero and the variance is 1. The middle panel shows the result of taking the average of $N=10$ random $\pm$1s over many repeated independent trials. Superimposed on this histogram is a Gaussian distribution with mean zero and variance $1/N$, as predicted by the Central Limit Theorem (1.50). The right panel shows the result of averaging $N = 100$ random variables.

Central Limit Theorem applies only for large N, this example shows that theorem is relevant even for $N = 10$. For reference, the middle panel of Figure 1.8 also shows a normal distribution evaluated from (1.50), using $\mu_X = 0$ and $\sigma_X = 1$ from (1.51) and (1.52). The histogram matches the predicted normal distribution fairly well. Repeating this experiment with $N = 100$ yields the histogram on the far right panel, which matches a normal distribution even better.

This example is an illustration of a *Monte Carlo technique*. A Monte Carlo technique is a computational procedure in which random numbers are generated from a prescribed population and then processed in some way. The technique is useful for solving problems in which the population is known but the distribution of some function of random variables from the population is difficult or impossible to derive analytically. In the example, the distribution of a sum of variables from a discrete distribution was not readily computable, but was easily and quickly estimated using a Monte Carlo technique.

An important property of the normal distribution is that a sum of independent normally distributed random variables also has a normal distribution.

Theorem 1.2 (Sum of Gaussian Variables is Gaussian) *Let X_1, X_2, ..., X_N be independent random variables with the following normal distributions:*

$$X_1 \sim \mathcal{N}(\mu_1, \sigma_1^2), \quad X_2 \sim \mathcal{N}(\mu_2, \sigma_2^2), \quad \dots \quad X_N \sim \mathcal{N}(\mu_N, \sigma_N^2). \tag{1.53}$$

Then the linear combination

$$Y = c_1 X_1 + c_2 X_2 + \dots + c_N X_N, \tag{1.54}$$

where c_1, c_2, ..., c_N are constants, is also normally distributed with mean

$$\mu_Y = c_1 \mu_1 + c_2 \mu_2 + \dots + c_N \mu_N, \tag{1.55}$$

and variance

$$\sigma_Y^2 = c_1^2 \sigma_1^2 + c_2^2 \sigma_2^2 + \dots + c_N^2 \sigma_N^2. \tag{1.56}$$

More concisely,

$$Y \sim \mathcal{N}\left(\sum_{i=1}^{N} c_i \mu_i, \sum_{i=1}^{N} c_i^2 \sigma_i^2 \right). \tag{1.57}$$

Comment

From our theorems about expectations, we already knew the mean and variance of *iid* variables. What is new in Theorem 1.2 is that if the X's are normally distributed, then Y is also normally distributed. In other words, we now know the distribution of Y.

Example 1.13 (Distribution of the Sample Mean of Gaussian Variables) ***Question:*** *Let X_1, X_2, ..., X_N be independent random variables drawn from the normal distribution $\mathcal{N}(\mu, \sigma_X^2)$. What is the distribution of the sample mean of these variables?*

Answer: *The sample mean is a sum of independent normally distributed random variables. Therefore, by Theorem 1.2, the sample mean also has a normal distribution. Moreover, the expectation of the sample mean was shown in Example 1.8 to be μ_X, and the variance of the sample mean was shown in Example 1.11 to be σ_X^2/N. Therefore, the sample mean is normally distributed as*

$$\hat{\mu}_X \sim \mathcal{N}\left(\mu_X, \frac{\sigma_X^2}{N} \right). \tag{1.58}$$

The distributions (1.50) and (1.58) are identical, but the latter is exact because the original variables were known to be normally distributed, whereas the former holds only for large N because the original variables were not necessarily normally distributed.

Recall that the sample variance (1.7) involves *squares* of a variable. Importantly, X and X^2 do not have the same distribution. The relevant distribution for squares of normally distributed random variables is the *chi-squared distribution*.

Theorem 1.3 (The χ^2 Distribution) *Let X_1, X_2, ..., X_N be independent and identically distributed random variables from the normal distribution $\mathcal{N}(\mu_X, \sigma_X^2)$. Then the variable*

$$Y^2 = \frac{(X_1 - \mu_X)^2 + (X_2 - \mu_X)^2 + ... + (X_N - \mu_X)^2}{\sigma_X^2} \tag{1.59}$$

has a chi-squared distribution with N degrees of freedom. This statement is expressed as

$$Y^2 \sim \chi_N^2. \tag{1.60}$$

A corollary of Theorem 1.3 is that if $Z_1, \ldots, Z_N$ are independent variables from a *standardized* normal distribution then

$$Z_1^2 + Z_2^2 + \cdots + Z_N^2 \sim \chi_N^2. \tag{1.61}$$

The chi-squared distribution is completely specified by one parameter, *degrees of freedom*. Although the chi-squared distribution is defined using population quantities μ_X and σ_X, which are usually unknown in practical applications, this distribution is extremely useful, as will be shown in various places in this book. Examples of the chi-squared distribution are shown in Figure 1.9. The explicit density function of χ_N^2 is

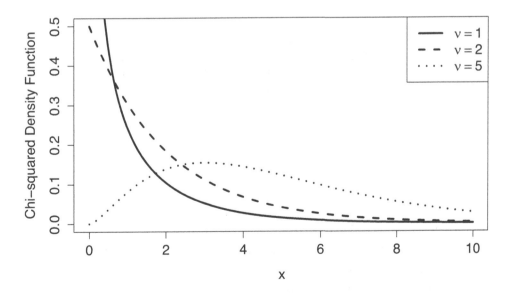

Figure 1.9 Illustration of the chi-squared distribution for three different values of the degrees of freedom.

$$p(x) = \begin{cases} \dfrac{x^{N/2-1}e^{-x/2}}{2^{N/2}\Gamma(N/2)}, & x > 0 \\ \\ 0, & \text{otherwise} \end{cases} \qquad (1.62)$$

where $\Gamma(\cdot)$ denotes the gamma function (a standard function in mathematics). Computations involving this distribution rarely require working with the explicit form (1.62). Instead, they can be performed using standard tables or statistical packages.

Selected Properties of the χ^2 Distribution

- $\mathbb{E}[\chi_N^2] = N$
- $\text{var}[\chi_N^2] = 2N$
- If $\chi_{N_1}^2$ and $\chi_{N_2}^2$ are two independent random variables with chi-squared distributions having N_1 and N_2 degrees of freedom, respectively, then $\chi_{N_1}^2 + \chi_{N_2}^2$ also has a chi-squared distribution with $N_1 + N_2$ degrees of freedom. This *additivity property* implies that the sum of any number of independent chi-squared variables is also chi-squared distributed, with the degrees of freedom equal to the sum of the degrees of freedom of the individual variables.

> **Theorem 1.4** (Distribution of the Sample Variance) *Suppose $X_1, X_2, \ldots, X_N$ are independent and identically distributed random variables drawn from normal distribution $\mathcal{N}(\mu, \sigma_X^2)$. Then the sample variance of these variables has the distribution*
>
> $$\frac{(N-1)\hat{\sigma}_X^2}{\sigma_X^2} \sim \chi_{N-1}^2. \qquad (1.63)$$

The reader may be curious as to why this χ^2 distribution involves $N-1$ degrees of freedom, rather than N. To understand this, note that the sample variance (1.7) involves the variables $X_1 - \hat{\mu}_X, X_2 - \hat{\mu}_X, \ldots, X_N - \hat{\mu}_X$. Importantly, these variables satisfy

$$\sum_{n=1}^{N}(X_n - \hat{\mu}_X) = \sum_{n=1}^{N} X_n - \sum_{n=1}^{N} \hat{\mu}_X = N\hat{\mu}_X - N\hat{\mu}_X = 0. \qquad (1.64)$$

This constraint holds for all realizations of $X_1, \ldots, X_N$ and does not depend on population or whether the variables are *iid*. The constraint is a simple consequence of the definition of the sample mean. Because of this constraint, the variables $(X_n - \hat{\mu}_X)$ are not independent, even if $X_1, \ldots, X_N$ are themselves independent. After all, if we know any $N-1$ values of $(X_n - \hat{\mu}_X)$, we know the N'th value exactly, hence the N'th value is not random. Because the variables are not independent, Theorem 1.3 cannot be invoked directly. The constraint (1.64) is a *linear* function of the random variables. The *degrees of freedom* is the number of random variables after removing the number of linear constraints. In general, the degrees of freedom v is determined as

$$v = N - K, \tag{1.65}$$

where

- N = number of independent random variables,
- K = number of linearly independent constraints on the random variables.

The term "linearly independent" is a linear algebra concept discussed in more detail in Chapter 7 (in particular, it does not imply "independence" in the probabilistic sense defined earlier). For the sample variance, there are N random variables of the form $X_n - \hat{\mu}_X$, but these variables satisfy constraint (1.64), hence $K = 1$ and $v = N - 1$.

A theorem that summarizes and extends some of the above theorems is the following:

> **Theorem 1.5** *Suppose X_1, X_2, ..., X_N are independent random variables drawn from the normal distribution $\mathcal{N}(\mu_X, \sigma_X^2)$. Then*
>
> *1 $\hat{\mu}_X \sim \mathcal{N}\left(\mu_X, \frac{\sigma_X^2}{N}\right)$.*
> *2 $(N-1)\hat{\sigma}_X^2 / \sigma_X^2 \sim \chi_{N-1}^2$.*
> *3 $\hat{\mu}_X$ and $\hat{\sigma}_X^2$ are independent.*

Note that items (1) and (2) have appeared in previous theorems, but item (3) is new. The Gaussian assumption is critical for this theorem – if the variables are not normally distributed, then none of the conclusions hold in general. For large N, the sample mean and variance tend to approach normal distributions, owing the Central Limit Theorem, and hence all three items become approximately true.

1.11 Independence versus Zero Correlation

If X and Y are independent, then $\rho_{XY} = 0$. However, the converse of this statement is false in general: $\rho_{XY} = 0$ does not generally imply independence of X and Y. A simple counter example is $Y = X^2$, where X is a random variable with an even density function (that is, $p(x) = p(-x)$), for then $\mathbb{E}[X] = 0$ and

$$\text{cov}[X, Y] = \mathbb{E}\left[(X - \mathbb{E}[X])\left(X^2 - \mathbb{E}[X^2]\right)\right]$$
$$= \int_{-\infty}^{\infty} x(x^2 - \mathbb{E}[X^2])p(x)dx = 0. \tag{1.66}$$

The covariance vanishes because the integrand is an odd function of x. Despite the vanishing covariance, the two variables are highly dependent: knowledge of X completely determines Y. The bottom right of Figure 1.3 illustrates this counter example. The correlation fails to indicate dependence here because X and Y are *nonlinearly* related, whereas the correlation quantifies the degree of *linear* relation between two variables.

For the bivariate normal distribution, zero correlation *does* imply independence.

1.12 Further Topics

This book covers a core set of statistical methods essential for climate research. Unfortunately, not all methods used in climate studies can be covered. Accordingly, at the end of each chapter, we briefly discuss related topics that were not covered in the main chapter.

This chapter considered only distributions related to the Gaussian distribution. However, some quantities, such as daily precipitation, involve *non-Gaussian* distributions. Some important non-Gaussian distributions in climate science include the Gamma, Beta, uniform, and Pareto distributions. This chapter also considered only continuous random variables that can take on values anywhere on the real number line. *Discrete* random variables, which take on only a countable number of values, are also important in climate science. Examples of discrete random variables include: number of category 5 hurricanes in a year, number of rainy days in a month, and number of blocking events during winter. Discrete distributions that commonly arise in climate applications include the Bernoulli, Poisson, multinomial, and hypergeometric distributions.

When estimating population mean and variance, it is natural to use sample mean and variance. A basic question one can ask is: are these the "best" estimators of the mean and variance? For instance, are they the most accurate? If not, what are the best estimators? The answers to these questions form the basis of *estimation theory*. One of the most powerful tools in estimation theory is *maximum likelihood estimation* (MLE), which provides estimates of population parameters when the statistical model has been specified. For instance, when the distribution is Gaussian, MLE tells us that the sample mean is the appropriate estimator of the population mean. However, for non-Gaussian distribution, the sample mean may not be the MLE estimator. The concepts of *consistency, asymptotic normality,* and *efficiency* play an important role in this approach. Other approaches to estimation theory include Bayes estimation and minimax theory.

For additional introductory material on probability and statistics, Spiegel et al. (2000) is highly recommended for a concise introduction to statistics with effective examples. Other recommended introductory texts include Johnson and Bhattacharyya (1992) and Mendenhall et al. (1986). For similar material with meteorological/oceanographic applications, we recommend von Storch and Zwiers (1999) and Wilks (2011).

1.13 Conceptual Questions

1 What is the difference between the mean, median, variance, and quantile?
2 What is the difference between a sample and a population?
3 If you take the average of N random numbers from the same distribution, what can you say about the expected value? Its variance? Its distribution?
4 If you average N random variables, what happens to its variability as N increases?
5 What does the correlation measure?
6 What does a negative correlation coefficient indicate?
7 What is the Central Limit Theorem?
8 What is the degrees of freedom and how do you calculate it?
9 What is a joint distribution? Marginal distribution? Conditional distribution?
10 Explain how the correlation between two variables can vanish yet the variables can be strongly related to each other.
11 If two variables are independent, what is the variance of their sum?
12 What is an estimator? Give an example. What is an unbiased estimator?
13 What is the (approximate) probability that a normally distributed random variable lies within two standard deviations of the mean?
14 If Z is from a standardized normal, what is the distribution of Z^2?

2

Hypothesis Tests

> The problem is to determine the rule which, for each set of values of the observations, specifies what decision should be taken.[1]
>
> *Lehman and Romano (2005)*

The previous chapter considered the following problem: given a distribution, deduce the characteristics of samples drawn from that distribution. This chapter goes in the opposite direction: given a random sample, *infer* the distribution from which the sample was drawn. It is impossible to infer the distribution *exactly* from a finite sample. Our strategy is more limited: we *propose a hypothesis* about the distribution, then decide whether or not to accept the hypothesis based on the sample. Such procedures are called *hypothesis tests*. In each test, a *decision rule* for deciding whether to accept or reject the hypothesis is formulated. The probability that the rule gives the wrong decision when the hypothesis is true leads to the concept of a *significance level*. In climate studies, perhaps the most common questions addressed by hypothesis test are whether two random variables:

- have the same *mean*.
- have the same *variance*.
- are *independent*.

This chapter discusses the corresponding tests for normal distributions, which are known as:

- the t-test (or the difference-in-means test).
- the F-test (or the difference-in-variance test).
- the correlation test.

---

[1] Reprinted by permission from Springer Nature: Springer *Testing Statistical Hypotheses* by E. L. Lehmann and J. P. Romano, 2005. Page 3.

2.1 The Problem

Most people learn in elementary school that the scientific method involves formulating a hypothesis about nature, developing consequences of that hypothesis, and then comparing those consequences to experiment. The initial stages of developing a scientific theory were beautifully described by Richard Feynman, Nobel prize winner in physics:

The principle of science, the definition, almost, is the following: The test of all knowledge is experiment. Experiment is the sole judge of scientific "truth". But what is the source of knowledge? Where do the laws that are to be tested come from? Experiment, itself, helps to produce these laws, in the sense that it gives us hints. But also needed is imagination to create from these hints the great generalizations – to guess at the wonderful, simple, but very strange patterns beneath them all, and then to experiment to check again whether we have made the right guess.[2]

This chapter discusses procedures for testing a hypothesis *after it has been formulated*. The much harder question of how to formulate the initial hypothesis will not be discussed, as it is too mysterious.

To test a scientific hypothesis, the hypothesis must make a testable prediction about the observed universe. To illustrate, consider the following scientific hypotheses:

- Increasing greenhouse gas concentrations will warm the planet.
- Adding fertilizer will enhance plant growth.
- Drinking sugary sodas leads to obesity.
- Chicken soup reduces the severity of a cold.

Each of these hypotheses makes a prediction of the form "if C changes, then E changes." (The letters C and E correspond to "cause" and "effect," but we avoid these terms because causation requires more than dependence.) A natural idea, then, is to observe E while holding C fixed, and then to hold C *at a different value* and observe E again. However, no two experiments yield *exactly* the same measurement because reality involves *randomness*. Because E varies even when C does not, differences in E cannot always be attributed to differences in C. To attribute changes in E to changes in C, such changes must be *significantly larger* than the changes that occur in E with no change in C. Thus, our goal is to decide whether the changes in E that are attributable to C are much larger than the changes that occur in E with no change in C. The extent to which E can change with constant C will be quantified using *multiple independent observations of E with C fixed*.

---

[2] From page 1-1 of Feynman et al. (1977), *Feynman Lectures on Physics* by Richard P. Feynman, copyright ©2000. Reprinted by permission of Basic Books, an imprint of Hachette Book Group, Inc.

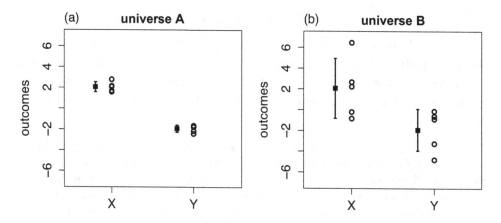

Figure 2.1 Results of experiments in universe A (a) and universe B (b). The circles show five outcomes for X and five outcomes for Y. The black square shows the sample mean for each condition, and the error bars show the mean plus/minus one standard deviation for each condition.

It is instructive to consider a concrete example. Consider Figure 2.1a, which shows the result of performing an experiment five times in universe A while holding C fixed, yielding outcomes $X_1, \ldots, X_5$, and then performed five more times with C fixed at a different value, yielding outcomes $Y_1, \ldots, Y_5$. As can be seen, the X-values are much closer to each other than they are to the Y-values. It would be difficult to argue that the Y-values are drawn from exactly the same population as the X-values. Therefore, these experiments support the claim that C influences the population. Now suppose we travel from our universe, A, to an alternate universe, B, and repeat the same experiments. The outcomes in universe B are shown in Figure 2.1b. Unlike universe A, the values X and Y overlap considerably. It is not unreasonable to suggest that the X- and Y-values were drawn from the *same* population, and that the differences between X and Y are simply a result of random chance. Our goal is to assess the plausibility of these explanations *quantitatively*.

To anticipate how this might be done, consider Figure 2.1 again. Most people would agree that data from universe A provide more compelling support for an influence of C than for B. Why? You might answer that the change in means is larger in A than B, but Table 2.1 reveals that this is not the case: The difference in means is *exactly* the same in the two universes (the samples were contrived to make this happen). In contrast, the *fluctuations about the mean* (that is, standard deviations) are larger in B than in A. This suggests that we base our conclusion on the *ratio* of the difference in means relative to some measure of the fluctuations about the mean, rather than the absolute difference in means.

Table 2.1. *Sample means and standard deviations of the experimental outcomes shown in Figure 2.1. The last column gives the value of t_0, defined later in (2.12).*

	Mean of X	Mean of Y	Stdv of X	Stdv of Y	t_0
universe A	2.06	-1.93	0.48	0.33	15.27
universe B	2.06	-1.93	2.88	2.01	2.54

2.2 Introduction to Hypothesis Testing

A basic paradigm in statistical analysis is to assume that observations are the output of a *random number generator*. A random number generator spits out random numbers from a particular probability distribution. Let $X_1, \ldots, X_{N_X}$ be random numbers from density p_X, and let $Y_1, \ldots, Y_{N_Y}$ be random numbers from density p_Y. The densities p_X and p_Y describe the populations under two separate conditions C, hence the scientific claim that C influences the experiments is tantamount to the claim that p_X differs from p_Y. However, we do not know p_X and p_Y; we only know samples drawn from those distributions. Our goal is to decide if p_X differs from p_Y using samples drawn from the respective distributions.

There are two possibilities: either the two distributions are the same or they are not. The hypothesis that the populations are the same is called the *null hypothesis* and is denoted H_0:

$$H_0^* : \quad p_X = p_Y. \tag{2.1}$$

The term "null" emphasizes that the hypothesis is the default assumption that condition C has no effect on the population. The asterisk * is used to distinguish the hypothesis from the more narrow hypothesis discussed next. The hypothesis that the distributions are not equal is called the *alternative hypothesis* and denoted H_A:

$$H_A^* : \quad p_X \neq p_Y. \tag{2.2}$$

The samples $X_1, \ldots, X_{N_X}$ are assumed to be *iid* as p_X, and $Y_1, \ldots, Y_{N_Y}$ are *iid* as p_Y.

If we know something about the population, then it is advantageous to incorporate that information into our analysis. Loosely speaking, we want to avoid "wasting" the data to confirm things we already know about the population, and instead use it efficiently to infer things we do not know about the population. Therefore, instead of allowing p_X and p_Y to be *any* distribution, we suppose that we know that the distributions are from a particular family, namely a *Gaussian* distribution. Even if the population is non-Gaussian, the test discussed here remains valid for large N_X and N_Y primarily because of the Central Limit Theorem. An advantage of a normal distribution is that it is specified completely by *two* quantities, namely the mean and variance. Let μ_X and σ_X^2 denote the mean and variance of p_X, and let μ_Y and σ_Y^2 denote the mean and variance of p_Y. These are known as *population parameters* because they describe properties of the population. For normal populations, the null hypothesis H_0^* can be expressed equivalently as

$$H_0: \quad \mu_X = \mu_Y \quad \text{and} \quad \sigma_X = \sigma_Y. \tag{2.3}$$

To investigate if H_0 is true, we now only need to decide if the means are equal and if the variances are equal. Naturally, we need to estimate these quantities. We use the estimators discussed in Chapter 1. Specifically, the mean and variance of p_X are estimated as

$$\hat{\mu}_X = \frac{1}{N_X} \sum_{n=1}^{N_X} X_n \quad \text{and} \quad \hat{\sigma}_X^2 = \frac{1}{N_X - 1} \sum_{n=1}^{N_X} (X_n - \hat{\mu}_X)^2. \tag{2.4}$$

Similar expressions give estimates for the mean and variance of p_Y, which are denoted $\hat{\mu}_Y$ and $\hat{\sigma}_Y^2$. Numerical values of these estimates for the data shown in Figure 2.1 are given in Table 2.1.

The most straightforward approach is to test equality of means and equality of variances *separately*. The test for equality of variance makes no assumptions about the means. After all, the sample variance depends on the difference $X_n - \hat{\mu}_X$, which does not depend in any way on μ_X; e.g., $\mathbb{E}[X_n - \hat{\mu}_X] = 0$. Therefore, tests based on sample variances are *noncommittal* regarding the means, so it makes sense to test equality of variance first. If a difference in variance is detected, then we stop and declare detection of a difference in distributions. If no difference in variance is detected, then we proceed to test equality of means *assuming the populations have the same variance*. This assumption is common when testing differences in means. Unequal population variances could be allowed, but the associated test, known as the *Behrens–Fisher problem* (Scheffé, 1970), involves more complicated sampling distributions and involves subtleties that we prefer to avoid.

In the climate literature, equality-of-means tests are far more common than equality-of-variance tests. Accordingly, we postpone discussion of equality of variances to Section 2.4.2 and focus on the question of whether the population means differ, assuming no difference in variance. In this case, our alternative hypothesis is

$$H_A: \quad \mu_X \neq \mu_Y \quad \text{and} \quad \sigma_X = \sigma_Y. \tag{2.5}$$

A natural idea is to examine the difference in sample means, $\hat{\mu}_X - \hat{\mu}_Y$. How do we use the observed value of $\hat{\mu}_X - \hat{\mu}_Y$ to make a judgment about the null hypothesis? Answering this question requires thinking about the problem from a broader perspective. Imagine that other investigators collect X- and Y-samples *independently*.

Each investigator will obtain a different value of $\hat{\mu}_X - \hat{\mu}_Y$. Our value of $\hat{\mu}_X - \hat{\mu}_Y$ is but one realization out of many from this army of investigators. Imagine collecting these values and plotting a histogram to gain an idea of the distribution of $\hat{\mu}_X - \hat{\mu}_Y$. The distribution of $\hat{\mu}_X - \hat{\mu}_Y$ is known as the *sampling distribution*. Our goal is to derive the sampling distribution of $\hat{\mu}_X - \hat{\mu}_Y$ for a universe in which the null hypothesis H_0 is true, and then see where our particular value of $\hat{\mu}_X - \hat{\mu}_Y$ lies relative to this distribution.

The distribution of $\hat{\mu}_X - \hat{\mu}_Y$ can be computed from the methods discussed in Chapter 1. Specifically, from Example 1.13, the distributions of $\hat{\mu}_X$ and $\hat{\mu}_Y$ are

$$\hat{\mu}_X \sim \mathcal{N}\left(\mu_X, \frac{\sigma_X^2}{N_X}\right) \quad \text{and} \quad \hat{\mu}_Y \sim \mathcal{N}\left(\mu_Y, \frac{\sigma_Y^2}{N_Y}\right). \tag{2.6}$$

Incidentally, if N_X and N_Y are large, these distributions would be obtained even if X and Y came from non-Gaussian distributions (see Central Limit Theorem 1.1, particularly (1.50)). By Theorem 1.2, the distribution of the *difference* in sample means is

$$\hat{\mu}_X - \hat{\mu}_Y \sim \mathcal{N}\left(\mu_X - \mu_Y, \frac{\sigma_X^2}{N_X} + \frac{\sigma_Y^2}{N_Y}\right). \tag{2.7}$$

By Example 1.3, the difference in sample means can be expressed in standardized form as

$$\frac{(\hat{\mu}_X - \hat{\mu}_Y) - (\mu_X - \mu_Y)}{\sqrt{\frac{\sigma_X^2}{N_X} + \frac{\sigma_Y^2}{N_Y}}} \sim \mathcal{N}(0, 1). \tag{2.8}$$

Under either H_0 or H_A, we have $\sigma_X = \sigma_Y = \sigma$, hence the previous expression simplifies to

$$Z = \frac{(\hat{\mu}_X - \hat{\mu}_Y) - (\mu_X - \mu_Y)}{\sigma\sqrt{\frac{1}{N_X} + \frac{1}{N_Y}}} \sim \mathcal{N}(0, 1). \tag{2.9}$$

The variable Z is of the form

$$Z = \frac{\text{observed} - \text{hypothesized}}{\text{standard error}}. \tag{2.10}$$

The "observed" quantity is $\hat{\mu}_X - \hat{\mu}_Y$ and the "hypothesized" quantity is $\mu_X - \mu_Y$. "Standard error" refers to the standard deviation of $\hat{\mu}_X - \hat{\mu}_Y$. Z is an example of a *signal-to-noise ratio*, which arises explicitly or implicitly in all hypothesis tests. Z describes the difference between observed and hypothesized value in units of the standard error. For instance, $Z = 3$ means that the observed value is three standard errors away from the hypothesized value.

Unfortunately, Z cannot be evaluated because σ is an unknown population parameter. A natural idea is to substitute an estimate of σ into the expression for Z. Because the variances are equal under both H_0 and H_A, two natural estimates of σ^2 are $\hat{\sigma}_X^2$ and $\hat{\sigma}_Y^2$, but choosing only one of those estimates while ignoring the other is wasteful. A more efficient use of data is to combine those estimates into a single (presumably better) estimate. The appropriate combination is the *pooled variance* $\hat{\sigma}_P^2$, as defined in (2.20). If the sample sizes are equal, then the pooled estimate is simply the average of $\hat{\sigma}_X^2$ and $\hat{\sigma}_Y^2$. Substituting $\hat{\sigma}_P$ for σ in the expression for Z yields the quantity

$$T = \frac{(\hat{\mu}_X - \hat{\mu}_Y) - (\mu_X - \mu_Y)}{\hat{\sigma}_P \sqrt{\frac{1}{N_X} + \frac{1}{N_Y}}}. \tag{2.11}$$

If the null hypothesis (2.3) is true, then $\mu_X = \mu_Y$ and (2.11) becomes

$$T_0 = \frac{\hat{\mu}_X - \hat{\mu}_Y}{\hat{\sigma}_P \sqrt{\frac{1}{N_X} + \frac{1}{N_Y}}}. \tag{2.12}$$

T_0 depends *only* on sample quantities and therefore can be evaluated. The values of T_0 for the data shown in Figure 2.1 are given in Table 2.1. For instance, for universe B, the value is

$$t_0 = \frac{2.06 - (-1.93)}{\sqrt{\frac{2.88^2 + 2.01^2}{2}} \sqrt{\frac{1}{5} + \frac{1}{5}}} \approx 2.54. \tag{2.13}$$

The presence of $\hat{\sigma}_P$ in (2.12) implies that T_0 is a *ratio* of random variables, which generally does not have a normal distribution. In fact, when H_0 is true, T_0 has a distribution known as the *t-distribution*. The t-distribution depends on the sum of the degrees of freedom of $\hat{\sigma}_X$ and $\hat{\sigma}_Y$, namely $N_X + N_Y - 2$. For the example in Section 2.1, $N_X = N_Y = 5$, which gives 8 degrees of freedom. The corresponding t-distribution when H_0 is true is shown in Figure 2.2. The t-distribution looks similar to a normal distribution, but it is not quite the same (it approaches the normal distribution for large degrees of freedom).

What happens to the distribution of T_0 if H_A is true? Comparing (2.11) and (2.12) shows that T and T_0 are related as

$$T_0 = T + \frac{\mu_X - \mu_Y}{\hat{\sigma}_P \sqrt{\frac{1}{N_X} + \frac{1}{N_Y}}}. \tag{2.14}$$

T has a t-distribution regardless of whether H_0 and H_A is true. The last term on the right in (2.14) vanishes if H_0 is true, but it is nonzero if H_A is true. As a concrete example, suppose $\mu_X - \mu_Y = 4$. Then, (2.14) shows that the mean of T_0

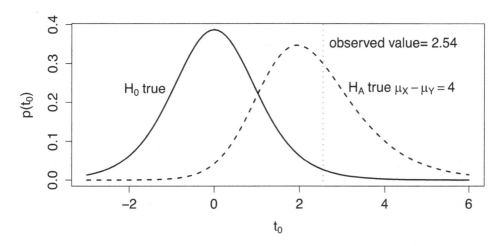

Figure 2.2 Density function for the t-distribution with 8 degrees of freedom when H_0 is true (solid) and when H_A is true (dashed). The value of t_0 for the data shown in Figure 2.1 for universe B is indicated by the vertical dotted line. The distribution under H_A is based on noncentrality parameter 2.1.

under H_A is *shifted to the right* relative to the mean under H_0. This distribution is shown in Figure 2.2 (drawing the latter distribution requires knowing the *noncentral* t-distribution, but this distribution is not needed for the test discussed here, hence we do not discuss it further).

Figure 2.2 summarizes what we have so far: the distribution of T_0 under hypotheses H_0 and H_A, and the specific value of t_0 obtained from the experiments for universe B. How can these results be used to decide whether to accept or reject the null hypothesis? To answer this question, let us return to our imaginary army of investigators, each of whom is repeating the experiments independently. Suppose we are the superintendent who must give instructions to each investigator about how to accept or reject the null hypothesis. Importantly, the instructions must be based only on the sample seen by the investigator and must be given *before* any data is collected. Also, there is no basis to distinguish one investigator over another, so the instructions given to each investigator must be the *same*. This instruction is known as a *decision rule*. To formulate a decision rule, all possible values of a statistic are divided into two groups, which are known as the *acceptance* and *rejection* regions. If the observed value falls in the acceptance region, then the investigator should accept the null hypothesis, otherwise the investigator should reject the null hypothesis.

How should the acceptance and rejection regions be defined? Answering this question requires considering the alternative hypothesis more carefully. The alternative hypothesis (2.5) merely states that $\mu_X - \mu_Y$ is nonzero, it does not specify the *sign* of the difference. According to (2.14), if $\mu_X - \mu_Y$ is either positive or negative, then the mean of T_0 is shifted *away from zero*, suggesting that H_0 should be rejected for large *absolute* values of T_0. Accordingly, the decision rule is of the form

$$\text{Reject } H_0 \text{ if } |T_0| > t_C \qquad\qquad (2.15)$$
$$\text{Accept } H_0 \text{ if } |T_0| < t_C,$$

for some constant t_C, known as the *critical value*. How should the critical value t_C be chosen? The answer depends on the error rate that we are willing to tolerate. No decision rule will give the correct answer all the time, but we can control the probability of certain errors through our choice of t_C. If the null hypothesis is true and the statistic falls in the rejection region, then some fraction of our investigators will reject the null hypothesis and therefore draw the wrong conclusion. This error is termed a Type I error. (Type II errors are discussed in Section 2.6.) The probability of a Type I error – i.e., rejecting the null hypothesis when it is true – is termed the *significance level*, often denoted by α. This probability equals the area under the density function in the region $|T_0| > t_C$. Thus, specifying t_C is equivalent to specifying α, the probability of committing a Type I error.

In scientific studies, a standard choice for the significance level is 5%. This choice has achieved a mythical status, but it is merely the legacy of a choice made a long time ago for compiling statistical tables when computers were not readily available. For our specific example with 8 degrees of freedom, the value of t_C such that

$$\mathcal{P}\left(|T_0| > t_C\right) = 5\%, \qquad\qquad (2.16)$$

is $t_C = 2.31$ (this value is obtained from a table or a statistical software). The corresponding rejection and acceptance regions are indicated in Figure 2.3. The actual value obtained in universe B is $t_0 = 2.54$. Since $|t_0| > t_C$, the decision rule tells us to reject the null hypothesis. This procedure is known as *Student's t-test*, or simply the *t-test*.

According to the decision rule, H_0 is rejected as long as $|t_0| > t_C$. No distinction is made between "close calls," where t_0 is close to t_C, versus "out of the park," where t_0 is far from t_C. A more informative conclusion is to state the *p-value*. The p-value is the probability of obtaining the observed value, or a more extreme value, of a test statistic when the null hypothesis is true. For our example with 8 degrees of freedom, the probability $|T_0| > 2.54$ is 3.4%. Thus, the p-value of t_0 for our experiments is 3.4% (again, this value is obtained from a table or a statistical software package). In general, if the null hypothesis is rejected, then the p-value is *less* than the chosen significance level.

In universe A, the numerical values give $t_0 = 15.27$, which far exceeds the critical value of $t_C = 2.31$ and therefore is strong evidence of a difference in means, consistent with our initial intuition. In fact, the corresponding p-value is about 1 in 3 million.

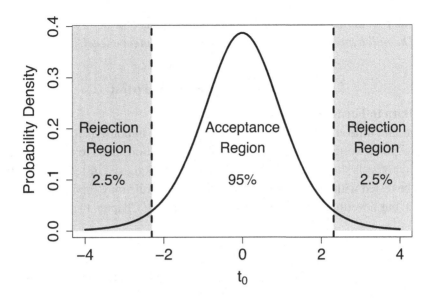

Figure 2.3 Acceptance and rejection regions corresponding to a t-distribution with 8 degrees of freedom for a two-sided significance test at the 5% level.

If the statistic falls in the rejection region, then the procedure is to reject H_0. In this case, it is incorrect to say "the populations means *are* different." After all, there is a 5% chance the statistic could have fallen in the rejection region even if the populations were the same. Also, it is incorrect to claim "the probability that the null hypothesis is true is 5%." In fact, the probability that the null hypothesis is correct is either 100% or 0% (either it is true or not). The fact that a statistic falls in the rejection region simply means that the observed value is unlikely – but not impossible – under the null hypothesis. Similarly, if the statistic falls in the acceptance region, avoid claiming the null hypothesis is true. The decision to reject is merely a statement about the rarity of the data when the null hypothesis is true.

Example 2.1 (Expressing the Decision) *Rejection of the null hypothesis at $\alpha = 5\%$ can be expressed in different ways:*

- *"The value of the statistic is significant at the 5% level."*
- *"The null hypothesis is rejected at the 5% significance level."*
- *"The statistic lies outside the range of likely values under the null hypothesis."*
- *"A difference in means has been detected at the 5% significance level."*

The following are **incorrect** *ways of characterizing the rejection:*

- *"The null hypothesis is false."*
- *"The alternative hypothesis is true."*

- *"The probability that the null hypothesis is true is 5%."*
- *"The null hypothesis is rejected at the 95% confidence level."*

2.3 Further Comments on the t-test

Using Data to Formulate Hypotheses

We have said little about how to formulate a scientific hypothesis for testing. However, one point worth emphasizing is that the observations used to formulate a scientific hypothesis should not be *re-used* to test the hypothesis. Instead, a hypothesis should be tested with *independent* observations. This rule may seem too obvious to mention, but nevertheless it is violated quite often (see Chapter 11).

Confounding Factors

An implicit assumption in our discussion was that an "experiment" can be performed in which one thing is varied while all other things are held fixed. This ideal experiment can never be realized exactly. For the real climate system, repeated controlled experiments are not even possible. One might try to approximate this ideal by separating observations according to some condition C, but the climate system is constantly fluctuating and variables that have been ignored will not be constant and may affect the observation. Such extraneous factors are known as *confounding factors*. Confounding factors need to be taken into account to avoid invalid inferences, usually by further subdividing the observations, or constructing models that take into account confounding factors.

Test Statistic

The t statistic is a special case of a test statistic.

> **Definition 2.1** (Test Statistic) *A test statistic is a function of one or more random variables that does not depend on any unknown parameter:*

1 $\sum_i (X_i - \mu)$ is not a test statistic since the population mean μ is unknown.
2 $\hat{\mu}_X = \frac{1}{N} \sum_i X_i$ is a test statistic.
3 $\hat{\sigma}_X^2 = \frac{1}{N-1} \sum_i (X_i - \hat{\mu}_X)^2$ is a test statistic.
4 Z defined in (2.9) is not a test statistic because μ_X, μ_Y, and σ are unknown.
5 T_0 defined in (2.12) is a test statistic.

Central Limit Theorem

Although we assumed the populations had normal distributions, the test gives reasonable results even if this is not the case. One reason for this is the Central Limit Theorem. For the hypothetical data shown in Figure 2.1, the sample size is 10. Is 10 sufficiently large to invoke the central limit theorem? This question is difficult to answer because the Central Limit Theorem provides no guidance on the sample size needed for it to apply. Even if it did, the condition would depend on the degree to which the population differs from a Gaussian, which is unknown (otherwise we would not be asking this question). For univariate data, one can plot histograms of the data and see if any obvious departures from normality exist, although for small sample sizes the histogram will be noisy. Objective procedures for testing normality include the *Shapiro–Wilk test, Lillifors test*, and the *Anderson–Darling test*.

Independence Assumption
In deriving the sampling distribution, we assumed that each X was independent of the others, each Y was independent of the others, and the X's and Y's were independent of each other. Is independence a good assumption? Usually not in climate studies! Most climate variables are correlated over sufficiently short time scales. For example, today's high temperature is a good predictor of tomorrow's high temperature. This means that local air temperature is not independent over time scales of days. For most meteorological variables, the value at any geographic location appears to be uncorrelated with itself after two years, provided trends and annual and diurnal cycles have been subtracted out (one year is not long enough because some ENSO indices are significantly correlated beyond one year; methods for removing annual cycles and trends will be discussed in Chapter 8). Thus, one could analyze samples separated by one year, but at the expense of discarding data. A better approach is to account explicitly for serial correlation (see DelSole and Tippett (2020) for an example of such an approach).

The Pooled Variance
Under the null hypothesis (2.3), the population variances are the same. It is natural to estimate this variance using some combination of sample variances

$$\hat{\sigma}_X^2 = \frac{1}{N_X - 1} \sum_{i=1}^{N_X} (X_i - \hat{\mu}_X)^2 \quad \text{and} \quad \hat{\sigma}_Y^2 = \frac{1}{N_Y - 1} \sum_{i=1}^{N_Y} (Y_i - \hat{\mu}_Y)^2. \quad (2.17)$$

From Theorem 1.4, we know that these quantities have the following distributions:

$$\frac{(N_X - 1)\hat{\sigma}_X^2}{\sigma^2} \sim \chi_{N_X-1}^2 \quad \text{and} \quad \frac{(N_Y - 1)\hat{\sigma}_Y^2}{\sigma^2} \sim \chi_{N_Y-1}^2. \quad (2.18)$$

These quantities are independent and thus their sum has a chi-squared distribution with degrees of freedom equal to the sum of the individual degrees of freedom:

$$\frac{(N_X - 1)\hat{\sigma}_X^2 + (N_Y - 1)\hat{\sigma}_Y^2}{\sigma^2} \sim \chi_{N_X+N_Y-2}^2. \quad (2.19)$$

Using the fact $\mathbb{E}[\chi_v^2] = v$, it is a simple exercise to show that the normalized quantity

$$\hat{\sigma}_p^2 = \frac{(N_X - 1)\hat{\sigma}_X^2 + (N_Y - 1)\hat{\sigma}_Y^2}{N_X + N_Y - 2}, \quad (2.20)$$

is an *unbiased estimate of the variance* (that is, $\mathbb{E}[\hat{\sigma}_p^2] = \sigma^2$). The quantity $\hat{\sigma}_p^2$ is termed the *pooled estimate* of the variance. The pooled estimate has distribution

$$(N_X + N_Y - 2)\left(\frac{\hat{\sigma}_p^2}{\sigma^2}\right) \sim \chi_{N_X+N_Y-2}^2. \tag{2.21}$$

The population variance of $\hat{\sigma}_p^2$ is always less than that of either $\hat{\sigma}_X^2$ or $\hat{\sigma}_Y^2$; hence the pooled estimate has less uncertainty and is a more precise estimate of σ^2.

Theorem 2.1 (The t-distribution) *Let Z and W be independent random variables, where Z is normally distributed as $\mathcal{N}(0,1)$ and W is chi-squared distributed with v degrees of freedom. Then the random variable*

$$T = \frac{Z}{\sqrt{\frac{W}{v}}} \tag{2.22}$$

has a t-distribution with v degrees of freedom. This distribution is also known as a Student's t-distribution and denoted t_v. Properties of the t-distribution include:

1 $\mathbb{E}[T] = 0$
2 $\operatorname{var}[T] = v/(v-2)$
3 *as $v \to \infty$, $T \sim \mathcal{N}(0,1)$*

Theorem 2.1 can be used to show that T_0 defined in (2.12) has a t-distribution. Specifically, under the null hypothesis, (2.3) and (2.9) imply

$$Z = \frac{\hat{\mu}_X - \hat{\mu}_Y}{\sigma\sqrt{\frac{1}{N_X} + \frac{1}{N_Y}}} \sim N(0,1). \tag{2.23}$$

In addition, we may identify W with (2.21) and therefore $v = N_X + N_Y - 2$,

$$W = (N_X + N_Y - 2)\left(\frac{\hat{\sigma}_p^2}{\sigma^2}\right) \sim \chi_{N_X+N_Y-2}^2. \tag{2.24}$$

Furthermore, according to Theorem 1.5, $\hat{\mu}_X$ and $\hat{\sigma}_X^2$ are independent, and $\hat{\mu}_Y$ and $\hat{\sigma}_Y^2$ are independent. It follows that Z and W are independent and therefore satisfy the assumptions of Theorem 2.1. Therefore, substituting (2.23) and (2.24) into Theorem (2.1) yields

$$T_0 = \frac{\hat{\mu}_X - \hat{\mu}_Y}{\hat{\sigma}_p\sqrt{\frac{1}{N_X} + \frac{1}{N_Y}}} \sim t_{N_X+N_Y-2} \quad \text{if } H_0 \text{ is true.} \tag{2.25}$$

Note that T_0 depends only on sample quantities – that is, T_0 is a statistic (e.g., σ has cancelled out). Moreover, the sampling distribution of T_0 is independent of the unknown population parameters σ^2, μ_X, and μ_Y. This result is amazing! It implies that we can test a hypothesis about the population *without having to know any of the population parameters.*

Definition 2.2 (Upper-tail critical value for a t-distribution) *The upper-tail critical value is the number $t_{\alpha, v}$ that has an area α to the right of it, for a t-distribution with v degrees of freedom. Equivalently, it is the number $t_{\alpha, v}$ that satisfies*

$$P\left(t > t_{\alpha, v}\right) = \alpha, \tag{2.26}$$

for a t-distribution. Conversely, the lower-tail critical value is the number that has area α to the left of it. The t-distribution is symmetric, hence the lower-tail critical value is $t_{1-\alpha, v} = - t_{\alpha, v}$. In general, critical values specify the boundaries of the rejection region.

One-tailed and Two-tailed Tests
One subtle concept needs to be clarified. There exist situations in which alternative hypotheses specify the sign of $\mu_X - \mu_Y$. For instance, increasing greenhouse gas concentrations will cause global *warming*; drinking sugary sodas will *increase* your chance of obesity; adding fertilizer will *enhance* plant growth. In such cases, the rejection region lies only in the positive or negative tail of the distribution but not both. Such decision rules are termed *one-tailed*. In contrast, decision rule (2.15) is said to be *two-tailed* because the rejection region lies in both the positive and negative tails. In this case, the decision rule is related to the critical value for the t-distribution as

$$\text{Reject } H_0 \text{ if } |T_0| > t_{\alpha/2} \tag{2.27}$$
$$\text{Accept } H_0 \text{ if } |T_0| < t_{\alpha/2}.$$

Such a decision rule is illustrated in Figure 2.3. In practice, a two-tailed test is generally recommended even if the alternative hypothesis implies a one-tailed test, because a two-tailed test is more conservative and, in practice, a significant difference in the direction contrary to expectations would be worth identifying.

2.4 Examples of Hypothesis Tests

This section briefly outlines other hypothesis tests that are common in climate studies.

2.4.1 Difference-in-Means Test

Although the difference-in-means test has been discussed in Sections 2.2 and 2.3, the description is spread over several pages. For convenience, this test is summarized concisely here. The difference-in-means test is most appropriate when $X_1, \ldots, X_{N_X}$

are independent and normally distributed random variables, and $Y_1, \ldots, Y_{N_Y}$ are independent and normally distributed random variables. The null hypothesis is that the two normal distributions have the same mean and variance. Let the sample mean and variance of the X's be $\hat{\mu}_X$ and $\hat{\sigma}_X^2$, and those for Y be $\hat{\mu}_Y$ and $\hat{\sigma}_Y^2$. If the null hypothesis is true, then the test statistic T_0 defined in (2.12) has a t-distribution with $N_X + N_Y - 2$ degrees of freedom. The decision rule with a $100\alpha\%$ significance level is to reject the null hypothesis when

$$
\begin{aligned}
T_0 &> t_{\alpha, N_X+N_Y-2} & &\text{one-tailed test for } \mu_X > \mu_Y \\
T_0 &< t_{\alpha, N_X+N_Y-2} & &\text{one-tailed test for } \mu_X < \mu_Y \qquad (2.28)\\
|T_0| &> t_{\alpha/2, N_X+N_Y-2} & &\text{two-tailed test.}
\end{aligned}
$$

Example 2.2 (Test Equality of Means) *A person claims that ENSO influences the winter-mean temperature in a particular city. To investigate this claim, a scientist collects the winter-mean temperature for the past 20 El Niños and the past 20 La Niñas (giving a total of 40 numbers). For the 20 El Niño years, the sample mean is 280K and the standard deviation is 3K. For the 20 La Niña years, the sample mean is 275K and standard deviation is 4K. Do these data support the claim?*

To investigate the claim, the scientist tests the null hypothesis that the population mean during El Niño equals the population mean during La Niña. Since the sample variances are $\hat{\sigma}_X^2 = 9K^2$ and $\hat{\sigma}_Y^2 = 16K^2$, the pooled variance is

$$
\hat{\sigma}_p^2 = \frac{(N_X - 1)\hat{\sigma}_X^2 + (N_Y - 1)\hat{\sigma}_Y^2}{N_X + N_Y - 2} = \frac{19 * 9 + 19 * 16}{20 + 20 - 2} = 12.5.
$$

The difference in sample means is $\hat{\mu}_X - \hat{\mu}_Y = 280K - 275K = 5K$. Therefore, the test statistic (2.12) is

$$
T_0 = \frac{5}{\sqrt{12.5}\sqrt{\frac{1}{20} + \frac{1}{20}}} \approx 4.5.
$$

The claim is noncommittal regarding the direction of change, so a two-tailed test is performed. The critical value for a two-tailed significance test at the 5% level for a t-distribution with 38 degrees of freedom is 2.0 (this value is obtained by looking it up in a table or from a statistical software package). Since 4.5 > 2.0, the null hypothesis is rejected at the 5% significance level. Thus, the data support the claim.

2.4.2 Difference in Variance Test (also called the F-test)

Recall that the null hypothesis (2.3) assumes that the *variances* are equal. The natural statistic for testing equality of variances is the *ratio* of variances. The sampling distribution of a variance ratio is given by the following theorem.

Theorem 2.2 (Distribution of a Ratio of Chi-Squared Random Variables) *Let V_1 and V_2 be independent random variables that are chi-squared distributed with ν_1 and ν_2 degrees of freedom, respectively. Then the random variable*

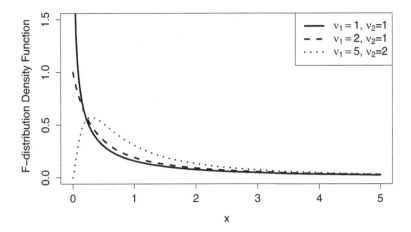

Figure 2.4 Illustration of the F-distribution for three different values of the degrees of freedom v_1 and v_2.

$$F = \frac{V_1/v_1}{V_2/v_2} \tag{2.29}$$

has a distribution termed the F-distribution with v_1 and v_2 degrees of freedom. Examples of the F-distribution are illustrated in Figure 2.4.

To apply this theorem, we assume that the X's and Y's are independent and normally distributed. Then, by Theorem 1.4, the sample variances $\hat\sigma_X^2$ and $\hat\sigma_Y^2$ satisfy (2.18), which leads to the identifications

$$V_1 = \frac{(N_X - 1)\hat\sigma_X^2}{\sigma_X^2}, \quad v_1 = N_X - 1, \quad \text{and} \quad \frac{V_1}{v_1} = \frac{\hat\sigma_X^2}{\sigma_X^2}. \tag{2.30}$$

Similarly,

$$\frac{V_2}{v_2} = \frac{\hat\sigma_Y^2}{\sigma_Y^2}. \tag{2.31}$$

Therefore, Theorem 2.2 implies that

$$\frac{\hat\sigma_X^2}{\hat\sigma_Y^2} \frac{\sigma_Y^2}{\sigma_X^2} \sim F_{N_X-1, N_Y-1}. \tag{2.32}$$

Under the null hypothesis $\sigma_X = \sigma_Y$, the theorem tells us that the test statistic

$$\hat{F} = \frac{\hat\sigma_X^2}{\hat\sigma_Y^2} \tag{2.33}$$

has an F-distribution. Here, a significant difference in variance corresponds to a value of F *different from one*. This test is called the F-test.

Definition 2.3 (Upper-tail critical value for an F-distribution) *The upper-tail critical value is the number F_{α, v_X, v_Y} that has an area α to the right of it, for an F-distribution with v_X and v_Y degrees of freedom. Equivalently, it is the number F_{α, v_X, v_Y} that satisfies*

$$P\left(F > F_{\alpha, v_X, v_Y}\right) = \alpha, \tag{2.34}$$

for an F-distribution. Conversely, the lower-tail critical value is the number that has an area α to the left of it. Because swapping the X and Y labels leads to the reciprocal of F, the critical value satisfies $F_{1-\alpha, v_X, v_Y} = 1/F_{\alpha, v_X, v_Y}$.

The two-tailed version of the F-test is to reject the null hypothesis if

$$\hat{F} < F_{1-\alpha/2, N_X-1, N_Y-1} \quad \text{or} \quad \hat{F} > F_{\alpha/2, N_X-1, N_Y-1}. \tag{2.35}$$

Of course, the labels X and Y are arbitrary, so whatever decision is made regarding the null hypothesis should not depend on the arbitrary assignment of labels. Swapping the X and Y labels has the effect of taking the reciprocal of the original $\hat{F}$. Moreover, as described in Theorem (2.2), the F-distribution satisfies the identity

$$F_{1-\alpha, v_X, v_Y} = \frac{1}{F_{\alpha, v_Y, v_X}}. \tag{2.36}$$

The implication of these properties is that one can simply assign the label X to the variable with the larger sample variance, ensuring $\hat{F} \geq 1$, and then use the decision rule

$$\text{Reject } H_0 \text{ if } \hat{F} > F_{\alpha/2, N_X-1, N_Y-1} \tag{2.37}$$

$$\text{Accept } H_0 \text{ if } \hat{F} < F_{\alpha/2, N_X-1, N_Y-1}.$$

This test is equivalent to a two-tailed test even though it takes the form of a one-tailed test. The equivalence depends on assigning the X and Y labels after looking at the data.

Importantly, if a null hypothesis involves multiple assumptions (as it often does), then violation of any one of those assumptions could lead to rejection of the null hypothesis. For instance, recall that the F-test assumes Gaussian distributions for X and Y. Unlike the t-test, the F-test is sensitive to departures from normality. Specifically, a large F-value that is rare when X and Y have the same normal distribution may not be rare when X and Y have the same non-normal distribution. As a result, the F-test may reject a null hypothesis because the distributions are non-Gaussian, rather than because the variances differ. Thus, when the F-test rejects the null hypothesis, the claim that the variances differ should be supported by the fact that the data plausibly satisfies the other assumptions in the null hypothesis, particularly the assumption of normality and independence.

Example 2.3 (Difference-in-Variance Test) *A person claims that ENSO influences the variability of the winter-mean temperature in a particular city. To investigate this claim, a scientist collects the winter-mean temperature for the past 20 El Niños* *and for the past 20 La Niñas (giving a total of 40 numbers). The standard deviations are 3K for the El Niño periods and 4K during the La Niña periods. Do these data support the claim?*

The above claim can be investigated by testing the hypothesis that the population variances are the same during El Niño and La Niña. The sample variances are $\hat{\sigma}_Y^2 = 9K^2$ and $\hat{\sigma}_X^2 = 16K^2$, where X labels the variable with larger variance. The test statistic is

$$\frac{\hat{\sigma}_X^2}{\hat{\sigma}_Y^2} = \frac{16}{9} \sim 1.78.$$ (2.38)

The claim does not assert a direction to the dependence, so a two-tailed test is performed. The upper 97.5% critical value for an F distribution with 19 and 19 degrees of freedom is 2.5 (this value is obtained by looking it up in a table or from a statistical software package). Because $1.78 < 2.5$, the null hypothesis is not rejected. That is, the ratio of variances is not significantly different from one. Thus, the data do not support the claim.

2.4.3 The Correlation Test

Another assumption in the null hypothesis is that the samples were *independent*. Thus, to complete our analysis, we need a test for independence. The question of independence arises frequently in climate studies, so this test is important for its own sake. The test discussed here is called the *correlation test*. The basic idea relies on a simple fact from Chapter 1, namely that if X and Y are independent, then their correlation coefficient vanishes. Conversely, if the correlation coefficient does not vanish, then X and Y are not independent. A sample estimate of the correlation coefficient is given by

$$\hat{\rho} = \frac{\sum\limits_{j=1}^{N} \left(X_j - \hat{\mu}_X\right)\left(Y_j - \hat{\mu}_Y\right)}{\sqrt{\left(\sum\limits_{j=1}^{N} \left(X_j - \hat{\mu}_X\right)^2\right)\left(\sum\limits_{k=1}^{N} \left(Y_k - \hat{\mu}_Y\right)^2\right)}}.$$ (2.39)

Just as in the t-test, we postulate a null hypothesis and then use the data to test that null hypothesis. Here, the appropriate null hypothesis is that the population correlation coefficient vanishes; that is, $\rho = 0$. To test this hypothesis, we need to know the sampling distribution of $\hat{\rho}$ under the null hypothesis. This distribution is given by the following theorem.

Theorem 2.3 (Distribution of the Correlation Coefficient) *Assume $X_1, \ldots, X_N$ are independent samples from a normal distribution, $Y_1, \ldots, Y_N$ are independent samples from another normal distribution, and the X's and Y's are independent of each other. Let $\hat{\rho}$ be their sample correlation defined in (2.39). Then, the statistic*

$$T = \frac{\hat{\rho}\sqrt{N-2}}{\sqrt{1-\hat{\rho}^2}}, \qquad (2.40)$$

has a t-distribution with $N-2$ degrees of freedom.

Note that the number of degrees of freedom is $N-2$. The reason for this will be discussed in Chapter 9. Also, it is generally recommended that one apply a two-tailed test, for reasons discussed in Section 2.3. Then, the decision rule is (2.27) with $t_C = t_{\alpha/2, N-2}$.

Theorem 2.3 expresses the sampling distribution of $\hat{\rho}$ in terms of T. Often, it is more convenient to express the decision rule in terms of $\hat{\rho}$ directly. Using (2.40), the decision rule $|T| < t_{\alpha/2, N-2}$ can be manipulated algebraically into the equivalent decision rule

$$|\hat{\rho}| < \rho_{\alpha, N}, \qquad (2.41)$$

where

$$\rho_{\alpha, N} = \frac{t_{\alpha/2, N-2}}{\sqrt{N-2+t_{\alpha/2, N-2}^2}}. \qquad (2.42)$$

Therefore, the decision rule for the hypothesis $\rho = 0$ is

$$\text{Reject } H_0 \text{ if } |\hat{\rho}| > \rho_{\alpha, N} \qquad (2.43)$$
$$\text{Accept } H_0 \text{ if } |\hat{\rho}| < \rho_{\alpha, N}.$$

Thus, there are two equivalent procedures for the correlation test: evaluate (2.40) and use the decision rule (2.27), or evaluate $\hat{\rho}$ and use the decision rule (2.43). The second approach often is more convenient because the 5% critical value for the correlation coefficient can be approximated by the simple formula

$$\rho_{0.05, N}^{\text{approx}} = \frac{2}{\sqrt{N}}. \qquad (2.44)$$

A comparison between exact and approximate critical values for a 5% significance test is shown in Table 2.2 and reveals that the approximation is good even for small sample sizes.

One last point deserves to be mentioned. Although independence implies vanishing correlation, the converse of this statement is not true; that is, two variables may be dependent even if their correlation coefficient vanishes. A more precise description of the correlation test is that it tests for a *linear* dependence between variables.

Example 2.4 (Test for Vanishing Correlation Coefficient) *A person claims that ENSO influences the winter-mean temperature in a particular city. To investigate this claim, the scientist collects the Niño 3.4 index and winter-mean temperature for the last 20 years. The sample correlation between the 20 pairs of Niño 3.4 indices and temperature values is 0.4. Does this correlation support the claim?*

Table 2.2. *Comparison between exact and approximate critical values of the correlation coefficient for a 5% significance test, as a function of sample size. The exact critical values are derived from (2.42) while the approximate values are derived from (2.44).*

Sample Size N	Exact $\rho_{0.05, N}$	Approx. $2/\sqrt{N}$	Sample Size N	Exact $\rho_{0.05, N}$	Approx. $2/\sqrt{N}$
5	0.88	0.89	30	0.36	0.37
10	0.63	0.63	35	0.33	0.34
15	0.51	0.52	40	0.31	0.32
20	0.44	0.45	45	0.29	0.30
25	0.40	0.40	50	0.28	0.28

The claim can be assessed by testing the null hypothesis that the population correlation between ENSO and city temperature vanishes. The test statistic (2.40) for this study is

$$T = \frac{0.4\sqrt{20 - 2}}{\sqrt{1 - 0.4^2}} \approx 1.85.$$

The claim made no assertion about the direction of change, so a two-tailed test is performed. The critical value for a t-distribution with 18 degrees of freedom for a two-tailed significance test at the 5% level is 2.1 (this value is obtained by looking it up in a table or from a statistical software package). Since $|1.85| < 2.1$, the null hypothesis is not rejected – the observed correlation is not significantly different from zero. Thus, the data do not support the claim. Equivalently, we may evaluate (2.42) to obtain the critical value

$$\rho_c = \frac{2.1}{\sqrt{20 - 2 + 2.1^2}} \approx 0.44.$$

This value is nearly the same as derived from (2.44): $2/\sqrt{20} \approx 0.44$. Since $|0.4| < 0.44$, the null hypothesis is not rejected, consistent with the conclusion reached earlier.

2.5 Summary of Common Significance Tests

	difference in means	independence	difference in variance
null hypothesis	$\mu_X = \mu_Y$	$\rho_{XY} = 0$	$\sigma_X = \sigma_Y$
test statistic	$T_0 = \dfrac{\hat{\mu}_X - \hat{\mu}_Y}{\hat{\sigma}_P\sqrt{\frac{1}{N_X} + \frac{1}{N_Y}}}$	$T = \dfrac{\hat{\rho}\sqrt{N - 2}}{\sqrt{1 - \hat{\rho}^2}}$	$F = \dfrac{\hat{\sigma}_X^2}{\hat{\sigma}_Y^2} (\hat{\sigma}_X^2 > \hat{\sigma}_Y^2)$
distribution under H_0	$t_{N_X + N_Y - 2}$	t_{N-2}	$F_{N_X - 1, N_Y - 1}$
rejection region (two-tailed)	$|T_0| > t_{\alpha/2, N_X + N_Y - 2}$	$|T| > t_{\alpha/2, N-2}$	$F > F_{\alpha/2, N_X - 1, N_Y - 1}$

where,

- $X_1, X_2, \ldots, X_{N_X}$ are *iid* as $\mathcal{N}\left(\mu_X, \sigma_X^2\right)$.
- $Y_1, Y_2, \ldots, Y_{N_Y}$ are *iid* as $\mathcal{N}\left(\mu_Y, \sigma_Y^2\right)$.
- ρ_{XY} is the population correlation coefficient between X and Y.
- $\alpha =$ significance level.
- $\hat{\sigma}_X^2$ and $\hat{\sigma}_Y^2$ are unbiased estimates of the variance of X and Y, respectively.
- $\hat{\mu}_X$ and $\hat{\mu}_Y$ are unbiased estimates of the mean of X and Y, respectively.
- $\hat{\rho}$ is a sample estimate of the correlation coefficient between X and Y.

2.6 Further Topics

One might wonder whether the tests discussed in this chapter are the best available. For instance, perhaps some other procedure has a higher probability of giving correct decisions. When deciding to reject a hypothesis, there are two types of errors: a hypothesis may be rejected when it is true, or it may be accepted when it is false. The former is known as a Type I error and has been the focus of the present chapter. The latter is known as a Type II error and its probability is often denoted by β. The *power* of a test is the probability of (correctly) rejecting a null hypothesis when it is false, which equals $1 - \beta$. A milestone in hypothesis testing is the *Neyman–Pearson Lemma*, which gives the test that maximizes the power of a decision rule subject to a pre-specified Type I error rate. The Neyman–Pearson Lemma assumes that all parameters of the null and alternative hypothesis distributions are specified, which is rarely the case in practice (for instance, the t-test requires estimating a variance and the alternative hypothesis requires specifying $\mu_X - \mu_Y$). For more complex hypotheses involving unknown population parameters, the Likelihood Ratio Test is a standard approach. For many cases of practical interest, the Likelihood Ratio Test gives optimal tests for sufficiently large sample sizes. This procedure often suggests a test statistic in situations where it may be difficult to even guess a suitable test statistic. All of the tests discussed in this chapter can be derived from the Likelihood Ratio Test.

In Chapter 9, we will discuss procedures for testing hypotheses regarding parameters in a *regression model*. Interestingly, the difference-in-means test and the correlation test can be derived from this framework. The regression model framework is extremely general and provides a unifying framework for testing a variety of different hypotheses.

Another method of inference is *nonparameteric* or *distribution-free* methods, which are discussed in Chapter 4. Nonparametric methods make fewer assumptions about the population and therefore are simpler to understand and have wider applicability than parametric methods. Of course, this greater generality comes at a cost: Nonparametric tests often have less power than parametric tests when the parametric tests are appropriate. On the other hand, some nonparametric tests are *almost* as powerful as the parametric tests. Nonparameteric methods are often based on ranks rather than numerical values.

These methods are concerned with testing one hypothesis. On the other hand, climate science deals with complex data that often require testing several hypotheses simultaneously. This topic is called the *multiple testing problem*. Chapter 11 discusses techniques for testing multiple hypotheses while imposing a significance level over the whole sequence of tests. Techniques for multiple testing that take into account spatial correlations in the data are called *field significance tests* and are discussed in Chapter 13.

2.7 Conceptual Questions

1 What is a p-value?
2 What is a null hypothesis?
3 What is a significance level?
4 How do you state the results of a significance test?
5 What is wrong with saying "The probability that the null hypothesis is true is 5%"?
6 What is a rejection region? How does one define a rejection region?
7 What is a test statistic?
8 Why does the t-test use the t-distribution instead of a normal distribution?
9 If the null hypothesis $\mu_X = \mu_Y$ is false, what can you say about the distribution of T_0?
10 If N_X and N_Y are large, what is the critical value for 5% significance of a t-distribution?
11 When do you use a one-tailed test? When do you use a two-tailed test?
12 If $N = 25$, is the correlation $\hat{\rho} = 0.5$ significant? (Do this in your head with no paper.)
13 Does statistically significant mean *physically* significant? Why or why not?
14 When do you use a chi-squared distribution? t-distribution? F-distribution?

3

Confidence Intervals

When possible, quantify findings and present them with appropriate indicators of measurement error or uncertainty (such as confidence intervals). Avoid relying solely on statistical hypothesis testing, such as P values, which fail to convey important information about effect size and precision of estimates.[1]

International Committee of Medical Journal Editors (2019)

A major goal in statistics is to make inferences about a population. Typically, such inferences are in the form of estimates of *population parameters*; for instance, the mean and variance of a normal distribution. Estimates of population parameters are imperfect because they are based on a finite amount of data. Therefore, when reporting the estimated value of a population parameter, it is helpful to report its uncertainty. In fact, the estimate itself is almost meaningless without an indication about how far the estimate may be from the population value. A *confidence interval* provides a way to quantify uncertainty in parameter estimates. A confidence interval is a random interval that encloses the population value with a specified probability. Confidence intervals are related to hypothesis tests about population parameters. Specifically, for a given hypothesis about the value of a parameter, a test at the 5% significance level would accept the hypothesis if the 95% confidence interval contained the hypothesized value. While hypothesis tests merely give a binary decision – accept or reject – confidence intervals give a sense of whether the decision was a "close call" or a "miss by a mile." This chapter constructs a confidence interval for a difference in means, a ratio of variances, and a correlation coefficient. These confidence intervals assume the population has a normal distribution. If the population is not Gaussian, or the quantity being inferred is complicated, then *bootstrap* methods offer an important alternative approach, as discussed at the end of this chapter.

---

[1] Reprinted by permission from International Committee of Medical Journal Editors (2019).

3.1 The Problem

Suppose an experiment is performed under two different conditions. Let $\hat{\mu}_X$ and $\hat{\mu}_Y$ be the average of the data collected under the two conditions. Then, $\hat{\mu}_X - \hat{\mu}_Y$ is an estimate of the difference in population means $\mu_X - \mu_Y$. However, the estimate $\hat{\mu}_X - \hat{\mu}_Y$ is computed from a finite amount of data and therefore differs from $\mu_X - \mu_Y$ by a random error. The error may be so large that even the sign of the difference is uncertain. How can this uncertainty be quantified? One approach is based on hypothesis testing. In Chapter 2, we learned a test for the hypothesis that the difference in means *equals zero*. What if we want to test other hypothetical values, such as $\mu_X - \mu_Y = 2$? More generally, can we identify the range of values of $\mu_X - \mu_Y$ that would not be rejected based on the data? The *confidence interval* addresses this kind of question.

3.2 Confidence Interval for a Difference in Means

We begin by deriving a confidence interval for a difference in means. Let $X_1, \ldots, X_{N_X}$ be independently and identically distributed (*iid*) as a normal distribution with mean μ_X and variance σ^2, and let $Y_1, \ldots, Y_{N_Y}$ be *iid* as a normal distribution, but with mean μ_Y and variance σ^2. The two populations have the same variance but possibly different means. An estimate of the difference in population means is the difference in sample means $\hat{\mu}_X - \hat{\mu}_Y$. Using Example 1.13, the sample means have distributions

$$\hat{\mu}_X \sim \mathcal{N}\left(\mu_X, \frac{\sigma^2}{N_X}\right) \quad \text{and} \quad \hat{\mu}_Y \sim \mathcal{N}\left(\mu_Y, \frac{\sigma^2}{N_Y}\right), \tag{3.1}$$

and their difference has distribution

$$\hat{\mu}_X - \hat{\mu}_Y \sim \mathcal{N}\left(\mu_X - \mu_Y, \sigma^2\left(\frac{1}{N_X} + \frac{1}{N_Y}\right)\right). \tag{3.2}$$

Standardizing this variable leads to

$$\frac{\hat{\mu}_X - \hat{\mu}_Y - (\mu_X - \mu_Y)}{\sigma\sqrt{\frac{1}{N_X} + \frac{1}{N_Y}}} \sim N(0,1). \tag{3.3}$$

Using the results of Section 2.3, the population variance σ^2 can be estimated using the pooled estimate of the variance (2.20):

$$\hat{\sigma}_p^2 = \frac{(N_X - 1)\hat{\sigma}_X^2 + (N_Y - 1)\hat{\sigma}_Y^2}{N_X + N_Y - 2}, \tag{3.4}$$

Section 2.3 shows that the quantity

$$t = \frac{\hat{\mu}_X - \hat{\mu}_Y - (\mu_X - \mu_Y)}{\hat{\sigma}_p \sqrt{\frac{1}{N_X} + \frac{1}{N_Y}}}, \tag{3.5}$$

has a t distribution with $N_X + N_Y - 2$ degrees of freedom. Therefore,

$$P\left(-t_{\alpha/2} \leq t \leq t_{\alpha/2}\right) = 1 - \alpha, \tag{3.6}$$

where $t_{\alpha/2}$ is the value of t that has area $\alpha/2$ to the right (see Definition 2.2; the degrees of freedom is $N_X + N_Y - 2$, but the corresponding subscript is suppressed to simplify the notation). Note that t involves the difference in population means $\mu_X - \mu_Y$. Our goal is to write a probability statement involving just this quantity. To simplify notation, define

$$\hat{\sigma}_{X-Y} = \hat{\sigma}_p \sqrt{\frac{1}{N_X} + \frac{1}{N_Y}}, \tag{3.7}$$

which is the *standard error* of $\hat{\mu}_X - \hat{\mu}_Y$. Then, the parameter t becomes

$$t = \frac{\hat{\mu}_X - \hat{\mu}_Y - (\mu_X - \mu_Y)}{\hat{\sigma}_{X-Y}}. \tag{3.8}$$

Substituting (3.8) into (3.6) and manipulating gives

$$1 - \alpha = P\left(-t_{\alpha/2} \leq \frac{\hat{\mu}_X - \hat{\mu}_Y - (\mu_X - \mu_Y)}{\hat{\sigma}_{X-Y}} < t_{\alpha/2}\right)$$

$$= P\left(-t_{\alpha/2}\hat{\sigma}_{X-Y} \leq \hat{\mu}_X - \hat{\mu}_Y - (\mu_X - \mu_Y) < t_{\alpha/2}\hat{\sigma}_{X-Y}\right)$$

$$= P\left(-\hat{\mu}_X - \hat{\mu}_Y - t_{\alpha/2}\hat{\sigma}_{X-Y} \leq -(\mu_X - \mu_Y) < -\hat{\mu}_X - \hat{\mu}_Y + t_{\alpha/2}\hat{\sigma}_{X-Y}\right)$$

$$= P\left(\hat{\mu}_X - \hat{\mu}_Y + t_{\alpha/2}\hat{\sigma}_{X-Y} \geq (\mu_X - \mu_Y) > \hat{\mu}_X - \hat{\mu}_Y - t_{\alpha/2}\hat{\sigma}_{X-Y}\right),$$

which we finally write as

$$P\left(\hat{\mu}_X - \hat{\mu}_Y - t_{\alpha/2}\hat{\sigma}_{X-Y} < \mu_X - \mu_Y \leq \hat{\mu}_X - \hat{\mu}_Y + t_{\alpha/2}\hat{\sigma}_{X-Y}\right) = 1 - \alpha. \tag{3.9}$$

In this expression, the population parameters occur in the middle while the (known) sample quantities occur at the upper and lower limits. The upper and lower limits define the *upper and lower confidence limits*, and the interval

$$\left[\hat{\mu}_X - \hat{\mu}_Y - t_{\alpha/2}\hat{\sigma}_{X-Y}, \hat{\mu}_X - \hat{\mu}_Y + t_{\alpha/2}\hat{\sigma}_{X-Y}\right], \tag{3.10}$$

is said to be the $100(1 - \alpha)\%$ *confidence interval* for $\mu_X - \mu_Y$. This confidence interval is often expressed as

$$\hat{\mu}_X - \hat{\mu}_Y \pm t_{\alpha/2}\hat{\sigma}_{X-Y}. \tag{3.11}$$

3.3 Interpretation of the Confidence Interval

The interpretation of (3.9) is very tricky. To illustrate, suppose data are collected and the 95% confidence interval is found to be [5,7]. Does expression (3.9) imply that the true difference in mean $\mu_X - \mu_Y$ has a 95% probability of being in the interval [5,7]? No! The parameter $\mu_X - \mu_Y$ is not a random number; it is a population parameter. For the sake of argument, suppose $\mu_X - \mu_Y = 2$. Then, we are asking "what is the probability that 2 is in the interval [5,7]?" The answer is zero! More generally, $\mu_X - \mu_Y$ is unknown, but it is still a fixed number, so the probability that it is in the interval is either 0 or 1 – either it is or it is not. Instead, it is the *end points* of the confidence interval that are random and move around from sample to sample. The proper interpretation of (3.9) is that the *random interval* (3.10) will include the population parameter $\mu_X - \mu_Y$ in 95% of the cases when *the same procedure is repeated on independent samples*. Thus, the uncertainty in the estimate of $\mu_X - \mu_Y$ is quantified by the confidence interval in the sense that if the analysis is repeated M times, approximately $0.95M$ intervals would contain $\mu_X - \mu_Y$, while $0.05M$ intervals would not bracket $\mu_X - \mu_Y$. This situation is illustrated in Figure 3.1.

For the above reasons, one should not speak of "the probability that the population parameter falls in the interval." Rather, one should speak of "the probability that *this procedure for computing intervals* will contain the population parameter."

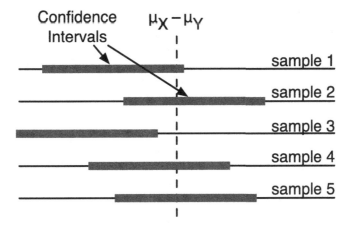

Figure 3.1 Illustration of confidence intervals for the difference in population means $\mu_X - \mu_Y$. The population parameter $\mu_X - \mu_Y$ is fixed while the confidence interval changes randomly from sample to sample. In this figure, the confidence interval includes the population parameter four out of five times (only sample 3 does not include the population parameter). It is incorrect to say that there is a 95% probability that the population parameter falls in a sample interval. Rather, we say that, before data is taken, there is a 95% probability that the interval encloses the population parameter.

As an analogy, consider throwing a net in a lake to capture fish. With enough repetitions, one might estimate that the net has a 95% chance of capturing a fish. Here, the 95% refers to a property of the net before it is thrown. After the net is thrown, the fish is either in the net or not; there is no fractional probability that the fish is in the net. Similarly, the 95% in a confidence interval refers to a property of the interval before any data are taken. Confidence intervals lose their probabilistic interpretation as soon as numbers are substituted. A more intuitive framework for quantifying uncertainty is Bayesian analysis, as discussed in Section 3.7.

Another interpretation follows from the fact that confidence intervals are closely related to hypothesis tests. To see this, recall that the decision rule to accept the null hypothesis $\mu_X = \mu_Y$ is $|t| < t_{\alpha/2}$, where

$$t = \frac{\hat{\mu}_X - \hat{\mu}_Y}{\hat{\sigma}_{X-Y}}. \tag{3.12}$$

The decision rule $|t| < t_{\alpha/2}$ is equivalent to

$$\hat{\mu}_X - \hat{\mu}_Y - t_{\alpha/2}\hat{\sigma}_{X-Y} < 0 < \hat{\mu}_X - \hat{\mu}_Y + t_{\alpha/2}\hat{\sigma}_{X-Y}. \tag{3.13}$$

Note that the left- and right-hand sides of this equation are identical to confidence interval (3.10). Thus, the criterion for accepting the null hypothesis $\mu_X - \mu_Y = 0$ is equivalent to the criterion that the confidence interval for $\mu_X - \mu_Y$ encloses zero. Conversely, if the null hypothesis is rejected, then the confidence interval will not enclose zero. These situations are illustrated in Figure 3.2. More generally, for a 95% confidence interval, values in the interval would be accepted by the corresponding hypothesis test at the 5% significance level. Conversely, values outside the interval would be rejected at the 5% significance level.

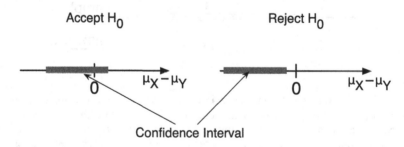

Figure 3.2 Illustration of the relation between hypothesis testing and confidence intervals: If the null hypothesis is accepted, then the confidence interval encloses zero; if the null hypothesis is rejected, then the confidence interval will not enclose zero.

3.4 A Pitfall about Confidence Intervals

Confidence intervals are often used to compare estimates of the population mean, but the conclusions derived from such comparisons may be inconsistent with the difference-in-means test. To illustrate, suppose $X_1, \ldots, X_N \overset{iid}{\sim} \mathcal{N}(\mu, \sigma^2)$, hence

$$\hat{\mu} \sim \mathcal{N}(\mu, \sigma^2/N). \tag{3.14}$$

Hypothesis tests for the mean are based on

$$\frac{\text{observed} - \text{hypothesized}}{\text{standard error}} = \frac{\hat{\mu} - \mu}{\sigma/\sqrt{N}} \sim \mathcal{N}(0, 1). \tag{3.15}$$

Replacing σ^2 by the sample estimate $\hat{\sigma}^2$ leads to the t-distribution

$$t = \frac{\hat{\mu} - \mu}{\hat{\sigma}/\sqrt{N}} \sim t_{N-1}. \tag{3.16}$$

Substituting this into (3.6) and following the same steps as before gives the $(1 - \alpha)$ 100% confidence interval for the mean μ_X as

$$\left(\hat{\mu} - t_{\alpha/2, N-1} \frac{\hat{\sigma}}{\sqrt{N}}, \hat{\mu} + t_{\alpha/2, N-1} \frac{\hat{\sigma}}{\sqrt{N}} \right). \tag{3.17}$$

Consider the example shown in Figure 3.3, which shows the values drawn from two different populations, labeled X and Y. Also shown are the 95% confidence intervals for μ_X, μ_Y, and their difference. One can see that the confidence intervals for the means *overlap with each other*. It is tempting to conclude from this that the means of the two samples are not significantly different. However, the corresponding *t*-test rejects the hypothesis that the means are equal, as can be seen by the fact that the confidence interval for the difference-in-means does not include zero. Thus, overlapping confidence intervals do not necessarily imply that the hypothesis of a difference in population parameter can be rejected. However, it turns out that if confidence intervals for the mean do not overlap, then the corresponding *t*-test always will reject the hypothesis of equal means.

3.5 Common Procedures for Confidence Intervals

Confidence intervals for other population parameters follow the same basic procedure:

1 Select an estimator $\hat{\gamma}$ for the population parameter γ of interest (for example, the previous section used $\hat{\mu}_X - \hat{\mu}_Y$ to estimate $\mu_X - \mu_Y$).
2 Specify the probability distribution of the estimator (in the previous section, for example, this distribution was embodied in (3.5)).

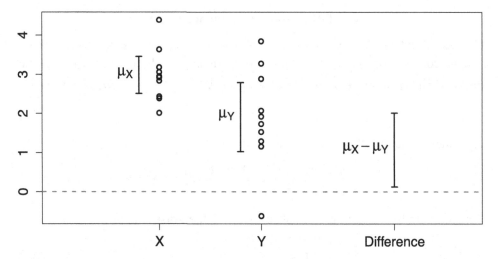

Figure 3.3 A sample drawn from $\mathcal{N}(3.2, 1)$ ("X") and from $\mathcal{N}(2, 1)$ ("Y") (open circles), and the corresponding 95% confidence intervals for the population mean of X, population mean of Y, and their difference (error bars). The horizontal dashed line shows zero, for reference.

3 Construct a confidence interval of the form $P(L < \gamma < U) = 1 - \alpha$, where the confidence limits, L and U, are sample quantities. Note the interval is random because its limits are sample quantities.

4 Evaluate the lower and upper confidence limits L and U from data.

5 After the confidence limits L and U have been computed from data, it is no longer sensible to speak about the probability of the interval containing the population parameter. Either the population parameter is in the interval or it is not. There is no "probability of being in the interval." Instead, $1 - \alpha$ is the proportion of intervals that would be expected to contain γ when the procedure is applied repeatedly to independent samples.

Example 3.1 (Confidence Interval for a Difference in Means) *A scientist wants to investigate whether ENSO influences the mean winter temperature in a particular city. To investigate this question, the scientist collects the mean winter temperature for the past 20 El Niños and the past 20 La Niñas (giving a total of 40 numbers). For the 20 El Niño years, the sample mean is 280K and the standard deviation is 3K. In contrast, for the 20 La Niña years, the sample mean is 275K and standard deviation is 4K. What is the 95% confidence interval for the difference in temperatures between El Niño and La Niña?*

The natural estimator for a difference in population means is the difference in sample means: $\hat{\mu}_X - \hat{\mu}_Y = 280K - 275K = 5K$. The upper and lower confidence limits for a difference in means was derived in (3.10), and requires the pooled variance. The sample variances are $\hat{\sigma}_X^2 = 9K^2$ and $\hat{\sigma}_Y^2 = 16K^2$. Therefore, the pooled variance is

$$\hat{\sigma}_p^2 = \frac{(N_X - 1)\hat{\sigma}_X^2 + (N_Y - 1)\hat{\sigma}_Y^2}{N_X + N_Y - 2} = \frac{19 * 9 + 19 * 16}{20 + 20 - 2} = 12.5,$$

and

$$\hat{\sigma}_{X-Y} = \hat{\sigma}_p\sqrt{\frac{1}{N_X} + \frac{1}{N_Y}} = \sqrt{12.5}\sqrt{\frac{1}{20} + \frac{1}{20}} \approx 1.12.$$

The critical value for a two-tailed significance test at the 5% level for a t distribution with $40 - 2 = 38$ degrees of freedom is 2.0 (this value is obtained by looking it up in a table or computing it from a statistical software package). Therefore, the upper and lower confidence limits for a difference in means are

$$\hat{\mu}_X - \hat{\mu}_Y \pm t_{\alpha/2, N_X+N_Y-2}\hat{\sigma}_{X-Y} \sim 5 \pm 2.0 * 1.12$$
$$\sim 5 \pm 2.24$$
$$\sim (2.77, 7.24).$$

Note that the confidence interval does not include zero. Thus, a hypothesis test would reject the null hypothesis of equal means, which is consistent with Example 2.2.

Example 3.2 *Referring to Example 3.1, which of the following statements is correct?*

1 *"There is a 95% chance that the true population parameter is between 2.77 and 7.24."*
2 *"The interval 2.77 and 7.24 has a 95% chance of bracketing the population parameter."*
3 *"If this procedure were repeated 100 times on independent samples, about 95 of them would overlap with this interval."*
4 *"If this procedure were repeated many times, the true population mean will be in the confidence interval 95% of the time. (2.77, 7.24) is one of those intervals."*

Answer: *Only (4) is a correct statement. Statement (1) is incorrect because neither the interval (2.77, 7.24) nor the population parameter are random (although the population parameter is unknown to us). Technically, the population parameter is either in the interval or not, so the probability that it is in the interval is 0% or 100%. Statement (2) is incorrect for the same reason. Statement (3) is incorrect because there is no guarantee about how confidence intervals overlap when computed from independent samples.*

3.5.1 Confidence Interval for the Correlation Coefficient

Exact confidence intervals for the correlation are difficult to obtain because the distribution of the sample correlation is a complicated function of the population parameters. The exact distribution of the sample correlation was derived by Ronald

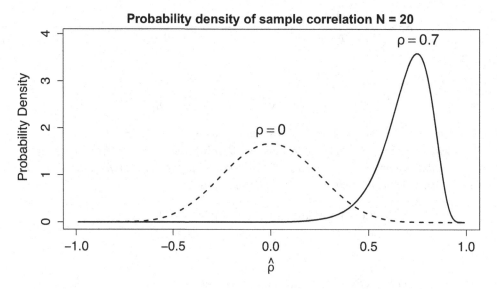

Figure 3.4 Exact probability density function of the sample correlation for $N = 20$ for two values of the population correlation, namely $\rho = 0$ (dashed) and $\rho = 0.7$ (solid).

Fisher in 1915 by ingenious mathematical arguments (Fisher, 1915). The distribution will not be presented here – it is available in standard multivariate texts (Anderson, 1984; Muirhead, 2009) – but we do show in Figure 3.4 the density function for two choices of the population correlation. One can see that the distribution changes shape as ρ increases from 0 to 1, which is exactly to be expected given that the sample correlation cannot exceed 1. Importantly, the variance of the distribution decreases as the population correlation increases. As a result, uncertainty depends on the population, which leads to an awkward confidence interval. However, Fisher (1921) further showed that if a sample correlation coefficient $\hat{\rho}$ is computed from N pairs of random variables with correlation ρ, then for large N, the statistic

$$z = \frac{1}{2} \ln \left(\frac{1 + \hat{\rho}}{1 - \hat{\rho}} \right), \tag{3.18}$$

has an approximately normal distribution with mean

$$\mu_Z = \frac{1}{2} \ln \left(\frac{1 + \rho}{1 - \rho} \right), \tag{3.19}$$

and variance

$$\sigma_Z^2 = \frac{1}{N - 3}. \tag{3.20}$$

Importantly, the variance depends only on the sample size and is independent of the population correlation ρ. This transformation is known as *Fisher's Transformation* and is an example of a *variance-stabilizing transformation*, which is a class of transformation that transforms a random variable into another one whose variance is independent of population parameters. It follows from Example 1.3 that

$$\frac{Z - \mu_Z}{\sigma_Z} \sim \mathcal{N}(0, 1). \tag{3.21}$$

Thus, an approximate $(1 - \alpha)100\%$ confidence interval for μ_Z can be derived from (1.47), which can be manipulated into the form

$$P\left(z - \sigma_Z z_{\alpha/2} \leq \mu_Z < z + \sigma_Z z_{\alpha/2}\right) = 1 - \alpha. \tag{3.22}$$

The confidence limits $z \pm \sigma_Z z_{\alpha/2}$ can then be inverted to derive a confidence interval for ρ. The procedure is illustrated in the next example.

Example 3.3 (Confidence Interval for a Correlation Coefficient) *The correlation coefficient between 20 pairs of forecasts and verifying observations is 0.4. What is the 95% confidence interval for the correlation coefficient?*
  For this problem, $N = 20$ and Fisher's Transformation yields

$$z = \frac{1}{2} \ln \left(\frac{1 + 0.4}{1 - 0.4}\right) = \frac{1}{2} \ln(2.3) \approx 0.42.$$

For $\alpha = 5\%$, $z_{\alpha/2} \approx 1.96$ (see Table 1.1) and (3.22) gives the 95% confidence interval

$$z \pm \sigma_Z z_{\alpha/2} = 0.42 \pm \frac{1.96}{\sqrt{20 - 3}} \approx 0.42 \pm 0.475 = (-0.06, 0.90).$$

This gives the confidence interval in terms of z, but we want the interval expressed in terms for ρ. To do this, we invert (3.19) for ρ:

$$z = \frac{1}{2} \ln \left(\frac{1 + \rho}{1 - \rho}\right) \quad \longrightarrow \quad \exp(2z) = \frac{1 + \rho}{1 - \rho} \quad \longrightarrow \quad \rho = \frac{e^{2\mu_Z} - 1}{e^{2\mu_Z} + 1}. \tag{3.23}$$

Substituting numerical values gives

$$\hat{\rho}_{lower} = \frac{\exp(-2 * 0.06) - 1}{\exp(-2 * 0.06) + 1} \approx -0.05$$

$$\hat{\rho}_{upper} = \frac{\exp(2 * 0.90) - 1}{\exp(2 * 0.90) + 1} \approx 0.72.$$

Thus, an approximate 95% confidence interval for ρ is $(-0.05, 0.772)$. Note the relatively large width of the interval. This interval includes zero, implying that the null hypothesis is not rejected, which is consistent with Example 2.4. Potentially, these two results could be inconsistent because the hypothesis test in Example 2.4 is exact whereas Fisher's Transformation is only approximate. If the exact distribution were used, then there would be no discrepancy. In practice, such inconsistencies occur

Table 3.1. *Correlation between Niño 3.4 and the all-India*
summer monsoon rainfall.

Period	Length	Correlation	Z
1951–1975	25 years	−0.62	−0.72
1976–2006	31 years	−0.43	−0.46

when the confidence limit is close to zero or the sample value is close to the critical
value. In the present case in which the lower limit is −0.05, this is close enough to zero
that one should perform the exact hypothesis test to be certain that the correlation is
statistically significant.

Fisher's Transformation also allows us to construct a confidence interval for a
difference in correlation. The basic procedure is illustrated in the next example.

Example 3.4 *A scientist wants to know if the relation between Indian summer*
monsoon rainfall and ENSO has changed in recent decades. To investigate this ques-
tion, the scientist computes the correlation between indices of monsoon rainfall and
ENSO over two different periods. The results are shown in Table 3.1. Did the ENSO–
monsoon relation change?

To address this question, the scientist tests the null hypothesis of no difference in
correlation. After applying Fisher's Transformation, the difference in z has distribu-
tion

$$Z_1 - Z_2 \sim \mathcal{N}(\mu_1 - \mu_2, \sigma_1^2 + \sigma_2^2). \tag{3.24}$$

The standardized version is therefore

$$\frac{Z_1 - Z_2 - (\mu_1 - \mu_2)}{\sqrt{\sigma_1^2 + \sigma_2^2}} \sim \mathcal{N}(0, 1), \tag{3.25}$$

and the $(1 - \alpha)100\%$ confidence interval is

$$(Z_1 - Z_2) \pm z_{\alpha/2}\sqrt{\sigma_1^2 + \sigma_2^2}. \tag{3.26}$$

From Table 3.1, $Z_1 - Z_2 = -0.72 + 0.46 = -0.26$, and from (3.20) the variance is

$$\sigma_1^2 + \sigma_2^2 = \frac{1}{25 - 3} + \frac{1}{31 - 3} \approx 0.40. \tag{3.27}$$

Thus, the 95% confidence interval in Z is

$$-0.26 \pm 1.96 * \sqrt{0.4} = (-1.5, 0.98). \tag{3.28}$$

Transforming this interval to ρ using (3.23) gives the 95% confidence interval

$$(-0.91, 0.75). \tag{3.29}$$

> *Since the interval includes zero, the scientist accepts the hypothesis of no change in correlation. Note the extremely wide confidence interval. The large width occurs because variance is additive when the difference of two correlations is computed.*

This test assumes that the sample used to compute one correlation is independent of the sample used to compute the second correlation. That is, if the samples are

$$\text{sample 1}: \quad \{(X_1, Y_1), (X_2, Y_2), \dots, (X_N, Y_N)\} \tag{3.30}$$

$$\text{sample 2}: \quad \{(X_1^*, Y_1^*), (X_2^*, Y_2^*), \dots, (X_{N*}^*, Y_{N*}^*)\}, \tag{3.31}$$

and $\hat{\rho}$ is computed from sample 1, and $\hat{\rho}^*$ is computed from sample 2, the test assumes that sample 1 and sample 2 are independent. This assumption is not true in some climate studies. For instance, correlation is often used to measure the skill of a forecast model. A frequent question is whether one forecast model is more skillful than another. To address this question, the correlation between forecast and observations is computed *over the same period*. Because the correlation is computed using the same observations, one of the variables is the same, say $X_1 = X_1^*, \dots, X_N = X_N^*$, hence the samples are not independent, and in fact the test is biased. Methods for measuring differences in skill that avoid this bias are discussed in DelSole and Tippett (2014) and Siegert et al. (2016).

3.5.2 Confidence Interval for Variance Ratios

The confidence interval for a variance ratio can be derived from the difference-in-variance test discussed in Chapter 2. As discussed in Section 2.4.2, if the populations have normal distributions with variances σ_X^2 and σ_Y^2, and N_X and N_Y random samples from the two populations yield sample variances $\hat{\sigma}_X^2$ and $\hat{\sigma}_Y^2$, respectively, then the ratio

$$F = \frac{\hat{\sigma}_X^2 / \sigma_X^2}{\hat{\sigma}_Y^2 / \sigma_Y^2}, \tag{3.32}$$

has an F distribution with $N_X - 1$ and $N_Y - 1$ degrees of freedom. A two-tailed test yields

$$1 - \alpha = p\left(F_{1-\alpha/2, N_X-1, N_Y-1} \le \frac{\hat{\sigma}_X^2}{\hat{\sigma}_Y^2} \frac{\sigma_Y^2}{\sigma_X^2} \le F_{\alpha/2, N_X-1, N_Y-1}\right). \tag{3.33}$$

Following the same kind of mathematics as used to derive a confidence interval in Section 3.2, we can construct the $100(1 - \alpha)\%$ equal-tail confidence interval of σ_X^2 / σ_Y^2 as

$$\left(\frac{1}{F_{\alpha/2, N_X-1, N_Y-1}} \frac{\hat{\sigma}_X^2}{\hat{\sigma}_Y^2}, \frac{1}{F_{1-\alpha/2, N_X-1, N_Y-1}} \frac{\hat{\sigma}_X^2}{\hat{\sigma}_Y^2}\right). \tag{3.34}$$

Similarly, a $100(1 - \alpha)\%$ confidence interval for the ratio of *standard deviations* σ_X/σ_Y follows from (3.33):

$$\left(\frac{1}{\sqrt{F_{\alpha/2, N_X-1, N_Y-1}}} \frac{\hat{\sigma}_X}{\hat{\sigma}_Y}, \frac{1}{\sqrt{F_{1-\alpha/2, N_X-1, N_Y-1}}} \frac{\hat{\sigma}_X}{\hat{\sigma}_Y} \right). \tag{3.35}$$

Example 3.5 (Confidence Interval for a Variance Ratio) *A scientist wants to investigate whether ENSO influences the variability of the mean winter temperature in a particular city. To investigate this question, the scientist collects the mean winter temperature for the past 20 El Niños and the past 20 La Niñas (giving a total of 40 numbers). For the 20 numbers in the El Niño group, the unbiased standard deviation was 3K. In contrast, for the 20 numbers in the La Niña group, the unbiased standard deviation was 4K. What is the 95% confidence interval for the ratio of variances between El Niño and La Niña?*

The sample variances are $\hat{\sigma}_X^2 = 9K^2$ and $\hat{\sigma}_Y^2 = 16K^2$, hence

$$\frac{\hat{\sigma}_X^2}{\hat{\sigma}_Y^2} = 9/16 \sim 0.56. \tag{3.36}$$

To construct a 95% confidence interval, we compute the critical values for an F distribution with 19 and 19 degrees of freedom. The 2.5% and 97.5% critical values are 0.40 and 2.53, respectively. It follows that the 95% confidence interval for the variance ratio is

$$\left(\frac{0.56}{2.53}, \frac{0.56}{0.4} \right) \sim (0.22, 1.4). \tag{3.37}$$

Since the interval includes one, a test for equality of variances would not reject the hypothesis of equal variances, consistent with Example 2.3. Note also the large difference in the upper and lower limits: they differ by over a factor of 6!

3.6 Bootstrap Confidence Intervals

The previously mentioned procedures assume that the population is Gaussian and that the parameter of interest is the mean, variance, or correlation. Can confidence intervals be constructed for non-Gaussian distributions or more complicated estimators? A procedure called the *bootstrap* offers a way. The bootstrap is an example of a *resampling* procedure in which samples are drawn randomly from the data set *with replacement* to construct an empirical distribution of a statistic. The bootstrap makes no assumptions about the shape of the population distribution and can handle any statistic that can be computed numerically. It does, however, assume the data are *iid*. The technique is very different from those discussed so far and has a considerable literature. In this section, we give a brief introduction to the technique.

3.6.1 The Method

Suppose $T(X_1, \ldots, X_N)$ is an estimator for some population quantity. For instance, $T(\cdot)$ could be an estimator for the population mean (that is, the sample mean). Let $X_1, \ldots, X_N$ be *iid* random variables. The bootstrap method proceeds as follows. Starting from the original sample $X_1, \ldots, X_N$, randomly select one element from this sample and call it X_1^*. Next, select a random element *from the original sample of N elements* and call it X_2^*. Importantly, X_2^* could equal X_1^* because the *same* element might have been chosen by chance. This type of sampling is called *sampling with replacement*, because whatever element is selected is returned to the original set before the next draw. Repeating this N times yields the *bootstrap sample*, from which T_1^* can be computed:

$$\text{bootstrap sample 1}: \quad X_1^*, X_2^*, \ldots, X_N^* \quad \longrightarrow \quad T_1^* = T(X_1^*, X_2^*, \ldots, X_N^*).$$
(3.38)

Then, this *entire* procedure is repeated B times, generating the estimators T_1^*, $T_2^*, \ldots, T_B^*$. The bootstrap estimates $T_1^*, T_2^*, \ldots, T_B^*$ can then be used to infer properties of the confidence interval (e.g., the 2.5% and 97.5% percentiles). B should be chosen to be large enough that the confidence interval does not depend on B (e.g., 10,000, or 1,000,000 if necessary).

> **Example 3.6** (Difference in Means) *A scientist wants to investigate whether ENSO influences the mean winter temperature in a particular city. To investigate this question, the scientist collects the mean winter temperature for the past 20 El Niños and the past 20 La Niñas (giving a total of 40 numbers). The resulting numbers (in degrees Kelvin) are shown in Table 3.2 in ascending order. The mean and standard deviation of the 20 numbers in each group are indicated in the right column. What is the 95% confidence interval for the difference in temperatures between El Niño and La Niña?*
>
> *To estimate a confidence interval based on the bootstrap method, we first draw 20 random samples with replacement from the La Niña data. An example of such a sample is shown in the table at the end of this example. It can be seen that several values*

Table 3.2. *Temperatures associated with warm and cold ENSO events for Example 3.6.*

La Niña	272.54	273.26	273.34	273.53	273.77	273.79	**mean: 276.74K**
	274.44	275.44	275.82	276.36	276.84	277.12	**SD: 3.1K**
	277.19	277.6	278.37	280	280.22	280.38	
	282.27	282.59					
El Niño	274.24	274.66	274.87	275.49	276.04	276.44	**mean: 279.31K**
	277.76	279.29	279.35	279.52	279.58	279.85	**SD: 3.5K**
	280.06	280.26	280.28	282.17	282.58	282.66	
	282.83	288.2					

are repeated, as anticipated. Also, the mean and standard deviation of the bootstrap sample differ slightly from the original data. This procedure is then repeated 100,000 times to generate 100,000 bootstrap estimates of the mean. The same procedure is applied to the El Niño data to generate 100,000 bootstrap estimates of the mean for El Niño conditions. Then, the difference between the El Niño and La Niña bootstrap estimated means are computed to generate 100,000 differences in means. The 2.5% and 97.5% percentiles of these 100,000 differences are then identified and found to be (−4.56, − 0.59). Since this confidence interval excludes 0, we reject the null hypothesis of equal means. For comparison, the 95% confidence interval derived from t-distribution is (−4.66, −0.46), which is similar to the bootstrap confidence interval, confirming the reasonableness of the bootstrap.

La Nina	272.54	273.26	273.34	273.34	273.34	274.44	**mean: 276.78K**
(bootstrap)	274.44	274.44	276.36	277.12	277.19	277.19	**SD: 3.1K**
	277.6	277.6	278.37	280	280.22	280.22	
	282.27	282.27					

This example shows that the bootstrap gives confidence intervals close to that based on the t-distribution. If that is the case, then why care about the bootstrap? The reason is that the bootstrap is much more general – it can be applied when the population is non-Gaussian or the estimator $T(\cdot)$ is a complicated function of the sample.

Example 3.7 (Confidence Interval for Mean Precipitation) *An engineer is designing a drainage system and needs to estimate the mean precipitation over a city. The annual average precipitation over the city for the past ten years is (in mm/day):*

0.0988	0.5550	0.2070	0.3470	3.9500	0.0221	0.2090	0.1620	0.0486	0.0163

If one were to adopt a Gaussian approximation, then one would use (3.17) to compute the confidence interval. The sample mean and standard deviation of the above data are 0.56 and 1.2, respectively, which gives the 95% confidence interval (−0.30, 1.42) for mean precipitation. Unfortunately, the lower limit of this interval is negative, which is unphysical (precipitation rate cannot be negative). In contrast, a bootstrap estimate based on 1,000 bootstrap samples gives a 95% confidence interval of (0.11, 1.37), which has positive upper and lower limits. Indeed, the bootstrap always gives positive values because it is based on resampling the original data, each of which is positive.

The ability to estimate confidence intervals in situations in which theory is complicated or nonexistent is the great advantage of the bootstrap.

3.6.2 Caveats about the Bootstrap

The accuracy of the bootstrap has few guarantees for small samples. For Gaussian populations, the bootstrap often converges to standard confidence intervals for sample sizes of 20 or more. Bootstrap confidence intervals are known to be inaccurate for medians or other quantiles, for which special methods are needed. Bootstrap intervals based on percentiles are most accurate when the distribution is symmetrical and the tails decay rapidly to zero; that is, when the distribution is near normal. When the distribution is asymmetric, the derived interval often does not match exact confidence intervals. For this reason, bootstrap confidence intervals for variances or ratio of variances are not accurate. A refinement known as the bias-corrected and accelerated bootstrap interval substantially improves over the percentile method with relatively little extra numerical computation. These and other issues with the bootstrap are discussed at an accessible level in Efron and Tibshirani (1994) and Good (2006).

3.7 Further Topics

Just as for hypothesis testing, one may wonder whether alternative confidence intervals exist and whether these might be preferable in some circumstances. This question is the focus of *estimation theory* and has a considerable literature. A basic consideration in estimation theory is the variance of an estimator: estimators with smaller variance are considered better. The Cramer–Rao inequality gives a lower bound on the variance of an estimator, so any estimator that achieves this bound is the "best." In some cases, a biased estimator – that is, an estimator that does not give the true population parameter on average – can have *much* smaller variance than an unbiased estimator. Consequently, there are a wide variety of techniques for producing biased estimators, for example, *ridge regression*.

The framework adopted in this chapter is said to be *Frequentist*, which means that a probability is interpreted as the frequency with which an event occurs in an infinite ensemble of realizations. Under this framework, a population parameter is not random and a confidence interval has the unintuitive interpretation discussed in Section 3.3. An alternative framework that is often more intuitive is Bayesian inference, which interprets probability distributions as quantifying "degrees of belief." For instance, the statement "the probability that Mr. X committed murder" cannot be checked by repeated experiments, so the Frequentist interpretation is unnatural, but most people nevertheless have an intuitive understanding of this statement. In a Bayesian framework, one can assign distributions to population parameters and speak about the probability that the parameter lies in an interval, (such an interval is called a "credible interval"). The Bayesian framework gives consistent results as long as belief measures are updated in a manner consistent with the axioms of probability. Although the updates follow the same rules, different people might *start* with a different *prior distribution*. The prior distribution expresses one's belief *before the data are available*. If one knows "nothing" about the quantity beforehand, then one can adopt an *uninformative prior* to reflect that lack of prior knowledge. After updating a prior distribution based on new data, the resulting distribution is called the *posterior* distribution. This framework is the basis of data assimilation discussed in Chapter 20. Because different people may start with different priors, they may end with different posteriors. Nevertheless, the posterior distributions converge to the same distribution in the limit of large sample size. Because the prior distribution is subjective, the results of Bayesian analysis may depend on this subjectivity, in contrast to the Frequentist framework, which does not depend on a prior distribution. However, this feature of Bayesian analysis is argued to be an advantage, since it allows prior information or relative ignorance to be incorporated in the analysis, something that is impossible in the Frequentist framework. Unfortunately, the proper posterior in a given situation can be difficult to calculate. This barrier is being removed rapidly with modern computers, especially with advances in *Markov Chain Monte Carlo*. The merits of Bayesian vs. Frequentist methods have been fiercely debated, but there is general concensus that the debate is over and has been decided in favor of Bayesian inference. Despite this, the Frequentist framework has been chosen in this book because it is simpler and by far the most common framework used in most climate studies. For further information about Bayesian analysis, we recommend Gelman et al. (2004) for an introduction, and Berger (1985) for more advanced reading.

3.8 Conceptual Questions

1 In what way is a 95% confidence interval more informative than a hypothesis test?
2 How can you use a confidence interval to infer what would happen if you were to perform a hypothesis test?
3 True or False: The probability that a 95% confidence interval contains the true value is 95%.
4 Does a confidence interval express the same information as a p-value?
5 Given a 95% confidence interval, to what does the 95% refer?
6 In your own words, describe how to compute a bootstrap confidence interval.
7 When should you use the bootstrap method?

4

Statistical Tests Based on Ranks

> It would not be a tool for creating new knowledge if a statistical proce-
> dure were to depend on assumptions that have to be taken for granted,
> or that cannot be verified by data themselves, or that cannot be justified
> on some other logical grounds.[1]
>
> *Rao (1973)*

The hypothesis tests discussed in the previous chapters are *parametric*. That is, the procedures assume samples come from a prescribed family of distributions, leaving only the *parameters* of the distribution open to question. For instance, a univariate Gaussian distribution is characterized by two parameters, the mean and variance, and hypotheses are expressed in terms of those parameters (e.g., see Chapter 2). This chapter discusses a class of procedures known as *nonparametric statistics*, or *distribution-free* methods, that make fewer assumptions. For some hypotheses, nonparametric tests are *almost* as powerful as parametric tests, hence some statisticians recommend nonparametric methods as a first choice. Nonparametric methods can be applied even if the data are in the form of ranks rather than numerical values. This chapter discusses the following nonparametric tests:

- Wilcoxon rank-sum test, a nonparametric version of the *t*-test.
- Kruskal–Wallis test, a nonparametric version of Analysis of Variance (see Chapter 18).
- a nonparametric version of the F-test, based on medians.
- Spearman's rank correlation, a nonparametric version of the correlation test.

This chapter assumes familiarity with the hypothesis testing framework discussed in Chapter 2, particularly the concepts of null hypothesis, decision rule, and significance level.

---

[1] Reprinted by permission from John Wiley & Sons, *Linear Statistical Inference and its Applications*, chapter 7, Theory of Statistical Inference, Rao (1973), p499.

4.1 The Problem

How will precipitation change in response to global warming? To investigate this question, a scientist runs a comprehensive climate model twice, once with current greenhouse gas concentrations, then again with doubled concentration. Results of these experiments are shown in Figure 4.1. The response to changing greenhouse gas concentrations is estimated by computing the difference between the mean precipitation of each simulation. However, the simulations contain random weather variability and therefore the associated sample means are noisy. As a result, the estimated change in precipitation might be caused by random sampling variability rather than a physical change. To investigate this possibility, the scientist decides to test the hypothesis that greenhouse gases have no effect on precipitation. Methods for doing this have been discussed in Chapter 2, but those methods assume that the population is Gaussian. Unfortunately, precipitation rate is clearly non-Gaussian, as indicated in Figure 4.1 (e.g., the histograms are far from symmetric). Therefore, the null hypothesis of Gaussian distributions is obviously wrong, and therefore rejecting

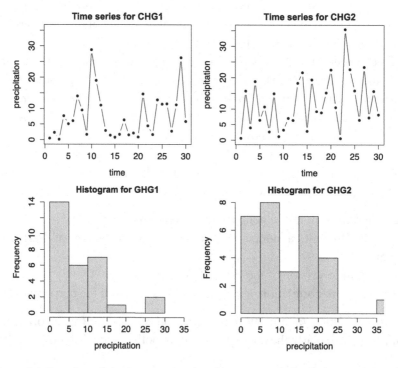

Figure 4.1 Local precipitation rate simulated by a general circulation model under two levels of greenhouse gas concentrations, labeled GHG1 and GHG2. The top row shows time series of the precipitation rate for the two cases and the bottom row shows the corresponding histograms.

this hypothesis would not be informative. The purpose of this chapter is to discuss tests that assume independent data, but otherwise make no assumptions about the distribution under the null hypothesis.

4.2 Exchangeability and Ranks

How can a hypothesis about a population be tested without making any assumptions about the shape of the distribution? The approach is surprisingly simple.

Suppose X and Y are continuous random variables drawn independently from the same population. Since X and Y are independent, their joint distribution satisfies

$$p_{XY}(x, y) = p_X(x)p_Y(y). \tag{4.1}$$

Also, since X and Y are drawn from the same populations, the distributions on the right are identical; i.e., $p_X = p_Y$. As a result, the x and y arguments on the right may be swapped without changing the joint distribution. In other words, the joint distribution remains the same after swapping x and y. More generally, if N variables are independent and identically distributed, then the joint distribution is invariant to any permutation of the variable labels. Random variables with this property are termed *exchangeable*. Exchangeability is a very powerful concept that will be used repeatedly in the discussion that follows.

To illustrate the usefulness of exchangeability, consider two variables X and Y that are exchangeable. What is the probability that $X > Y$? Whatever the answer, it cannot depend on which variable is labeled "X" or "Y," because no basis for labeling has been defined. Therefore, the probability of $X > Y$ must be identical to the probability of $X < Y$:

$$P(X > Y) = P(X < Y). \tag{4.2}$$

This expression illustrates the use of exchangeability: the probability on the left-hand side is simply the probability on the right-hand side after variables have been swapped. Recall that the total probability of all possible events equals one, thus

$$1 = P(X > Y) + P(X < Y) + P(X = Y). \tag{4.3}$$

However, for continuous random variables, $P(X = Y) = 0$ (see (1.14)), hence

$$1 = P(X > Y) + P(X < Y). \tag{4.4}$$

The identities (4.2) and (4.4) constitute two equations for two unknowns, and thus can be solved exactly. The solution is

$$P(X > Y) = P(X < Y) = 1/2. \tag{4.5}$$

In this way, the *exact* probability has been derived simply by assuming exchangeability. Moreover, the exact solution is true *regardless of the distribution of X and Y*!

Now instead of two variables, suppose there are three exchangeable variables X, Y, Z. What is the probability of $X > Y > Z$? Because the variables are exchangeable, the probability of this inequality equals the probability of any other inequality in which the variables are permuted. There are exactly six different permutations, namely

$$X > Y > Z, \quad Y > X > Z, \quad Z > X > Y,$$
$$X > Z > Y, \quad Y > Z > X, \quad Z > Y > X. \tag{4.6}$$

Since the variables are continuous, the probability of any two variables being equal has probability zero. Thus, these inequalities exhaust all possibilities, and hence their probabilities must sum to one. It follows that each ordering must have probability 1/6, hence

$$P(X > Y > Z) = 1/6. \tag{4.7}$$

More generally, if we have N continuous, exchangeable variables $X_1, X_2, \ldots, X_N$, then

$$P(X_1 > X_2 > \cdots > X_N) = \frac{1}{N!}. \tag{4.8}$$

Again, this equality holds regardless of the population distribution.

In general, objects that are arranged in order according to some criterion are said to be *ranked* with respect to that criterion. The complete list of ranked objects is known as a ranking. The rank of an object is an integer that indicates its relative position in the ranking.

The basic idea is to test hypotheses using ranks rather than numerical values. If variables are exchangeable, then any particular ranking is just as probable as any other ranking. Also, a test statistic that is a function of ranks has a sampling distribution that can be computed simply by counting the number of arrangements that give a particular value of the statistic and then dividing by the total number of arrangements. Astonishingly, the sampling distribution is independent of the particular distribution from which the variables are drawn. In practice, enumeration of all possible values of the statistic becomes unmanageable even for small sample sizes. If, however, the statistic is a sum of ranks, then in the limit of large sample size, a special version of the Central Limit Theorem can be invoked to derive the sampling distribution. These ideas will be exploited repeatedly in this chapter.

An attractive feature of rank-based tests is that ranks are not sensitive to *outliers*. Outliers refer to data values that are far from the other data values. Outliers can arise from faulty measurements and in that case we would hope that our analysis would not be unduly sensitive to such outliers. Suppose $X_1, \ldots, X_{N-1}$ are close to one, while X_N is very large (e.g., 1,000). If we were to perform the t-test discussed in Chapter 2, the t-statistic would be very sensitive to whether X_N is included or not. As a general rule, a conclusion that depends on a single data value is suspect. In contrast, the rank of X_N would be N regardless of by how much X_N exceeds other sample values. Hence, a rank-based test is not sensitive to the *value* of X_N and is said to be *robust* to outliers.

In the previous discussion, *ties* were ignored because their probability is zero for continuous variables. If, however, variables are discrete or ranks are based on discrete categories, then ties are common. Ties must be handled with special care in nonparametric statistics. In the procedures discussed next, ties can be accommodated by assigning a rank to each variable that equals the average of the ranks that would have been assigned to the variables had there been no ties. For simplicity, we assume that ties do not occur.

4.3 The Wilcoxon Rank-Sum Test

This section describes the Wilcoxon Rank-Sum Test, which may be considered the nonparametric alternative of the t-test discussed in Chapter 2. The test was first proposed by Wilcoxon (1945) and developed further by Mann and Whitney (1947). Accordingly, this test is sometimes called the Mann–Whitney test.

Consider the results of the GCM experiment shown in Figure 4.1. For notational convenience, let the precipitation rates for the GHG1 simulation be denoted $X_1, \ldots, X_{N_X}$, and those for GHG2 be $Y_1, \ldots, Y_{N_Y}$. The objective is to test whether the samples were drawn from the same distribution. Wilcoxon's idea was to combine the $N_X + N_Y$ samples and rank them in order from smallest to largest (ties are handled as described at the end of Section 4.2). If the samples were drawn from the same distribution, then the ranks assigned to the X's and Y's should be spread uniformly over the possible ranks. In contrast, if the distribution of Y is shifted to the right relative to the distribution of X, then the ranks assigned to the Y's generally should be higher than the ranks assigned to the X's. The actual ranks assigned to the variables are illustrated in Figure 4.2. The figure suggests that Y tends to be assigned high ranks whereas X tends to be assigned low ranks. To test this hypothesis objectively, a measure of these tendencies is needed. Wilcoxon suggested comparing the *sum of the ranks* assigned to the X and Y variables. The rank sums are shown in the right margin of Figure 4.2 and confirm that the rank sum for Y is larger than that of X.

To see how to construct a hypothesis test, it is instructive to consider a much smaller example. Accordingly, consider the case $N_X = N_Y = 3$, where the data may be listed as

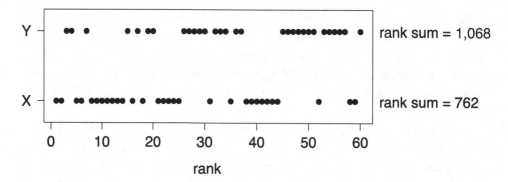

Figure 4.2 Schematic indicating which variable (X or Y) is assigned to each rank when both samples are combined and ordered from smallest to largest. The X and Y data are illustrated in Figure 4.1, and the respective rank sums are shown in the right margin.

$$X_1 \ \ X_2 \ \ X_3 \ \ Y_1 \ \ Y_2 \ \ Y_3. \tag{4.9}$$

There are six numbers, so the ranks to be assigned are $1, 2, \ldots, 6$. The Wilcoxon Rank-Sum test requires summing the ranks assigned to X (or Y, it does not matter which variable is chosen). There are 20 possible sums for the ranks of X_1, X_2, X_3:

$$
\begin{array}{llll}
1 + 2 + 3 = 6 & 1 + 3 + 5 = 9 & 2 + 3 + 4 = 9 & 2 + 5 + 6 = 13 \\
1 + 2 + 4 = 7 & 1 + 3 + 6 = 10 & 2 + 3 + 5 = 10 & 3 + 4 + 5 = 12 \\
1 + 2 + 5 = 8 & 1 + 4 + 5 = 10 & 2 + 3 + 6 = 11 & 3 + 4 + 6 = 13 \\
1 + 2 + 6 = 9 & 1 + 4 + 6 = 11 & 2 + 4 + 5 = 11 & 3 + 5 + 6 = 14 \\
1 + 3 + 4 = 8 & 1 + 5 + 6 = 12 & 2 + 4 + 6 = 12 & 4 + 5 + 6 = 15.
\end{array}
$$

If X and Y are exchangeable, then there is no basis for identifying one of these sums as more probable than another. Therefore, each one is equally likely and the probability that the rank sum equals a particular value is simply the number of counts of that value divided by 20. The count statistics are given next and illustrated in Figure 4.3.

rank sum	count	rank sum	count
6	1	11	3
7	1	12	3
8	2	13	2
9	3	14	1
10	3	15	1

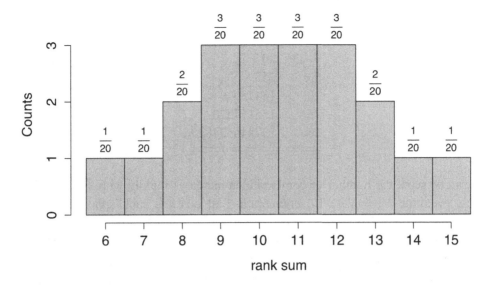

Figure 4.3 Counts (y-axis) and probabilities (labels at the top of histogram rectangles) of all possible rank sums for X for $N_X = N_Y = 3$.

For a two-tailed test, the possible p-values are

$$P(R_X \le 6 \text{ or } R_X \ge 15) = 2/20 = 10\%$$
$$P(R_X \le 7 \text{ or } R_X \ge 14) = 4/20 = 20\%$$
$$P(R_X \le 8 \text{ or } R_X \ge 13) = 8/20 = 40\%$$
$$P(R_X \le 9 \text{ or } R_X \ge 12) = 14/20 = 70\%$$
$$P(R_X \le 10 \text{ or } R_X \ge 11) = 20/20 = 100\%$$

If we were to choose the significance level to be 5%, then the hypothesis would never be rejected because the rarest event in this situation has a *ten* percent chance of occurring.

In the case of general N_X and N_Y, the total number of possible rank sums is a standard result in combinatorics. Specifically, the number of ways N_X things may be chosen from $N = N_X + N_Y$ things is

$$\binom{N}{N_X} = \frac{N!}{(N - N_X)! \, N_X!} = \frac{N(N - 1)\ldots(N - N_X + 1)}{N_X(N_X - 1)\ldots 1}. \tag{4.10}$$

Values of this number for various choices of $N_X = N_Y$ are shown next.

N_X	$\binom{N}{N_X}$
3	20
5	252
10	184756
15	155117520
20	137846528820

As can be seen, the number of combinations increases rapidly with sample size. Thus, even for $N_X = N_Y = 10$, enumerating all possible rank sums would be laborious.

Can the probabilities be estimated by simpler methods? One might think the Central Limit Theorem could be invoked here because the rank-sum is a sum. However, the Central Limit Theorem assumes that the variables in the sum are independent, but this is not the case for the rank sum. For instance, if the rank "3" appears as one of the terms in a rank sum, then it cannot appear in any other term in the rank sum. This means that the individual terms are not independent, as required by the Central Limit Theorem. Nevertheless, there is *another* Central Limit Theorem due to Erdös and Rényi that applies to the rank sum when the sample sizes are large. This theorem states that under certain (mathematically technical) conditions, the rank sum is approximately Gaussian for sufficiently large sample size N_X and N_Y. To apply this theorem, we need only determine the mean and variance of the rank sum. Let the rank sum of X be denoted R_X:

$$R_X = \sum_{n=1}^{N_X} \text{rank}\,[X_n], \tag{4.11}$$

where the ranking is based on the combined X and Y sample. It is shown in Section 4.9 that if the variables are exchangeable, then

$$\mathbb{E}\,[R_X] = \frac{N_X(N_X + N_Y + 1)}{2} \tag{4.12}$$

$$\text{var}\,[R_X] = \frac{N_X N_Y(N_X + N_Y + 1)}{12}. \tag{4.13}$$

These formulas are general. Thus, for exchangeable variables and large sample sizes,

$$z = \frac{R_X - \mathbb{E}[R_X]}{\sqrt{\text{var}[R_X]}} \sim \mathcal{N}(0, 1). \tag{4.14}$$

Following standard hypothesis test methodology (see Chapter 2), the null hypothesis would be rejected at the $\alpha 100\%$ significance level if the computed value of $|z|$ exceeds $z_{\alpha/2}$.

The normal approximation often can be improved by including a *continuity correction*, which adjusts for the use of a continuous distribution to approximate a discrete distribution. Basically, the continuity correction shifts z by a fraction toward the center. More precisely, the continuity correction is to subtract 1/2 from the sample rank-sum when it exceeds the expected mean, and to add 1/2 when it is less than the expected mean. That is,

$$z_{cor} = \frac{R_X - \mathbb{E}[R_X] + C}{\sqrt{\text{var}[R_X]}} \quad \text{where } C = \begin{cases} -0.5 & \text{for } R_X > \mathbb{E}[R_X] \\ 0.5 & \text{for } R_X < \mathbb{E}[R_X] \end{cases}. \quad (4.15)$$

For the case $N_X = N_Y = 3$ discussed previously, $\mathbb{E}[R_X] = 10.5$ and $\text{var}[R_X] = 5.25$, and the p-values computed from (4.15) are as follows.

criterion	exact P	estimated P
$R_X \le 6$ or $R_X \ge 15$	0.1	0.08
$R_X \le 7$ or $R_X \ge 14$	0.2	0.19
$R_X \le 8$ or $R_X \ge 13$	0.4	0.38
$R_X \le 9$ or $R_X \ge 12$	0.7	0.66
$R_X \le 10$ or $R_X \ge 11$	1	1

As can be seen, the estimated p-value correctly gets the first digit even though the sample size is very small. This would not happen without the continuity correction. For the data shown in Figure 4.2, the sample size is $N_X = N_Y = 30$, hence

$$\mathbb{E}[R_X] = 915, \quad \sqrt{\text{var}[R_X]} \approx 67.6 \quad \text{and} \quad R_X = 762, \quad (4.16)$$

which gives $z_{cor} \approx -2.25$. For a significance level $\alpha = 5\%$, the critical value of $z_{\alpha/2}$ is 1.96. Since $|z| > 1.96$, the hypothesis that the samples were drawn from the same distribution is rejected. Furthermore, the negative value of z implies that X tends to have lower ranks than Y, which is consistent with the impression from Figure 4.2.

> **Example 4.1** (Wilcoxon Rank-Sum Test) *A scientist decides to investigate if ENSO influences winter precipitation in the southeast United States. To investigate this question, the scientist collects the mean winter precipitation for the past four positive ENSO events and the past four negative ENSO events (giving a total of eight numbers).[2] The anomalies are listed in Table 4.1. Do these data support the hypothesis that ENSO influences the mean winter precipitation?*

[2] data from www.ncdc.noaa.gov and www.esrl.noaa.gov/psd/data/correlation/nina34.data

Table 4.1. *JFM precipitation anomalies over the southeast United States during the most extreme warm and cold ENSO events during 1950–2016. The left table shows the anomalies while the right table shows the corresponding ranks.*

Sample	Warm	Cold		Sample	Warm	Cold
1	3.16	3.96		1	2	4
2	4.43	4.41		2	7	6
3	3.06	3.9		3	1	3
4	4.17	5.23		4	5	8
mean	3.7	4.37		sum	15	21
st. dev.	0.7	0.61				

The scientist addresses the above question by testing the hypothesis that the samples are exchangeable using the Wilcoxon rank-sum test. To perform this test, the data are collected into a single group and ranked from smallest to largest. The results are shown on the right in Table 4.1. The positive events were assigned ranks 2, 7, 1, 5. Letting X refer to positive events, the sum of the ranks for the positive ENSO events is

$$R_X = 2 + 7 + 1 + 5 = 15. \qquad (4.17)$$

In this example, $N_X = N_Y = 4$, implying

$$\mathbb{E}[R_X] = \frac{4 * (4 + 4 + 1)}{2} = 18 \quad and \quad \text{var}[R_X] = \frac{4 * 4 * (4 + 4 + 1)}{12} = 12.$$

Since $R_X < \mathbb{E}[R_X]$, the continuity correction is $C = +0.5$. Thus, (4.15) is

$$z_{cor} \approx \frac{15 - 18 + 0.5}{\sqrt{12}} \approx -0.72. \qquad (4.18)$$

The threshold value for 5% significance is $z_{0.05/2} = 1.96$. Since $|z_{cor}| < 1.96$, the hypothesis of equal populations is not rejected. For such a small sample size, one generally should use the exact p-values available in standard tables. The normal distribution was used in this example strictly to illustrate the computations.

4.4 Stochastic Dominance

A subtlety about the Wilcoxon Rank-Sum test, and indeed all hypothesis tests, is that the test is designed to detect a specific kind of departure from the null hypothesis. To clarify this issue, let $F_X(\cdot)$ and $F_Y(\cdot)$ denote the cumulative distribution functions for X and Y, respectively. Then, rank-based tests often emerge when testing the hypothesis

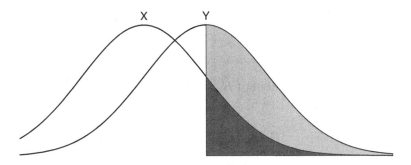

Figure 4.4 An example in which Y is stochastically dominant over X.

$$H_0 : F_Y(z) = F_X(z), \tag{4.19}$$

against the alternative

$$H_A : F_Y(z) > F_X(z). \tag{4.20}$$

If H_A is true, then Y is said to be *stochastically dominant* over X. This hypothesis is equivalent to the hypothesis

$$H_A : P(Y > z) > P(X > z), \tag{4.21}$$

which states that for any given value of z, Y has a larger probability of exceeding z than X. An example in which Y is stochastically dominant over X is illustrated in Figure 4.4. A special case of stochastic dominance occurs if Y is shifted to the right relative to X, that is, $p_Y(z) = p_X(z - \Delta)$. The parameter Δ is called a *location parameter*. Importantly, stochastic dominance is preserved under *any* continuous, strictly increasing function. That is, if $f(\cdot)$ is a continuous, strictly increasing function, and Y is stochastically dominant to X, then $f(Y)$ is stochastically dominant to $f(X)$. Strictly increasing functions include a wide class of functions, including $f(x) = x^3$ and $f(x) = \log(x)$. Because stochastic dominance is preserved for *all* continuous, strictly increasing functions $f(\cdot)$, the associated test cannot depend on the specific values of the data, and instead must depend only on certain order relations among the data; for instance, the ranks. The Wilcoxon rank-sum test is optimal only for a rather restrictive situation (the details are not important here; see Lehman and Romano, 2005), but is effective for situations in which it is not optimal. Accordingly, the Wilcoxon rank-sum test is often described as a test for *stochastic dominance*.

4.5 Comparison with the t-test

The Wilcoxon rank-sum test is often cited as the nonparametric counterpart to the t-test discussed in Chapter 2. Both tests effectively consider the same hypotheses

		Reality	
		H_0 true	H_0 false
Decision	**Accept H_0**	correct conclusion	Type II Error probability $= \beta$
	Reject H_0	Type I Error probability $= \alpha$	correct conclusion power $= 1 - \beta$

Figure 4.5 Schematic illustrating Type-I and Type-II errors and the respective probabilities α and β.

(4.19) and (4.20), which raises the question as to which is better. To quantify their relative performance, recall that two types of errors are possible when deciding between H_0 and H_A: The null hypothesis can be rejected when it is true (Type-I error), or it can be accepted when it is false (Type-II error). These errors are summarized in Figure 4.5. The discussion so far has emphasized significance levels, which quantify type-I errors. The other error is the type-II error of accepting the null hypothesis when it is false, the probability of which is usually denoted β. The *power* of a test is defined as the probability of (correctly) rejecting the null hypothesis when it is false, which is $1 - \beta$. In general, we prefer the most powerful test under plausible alternative hypotheses. For a given type-1 error rate and given difference in distribution, the sample size required for a test to obtain a certain type-2 error can be determined. The ratio of the sample sizes for the two tests is then called the *relative efficiency*. The relative efficiency depends on sample size and the details of how two distributions differ, but in some cases it converges to a constant in the limit of large sample size. This constant is called the *asymptotic relative efficiency* (A.R.E.).

For normal distributions, the A.R.E. of the Wilcoxon rank-sum test compared to the t-test is 0.955 (computation of this number requires methods beyond the scope of this book; see Mood, 1954). This number means that the Wilcoxon rank-sum test requires about 5% more samples to achieve the same level of performance than the t-test. However, when the distribution is non-Gaussian, the A.R.E. can be higher (even infinity). Also, if two distributions differ only by a location parameter, the A.R.E. is always greater than 0.864 (Hodges and Lehmann, 1956). Because the Wilcoxon rank-sum test is only slightly less inefficient than the t-test under a Gaussian distribution, its greater simplicity and broader applicability lead many statisticians to recommend the Wilcoxon rank-sum test as the primary test of the hypothesis that the means of two populations are equal.

4.6 Kruskal–Wallis Test

The Wilcoxon rank-sum test for comparing two groups can be extended to more than two groups. This extension is known as the Kruskal–Wallis Test and is often cited as the nonparametric version of Analysis of Variance (Analysis of Variance is the generalization of the t-test to more than two group means, as explained in Chapter 18).

To explain the Kruskal–Wallis test, consider an experiment that is conducted E times under a fixed condition labeled $c = 1$, and then again under a different (but fixed) condition labeled $c = 2$, and so on for a total of C conditions. The outcome of all experiments can be summarized by the matrix Y_{ec}, where $c = 1, 2, \ldots, C$ denotes the conditions, and $e = 1, 2, \ldots, E$ denotes the trial while holding condition c fixed. If condition c has no impact on the distribution, then each sample is drawn from the same distribution and Y_{ec} is exchangeable. Thus, following the same logic as for the Wilcoxon rank-sum test, all samples are pooled together and then ranked. The rank sum for condition c is

$$R_c = \sum_{e=1}^{E} \text{rank}\,[Y_{ec}], \tag{4.22}$$

where the rank is based on the combined EC sample values. If Y_{ec} is exchangeable, then the rank sums are equally likely and therefore tend to be close to each other, whereas if some populations are stochastically dominant relative to the others, then some of the rank-sums will be larger than others. These considerations motivate selecting a quantity that measures differences between the rank sums. A natural measure is variance, since this measure vanishes when all the variables are equal, and is large when there are large differences between variables. Thus, we use the measure

$$\chi^2 = \frac{1}{\nu} \sum_{c=1}^{C} (R_c - E[R_c])^2, \tag{4.23}$$

where ν is an appropriate constant. If variables are exchangeable, the expected rank for fixed c is the same as that considered in Section 4.3, where X denotes variables for one condition c, and Y is everything else, hence $N_X = E$, so $N_Y = EC - E$. Thus,

$$\mathbb{E}\,[R_c] = \frac{E(EC + 1)}{2}. \tag{4.24}$$

For large sample sizes, a version of the Central Limit Theorem implies that the rank-sums are normally distributed. In this case, it has been shown that (4.23) has a chi-square distribution with $C - 1$ degrees of freedom when $\nu = E(EC)(EC + 1)/12$, hence

Table 4.2. *Precipitation in the southeast United States for January–March (JFM), April–June (AMJ), July–September (JAS), and October–December (OND), during selected years.*

Year	JFM	AMJ	JAS	OND	Year	JFM	AMJ	JAS	OND
2012	3.23	4.26	5.51	2.77	2012	3	10	18	2
2013	4.41	5.46	6.2	3.51	2013	12	17	20	5
2014	3.92	4.93	4.74	3.73	2014	7	15	13	6
2015	3.46	4.27	5.11	5.94	2015	4	11	16	19
2016	3.96	4.21	4.79	2.69	2016	8	9	14	1
median	3.92	4.27	5.11	3.51	sum	34	62	81	33

$$\chi^2 = \frac{12}{E(EC)(EC+1)} \sum_{c=1}^{C} (R_c - E[R_c])^2 \sim \chi_{C-1}^2. \tag{4.25}$$

The case $C = 2$ reduces to the Wilcoxon rank-sum test. The Kruskal–Wallis test can also be generalized to unequal ensemble sizes.

Example 4.2 (Kruskal–Wallis Test) *Does precipitation in the southeast United States change with season? To investigate this question, a scientist collects data for precipitation in the southeast United States for four different seasons. The results are shown in the left table in Table 4.2. If precipitation varied with season, then it would not be exchangeable. Therefore, the scientist tests the hypothesis that precipitation data are exchangeable using the Kruskal–Wallis test. In this test, all the data are pooled together and then ranked. The result of this ranking is shown in the right table in Table 4.2. For instance, this latter table shows that the smallest value (rank 1) is 2.69 while the largest value (rank 20) is 6.2. What are E and C? E denotes the number of "repetitions" under fixed physical condition, while C denotes the number of physical conditions. In the present context, the physical condition that would cause changes is solar insolation, while the repetitions are the years (i.e., each year feels the same solar insolation). Thus, $C = 4$ and $E = 5$. Furthermore, the sum of the ranks for fixed physical condition is the sum of ranks for fixed season, yielding 34, 62, 81, and 33. If precipitation were exchangeable, then the expected rank sum would be (4.24), which is*

$$\mathbb{E}[R_c] = \frac{5 * (5 * 4 + 1)}{2} = 52.5.$$

Therefore, the chi-squared statistic (4.25) is

$$\chi^2 = 12 \left(\frac{(34 - 52.5)^2 + (62 - 52.5)^2 + (81 - 52.5)^2 + (33 - 52.5)^2}{5 * (5 * 4) * (5 * 4 + 1)} \right) \approx 9.29.$$

The 5% critical value of the chi-squared distribution with $C - 1 = 3$ degrees of freedom is 9.49, which is larger than the sample value 9.29. Since $\chi^2 < \chi_\alpha^2$, stochastic dominance cannot be rejected. Thus, the scientist was not able to detect a seasonal *cycle in seasonal mean precipitation in the southeastern United States. Note that this conclusion is based on five years of data and may not hold for a larger sample size. A small sample size was chosen strictly for illustration purposes.*

4.7 Test for Equality of Dispersions

The F-test was discussed in Chapter 2 as a test for differences in variance. Unfortunately, the F-test is sensitive to departures from normality (Box, 1953). One reason for this sensitivity is that variance involves the square of a variable, and squares lead to long-tailed distributions that are sensitive to the tail of the original population. An alternative measure of dispersion that has attractive properties for a wide class of non-Gaussian distributions is the *median absolute deviation*, defined as

$$\text{MAD}[X] = \text{median}\left[|X - \text{median}[X]|\right]. \tag{4.26}$$

Consider the random variables

$$\Delta X = |X - \text{median}[X]| \quad \text{and} \quad \Delta Y = |Y - \text{median}[Y]|. \tag{4.27}$$

The basic idea is that if Y has greater dispersion than X, then ΔY should be stochastically dominant over ΔX. Thus, applying the Wilcoxon rank-sum test to ΔX and ΔY provides a test for equality of dispersion. In practice, the population medians are not known and are approximated by sample medians, giving the statistics

$$\Delta X^* = |X - \hat{m}_X| \quad \text{and} \quad \Delta Y^* = |Y - \hat{m}_Y|, \tag{4.28}$$

where $\hat{m}_X$ and $\hat{m}_Y$ are the sample medians of X and Y. It turns out that applying the Wilcoxon rank-sum test to ΔX^* and ΔY^* yields results that depend somewhat on the population. Fortunately, this parametric dependence is small for large sample size.

It has been found that the sum of the *squared* ranks,

$$R'_X = \sum_{n=1}^{N_X} \left(\text{rank}[\Delta X_n^*]\right)^2, \tag{4.29}$$

leads to a more powerful test. The resulting test is called the *squared-rank test*.

As in the Wilcoxon rank-sum test, a version of the Central Limit Theorem can be invoked. The mean and variance of the sum square ranks is

$$\mathbb{E}[R'_X] = N_X \frac{(N+1)(2N+1)}{6} \tag{4.30}$$

$$\text{var}[R'_X] = \frac{N_X(N - N_X)}{N - 1}\left(\frac{1}{N}\sum_{i=1}^{N}\left(i^2 - \frac{(N+1)(2N+1)}{6}\right)^2\right), \tag{4.31}$$

Table 4.3. *Deviations of precipitation in the southeast United States for April–June (AMJ) and October–December (OND) (left table), derived from Table 4.2, and the corresponding ranks (right table). Deviation is defined as the value minus the median value.*

Year	AMJ	OND		Year	AMJ	OND
2012	−0.01	−0.74		2012	3	7
2013	1.18	0		2013	9	1.5
2014	0.66	0.21		2014	6	5
2015	0	2.42		2015	1.5	10
2016	−0.06	−0.83		2016	4	8
median	0	0		sum square	144.25	240.25

which is derived in Section 4.9. Accordingly, the standardized variable has distribution

$$z = \frac{R'_X - \mathbb{E}[R'_X] + C}{\sqrt{\text{var}[R'_X]}} \sim \mathcal{N}(0, 1), \tag{4.32}$$

where C is the appropriate continuity correction. H_0 is rejected when $|z| > z_{\alpha/2}$.

The test for equality of variance for more than two populations proceeds in the same way as the Kruskal–Wallis test, based on an analogous χ^2 statistic. Another nonparametric procedure for testing equality of variance is *Levene's test*, which is effectively Analysis of Variance applied to residuals about the median.

Example 4.3 (Equality of Dispersion Test) *Does the variability of precipitation in the southeast United States vary between spring and fall? To investigate this question, a scientist collects data for precipitation in the southeast United States for spring and fall. The results are shown in the left table in Table 4.2. The deviation – that is, the difference between the value and the median value – for each season are shown in the left table of Table 4.3. The median deviation vanishes by construction (see bottom line in the table). To apply the equality of dispersion test, absolute deviations for both seasons are pooled together and then ranked. The resulting ranks are shown in the right table of Table 4.3. For instance, this latter table shows that the largest value (rank 10) is 2.42. The sum square ranks are shown on the bottom line of the right table and have the values 144.25 and 240.25. If precipitation were exchangeable, then the expected sum square rank would be (4.30), whose value is*

$$\mathbb{E}[R'_X] = 5 * \frac{(10 + 1)(2 * 10 + 1)}{6} = 192.5. \tag{4.33}$$

Similarly, the variance of the sum square rank is

$$\text{var}[R'_X] = \frac{5 * (10 - 5)}{(10 - 1)} \frac{1}{10} \sum_{i=1}^{10} \left(i^2 - 192.5\right)^2 \approx 2919.6. \tag{4.34}$$

Therefore, the standardized statistic is

$$z = \frac{R'_X - \mathbb{E}[R'_X]}{\sqrt{\text{var}[R'_X]}} = \frac{144.25 - 192.5}{\sqrt{2919.6}} \approx -0.89. \tag{4.35}$$

The 5% critical value for a normal distribution is $z_{\alpha/2} = 1.96$. Since, $|z| < z_{\alpha/2}$, the equal dispersion hypothesis cannot be rejected. No change in variability from spring to fall precipitation is detected.

4.8 Rank Correlation

Chapter 2 discussed the correlation test for independence of normally distributed variables. This section presents a nonparametric version of this test. Consider N pairs of variables

$$(X_1, Y_1), \quad (X_2, Y_2), \quad \dots, \quad (X_N, Y_N). \tag{4.36}$$

Let rank$[X_n]$ denote the rank of X_n among the X-variables, and let rank$[Y_n]$ denote the rank of Y_n among the Y-variables. The correlation between ranks is known as *Spearman's Rank Correlation Coefficient* and is defined as

$$\hat{\rho}_{Sp} = \frac{\sum_{n=1}^{N} (\text{rank}[X_n] - \mathbb{E}[\text{rank}[X_n]]) (\text{rank}[Y_n] - \mathbb{E}[\text{rank}[Y_n]])}{\sqrt{\text{var}[\text{rank}[X_n]] \, \text{var}[\text{rank}[Y_n]]}}. \tag{4.37}$$

This expression can be simplified because the mean and variance are known. After all, the ranks $1, 2, \dots, N$ must occur exactly once and with equal probability. Thus,

$$\mathbb{E}[\text{rank}[X_n]] = \sum_{i=1}^{N} i \, P(\text{rank}[X_n] = i) = \frac{1}{N} \sum_{i=1}^{N} i = \frac{N+1}{2}. \tag{4.38}$$

The mean of ranks of Y also equals $(N + 1)/2$. The variance of the ranks is also known for the same reason and is given by

$$\text{var}[\text{rank}[X_n]] = \mathbb{E}\left[\left(\frac{1}{N} \sum_{n=1}^{N} \text{rank}[X_n] - \mathbb{E}[\text{rank}[X_n]] \right)^2 \right] = \frac{N(N^2 - 1)}{12}, \tag{4.39}$$

and similarly for the variance of rank$[Y_n]$. Substituting these quantities into (4.37) gives

$$\hat{\rho}_{Sp} = \frac{\sum_{n=1}^{N} \left(\text{rank}[X_n] - \frac{N+1}{2} \right) \left(\text{rank}[Y_n] - \frac{N+1}{2} \right)}{N(N^2 - 1)/12}. \tag{4.40}$$

It can be shown that Spearman's Rank Correlation can also be computed as

$$\hat{\rho}_{Sp} = 1 - \frac{6 \sum_{n=1}^{N} d_n^2}{N(N^2 - 1)},\tag{4.41}$$

where $d_n = \text{rank}[X_n] - \text{rank}[Y_n]$. These expressions are valid if there are no ties.

The significance levels for Spearman's Rank Correlation can be found in widely available tables and from standard software packages. For large sample sizes, a version of the Central Limit Theorem can be invoked using the approximations

$$\mathbb{E}\left[\hat{\rho}_{Sp}\right] \approx 0 \quad \text{and} \quad \text{var}\left[\hat{\rho}_{Sp}\right] \approx \frac{1}{N-1}.\tag{4.42}$$

Thus, a significance test for Spearman's Rank Correlation can be performed by computing

$$z = \frac{\text{observed} - \text{hypothesized}}{\text{standard error}} = \hat{\rho}_{Sp}\sqrt{N-1},\tag{4.43}$$

and rejecting the null hypothesis if $|z| > z_{\alpha/2}$. Note that the 95% significance threshold for Spearman's correlation is $1.96/\sqrt{N-1} \approx 2/\sqrt{N}$, the same as in (2.44).

Spearman's Rank Correlation is a test for how well two variables are related through a *monotonic function*. A monotonic function is a function that is entirely nonincreasing or entirely nondecreasing. For example, $f(x) = x^3$ and $f(x) = \log(x)$ are monotonic functions. If X is a monotonic function of Y, then the ranks are linearly related and Spearman's Rank Correlation equals one. This fact is illustrated in Figure 4.6.

The parametric counterpart of the rank correlation is Pearson's correlation coefficient $\hat{\rho}$, defined in (1.29). If X and Y are independent and identically distributed normal random variables, then $\hat{\rho}_{Sp}$ has an A.R.E. of 0.912 relative to Pearson's correlation. This means that Spearman's Rank Correlation requires about 9% more samples to achieve the same level of performance than Pearson's correlation coefficient for normal distributions.

Example 4.4 (Spearman's Rank Correlation) *Does ENSO influence winter precipitation in the southeast United States? To investigate this question, a scientist collects precipitation data for the eight warmest and eight coldest ENSO events during 1950–2016. The resulting values are shown in the left table of Table 4.4. The ranks of precipitation values and ENSO values separately are shown in the right table of Table 4.4. The expected rank (4.38) is*

$$\mathbb{E}[\text{rank}[X_n]] = \frac{16+1}{2} = 8.5.\tag{4.44}$$

Table 4.4. *JFM precipitation anomalies in the southeast United State for the most extreme ENSO events during 1950–2016 (left table), and the corresponding ranks (right table).*

Year	Precip	Niño 3.4	Year	Precip	Niño 3.4
1974	4.08	−1.59	1974	8	1
1950	3.16	−1.57	1950	1	2
1971	4.76	−1.42	1971	12	3
1989	3.6	−1.41	1989	4	4
2008	3.65	−1.39	2008	5	5
2000	3.24	−1.3	2000	2	6
1976	3.49	−1.23	1976	3	7
1999	3.69	−1.19	1999	6	8
1987	5.46	1.22	1987	14	9
1973	5.26	1.22	1973	13	10
2010	4.43	1.36	2010	11	11
1958	4.35	1.55	1958	10	12
1992	4.25	1.8	1992	9	13
1983	5.64	2.02	1983	15	14
1998	6.57	2.05	1998	16	15
2016	3.96	2.28	2016	7	16

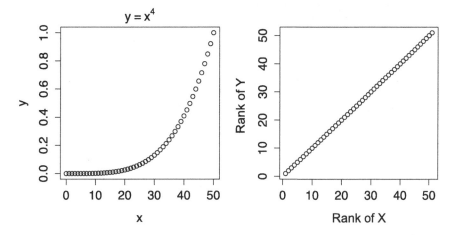

Figure 4.6 Illustration of the fact that if X and Y are monotonically related, then the ranks will be linearly related. The left panel shows points along the polynomial $Y = X^4$, and the right panel shows the rank of X versus rank of Y, where the X-variables are ranked among themselves and the Y variables are ranked among themselves.

Therefore, Spearman's rank correlation (4.40) is

$$\hat{\rho}_{Sp} = \frac{(8 - 8.5)(1 - 8.5) + \cdots + (7 - 8.5)(16 - 8.5)}{16 * (16^2 - 1)/12} \approx 0.553. \qquad (4.45)$$

The standardized value is therefore

$$z = \hat{\rho}_{Sp}\sqrt{N-1} \approx 0.553\sqrt{16-1} \approx 2.14. \tag{4.46}$$

The 5% critical value for a normal distribution is $z_{\alpha/2} = 1.96$. Since, $|z| > z_{\alpha/2}$, the independence hypothesis is rejected. Thus, the scientist concludes that ENSO influences winter precipitation in the southeastern United States at the 5% significance level.

An alternative approach to testing independence is based on *Kendall's Rank Correlation Coefficient* or *Kendall's tau coefficient*. Readers interested in this test may consult standard nonparametric texts (e.g., Conover, 1980). An advantage of Kendall's tau is that its distribution approaches a normal distribution very rapidly compared to Spearman's rho.

A common application of the rank correlation is the *Mann–Kendall Test* for a trend. To perform this test for X, we simply set the Y-variables to a linear function of time, such as

$$Y = \begin{pmatrix} 1 & 2 & \cdots & N \end{pmatrix}, \tag{4.47}$$

and perform a rank correlation test on X and Y. The corresponding Spearman's Rank Correlation measures the degree to which X is a monotonic function of time. In contrast, the corresponding Pearson Correlation measures the *linear* relation between X and time.

4.9 Derivation of the Mean and Variance of the Rank Sum

This section derives the mean and variance of the rank sum (4.11) under exchangeability. This material is mostly algebra and not essential for applying the rank sum test.

It will prove convenient to define the following variables:

$$Z_n = \text{rank}[X_n] \quad \text{and} \quad N = N_X + N_Y. \tag{4.48}$$

By definition, Z_n is a discrete random variable that can take on values $1, 2, \ldots, N$. Because the expectation is linear, it follows that

$$\mathbb{E}[R_X] = \mathbb{E}\left[\sum_{n=1}^{N_X} \text{rank}[X_n]\right] = \sum_{n=1}^{N_X} \mathbb{E}[Z_n] = N_X \mathbb{E}[Z_n], \tag{4.49}$$

where the last equality follows from the fact that the expected value of Z_n does not depend on n because of exchangeability. In addition, because X and Y are exchangeable, each value of Z_n is equally probable, thus

$$P(Z_n = i) = 1/N. \tag{4.50}$$

The expected rank is therefore

$$\mathbb{E}[Z_n] = \sum_{i=1}^{N} i P(Z_n = i) = \frac{1}{N} \sum_{i=1}^{N} i = \frac{N+1}{2}, \tag{4.51}$$

where the algebraic identity (A.3) was used. Substituting this into (4.49) yields

$$\mathbb{E}[R_X] = \frac{N_X(N+1)}{2}. \tag{4.52}$$

This proves (4.12).

The variance of the rank sum is

$$\text{var}[R_X] = \mathbb{E}\left[\left(\sum_{n=1}^{N_X} (Z_n - \mathbb{E}[Z_n])\right)^2\right] \tag{4.53}$$

$$= \sum_{n=1}^{N_X} \sum_{m=1}^{N_X} \mathbb{E}\left[(Z_n - \mathbb{E}[Z_n])(Z_m - \mathbb{E}[Z_m])\right] \tag{4.54}$$

$$= \sum_{n=1}^{N_X} \sum_{m=1}^{N_X} \text{cov}[Z_n, Z_m]. \tag{4.55}$$

Because the variables are exchangeable, the covariance equals a constant when $n \neq m$, and equals a different constant when $n = m$. There are N_X terms when $n = m$, and $N_X(N_X - 1)$ terms when $n \neq m$. Therefore,

$$\text{var}[R_X] = N_X(N_X - 1)\text{cov}[Z_n, Z_m]_{n \neq m} + N_X \text{var}[Z_n]. \tag{4.56}$$

The variance of the rank is

$$\text{var}[Z_n] = \sum_{i=1}^{N} (i - \mathbb{E}[Z_n])^2 P(Z_n = i) \tag{4.57}$$

$$= \frac{1}{N} \sum_{i=1}^{N} (i - \mathbb{E}[Z_n])^2 \tag{4.58}$$

$$= \frac{1}{N} \sum_{i=1}^{N} i^2 - (\mathbb{E}[Z_n])^2 \tag{4.59}$$

$$= \frac{1}{N} \frac{N(N+1)(2N+1)}{6} - \frac{(N+1)^2}{4} \tag{4.60}$$

$$= \frac{(N+1)(N-1)}{12}, \tag{4.61}$$

where (A.4) has been used. The covariance term is, by definition,

$$\text{cov}[Z_n, Z_m]_{n \neq m} = \sum_{\substack{i=1 \\ i \neq j}}^{N} \sum_{j=1}^{N} (i - \mathbb{E}[Z_n])(j - \mathbb{E}[Z_m])P(Z_n = i, Z_m = j). \quad (4.62)$$

The joint probability can be derived from the condition probability as

$$P(Z_n = i, Z_m = j) = P(Z_m = i | Z_n = j)P(Z_n = j) = \frac{1}{N-1}\frac{1}{N}, \quad (4.63)$$

where (4.50) has been used, and the conditional probability is $1/(N-1)$ because once Z_n is known, Z_m may equal any rank from 1 to N except the value of Z_n. Instead of evaluating the sum in (4.62), which skips over $i = j$, it is more convenient to sum over all i and j, and then subtract out the terms for which $i = j$. Accordingly, (4.62) becomes

$$\text{cov}[Z_n, Z_m]_{n \neq m}$$

$$= \frac{1}{N(N-1)} \left(\sum_{i=1}^{N} \sum_{j=1}^{N} (i - \mathbb{E}[Z_n])(j - \mathbb{E}[Z_n]) - \sum_{i=1}^{N} (i - \mathbb{E}[Z_n])^2 \right)$$

$$= \frac{1}{N(N-1)} \left(\left(\sum_{i=1}^{N} (i - \mathbb{E}[Z_n]) \right) \left(\sum_{j=1}^{N} (j - \mathbb{E}[Z_n]) \right) - \sum_{i=1}^{N} (i - \mathbb{E}[Z_n])^2 \right).$$

The first two sums on the right-hand side vanish individually, leaving

$$\text{cov}[Z_n, Z_m]_{n \neq m} = -\frac{1}{N(N-1)} \sum_{i=1}^{N} (i - \mathbb{E}[Z_n])^2 = -\left(\frac{1}{N-1} \right) \text{var}[Z_n]. \quad (4.64)$$

Substituting this into (4.56) gives

$$\text{var}[R_X] = -\frac{N_X(N_X - 1)}{N-1} \text{var}[Z_n] + N_X \text{var}[Z_n]$$

$$= \frac{1}{N-1} N_X(N - N_X) \text{var}[Z_n]. \quad (4.65)$$

Substituting (4.61) into (4.65) gives

$$\text{var}[R_X] = \frac{N_X(N - N_X)(N + 1)}{12}. \quad (4.66)$$

This proves (4.13).

Some tests use a *function of the ranks* rather than the ranks themselves. For instance, equality-of-variance tests often use the square of the rank. To account for this case, let

$$R_f = \sum_{n=1}^{N_X} f(Z_n), \qquad (4.67)$$

where $f(\cdot)$ is arbitrary function (e.g., in the equality-of-dispersion test, $f(x) = x^2$). Following the same arguments as previously, the mean of this statistic is

$$\mathbb{E}[R_f] = \sum_{n=1}^{N_X} \mathbb{E}[f(Z_n)] = N_X \mathbb{E}[f(Z_n)], \qquad (4.68)$$

where all terms in the sum are identical because of exchangeability. Since the ranks are exchangeable they are equally probable, hence

$$\mathbb{E}[f(Z_n)] = \sum_{i=1}^{N} f(i)P(Z_n = i) = \frac{1}{N} \sum_{i=1}^{N} f(i) = \overline{f}. \qquad (4.69)$$

Therefore, the expected value of R_f is

$$\mathbb{E}[R_f] = N_X \overline{f}. \qquad (4.70)$$

In the same way that (4.65) was derived, the variance of R_f is

$$\mathrm{var}[R_f] = \frac{1}{N-1} N_X(N - N_X) \, \mathrm{var}[f(Z_n)], \qquad (4.71)$$

where

$$\mathrm{var}[f(Z_n)] = \sum_{i=1}^{N} (f(i) - \mathbb{E}[f(Z_n)])^2 \, P(Z_n = i) \qquad (4.72)$$

$$= \frac{1}{N} \sum_{i=1}^{N} \left(f(i) - \overline{f} \right)^2 \qquad (4.73)$$

$$= \hat{\sigma}_f^2. \qquad (4.74)$$

Thus, the variance is

$$\mathrm{var}[R_f] = \frac{1}{N-1} N_X(N - N_X) \left(\frac{1}{N} \sum_{i=1}^{N} \left(f(i) - \overline{f} \right)^2 \right) \qquad (4.75)$$

In the special case $f(i) = i^2$,

$$\overline{f} = \frac{1}{N} \sum_{i=1}^{N} i^2 = \frac{(N+1)(2N+1)}{6}. \qquad (4.76)$$

4.10 Further Topics

This chapter discussed rank-based nonparametric methods. Another class of non-parametric methods, known as *resampling* methods, are based on drawing random samples *from the given sample* to make inferences about the population. An example of this is the *bootstrap* method discussed in Chapter 2, which gives confidence intervals. Another example are *permutation tests*, which perform significance tests. For instance, many null hypotheses are of the form that certain random variables are exchangeable, in the sense the distribution is invariant to rearrangements of the labels assigned to individual sample values. For instance, in the t-test, the distribution of X and Y under the null hypothesis is invariant to randomly reassigning the X- and Y- labels. Therefore, the sampling distribution of the test statistic can be derived by calculating the value of the statistic for all possible rearrangements of the labels. This procedure effectively permutes the sample, hence the name. A permutation test can give *exact* significance levels for real-world problems because it does not make any assumption about the distribution beyond exchangeability. This generality comes at a cost: permutation tests are computationally intensive. Introductions to resampling methods can be found in Efron and Tibshirani (1994), Good (2005), and Good (2006).

Another class of nonparametric methods are based on contingency tables. A contingency table is an array of counts or frequencies in matrix form, often used to summarize the outcomes of discrete events. For instance, a forecaster may use a contingency table to record the number of times a forecast system successfully predicted a weather event, such as a fog, storm, flood, rain, etc. Based on the resulting table, the forecaster may want to test whether the forecast and observations are independent (in which case the forecast has no skill). Contingency tables may also be used to test for differences in probabilities. A comprehensive text on contingency table methods is Agresti (2012).

Another class of nonparametric methods are used to test if samples are drawn from specific distributions. For instance, the tests discussed in earlier chapters, such as the t-test, F-test, and correlation test, *assume* samples come from a normal distribution. What if the assumption is wrong? Is there a way to test whether the sample came from a normal distribution? Andrey Kolmogorov proposed a very simple procedure for the special case in which the theoretical distribution is specified fully (e.g., the mean and variance of the normal distribution are known). Specifically, one computes the empirical cumulative distribution $F_N(x)$, which gives the fraction of samples that are less than or equal to x, and then computes the maximum absolute difference D between this and the theoretical cumulative distribution function $F(x)$ over all x. Kolmogorov proved that the distribution of D is independent of $F(x)$ for continuous $F(X)$ and gave its exact distribution (see Kolmogorov, 1992, for an English translation of the original paper). In practice, this test is not very useful because it assumes that $F(x)$ is known completely (i.e., all parameters are known exactly), but it inspired more useful tests, such as the Kolmogorov–Smirnov (KS) test, which tests if two random sample come from the same distribution. The KS test may be preferred over contingency table tests because the Kolmogorov-type tests are exact even for small sample sizes. Kolmogorov-type tests are based on the maximum absolute difference between two cumulative distribution functions. Another type of tests are based on a weighted integral of the *squared difference* between $F_N(x)$ and $F(x)$, the most well-known of which include the Anderson–Darling test and Cramér–von Mises tests. These latter tests are some of the most powerful tests for detecting departures from normality.

4.11 Conceptual Questions

1 What are some advantages of rank-based hypothesis tests?
2 What are some disadvantages of rank-based hypothesis tests?
3 When should you use the rank correlation rather than the Pearson correlation?
4 In your own words, describe the Wilcoxon Rank-Sum test.
5 Suppose the Wilcoxon Rank-Sum test says there is a statistically significant difference between two samples. What general aspect of the samples do you expect to be different?
6 What is the relative efficiency of a test?
7 What are some reasons why the Wilcoxon Rank-Sum test might be recommended instead of the t-test?
8 In your own words, describe the rank correlation test.
9 What is the advantage of a rank correlation test compared to a standard correlation test?
10 Describe a situation in which the rank correlation and Pearson's correlation give different answers. Which one would be "right"? Now think of a situation in which the answers are different but the other one might be considered "right."

5

Introduction to Stochastic Processes

If we observe at short equal intervals of time the departures of a simple harmonic pendulum from its position of rest, errors of observation
will cause superposed fluctuations ... But by improvement of apparatus
and automatic methods of recording, let us say, errors of observation
are practically eliminated. The recording apparatus is left to itself, and
unfortunately boys get into the room and start pelting the pendulum
with peas, sometimes from one side and sometimes from the other. The
motion is now affected, not by superposed fluctuations but by true disturbances, and the effect on the graph will be of an entirely different kind.
The graph will remain surprisingly smooth, but amplitude and phase will
vary continually.[1]

Yule (1927)

Climate data are correlated over sufficiently short spatial and temporal scales.
For instance, today's weather tends to be correlated with tomorrow's weather,
and weather in one city tends to be correlated with weather in a neighboring city.
Such correlations imply that weather events are *not independent*, and therefore the
"independent and identically distributed" assumption (*iid*) used in previous chapters
is very unrealistic for weather variables over short space or timescales. Similarly,
ocean variables tend to be correlated over a few months, hence the *iid* assumption is
not appropriate for ocean variables on monthly timescales. Dependencies in space
and time can be taken into account using the concept of *stochastic processes*. A
stochastic process is a collection of random variables indexed by a parameter, such
as time or space. A stochastic process is described by its moments (e.g., mean
and variance, which might depend on time), and by the degree of dependence
between two times, often measured by the *autocorrelation function*. This chapter
reviews these concepts and discusses common mathematical models for generating
stochastic processes, especially *autoregressive models*. The focus of this chapter is
on developing the language for describing stochastic processes. Challenges in estimating parameters and testing hypotheses about stochastic processes are discussed.

---

[1] Reprinted by permission from The Royal Society (U.K.), *Philosophical Transactions of the Royal Society of
London. Series A, Containing Papers of a Mathematical or Physical Character*, Yule (1927).

Niño 3.4 index (raw)

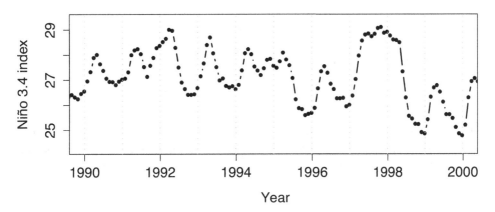

Figure 5.1 The observed monthly Niño 3.4 index for a selected 10-year period.

5.1 The Problem

A collection of observations sequenced in time is called a *time series*. This sequence in time can be denoted as $X_1, X_2, \ldots$, where the subscript refers to the order in time. As a concrete example, consider the monthly Niño 3.4 index shown in Figure 5.1. Let us try to predict X_{t+1} given the previous values $X_t, X_{t-1}, \ldots$. To do this, patterns in the data need to be identified. One pattern is that the peaks and valleys occur at nearly periodic intervals, reflecting the familiar seasonal cycle. However, the time series is not *exactly* periodic. Accordingly, we hypothesize that the data were generated by a model of the form

$$X_t = \mu_t + \epsilon_t, \tag{5.1}$$

where μ_t is a *deterministic* function of time and ϵ_t is a *random* variable. The deterministic term depends on hypotheses about the mechanisms that cause variations in X_t. For instance, diurnal and annual cycles are driven by well understood variations in solar insolation, so it is accepted practice to model these cycles with a periodic function of time. Other time series exhibit clear trends and oscillations over decades or centuries, but whether these variations should be included in μ_t is less clear. For instance, if the trends are caused by changes in greenhouse gas concentrations and aerosols, then it is accepted practice to model μ_t by a physics-based model, or, because of its smooth time variations, by a low-order polynomial in time with coefficients estimated from data. On the other hand, if these trends reflect slowly varying *random* components of the climate system, such as persistent ocean variations, then these trends should not be included in μ_t.

For the present example, we model μ_t as a periodic function of time as

$$\mu_t = a_0 + \sum_{k=1}^{2} \left(a_k \cos\left(\frac{2\pi kt}{12}\right) + b_k \sin\left(\frac{2\pi kt}{12}\right)\right), \tag{5.2}$$

where a_k and b_k are constant coefficients. This model has five parameters (a_0, a_1, a_2, b_1, b_2), which are estimated from past data using the method of least squares (see Chapter 8). The specific procedure is not important for the present discussion. Suppose this estimation yields $\hat{\mu}_t$. Then, the difference

$$X'_t = X_t - \hat{\mu}_t, \tag{5.3}$$

is termed the *anomaly* index. The anomaly index for Niño 3.4 is shown in Figure 5.2. Can the anomaly index be predicted? Again, we look for patterns. A distinctive feature of the anomaly index shown in Figure 5.2 is that it is a fairly smooth function of time. Smooth time series imply a dependence in time and are said to be *serially correlated*. To demonstrate this dependence more explicitly, consider Figure 5.3, which shows a scatter plot of the anomaly index plotted against its value the following month. This figure provides a basis for making predictions. For instance, if the current anomaly were $x_t = 2$, then a reasonable prediction from Figure 5.3 is that next month's anomaly is likely to be between 1 and 2.

Figure 5.3 shows that X_t and X_{t+1} are dependent regardless of the specific value of t. It follows that X_{t+1} and X_{t+2} are dependent, and so on in a chain-like fashion.

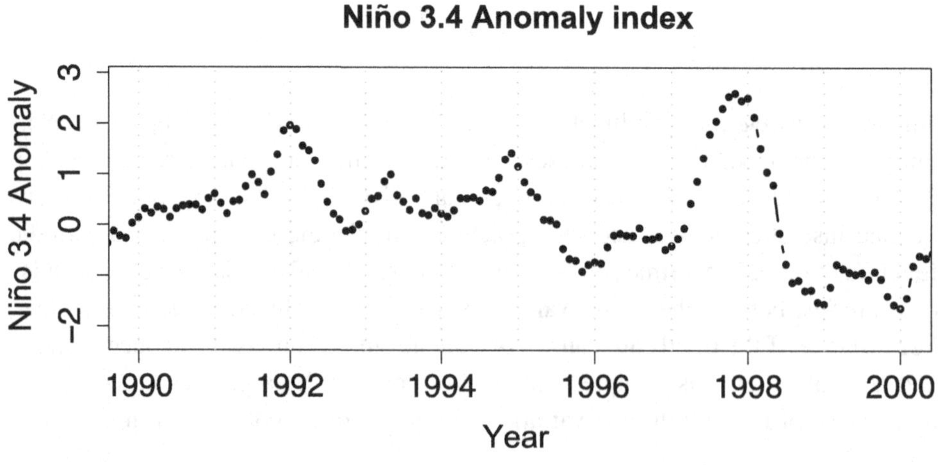

Niño 3.4 Anomaly index

Figure 5.2 Monthly mean Niño 3.4 anomaly index.

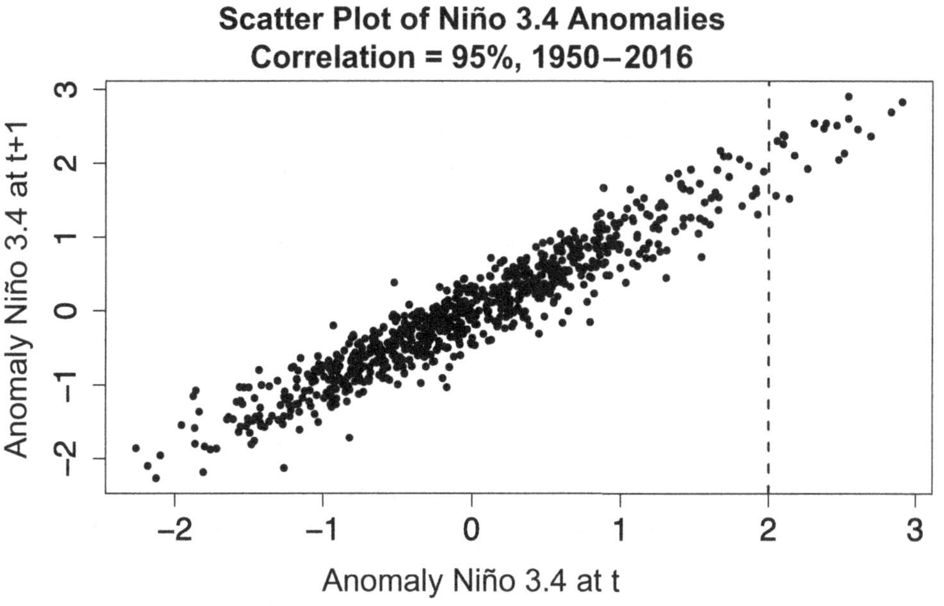

Scatter Plot of Niño 3.4 Anomalies
Correlation = 95%, 1950–2016

Figure 5.3 Scatter plot of two consecutive values of the Niño 3.4 anomaly. The text discusses $X_t = 2$ and is indicated by the dashed vertical line as a visual aid.

This fact raises the possibility that, given X_t, not only is X_{t+1} predictable but so are $X_{t+2}, X_{t+3}, \ldots$, to differing degrees. To investigate this possibility, the dependence between months needs to be quantified. A natural measure of linear dependence is the correlation coefficient (see Chapter 1). For example, the correlation of the data shown in Figure 5.3 is 0.95, which is close to one and thus indicates a strong month-to-month dependence. In the same way, the dependence on longer timescales can be quantified by computing the correlation between anomalies separated by τ months. The parameter τ is known as the *time lag*. The temporal dependence of a time series is measured by the autocovariance function.

Definition 5.1 (Sample Autocovariance Function) *Let $X_1, X_2, \ldots, X_N$ be a time series of length N. The sample autocovariance function is defined as*

$$\hat{c}(\tau) = \frac{1}{N} \sum_{t=1}^{N-|\tau|} \left(X_{t+|\tau|} - \hat{\mu}_{t+|\tau|} \right) \left(X_t - \hat{\mu}_t \right), \qquad (5.4)$$

for $\tau = 0, \pm 1, \pm 2, \ldots, \pm(N-1)$, where $\hat{\mu}_t$ is the estimated deterministic component.

This quantity differs from the sample covariance function defined in Chapter 1 in two ways. First, the sum is normalized by $1/N$, rather than by $N - 1$. Second, the definition computes differences relative to a single deterministic function $\hat{\mu}_t$, which is estimated using the *full* sample of length N, rather than using two separate samples, one from $t \in (1, N - \tau)$ and the other from $t \in (\tau + 1, N)$. The reason for defining (5.4) in this way cannot be fully explained now, but is related to improving accuracy and ensuring that the sample autocovariance function satisfies constraints that are satisfied by the population autocovariance function (e.g., positiveness of the power spectrum; see Chapter 6).

The *sample autocorrelation function* (ACF) is defined as

$$\hat{\rho}_\tau = \frac{\hat{c}(\tau)}{\hat{c}(0)}. \tag{5.5}$$

The autocorrelation function of the Niño 3.4 anomaly is shown in Figure 5.4. The ACF starts at one and decays smoothly to zero in 10 months, and then overshoots zero and becomes negative for another 20 months, giving the impression of a *damped oscillation*. These features foreshadow the characteristics of a linear prediction of the Niño 3.4 anomaly. For instance, the negative correlation at 12 months implies that a prediction of the Niño 3.4 anomaly one year from now would have a sign opposite to the current anomaly.

The ACF is a fundamental quantity in time series analysis for summarizing the statistical characteristics of a time series. More examples of its use will be presented later this chapter. A common question is whether a given autocorrelation differs significantly from zero. The following theorem is useful for answering this question.

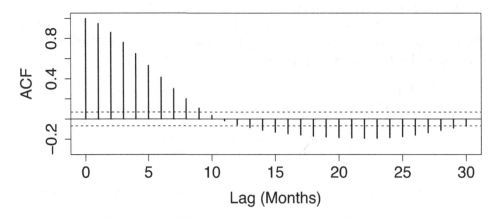

Figure 5.4 Autocorrelation function of the Niño 3.4 anomaly for 1950–2013. The horizontal dashed line shows an approximate 95% bound for independent samples from a Gaussian distribution.

Theorem 5.1 (Variance of the Sample Autocorrelation) *(Bartlett, 1946) Let* $X_1, \ldots, X_N$ *be independent random variables from a normal distribution. Then, for large N, the sample autocorrelation function (5.5) is normally distributed with zero mean and variance*

$$\text{var}\left[\hat{\rho}_{\tau \neq 0}\right] \approx \frac{1}{N}. \tag{5.6}$$

This theorem is proven in many texts on time series analysis (e.g., chapter 5 in Jenkins and Watts (1968); sec. 5.3.4 in Priestley (1981)). The horizontal dashed lines in Figure 5.4 show $\pm 1.96/\sqrt{N}$, which, by Theorem 5.1, gives an approximate 95% confidence interval for independent samples from a normal distribution. The figure shows that well over 5% of the sample correlations exceed the 95% bound, which leads us to reject the hypothesis of independence and decide that the time series is serially correlated.

Sometimes, deciding serial correlation based on figure such as Figure 5.4 is not straightforward. For example, if several lags are considered, then some would be expected to exceed the 95% bound by random chance even if the samples were *iid*. Furthermore, if a particular correlation happens to be large at a certain time lag, then neighboring sample correlations also tend to be large simply because they are derived from nearly the same data. Instead of checking each $\hat{\rho}_\tau$ individually, it is desirable to apply a single test that depends on, say, the first H autocorrelations $\hat{\rho}_1, \ldots, \hat{\rho}_H$. A natural statistic is N times the sum square autocorrelations, which, by Theorems 1.3 and 5.1, should have a chi-squared distribution with H degrees of freedom. However, Theorem 5.1 applies in an asymptotic sense and may not be good for small samples. Ljung and Box (1978) show that a better statistic is

$$Q = N(N+2) \sum_{\tau=1}^{H} \frac{\hat{\rho}_\tau^2}{N-\tau}. \tag{5.7}$$

For independent samples from a normal distribution, Q has a chi-squared distribution with H degrees of freedom. In the case of the sample autocorrelations shown in Figure 5.4, $Q \approx 3,000$, which for $H = 30$ degrees of freedom has a p-value $< 10^{-20}$, indicating that the ACF in Figure 5.4 is virtually impossible to obtain from *iid* samples (not surprisingly).

Having decided that the time series is serially correlated, the next natural question is *how* to make predictions. To make a prediction, a model is needed to describe how X_t is related to X_{t+1}. Ideally, this model would come from firmly established physical laws, but empirical laws inferred from past data could also be used. In addition, no model is perfect, so the prediction will be uncertain. The most complete description of this uncertainty is a probability distribution. Thus, a formalism is needed that explains how to make a prediction using a model and for describing the uncertainty in that prediction.

5.2 Stochastic Processes

Although climate variables behave randomly, this randomness exhibits certain distinctive characteristics in space and time. Describing these characteristics requires considering a *collection* of random variables. An ordered collection of random variables is known as a *stochastic process, random process,* or simply a *process*. Stochastic processes may be either continuous or discrete, but here only discrete processes are considered. A discrete stochastic process is denoted by X_t, where t is an indexing parameter that takes on discrete values $t = 1, 2, \ldots, N$. In this chapter, the parameter t is called "time." A sample from a stochastic process is called a *time series*. Because order is important, the distribution of a stochastic process at a single time step or single location is an incomplete description. Thus, unlike previous chapters, the statement $X_t \sim \mathcal{N}(\mu, \sigma^2)$ is incomplete because the *relation* between time steps is unspecified. Instead, the distribution of a stochastic process is described by a *joint* distribution at several different times, say $p(x_1, x_2, \ldots, x_N)$.

The conditional distribution given $X_1 = x_1$ follows from definition (1.12):

$$p(x_2, x_3, \ldots | x_1) = \frac{p(x_1, x_2, x_3, \ldots, x_N)}{p(x_1)}. \tag{5.8}$$

This distribution describes everything that is known about X_t given $X_1 = x_1$. One approach to visualizing this complex distribution is to imagine individual realizations of the stochastic process, as illustrated in Figure 5.5. Each curve shows a possible realization of the stochastic process. A collection of possible realizations of the stochastic process is called an *ensemble*. The observed system in which we live is but

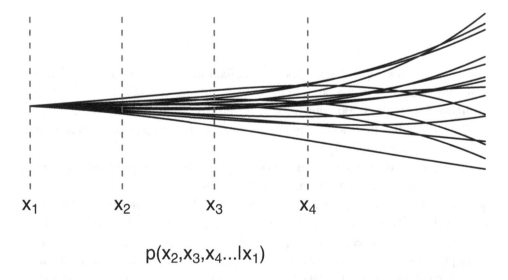

$$p(x_2, x_3, x_4 \ldots | x_1)$$

Figure 5.5 Possible trajectories of a stochastic process after X_1 is known.

one member of this ensemble. Importantly, the expectation of a stochastic process is an *ensemble average*, not a time average. The ensemble mean and variance of a stochastic process *can depend on time*:

$$\mathbb{E}[X_t] = \mu_t \quad \text{and} \quad \text{var}[X_t] = \mathbb{E}\left[(X_t - \mu_t)^2\right]. \tag{5.9}$$

For instance, the mean of X_2 is the mean of all ensemble members at time $t = 2$. Deterministic functions of time are constants under an expectation:

$$\mathbb{E}[f(t)X_t] = f(t)\mathbb{E}[X_t]. \tag{5.10}$$

These concepts provide the basis for the modern definition of climate (Leith, 1978; DelSole and Tippett, 2017). Early studies defined climate as the mean over a few decades (usually 30 years). However, this definition is problematic for a number of reasons. First, other statistical properties beyond the mean are important characteristics of the climate (e.g., variance and extremes). Second, this definition implies that climate depends on the averaging period. This dependence could be eliminated by using an extremely long averaging period, but the data required for such averaging are not available. Moreover, in the limit of an infinitely long period, climate would no longer depend on period, but then it also cannot change in time, and hence the concept of climate change cannot be defined.

To define climate, it would be hard to improve upon the definition given by Leith (1978):

One imagines an ensemble of Earths, each subjected to the same external influences, each with the same distribution of continents and oceans, and the weather on each going through normal fluctuations independently of the others. One next imagines that the distribution of states ... can change relatively slowly owing to changing external influences. The climate is then defined in terms of averages over this ensemble. It is immediately evident that this ensemble-averaged climate is an ideal theoretical concept inaccessible by direct observations ...[2]

Because climate is a theoretical construct, quantitative comparisons require some kind of model. The model could be empirical, e.g., a low-order polynomial in time, with parameters estimated from data. Alternatively, the model could be a comprehensive climate model that simulates weather variability based on physical laws. In the latter case, ensemble members are generated by integrating the climate model from a set of initial states that are consistent with observational information up to that time. Then, statistical properties of the climate distribution over any study period are estimated from the ensemble. In either case, the model constitutes a hypothesis whose reasonableness needs to be assessed.

---

[2] Adapted by permission from Springer Nature, *Nature*, Leith (1978).

Without further assumptions, a complete description of a stochastic process requires specifying the joint distribution at all possible times. Clearly, this is a difficult task. To make progress, the distribution needs to be constrained in some way. The most common simplifying assumptions are associated with the following processes:

- Markov Process
- Branching Process
- Birth and Death Process
- Diffusion Process

We consider only a subset of Markov processes. A Markov process is a process in which the distribution of the future given the present is the same as the distribution of the future given both the present and past. That is, a Markov process satisfies

$$p(\text{future} \mid \text{present}) = p(\text{future} \mid \text{present and past}). \tag{5.11}$$

Equivalently, given the present, the future and the past are independent. Dynamical systems are a special case of Markov processes in the sense that if the initial condition is specified completely, the future does not depend on the states before the initial condition.

Another important class of stochastic processes is *stationary* processes.

Definition 5.2 (Stationary) *A strictly stationary process is a process whose joint distribution is invariant with respect to a shift in time. That is, the joint probability satisfies*

$$p\left(x_{t_1}, x_{t_2}, \ldots, x_{t_K}\right) = p\left(x_{t_1+h}, x_{t_2+h}, \ldots, x_{t_K+h}\right) \tag{5.12}$$

for any collection of time steps $t_1, t_2, \ldots, t_K$, and for any shift h.

Example 5.1 *The expectation of a stationary stochastic process is independent of time:*

$$\mathbb{E}[X_t] = \mathbb{E}\left[X_{t+h}\right] = \ldots = \mathbb{E}\left[X_{t+2h}\right] = \mu \tag{5.13}$$

By the same reasoning, the variance of a stationary process is also independent of time:

$$\text{var}[X_t] = \mathbb{E}\left[(X_t - \mu)^2\right] = \sigma^2. \tag{5.14}$$

A *stationary Gaussian process* is a process in which every linear combination of variables has a stationary normal distribution. Examples of such processes will be given in Section 5.4. A process that is not stationary is called *nonstationary*. Most climate time series are nonstationary because they depend on calendar day and time-of-day: winter and summer temperatures have different distributions, as do day and night temperatures. Climate change is another form of nonstationarity in which the distribution depends on year.

Returning to the Niño 3.4 index, Figure 5.1 shows that this time series is nonstationary because of the strong annual cycle. However, after computing the anomaly $X'_t = X_t - \hat{\mu}_t$, which removed the annual cycle, no obvious deterministic time dependence is evident. (Actually, further analysis reveals that the *variance* of the anomaly depends on calendar month, but we ignore this aspect of nonstationarity.) Implicitly, the stationarity assumption was invoked when the scatter diagram in Figure 5.3 was used to characterize the month-to-month dependence of the Niño 3.4 anomaly. After all, this was done *without regard to the specific time t*. In effect, the pairs (X'_t, X'_{t+1}) were assumed to be drawn from a joint distribution $p(x_t, x_{t+1})$ that does not depend on time origin, and as a consequence, samples could be combined over all months to estimate a single 1-month lag correlation.

One of the most important characteristics of a stochastic process is the covariance between two different times $\text{cov}[X_{t+\tau}, X_t]$. If a stochastic process is stationary, then the covariance between any two time steps depends only on the difference in times. This fact can be seen by considering various covariances shifted in time:

$$\text{cov}\,[X_t, X_t] = \text{cov}\,[X_{t+1}, X_{t+1}] = \dots = \text{constant } 1$$

$$\text{cov}\,[X_{t+1}, X_t] = \text{cov}\,[X_{t+2}, X_{t+1}] = \dots = \text{constant } 2$$

$$\text{cov}\,[X_{t+2}, X_t] = \text{cov}\,[X_{t+3}, X_{t+1}] = \dots = \text{constant } 3.$$

Thus, for stationary processes, the covariance between any two time steps is denoted by

$$c(\tau) = \text{cov}\,[X_{t+\tau}, X_t], \tag{5.15}$$

where $c(\tau)$ is the (population) *autocovariance function*, and the parameter τ is termed the *time lag*. Note that in this notation $c(0) = \text{var}[X_t]$. The autocovariance function is estimated using the sample autocovariance function $\hat{c}(\tau)$, defined in Definition (5.1).

Example 5.2 *The autocovariance function of a stationary process is an even function of time-lag. To prove this, note that the autocovariance of a stationary process is invariant to shifts in time and does not depend on the order of the variables:*

$$c(-\tau) = \text{cov}[X_{t-\tau}, X_t] = \text{cov}[X_t, X_{t+\tau}] = \text{cov}[X_{t+\tau}, X_t] = c(\tau). \tag{5.16}$$

The correlation between two times is defined as

$$\rho_\tau = \text{cor}[X_{t+\tau}, X_t] = \frac{\text{cov}[X_{t+\tau}, X_t]}{\sqrt{\text{var}[X_{t+\tau}]\,\text{var}[X_t]}}. \tag{5.17}$$

If the stochastic process is stationary, two simplifications can be made. First, the variance is independent of time, so $\text{var}[X_{t+\tau}] = \text{var}[X_t] = c_0$. Second, the covariance depends only on the time lag. Accordingly, the correlation function of a stationary process is

$$\rho_\tau = \frac{c(\tau)}{c(0)}. \tag{5.18}$$

This function is called the *autocorrelation function*, or ACF. The ACF is estimated using the sample autocorrelation function $\hat{\rho}_\tau$ defined in (5.5).

An important special case of a stationary stochastic process is white noise.

Definition 5.3 (Stationary White Noise) *Stationary white noise is a stochastic process X_t such that any collection $\{X_{t_1}, X_{t_2}, \dots, X_{t_N}\}$ is iid. Thus, a stationary white noise process has the covariance function*

$$\text{cov}[X_t, X_s] = \begin{cases} 0 & \text{if } t \neq s \\ \sigma^2 & \text{if } t = s \end{cases}. \tag{5.19}$$

It follows that the autocorrelation function of stationary white noise is

$$\rho_\tau = \begin{cases} 1 & \text{if } \tau = 0 \\ 0 & \text{if } \tau \neq 0 \end{cases}. \tag{5.20}$$

If X_t also has a Gaussian distribution with zero mean and variance σ^2, then X_t is said to be Gaussian white noise and denoted

$$X_t \sim GWN(0, \sigma^2). \tag{5.21}$$

GWN is exactly equivalent to

$$X_1, \dots, X_N \overset{iid}{\sim} \mathcal{N}(0, \sigma^2). \tag{5.22}$$

A white noise process is said to have "no memory," in the sense that the value at time t gives no information about the value at any other time. *Stationary white noise is equivalent to the "independent and identically distributed" assumption used in previous chapters.*

5.3 Why Should I Care if My Data Are Serially Correlated?

Serial correlation is beneficial in forecasting because it provides the basis for predicting the future based on the present. However, serial correlation creates problems in hypothesis testing. Recall that Chapters 2–4 discussed procedures for making inferences about a population. Each one of those procedures assumed that the samples were *iid*. Unfortunately, if a stochastic process is serially correlated, then the *iid* assumption is incorrect and, as it turns out, those tests tend to be *biased* in a seriously misleading way.

To gain insight into the impact of serial correlation, consider the time mean $\overline{X}$. If X_t is a stationary process, then the variance of the time mean of $X_1, \ldots, X_N$ is

$$\text{var}[\overline{X}] = \mathbb{E}\left[\left(\overline{X} - \mathbb{E}[\overline{X}]\right)^2\right] = \mathbb{E}\left[\left(\frac{1}{N}\sum_{i=1}^{N}(X_i - \mathbb{E}[X_i])\right)\left(\frac{1}{N}\sum_{j=1}^{N}(X_j - \mathbb{E}[X_j])\right)\right]$$

$$= \frac{1}{N^2}\sum_{i=1}^{N}\sum_{j=1}^{N}\mathbb{E}\left[(X_i - \mathbb{E}[X_i])(X_j - \mathbb{E}[X_j])\right]$$

$$= \frac{1}{N^2}\sum_{i=1}^{N}\sum_{j=1}^{N}\text{cov}[X_i, X_j]. \tag{5.23}$$

If the process is stationary, then the covariance depends only on the difference in times. Furthermore, by definition of the ACF (5.18), we have

$$\text{cov}[X_i, X_j] = \text{var}[X_t]\rho_{i-j}. \tag{5.24}$$

Substituting this into (5.23) yields

$$\text{var}[\overline{X}] = \frac{\text{var}[X_t]}{N^2}\sum_{i=1}^{N}\sum_{j=1}^{N}\rho_{i-j}. \tag{5.25}$$

This expression can be simplified further by noting that the term ρ_0 in the double sum (5.25) occurs only when $i = j$, and this occurs N times. Next, the term ρ_1 occurs for the (i, j) pairs $(2, 1), (3, 2), \ldots, (N, N-1)$, and this occurs a total of $N-1$ times. Similarly, the term ρ_{-1} occurs $N-1$ times. Therefore, the terms ρ_1 and ρ_{-1} occur a total of $2(N-1)$ times. More generally, the terms ρ_τ and $\rho_{-\tau}$ occur $2(N-\tau)$ times. Therefore, the double sum can be expressed as a single sum over the difference parameter $\tau = i - j$:

$$\sum_{i=1}^{N}\sum_{j=1}^{N}\rho_{i-j} = N + 2\sum_{\tau=1}^{N-1}(N-\tau)\rho_\tau. \tag{5.26}$$

Substituting this equation in (5.25) gives

$$\text{var}\left(\overline{X}\right) = \text{var}[X_t]\frac{N_1}{N},\tag{5.27}$$

where the new parameter N_1 is

$$N_1 = 1 + 2\sum_{\tau=1}^{N-1} \rho_\tau \left(1 - \frac{\tau}{N}\right).\tag{5.28}$$

Many climate time series appear to be processes with *positive* ACF. In such cases, N_1 exceeds one (because all terms in the (5.28) are positive). Thus, the variance of the time mean given by (5.27) exceeds the variance associated with the case of *iid*. In other words, *serial correlation increases uncertainty in the sample mean*.

To understand the implications of (5.27), it may help to re-write the equation as

$$\text{var}\left(\overline{X}\right) = \frac{\text{var}[X_t]}{N_{\text{eff}}},\tag{5.29}$$

where

$$N_{\text{eff}} = \frac{N}{N_1}.\tag{5.30}$$

In the extreme case of a white noise process, $N_1 = 1$ (because ρ_τ vanishes for $\tau > 0$), and (1.45) is recovered, as it should since white noise is equivalent to *iid*. On the other hand, for serially correlated time series, N_{eff} is less than one. In a sense, then, serial correlation can be interpreted as *shrinking the effective sample size*. This qualitative impact of serial correlation holds for *any* estimator, not just the time mean. After all, if samples are *iid*, then each new sample provides new information. However, if the samples are correlated, then a new sample is somewhat *redundant* and not as informative as an independent sample. Indeed, in the extreme limit of perfect correlation, a new sample adds *no* new information.

As another example, consider the sample ACF. If the process is such that the autocorrelation after lag q vanishes, then it can be shown that (Cryer and Chan, 2010, eq. 6.1.11)

$$\text{var}[\hat{\rho}_{|\tau|>q}] = \frac{N_2}{N},\tag{5.31}$$

where

$$N_2 = 1 + 2\sum_{\tau=1}^{\infty} \rho_\tau^2.\tag{5.32}$$

Comparing (5.31)–(5.6) shows again that serial correlation *inflates* sampling variability.

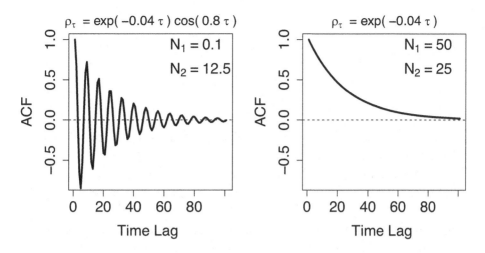

Figure 5.6 Illustration of two types of autocorrelation functions.

If a process is serially correlated, we often want to distinguish processes by their *time scale*. Timescale is the time over which a process becomes uncorrelated with itself. Atmospheric processes typically have a much shorter timescale than oceanic processes. Both N_1 and N_2 may define a measure of timescale. A key difference between these measures can be appreciated by considering the autocorrelations illustrated in Figure 5.6. The left panel shows an oscillatory autocorrelation while the right shows a purely decaying ACF. Both ACFs decay at the same rate, so the timescale for approaching zero correlation is the same. Nevertheless, the oscillatory ACF has timescale $N_1 = 0.1$, whereas the timescale for the monotonically decaying ACF is $N_1 = 50$! The reason N_1 differs so much between the two ACFs is that when the sum is taken, the oscillations in the former cancel before appreciable amplitude decay. Thus, N_1 can be a misleading measure of timescale for oscillatory ACFs. In contrast, N_2 differs from N_1 by summing over *squares* of the autocorrelation function. For the two autocorrelations shown in Figure 5.6, $N_2 = 12.5$ and 25 (as indicated in each panel). Because N_1 and N_2 are within a factor of two of each other for pure exponential decay, $2N_2$ may be used to measure timescale for either type of ACF.

To illustrate the impact of serial correlation on hypothesis testing, consider the *t*-test. Let X_t and Y_t be two independent stochastic processes from a population with mean μ and variance σ^2. If the sample size is N for both, then the time mean has distributions

$$\overline{X_t} \sim \mathcal{N}\left(\mu, \sigma^2 \frac{N_1}{N}\right) \quad \text{and} \quad \overline{Y_t} \sim \mathcal{N}\left(\mu, \sigma^2 \frac{N_1}{N}\right). \tag{5.33}$$

It follows that the difference in time means has the distribution

$$Z = \frac{\overline{X}_t - \overline{Y}_t}{\sigma\sqrt{\frac{2}{N}}} \sim \mathcal{N}(0, N_1). \tag{5.34}$$

The quantity Z is the same as (2.9) used in Chapter 2 to derive the t-test. When the stochastic process is *iid*, then $N_1 = 1$ and Z has a standardized normal distribution, as assumed in the t-test. However, if the ACF is positive for nonzero lags, then $N_1 > 1$, and the variance of Z is *inflated* relative to the *iid* case. That is, if the process is serially correlated, the difference in means has larger variance than assumed in the t-test.

The sampling distribution of the T statistic (2.25) can be estimated by Monte Carlo methods. As a concrete example, the top panel of Figure 5.7 shows a time series of length 50 generated by a mathematical model that can produce samples

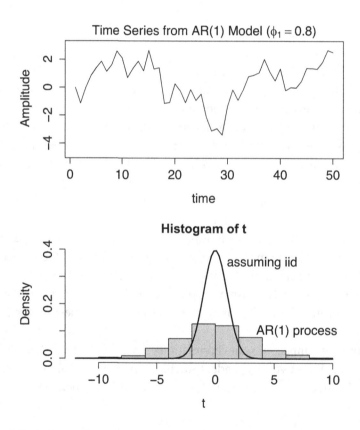

Figure 5.7 A time series generated by an AR(1) model with $\phi_1 = 0.8$ (top), and the distribution of the t-statistic (bottom) when samples are from the AR model (shaded histogram) and when the samples are *iid* (curve).

with a prescribed autocorrelation function (to be more precise, it is the AR(1) model discussed in Section 5.4). The T-statistic (2.25) is computed for this sample. Then, this procedure is repeated 1,000 times, yielding 1,000 T-values. A histogram of the resulting T-values is shown in the bottom panel of Figure 5.7. For comparison, the bottom panel also shows the t-distribution with 98 degrees of freedom, corresponding to an *iid* assumption. The figure shows that the distribution is much wider for serially correlated process than for *iid* processes. Thus, for serially correlated process, extreme T-values are more likely and the decision rule will have a larger probability of type-1 error than the nominal significance level α of the standard t-test.

An approach to accounting for serial correlation in the t-test is discussed in Section 5.8.

5.4 The First-Order Autoregressive Model

A popular class of models for serially correlated time series are *autoregressive models*. An autoregressive model represents the value at one time step as a linear combination of previous steps, plus white noise. A nice property of autoregressive models is that they generate a stochastic process with prescribed values of the ACF at a specified number of consecutive lags. For instance, the first-order autoregressive model, denoted AR(1), generates a stochastic process with a lag-1 ACF equal a specified value. The AR(1) is so common in climate science that it is worth reviewing all by itself. An AR(1) model is

$$X_t = \phi X_{t-1} + W_t + k, \tag{5.35}$$

where ϕ is a constant, W_t is a Gaussian white noise process with variance σ_W^2, and k is a constant. Thus, $W_1, W_2, \ldots, W_N$ are *iid*, but if $\phi \neq 0$, the resulting $X_2, \ldots, X_N$ are not *iid*. A process generated by this model is called an *AR(1) process*. One reason the AR(1) is popular is that it can be interpreted as a finite discretization of a "physical model"

$$\frac{dx}{dt} = ax + \text{noise}. \tag{5.36}$$

Such models often arise from simplified conservation laws (e.g., thermodynamic mixed-layers; Hasselmann, 1976). If $dx/dt \approx (X_t - X_{t-1})/\delta t$, and ax is identified with aX_{t-1}, then the resulting discretized version of (5.36) can be rearranged into the form (5.35).

Suppose the initial value at $t = 0$ is x_0 (here, a lower case symbol denotes a specific realization, whereas a capital letter denotes a random variable). What can be said about the random variable X_1? For instance, to what extent can we *predict*

X_1 knowing the initial state x_0? According to the model (5.35), $X_1 = \phi x_0 + W_1 + k$, hence the state at $t = 1$ is a constant $(\phi x_0 + k)$ plus a Gaussian random variable W_t, so X_1 has a Gaussian distribution. Therefore, a prediction of X_1 is described by a distribution. Because X_1 is Gaussian, it is sufficient to determine the mean and variance of X_1. This mean and variance can be interpreted as *conditioned on the initial state* $X_0 = x_o$. Specifically, the mean of X_1 conditioned on the value x_0 is

$$\mathbb{E}[X_1|x_o] = \mathbb{E}[(\phi x_0 + W_1 + k)|x_0] = \phi x_0 + k + \mathbb{E}[W_1] = \phi x_0 + k, \quad (5.37)$$

where we have used the fact that x_0 and k are constant under the conditional distribution, and that W_t has zero mean regardless of the conditional. The conditional variance is

$$\text{var}[X_1|x_0] = \text{var}[(\phi x_0 + W_1 + k)|x_0] = \text{var}[W_1] = \sigma_W^2, \quad (5.38)$$

where we have used the fact that adding a constant to a random variable does not alter its variance. It follows that the conditional distribution after one step is

$$p(x_1|x_0) \sim \mathcal{N}\left(\phi x_0 + k, \sigma_W^2\right). \quad (5.39)$$

A schematic of this one-step stochastic process is illustrated in Figure 5.8.

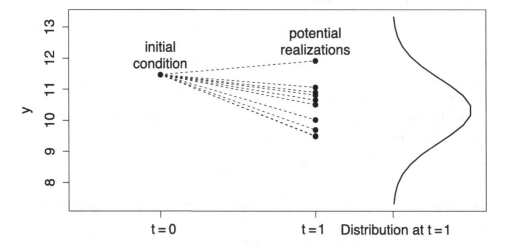

Figure 5.8 Schematic of a one-step AR(1) process. The initial condition at $t = 0$ is certain, but the next step at $t = 1$ is uncertain because of the random term in the AR(1) model (5.35). The distribution of the process after one step is illustrated on the far right-hand side, and possible realizations from this distribution are termed *ensembles* and indicated by dots at $t = 1$.

Next, consider the conditional distribution for $X_2, X_3, \ldots$. The general solution can be inferred by examining the first few iterations:

$$X_1 = \phi x_0 + W_1 + k$$
$$X_2 = \phi X_1 + W_2 + k = \phi \, (\phi x_0 + W_1 + k) + W_2 + k$$
$$= \phi^2 x_0 + (1 + \phi)k + (W_2 + \phi W_1)$$

$$\vdots$$

$$X_t = \phi^t x_0 + \sum_{j=0}^{t-1} \phi^j k + \sum_{j=0}^{t-1} \phi^j W_{t-j}. \tag{5.40}$$

According to (5.40), X_t is a sum of independent Gaussian processes plus a constant, hence it is itself normally distributed. Following the same procedure as before, the mean is

$$\mathbb{E}[X_t | x_0] = \phi^t x_0 + \sum_{j=0}^{t-1} \phi^j k, \tag{5.41}$$

and the variance is

$$\mathrm{var}[X_t | x_0] = \mathrm{var}\left[\sum_{j=0}^{t-1} \phi^j W_{t-i} \right] = \sum_{j=0}^{t-1} \sigma_W^2 \phi^{2j}, \tag{5.42}$$

where the last expression uses the fact that the W_t's are independent, and therefore the variance of a sum equals the sum of the variances. Since k and σ_W are constants, the sums involving them are merely geometric series and can therefore be simplified using standard summation formulas. Consolidating these results leads to

$$p(x_t | x_0) \sim \mathcal{N}\left(\phi^t x_0 + k \frac{1 - \phi^t}{1 - \phi}, \sigma_W^2 \frac{1 - \phi^{2t}}{1 - \phi^2} \right). \tag{5.43}$$

In general, solutions should be finite as $t \to \infty$. It is clear from this solution that a necessary condition for bounded solutions is that $|\phi| < 1$, otherwise the sums (5.41) and (5.42) diverge to infinity. Interestingly, the variance of (5.43) at each time t is independent of the initial condition x_0. This may or may not be realistic.

The evolution of the distribution of X_t is illustrated in Figure 5.9. Initially, the distribution is narrow because the initial condition is perfectly known. As time increases, the distribution shifts and spreads out because of the addition of random noise with each time step. Eventually, the accumulated variance stops growing, reflecting a balance between the increase in variance as a result of random noise

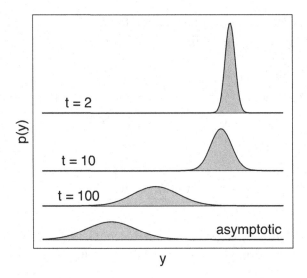

Figure 5.9 The distribution of Y_t at different times for an AR(1) process. The distributions have been offset in the vertical for clarity. The AR parameters are $\phi = 0.99$, $\sigma_W^2 = 1$, and $k = 0$. The initial condition is chosen at the (unlikely) large value of five standard deviations to exaggerate changes in distribution. Axis labels have been omitted to highlight the qualitative behavior.

and the decrease in variance as a result of the damping term ϕ. In the asymptotic limit $t \to \infty$, the distribution becomes

$$\lim_{t \to \infty} p(x_t|x_0) \sim \mathcal{N}\left(\frac{k}{1-\phi}, \frac{\sigma_W^2}{1-\phi^2}\right). \tag{5.44}$$

This distribution is stationary because the right-hand side depends on neither t nor x_0. If $(X_t, X_{t+\tau})$ are drawn from the stationary distribution, then numerous properties of the AR(1) process can be derived very simply, as illustrated next.

Example 5.3 (Mean and Variance of a Stationary AR(1) Process) *Suppose an AR(1) process is stationary. Taking the expectation of both sides of (5.35) gives*

$$\mathbb{E}[X_t] = \phi\mathbb{E}[X_{t-1}] + k, \tag{5.45}$$

where $\mathbb{E}[W_t] = 0$ has been used. For a stationary process, $\mathbb{E}[X_t] = \mathbb{E}[X_{t-1}]$. Substituting this relation and solving gives

$$\mathbb{E}[X_t] = \frac{k}{1-\phi}, \tag{5.46}$$

in agreement with (5.44). Similarly, taking the variance of both sides of (5.35) gives

$$\text{var}[X_1] = \text{var}[\phi X_{t-1} + W_t + k]. \tag{5.47}$$

The constant k can be dropped since it does not affect the variance. In addition, X_t and W_t are independent, so the variance of the sum equals the sum of variances:

$$\text{var}[X_t] = \phi^2 \, \text{var}[X_{t-1}] + \text{var}[W_t]. \tag{5.48}$$

This equation defines a balance equation for variance. Stationarity implies $\text{var}[X_t] = \text{var}[X_{t-1}]$. *Substituting this and solving for* $\text{var}[X_t]$ *gives*

$$\text{var}[X_t] = \frac{\text{var}[W_t]}{1 - \phi^2}, \tag{5.49}$$

in agreement with (5.44). Because $|\phi| < 1$, the balance equation (5.48) is stable, so any initial variance relaxes toward the asymptotic variance on a timescale determined by ϕ.

The most important quantity characterizing an AR(1) process is its ACF. For a stationary AR(1) process, the expectation does not depend on time, so let $\mathbb{E}[X_t] = \mu$. Subtracting the mean from both sides of (5.35) yields

$$X_t - \mu = \phi \, (X_{t-1} - \mu) + W_t + (\phi - 1)\mu + k. \tag{5.50}$$

The constant term $(\phi - 1)\mu + k$ vanishes because $\mu = k/(1 - \phi)$ (see (5.46)). Multiplying both sides by $X_{t-\tau} - \mu$ and taking expectations gives

$$\text{cov}[X_t, X_{t-\tau}] = \phi \, \text{cov}[X_{t-1}, X_{t-\tau}] + \text{cov}[W_t, X_{t-\tau}]. \tag{5.51}$$

An important observation is the following: the variable $X_{t-\tau}$ is independent of W_t for all $\tau \geq 1$. This fact follows from causality: X_t can only depend on *past* values of W_t, not future values of W_t (which are independent). This fact can be seen directly from the solution (5.40), which shows that X_t depends only on past values of $W_{t-1}, W_{t-2}, \ldots$. Thus, $\text{cov}[W_t, X_{t-\tau}] = 0$ for $\tau \geq 1$, and (5.51) reduces to

$$c(\tau) = \phi \, c(\tau - 1), \tag{5.52}$$

where we have used the fact that the covariances depend only on time lag. This recursive equation has the general solution $c(\tau) = \phi^\tau c(0)$. Since the autocovariance function is an even function of lead, we know immediately that the autocorrelation at arbitrary lead is

$$\rho_\tau = \frac{c(\tau)}{c(0)} = \phi^{|\tau|}. \tag{5.53}$$

Example 5.4 (Time Scales of an AR(1) Process) *For an AR(1) process with parameter ϕ_1, taking the limit $N \to \infty$ for N_1 gives*

$$N_1 = 1 + 2\sum_{\tau=1}^{\infty} \rho_\tau = 1 + 2\sum_{\tau=1}^{\infty} \phi_1^\tau = 1 + 2\left(\frac{1}{1 - \phi_1} - 1\right) = \frac{1 + \phi_1}{1 - \phi_1}. \tag{5.54}$$

Similarly, the value of N_2 is

$$N_2 = \sum_{\tau=-\infty}^{\infty} \rho_\tau^2 = 2\sum_{\tau=0}^{\infty} \rho_\tau^2 - 1 = 2\sum_{\tau=0}^{\infty} \phi_1^{2\tau} - 1 = 2\frac{1}{1-\phi_1^2} - 1 = \frac{1+\phi_1^2}{1-\phi_1^2}. \quad (5.55)$$

As an example, if $\phi = 0.7$, then $N_1 \approx 5.7$ and $N_2 \approx 3$. This means that the sampling variance of the mean will be over five times larger than if the process were white noise, and that the sampling variance of the ACF at large lags will be about three times larger than if the process were white noise. Both measures of timescale grow without bound as $\phi \to 1$.

To gain a sense of the kinds of time series that can be generated by an AR(1) model, a few realizations are shown in Figure 5.10. These realizations were derived by solving (5.35) for a specific sequence of noise terms $w_1, w_2, \ldots$. The left panels show individual realizations while the right panels show the corresponding sample and population ACFs. Comparison of the top two rows shows the result of increasing the value of ϕ: as ϕ approaches 1, the time series becomes "smoother" and the autocorrelation function takes a longer time to decay. The top row shows that the sample ACF fluctuates mostly within the 95% bound beyond the first four lags. However, compare the first and third rows, which show samples from exactly the same AR(1) process but for different sample sizes N. For small N (third row), the sample correlations not only exceed the 95% bound at large lags, but tend to persist, giving the impression of an oscillatory ACF. Indeed, one might feel that the correlations in the bottom right panel seem too organized to be dismissed as random. However, the population ACFs (5.53) are plotted in the right panels of Figure 5.10 as the thick, dashed curves. Comparison with the corresponding sample ACFs shows that the sample ACFs can sometimes differ substantially from the population ACF. These examples illustrate the difficulty in interpreting sample autocorrelation functions from short time series.

To make a prediction with an AR(1) model, the parameters k, ϕ, and σ_W^2 need to be specified. Since the anomaly time series has zero mean (by construction), we set $k = 0$. The parameter ϕ can be estimated from the ACF using (5.53). Although $\hat{\rho}_\tau$ and ρ_τ can differ substantially, the differences are relatively small at the first lag, suggesting the estimate

$$\hat{\phi} = \hat{\rho}_1. \quad (5.56)$$

As mentioned in relation to the scatter plot Figure 5.3, $\hat{\rho}_1 \approx 0.95$. Finally, the parameter σ_W can be estimated from the variance of the residuals $X'_t - \phi X'_{t-1}$, which is found to be $\hat{\sigma}_W = 1.66$. Substituting these estimates into the distribution (5.43) yields the forecast distribution. The forecast distribution for two selected initial conditions is shown in Figure 5.11. As anticipated, the spread of the forecast

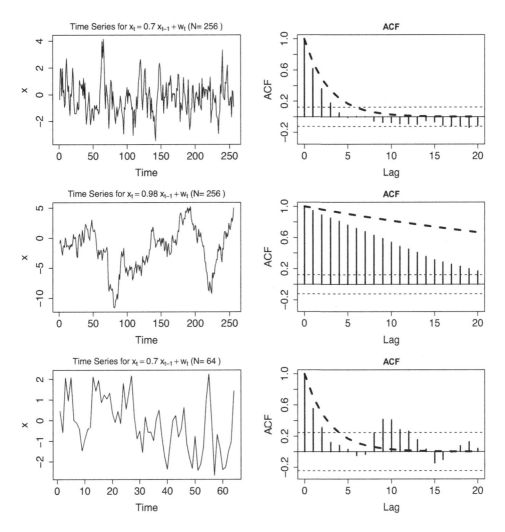

Figure 5.10 A realization of a stochastic process generated by an AR(1) model (left panels) and associated autocorrelation functions (right panels). The AR(1) and sample size are indicated in the title of the left panel. In the right panels, the vertical bars are the sample autocorrelation (derived from the time series shown in the immediate left panel), the thick dashed curve is the exact autocorrelation function, and the thin, horizontal dashed lines indicate the 95% bound $\pm 1.96/\sqrt{N}$. Differences between the sample ACF and exact ACF are a result of random sampling.

distribution increases with lead time, and the mean of the forecast distribution damps toward zero.

Consider the ACF for Niño 3.4, which is shown again in Figure 5.12. This figure also shows the ACF of an AR(1) process that matches the sample lag-1 ACF.

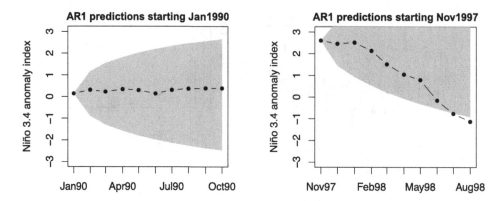

Figure 5.11 Probabilistic forecasts of the Niño 3.4 anomaly index using an AR(1) model starting from January 1990 (left) and November 1997 (right). The observed values are indicated by the curves with black dots, and the 95% interval for the forecast distribution is shown as the shaded region.

ACF of Niño 3.4 Anomalies 1950–2017

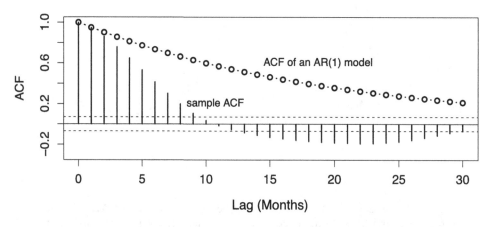

Figure 5.12 The sample ACF of the Niño 3.4 anomaly, as in Figure 5.4, but with the ACF of an AR(1) process with $\hat{\phi} = 0.95$ superimposed (curve with open circles).

Obviously, the AR(1) model is a poor model of the Niño 3.4 anomaly (however, it is a good model for most atmospheric variables). Moreover, note that the sample ACF is a damped oscillation that becomes *negative* after 10 months. A first-order autoregressive model cannot capture a damped oscillation because it stays positive for $\phi > 0$. These considerations suggest the need for a broader range of models for simulating stochastic processes.

5.5 The AR(2) Model

The next simplest autoregressive model is an AR(2) model, which is of the form

$$X_t = \phi_1 X_{t-1} + \phi_2 X_{t-2} + W_t + k, \qquad (5.57)$$

where ϕ_1 and ϕ_2 are constant parameters, W_t is a white noise process, and k is a constant. The general solution of an AR(2) process is fairly complicated and will not be derived here. Instead, we consider only stationary solutions, whose properties can be obtained by elementary methods. Specifically, using the same procedure as in the previous section, the following equation for the stationary covariances of the AR(2) model can be obtained:

$$\mathrm{cov}[X_t, X_{t-\tau}] = \phi_1 \, \mathrm{cov}[X_{t-1}, X_{t-\tau}] + \phi_2 \, \mathrm{cov}[X_{t-2}, X_{t-\tau}]. \qquad (5.58)$$

Dividing by the variance of X_t and using the fact that stationary covariances depend only on the difference in times gives

$$\rho_\tau = \phi_1 \rho_{\tau-1} + \phi_2 \rho_{\tau-2}. \qquad (5.59)$$

This equation that can be solved recursively provided the initial values ρ_0 and ρ_1 are known. By definition, $\rho_0 = 1$. Also, evaluating (5.59) for $\tau = 1$, and making use of the fact that ρ_τ is even function of τ, gives $\rho_1 = \phi_1 + \phi_2 \rho_1$, which implies

$$\rho_1 = \frac{\phi_1}{1 - \phi_2}. \qquad (5.60)$$

Using these initial values, the remaining values of ρ_τ can be solved recursively from (5.59). This procedure illustrates a general property of autoregression processes: all lagged correlations of an AR(p) model can be written in terms of the first p correlations.

The complete autocorrelation function of an AR(2) process depends only on the two parameters ϕ_1 and ϕ_2. Thus, by suitable choice of these two parameters, the autocorrelation function may be prescribed at two distinct lags. However, for some values of ϕ_1 and ϕ_2, solutions from (5.59) are unbounded. In Section 5.7, it is shown that bounded solutions occur only if the values of ϕ_1 and ϕ_2 lie in the triangle shown in Figure 5.13. That is, if one chooses values of ϕ_1 and ϕ_2 that are outside of the triangle, then solutions from (5.57) grow to infinity. Also, depending on the values of ϕ_1 and ϕ_2 inside the triangle, the autocorrelation function may decay monotonically to zero, or exhibit a *damped oscillation*. Examples of sample time series and autocorrelation functions for various second-order AR models are shown in the lower panel of Figure 5.13. The most important new behavior relative to the AR(1) is that the ACFs of an AR(2) can oscillate and change sign as they decay to zero.

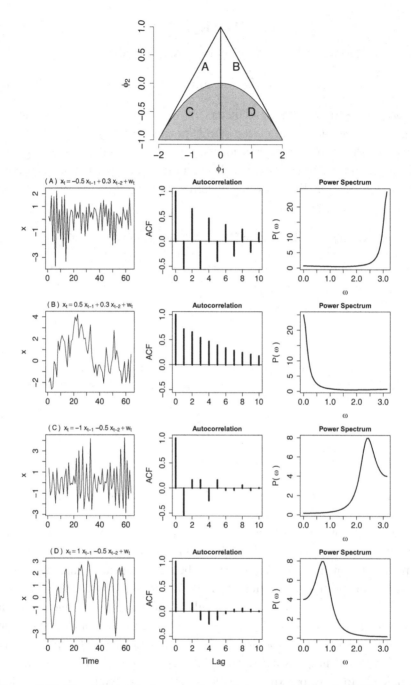

Figure 5.13 The top panel shows values of ϕ_1 and ϕ_2 in an AR(2) model that produce realizable autocorrelation functions. The shading shows values that produce oscillatory autocorrelation functions. In the lower panels, each row shows the following results from a specific AR(2) model: sample time series realization (left panel), the corresponding population autocorrelation function (middle panel), and corresponding power spectrum (far right panel; this quantity is explained in Chapter 6). The exact AR(2) model is indicated in the title of each left panel.

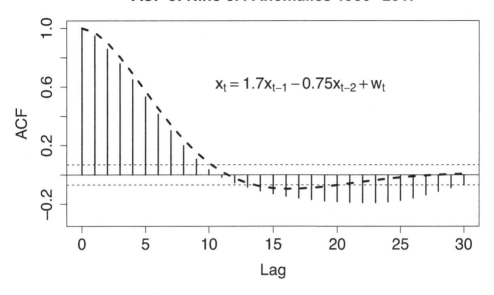

Figure 5.14 Autocorrelation function of the Niño 3.4 anomaly for 1950–2013 (bars) and the ACF for an AR(2) model with parameters chosen to fit the observed ACF (dashed). The fit is not based on optimization, but rather is chosen to highlight the oscillatory nature of the ACF. The horizontal dashed lines shows the approximate 95% bound for independent samples from a normal distribution.

An example of an AR(2) "fit" to the observed Niño 3.4 anomaly is shown in Figure 5.14. The fit is not perfect, but at least it captures the oscillatory nature of the observed ACF.

5.6 Pitfalls in Interpreting ACFs

In practice, the precise stochastic process that generated a time series is unknown; instead the process must be inferred from finite samples. Standard issues in such inferences include deciding the model and estimating its parameters, which are discussed in Chapters 8 and 10. In addition, there are important pitfalls in such inferences. Suppose one observes a sample autocorrelation that does not asymptote to zero for long lead times. It is natural to assume that the underlying process is stationary but has a long memory. However, persistent correlations could also arise from choosing a poor model of μ_t. In other words, persistent correlations can arise from *nonstationary* components in the time series.

To illustrate, Figure 5.15a shows three time series labeled A, B, and C. These time series were generated by a model of the form (5.1), where ϵ_t is *iid* and

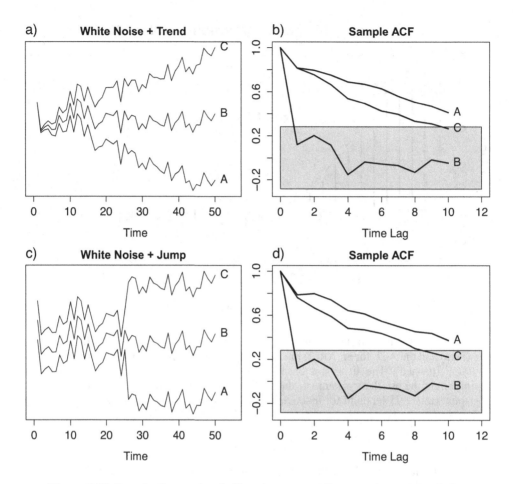

Figure 5.15 Sample time series (left) and corresponding sample autocorrelation functions (right). The top row shows results for three time series comprising white noise plus a straight line (known as a "trend"), and the bottom row shows results for three time series comprising white noise plus a "jump" (a function that is constant for the first half and a different constant for the second half of the period). The shaded region in the ACF plots shows the region for insignificance at the 5% level.

μ_t is a *linear function of time*. The corresponding sample ACFs are shown in Figure 5.15b. The time series marked B has no trend and the corresponding ACF behaves as expected: it is significant only at zero lag. However, the ACFs for the other time series decay much more slowly. This example shows that the sum of a trend plus *iid* noise can produce sample ACFs that decay slowly. As a second example, Figure 5.15c shows time series generated as the sum of *iid* noise plus a *step function*. The step function is a constant for the first half of the period and then a different constant for the second half. The corresponding ACFs for A and C, shown in 5.15d, decay slowly. Importantly, the ACFs produced by the step function are nearly indistinguishable to the ACFs generated by the trend. Thus, different models of μ_t can generate similar ACFs.

These examples illustrate three important principles: (1) jump discontinuities or slowly varying time series can inflate ACFs that otherwise would be small, (2) very different functions μ_t can generate similar ACFs over finite samples, and (3) if a sample ACF does not asymptote to zero, it is good practice to inspect the time series directly for nonstationary effects such as trends, periodicities, and discontinuities. Of course, trends, periodicities, and discontinuities could arise from stationary processes. In general, one must rely on physical understanding to decide whether a persistent ACF is a result of nonstationary effects, or to sampling a stationary process over a timescale that is short compared to the natural timescale of the process.

In climate applications, common examples of nonstationary effects include annual and diurnal cycles, trends resulting from climate change or model drift, and discontinuities resulting from changes in observation network. Nonstationary components should be removed from the time series before computing the ACF. For instance, given time series A in Figure 5.15a, one might postulate that μ_t is a linear function of time and subtract it from A. By subtracting a polynomial from the time series, the ACF tends to change from the slowly decaying function A to the quickly decaying function B in Figure 5.15b. Similarly, given time series A in Figure 5.15c, one might postulate a step function for μ_t and subtract it from A, producing the ACF B in Figure 5.15d. This illustrates a typical goal in time series analysis, namely to remove or transform the data until the only thing left over is *iid* noise.

5.7 Solutions of the AR(2) Model

In Section 5.5, we asserted that the parameters of an AR(2) had to lie within a certain triangle in (ϕ_1, ϕ_2) coordinates. A method for deriving this triangle is derived in this section. Moreover, the method discussed here can be generalized to determine solutions for arbitrary AR(p) models. The basic idea is to recognize that (5.59) is a linear, homogeneous difference equation that can be solved by the method of characteristic roots. The method of characteristic roots proceeds by assuming a solution of the form

$$\rho_\tau = a\zeta^\tau, \tag{5.61}$$

where a and ζ are unknown parameters. Substituting this into (5.59) and simplifying gives

$$\zeta^2 - \phi_1\zeta - \phi_2 = 0, \tag{5.62}$$

where we have divided out $a\zeta^{\tau-2}$ assuming $\zeta \neq 0$. This equation is called the *characteristic equation* of (5.59). The roots of this equation, called the *characteristic roots*, are

$$\zeta = \frac{\phi_1 \pm \sqrt{\phi_1^2 + 4\phi_2}}{2}. \tag{5.63}$$

It follows that if ζ is a characteristic root, then $\rho_\tau = a\zeta^\tau$ satisfies the difference equation (5.59). Since there are two characteristic roots, both roots will satisfy the difference equation. Moreover, since the difference equation is linear, any linear combination of the characteristic roots will satisfy the difference equation. Thus, the most general solution is

$$\rho_\tau = a_1\zeta_1^\tau + a_2\zeta_2^\tau, \tag{5.64}$$

where ζ_1 and ζ_2 denote the two solutions from (5.63).

However, not all solutions of the form (5.64) are autocorrelation functions. For instance, an autocorrelation function must lie in $[-1,1]$. This fact constrains ϕ_1 and ϕ_2. To understand the implications of this constraint, recall that the roots of a quadratic equation are either real or complex. If the roots are real and differ from each other, then as $\tau \to \infty$, (5.64) will be dominated by the larger root, which in turn will be bounded only if it is less than one. If the roots are complex, similar reasoning implies that their modulus must be less than one. These conditions can be summarized concisely as the characteristic roots must lie inside the unit circle in the complex ζ-plane.

The boundary between real and complex values of (5.63) is $\phi_1^2 + 4\phi_2 = 0$, which is an equation for a parabola in the (ϕ_2, ϕ_1) plane. In the case of real roots, the constraint $|\zeta_1| < 1$ implies $-1 \leq \zeta_1$ and $\zeta_1 \leq 1$. Substituting (5.63) into these inequalities gives

$$\phi_2 \leq 1 - \phi_1 \quad \text{and} \quad \phi_2 \leq 1 + \phi_1. \tag{5.65}$$

On the other hand, if the roots are complex, then the modulus of the roots is

$$\zeta\zeta^* = |\phi_2| \leq 1. \tag{5.66}$$

Inequalities (5.65) and (5.66) divide the (ϕ_1, ϕ_2) plane into a triangle, and the parabola $\phi_1^2 + 4\phi_2 = 0$ divides the triangle into real and complex roots, as illustrated in Figure 5.13.

5.8 Further Topics

An alternative class of models that are used in time series analysis are *moving average (MA) models*. A moving average model of order q is of the form

$$X_t = k + \sum_{i=0}^{q} \theta_i W_{t-i}, \tag{5.67}$$

where θ_i are model parameters. The model gets its name from the fact that it computes a "moving average" of white noise processes. The autocorrelation function of a MA(q) vanishes *exactly* beyond lag q. Estimation of moving average processes is often based on maximum likelihood methods.

The autoregressive model has several extensions that capture a wider variety of statistical behavior. One extension is to combine the moving average model with the autoregressive model. The resulting model is called a *autoregressive moving average model (ARMA)*. It turns out that an infinite-order AR model can capture a MA process, and an infinite-order MA model can capture an AR process. The philosophy of ARMA modeling is to choose the lowest possible orders of the AR and MA models to capture the desired probabilistic behavior. To capture certain common forms of nonstationary behavior, *autoregressive integrated moving average models (ARIMA)* have been formulated. These models simulate the behavior of *differences* between consecutive time steps. For instance, if a given process contains an additive linear function of time, then differencing the time series will remove the trend exactly. The trend can be put back into the time series by "integration," hence the name. More complicated nonstationary behavior, such as a quadratic growth in the time series, can be removed by further differencing. Another extension is to include multiple variables simultaneously, including their (linear) interactions. Such models are called *vector autoregressive models (VAR)*. Another extension is to allow the noise to vary in time. If the time dependence is a known deterministic function, then the probabilistic behavior can be determined using the techniques discussed in this chapter. In econometrics, the *autoregressive conditional heteroskedasticity (ARCH)* model assumes the variance of the noise term is state dependent. Another extension is to allow the model parameters to vary periodically in time, such as might be appropriate for capturing seasonal or diurnal variations. Such models are said to be *cyclostationary*. The Niño 3.4 index is a noteworthy example of such a time series (Torrence and Webster, 1998).

ARMA models appear to capture the statistical behavior of variables such as temperature and geopotential height fairly well, but do not capture variables such as daily precipitation. For instance, precipitation time series are characterized by sequences of consecutive zeros interrupted by periods of random precipitation activity. Katz (1977) proposed a *chain-dependent* probabilistic model for daily precipitation in which precipitation occurrence is simulated by a Markov chain that predicts the occurrence or nonoccurrence of precipitation, and the intensity of rainfall is a random draw from a specified (non-Gaussian) distribution. Subsequent studies showed that a first-order Markov chain, in which the probability of occurrence depends only on the previous day, exhibits shortcomings that can be improved by generalizing to second or higher order Markov chains (Chin, 1977).

The present chapter has focused on *discrete* stochastic processes. There exist *continuous* stochastic processes as well, although the mathematics of such models involve a new form of calculus since the concept of a deterministic limit does not apply. In fact, different versions of this calculus exist corresponding to different conventions for computing the limit. The two most important versions of this calculus are *Ito* and *Stratonovich* calculus. The *Fokker–Planck equation* or *Kolmogorov equations* are fundamental to continuous stochastic processes.

In this chapter, the main tool for characterizing a stochastic process was the autocorrelation function. Another important tool is the *power spectrum*, as discussed in Chapter 6. These tools are sufficient for characterizing stationary, Markov, Gaussian processes. Nonstationary, non-Markov, non-Gaussian processes require additional tools. For instance, some time series exhibit *long memory*: the degree of dependence across time lags decays more slowly than exponential. A measure of this dependence is the *Hurst exponent. Autoregressive fractionally integrated moving average (ARFIMA)* models have been developed to capture some types of long memory behavior. *Information theory* provides a new class of tools for quantifying the behavior of nonlinear or non-Gaussian processes.

To use stochastic models in practice, one has to *select* a model, *estimate* its parameters, and *assess the adequacy* of the model. A wide variety of models can capture the same statistical properties, so deciding the best model can be something of an art. Also, one important use of stochastic models is to *predict* future values. Box et al. (2008) is a widely respected textbook on these topics with substantial practical advice.

Autoregressive models provide a framework for accounting for serial correlation in hypothesis testing. For instance, consider testing a difference in means. The standard t-test assumes that time series are *iid*. To account for serial correlation, one might assume that each time series is an AR(1) process. The mean of an AR(1) process from (5.35) is related to the model parameters as $\mu = k/(1 - \phi)$ (see 5.46). Thus, if two AR(1) processes have the same values of (k, ϕ), then they must have the same means. A test for differences in AR models is discussed in DelSole and Tippett (2020).

5.9 Conceptual Questions

1 What is a stationary process?
2 What is the autocorrelation function?
3 A time average does not depend on time, since the time dimension has been averaged. How can the expectation of a stochastic process depend on time?

4 If a process is stationary, what can you say about the expectation? About the autocorrelation function?

5 What is Gaussian white noise? How is it related to the *iid* assumption?

6 How does serial correlation impact hypothesis tests?

7 What is a first-order autoregressive process? What kind of time series can it simulate? What kind of time series can it not simulate?

8 What is the autocorrelation function of a first-order autoregressive order process?

9 What is a second-order autoregressive process? What kind of time series can it simulate that a first-order model cannot?

10 In what way do trends and discontinuities in time series affect the sample autocorrelation function?

6

The Power Spectrum

… for autoregressive series, periodogram analysis is not only of no value
but may be dangerously misleading.

Kendall (1946)

This chapter introduces the power spectrum. The power spectrum quantifies how variance is distributed over frequencies and is useful for identifying periodic behavior in time series. The power spectrum is the Fourier Transform of the autocovariance function, and the autocovariance function is the (inverse) Fourier Transform of the power spectrum. As such, the power spectrum and autocovariance function offer two complementary *but mathematically equivalent* descriptions of a stochastic process. Basic characteristics of a stochastic process can be inferred directly from the shape of its power spectrum. This chapter introduces the discrete Fourier Transform, which is a procedure for decomposing any time series into a sum of periodic functions. Results from the discrete Fourier Transform can be summarized in a periodogram, which provides a starting point for estimating power spectra. Estimation of the power spectrum can be counterintuitive because the uncertainty in periodogram elements does not decrease with increasing sample size. To reduce uncertainty, periodogram estimates are averaged over a frequency interval known as the *bandwidth*. Trends and discontinuities in time series can lead to similar low-frequency structure despite very different temporal characteristics. Spectral analysis provides a particularly insightful way to understand the behavior of linear filters. Specifically, the power spectrum of a linearly filtered stochastic process can be factored into two terms, one which depends only on the filter, and the other which depends only on the spectrum of the original process. This factorization is illustrated by deriving the filtering properties of a running mean and the power spectrum of an arbitrary AR(p) process.

6.1 The Problem

The discovery of the ice ages constitutes one of the most fascinating detective stories in modern science (Krüger, 2013). At various times throughout history, villagers near glacial mountains (e.g., the Swiss Alps) claimed that erratically located boulders in their area were placed there by the movement of glaciers that were no longer there. Few geologists took this idea seriously for a long time, but by the 1880s the geological evidence became so overwhelming that most geologists accepted that glaciers covered enormous regions of the Earth in its ancient past. Unfortunately, the mechanism for these changes was a mystery. The astronomer John Herschel advanced the hypothesis that changes in the Earth's orbit caused ice ages, but later dismissed it on the grounds that such orbital changes were too weak to cause climatic changes. Changes in the earth's orbit can be computed from Newton's laws by taking into account the gravitational attraction of other bodies in the solar system, especially the moon, Jupiter and Saturn. Because celestial bodies in the solar system travel in near circles, the forces perturbing earth's orbit are approximately periodic, causing *cyclical* changes in earth's climate. James Croll further developed Herschel's theory and advanced the ice–albedo feedback to explain the amplification of the response to solar variations, but predictions from his theory, which assumed ice ages occurred during periods of high eccentricity, did not fit observations. Attempts to reconcile the uncertain geologic record with various aspects of earth's orbital changes were so discouraging that the theory fell into disrepute. Milutin Milankovic believed that previous studies were not decisive and decided to re-consider the problem in a mathematically rigorous manner. During the 1920s and 1930s, Milankovic computed changes in solar insolation caused by orbital changes for the past 650,000 years and argued that the most important factor was neither the total amount of sunlight nor the amount during winter, but rather the amount shining at high latitudes during summer. After all, high latitude winters are always below freezing, so temperature changes during winter have little impact on snow, whereas a significantly cooler summer results in less snow melt, thereby favoring more ice formation throughout the year. Milankovic identified three important cyclical movements of the earth, namely those of the eccentricity, obliquity, and apsidal precession, whose periods are around 100,000, 41,000, and 21,000 years, respectively. These cycles are now known as *Milankovic cycles*.

Numerous theories of ice ages have been advanced in the past, but Milankovic's theory is the only theory that *predicts* the frequencies of major glacial fluctuations and hence can be tested by examining observed periodicities in geological data. Historically, this comparison was thwarted by several difficulties, especially dating the chronology of geological data, but was eventually accomplished by Hays et al. (1976). To appreciate the difficulty with identifying Milankovic cycles, consider

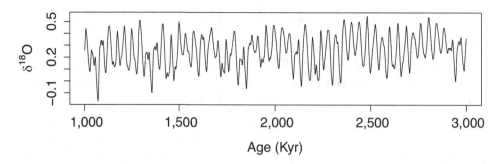

Figure 6.1 Time series of δ^{18}O from the study of Raymo et al. (2006).

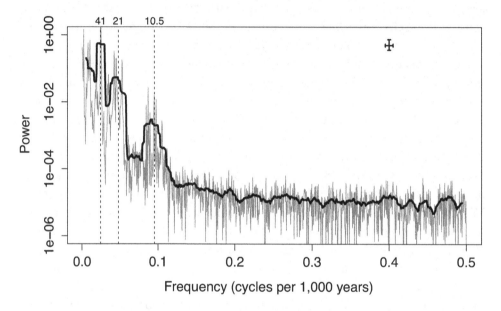

Figure 6.2 An estimate of the power spectrum (thick black curve) and associated periodogram (thin gray curve) of the δ^{18}O time series in Figure 6.1. The error bars in the upper right show the 90% confidence interval and bandwidth of the spectral average estimator (see Section 6.8). Vertical dashed lines show the frequencies corresponding periods of 41,000, 21,000, and 10,500 years.

the time series shown in Figure 6.1, which shows a quantity known as δ^{18}O, derived from ocean sediments. This quantity is a measure of the ratio of ^{18}O to ^{16}O, which in turn is a proxy for ice volume (Raymo et al., 2006). The time series suggest that δ^{18}O may be dominated by certain periodic functions, but which ones?

This question can be addressed using *power spectra*. A power spectrum shows how variance is distributed across frequencies. As an example, the power spectrum estimated from the δ^{18}O time series is shown in Figure 6.2. The x-axis shows frequencies in units of cycles-per-1,000-years and the y-axis shows an estimate of the power spectrum density. Note that the y-axis is on a logarithmic scale and therefore the power differs by orders of magnitude over frequency. Integrating the power spectrum over all frequencies yields the total variance. There are two sets of curves: a thin (noisy) curve showing the squared amplitude of each oscillation frequency, and a thick curve showing a smoothed version of the thin curve. The thick curve is the estimated power spectrum and can be interpreted as a kind of histogram in which data within "spectral bins" are pooled together to estimate the density within the bin. The range of frequencies that are averaged together to construct the smooth curve is indicated by the horizontal error bar in the upper right. This frequency interval is known as the *bandwidth* and variations of the smooth curve within the bandwidth are said to be "unresolved." The vertical error bar shows the 90% confidence interval, and variations of the smooth curve within the confidence interval are said to be insignificant.

A striking aspect of the power spectrum are the peaks at 41,000, 21,000, and 11,000 years. The fact that local peaks occur near Milankovic cycles (namely 41,000 and 21,000 years), and that the peaks rise well above a confidence interval relative to neighboring frequencies, strongly supports the presence of Milankovic cycles in the time series. On the other hand, no clear peak exists at 100,000 years, while a clear peak occurs at 10,500 years, which happens to be half the period of the 21,000-year Milankovic cycle. A goal of paleoclimate studies is to explain such discrepancies in a scientifically consistent manner.

The purpose of this chapter is to explain the concept of a power spectrum and how it can be estimated from finite time series.

6.2 The Discrete Fourier Transform

A well-known fact is that any discrete time series $X_1, X_2, \ldots, X_N$ can be represented as a *Discrete Fourier Transform* (DFT), which is a sum of sines and cosines of the form

$$X_t = \sum_{j=0}^{N/2} \left(A_j \cos(\omega_j t) + B_j \sin(\omega_j t) \right), \tag{6.1}$$

where N is assumed to be even, and the *Fourier frequencies* ω_j are

$$\omega_j = 2\pi j/N. \tag{6.2}$$

This decomposition is remarkable because it applies to *any* time series X_t. A Discrete Fourier Transform exists for odd N too, but in practice the transform is computed by the Fast Fourier Transform, which assumes even N, so the case of odd N is not practically relevant. The Fourier frequencies $\omega_1, \omega_2, \ldots$ are special because for these frequencies the sines and cosines are orthogonal, in the sense that they satisfy the following identities:

$$\sum_{t=1}^{N} \cos(\omega_j t) \cos(\omega_k t) = \begin{cases} 0 & \text{if } j \neq k \neq 0 \\ N/2 & \text{if } j = k \neq 0 \\ N & \text{if } j = k = (0 \text{ or } N/2) \end{cases} \tag{6.3}$$

$$\sum_{t=1}^{N} \sin(\omega_j t) \sin(\omega_k t) = \begin{cases} 0 & \text{if } j \neq k \neq 0 \\ N/2 & \text{if } j = k \neq 0 \\ 0 & \text{if } j = k = (0 \text{ or } N/2) \end{cases} \tag{6.4}$$

$$\sum_{t=1}^{N} \cos(\omega_j t) \sin(\omega_k t) = \sum_{t=1}^{N} \sin(\omega_j t) = 0 \quad \text{for all } j, k \tag{6.5}$$

$$\sum_{t=1}^{N} \cos(\omega_j t) = \begin{cases} 0 & \text{if } j \neq 0 \\ N & \text{if } j = 0 \end{cases}. \tag{6.6}$$

These identities may appear daunting, but in fact they are very simple: the sums vanish when $j \neq k$, and when cos and sin multiply each other, otherwise they equal N or $N/2$, depending on whether j equals N or not. An illustration of how these orthogonality relations arise from symmetry of trigonometric functions is given in Figure 6.3.

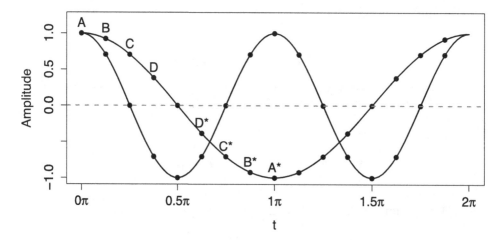

Figure 6.3 Illustration of how $\cos(t)$ and $\cos(2t)$ have vanishing inner product. Both trig functions are symmetric about π. However, $\cos[t]$ is *antisymmetric* about 0.5π while $\cos[2t]$ is symmetric about 0.5π. Thus, when the inner product is computed, the terms associated with A, B, C, D are cancelled by the terms associated with A*, B*, C*, D*.

These orthogonality relations allow A_j and B_j to be calculated directly from the time series X_t. For instance, to obtain A_k, assuming k is not an endpoint (that is, $k \neq 0$ and $k \neq N/2$), multiply (6.1) by $\cos(\omega_k t)$ and sum over t:

$$\sum_{t=1}^{N} X_t \cos(\omega_k t) = \sum_{j=0}^{N/2} \sum_{t=1}^{N} A_j \cos(\omega_j t) \cos(\omega_k t) + \sum_{j=0}^{N/2} \sum_{t=1}^{N} B_j \sin(\omega_j t) \cos(\omega_k t)$$

$$= \frac{N}{2} A_k. \tag{6.7}$$

Similarly, B_k can be derived by multiplying both sides of (6.1) by $\sin(\omega_k t)$ and summing over t. The result is that the coefficients for j not at the end points $(0, N/2)$ are given by

$$A_j = \frac{2}{N} \sum_{t=1}^{N} X_t \cos(\omega_j t) \quad \text{and} \quad B_j = \frac{2}{N} \sum_{t=1}^{N} X_t \sin(\omega_j t), \qquad j \neq (0, N/2).$$

$$\tag{6.8}$$

For practical calculations, the DFT is computed from a computational technique known as the Fast Fourier Transform (FFT), which is available in R, Python, Matlab, etc.

The end points of the DFT require separate discussion. For $j = 0$, $\omega_0 = 0$ and $\sin(\omega_0 t) = 0$, in which case B_0 is irrelevant and can be ignored. Also, $\cos(\omega_0 t) = 1$, in which case

$$A_0 = \frac{1}{N} \sum_{t=1}^{N} X_t. \tag{6.9}$$

Thus, the coefficient A_0 is simply the time mean of the data, denoted $\overline{X}$. Since N is even, $\sin(\omega_{N/2} t) = 0$ so $B_{N/2}$ is irrelevant and hence ignored. Moreover,

$$A_{N/2} = \frac{1}{N} \sum_{t=1}^{N} (-1)^t X_t. \tag{6.10}$$

The Discrete Fourier Transform is *invertible*, in the sense that substituting the coefficients from (6.8) and (6.10) into (6.1) returns the original time series *exactly*. Thus, there is no loss of information in transforming the time series into Fourier coefficients, since the coefficients can be used to derive the time series. This makes perfect sense because there are $N + 2$ total coefficients in the summation (6.1), but B_0 and $B_{N/2}$ are irrelevant, as discussed previously, leaving exactly N nonzero coefficients A_j and B_j. Thus, starting with a time series of length N, we obtain N nontrivial coefficients in the DFT.

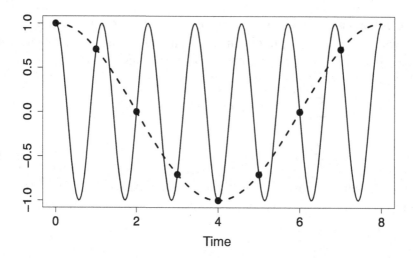

Figure 6.4 Illustration of aliasing. The functions $\cos(\pi t/4)$ (solid) and $\cos(7\pi t/4)$ (dashed) are evaluated in the time range $t = [0,8]$, and the integer points $t = 0, \ldots, 7$ are indicated as dots. The two cosine functions intersect at exactly the same value at the sample points $t = 0, \ldots, 7$.

An important limitation of the Discrete Fourier Transform is that it can resolve oscillations only up to frequency $\omega = \pi$, termed the *Nyquist frequency*. An illustration of this fact is shown in Figure 6.4, which shows two sinusoids sampled at the integer values $t = 0, 1, 2, \ldots$. Even though the two sinusoids differ, they cannot be distinguished *when sampled at the integer values of* t. Thus, given the values at $t = 0, 1, 2, \ldots$, it is impossible to decide whether the high- or low-frequency oscillation is present. This effect follows from standard trigonometric identities, and is known as *aliasing*. Aliasing means that oscillations with frequencies beyond the Nyquist frequency will be attributed to frequencies below the Nyquist frequency, giving a false impression of how variability is distributed by frequency. In this case, oscillations at frequencies larger than $\omega = \pi$ are said to be *aliased* into lower frequencies. To avoid significant aliasing, the sampling rate must be chosen to be high enough so that the process being sampled changes little in between sampling times.

The frequency ω has units of radians per time step. It is often more natural to measure frequency in cycles per time step: $f = \omega/(2\pi)$. In this case, the Fourier frequencies are

$$
\begin{array}{cccccll}
j & = & 1 & 2 & \cdots & N/2 & \text{Fourier index} \\
\omega_j & = & 2\pi/N & 4\pi/N & \cdots & \pi & \text{radians per time step.} \\
f_j & = & 1/N & 2/N & \cdots & 1/2 & \text{cycles per time step}
\end{array} \tag{6.11}
$$

In terms of cycles per time step, the Nyquist frequency is 1/2, or equivalently, the Nyquist period is 2 time steps per cycle. Thus, for daily data, the Nyquist period is 2 days; for monthly data, the Nyquist period is 2 months; for annual data the Nyquist period is 2 years. Periods shorter than the Nyquist period are aliased.

Some calculations in spectral analysis are greatly simplified by using complex variables. Complex variables have the property that $e^{i\omega t} = \cos(\omega t) + i \sin(\omega t)$, which in turn implies

$$2\cos(\omega t) = e^{i\omega t} + e^{-i\omega t} \quad \text{and} \quad 2i \sin(\omega t) = e^{i\omega t} - e^{-i\omega t}. \tag{6.12}$$

Most derivations presented in this chapter use trigonometric functions because trigonometric functions are more familiar to most readers. However, complex variables will be used to simplify certain calculations in certain parts of this chapter.

6.3 Parseval's Identity

An important relation exists between the DFT and the sample variance

$$\hat{\sigma}_X^2 = \frac{1}{N} \sum_{t=1}^{N} (X_t - \overline{X})^2, \tag{6.13}$$

where $\overline{X}$ is the time mean of X_t (we normalize by N instead of $N - 1$ to simplify certain expressions). Subtracting $\overline{X}$ from X_t is equivalent to dropping A_0 in (6.1), with no other change in the DFT. Therefore, the sample variance can be written as

$$\hat{\sigma}_X^2 = \frac{1}{N} \sum_{t=1}^{N} \left(\sum_{j=1}^{N/2} (A_j \cos(\omega_j t) + B_j \sin(\omega_j t)) \right)^2, \tag{6.14}$$

where the j index starts at one (instead of zero like (6.1)). Expanding the square and using orthogonality relations (6.3)–(6.6) to eliminate the cross-terms yields

$$\hat{\sigma}_X^2 = \frac{1}{2} \sum_{j=1}^{\frac{N}{2}-1} (A_j^2 + B_j^2) + A_{N/2}^2. \tag{6.15}$$

This equation, known as *Parseval's Identity*, reveals that the sample variance equals the sum of squared coefficients of the DFT. In effect, Parseval's Identity shows that variance may be *partitioned* into different oscillation frequencies.

6.4 The Periodogram

In Chapter 5, we learned that a measure of temporal dependence of a time series is the autocovariance function $c(\tau)$, whose sample estimate is

$$\hat{c}(\tau) = \frac{1}{N} \sum_{t=1}^{N-|\tau|} \left(X_{t+\tau} - \hat{\mu}_{t+\tau}\right)\left(X_t - \hat{\mu}_t\right),\tag{6.16}$$

where $-(N-1) \leq \tau \leq N-1$. The *periodogram* $\hat{p}(\omega)$ is defined as

$$\hat{p}(\omega) = \sum_{\tau=-(N-1)}^{N-1} \hat{c}(\tau)\cos(\omega\tau).\tag{6.17}$$

The periodogram can be evaluated for *all* frequencies. Because the autocovariance function is an even function of τ, it can be written equivalently as

$$\hat{p}(\omega) = \hat{c}(0) + 2\sum_{\tau=1}^{N-1} \hat{c}(\tau)\cos(\omega\tau).\tag{6.18}$$

The periodogram (6.17) can be evaluated for all frequencies, but only the frequencies $0 < \omega \leq \pi$ are needed. After all, (6.22) is periodic, that is, $\hat{p}(\omega+2\pi k) = \hat{p}(\omega)$ for any integer k. Also, the periodogram is an even function of ω, so without loss of generality only positive frequencies need to be considered. Consequently, the periodogram defined in the frequency range $(0, \pi]$ is sufficient to specify it at all other frequencies.

The sample autocovariance can be derived from the periodogram by inverting the transform. In this particular case, invertibility follows from the identity

$$\int_{-\pi}^{\pi} \cos(a\omega)\cos(b\omega)d\omega = \begin{cases} 0 & \text{if } a \neq b \\ \pi & \text{if } a = b \end{cases},\tag{6.19}$$

for any integers a and b. Thus, to derive the inverse transform, we multiply the periodogram (6.18) by $\cos(\omega s)$ and integrate over ω. The result is

$$\hat{c}(\tau) = \frac{1}{\pi} \int_{-\pi}^{\pi} \hat{p}(\omega)\cos(\omega\tau)d\omega.\tag{6.20}$$

The fact that the sample autocovariance and periodogram can be derived from each other implies that they have equivalent information content.

Evaluating (6.20) at $\tau = 0$ implies that

$$\hat{c}(0) = \frac{1}{\pi} \int_{-\pi}^{\pi} \hat{p}(\omega)d\omega.\tag{6.21}$$

Recalling the $\hat{c}(0)$ is the sample variance, this expression shows that sample variance equals the integral of the periodogram. In other words, the periodogram quantifies how *sample* variance is distributed over frequencies. We will soon see that the periodogram provides the basis for inferring how *population* variance is distributed over frequencies. The periodogram can be expressed in terms of X_t as follows (see Section 6.11 for the proof):

Definition 6.1 (The Periodogram) *Given a time series* $X_1, X_2, \ldots, X_N$, *the periodogram* $\hat{p}(\omega)$ *is defined as the function over* $\omega \in (0, \pi]$

$$\hat{p}(\omega) = \frac{1}{N}\left(\left(\sum_{t=1}^{N}(X_t - \hat{\mu}_t)\cos(\omega t)\right)^2 + \left(\sum_{t=1}^{N}(X_t - \hat{\mu}_t)\sin(\omega t)\right)^2\right), \tag{6.22}$$

where $\hat{\mu}_t$ *is an estimate of the mean of the process at fixed time (e.g., this term would include the time mean and estimated trends, annual cycles, diurnal cycles, etc). The periodogram can be evaluated at any frequency* ω, *not just the Fourier frequency* ω_j. *At the Fourier frequencies (6.2), the quantity* $\hat{p}(\omega_j)$ *is known as a* periodogram ordinate *and equals*

$$\hat{p}(\omega_j) = \begin{cases} \frac{N}{4}\left(A_j^2 + B_j^2\right) & \text{if } j \neq N/2 \\ NA_j^2 & \text{if } j = N/2, \end{cases} \tag{6.23}$$

where A_j *and* B_j *are defined in (6.8), and Parseval's Identity (6.15) becomes*

$$\hat{\sigma}_X^2 - \frac{1}{N}\left(2\sum_{j=1}^{\frac{N}{2}-1}\hat{p}(\omega_j) + \hat{p}(\omega_{N/2})\right). \tag{6.24}$$

6.5 The Power Spectrum

Replacing the sample autocovariance function $\hat{c}(\tau)$ by its population counterpart $c(\tau)$, and taking the limit $N \to \infty$ in (6.18), leads to the power spectrum:

Definition 6.2 (The Power Spectral Density Function) *The power spectral density function, or power spectrum, of a stationary process with autocovariance* $c(\tau)$ *is defined as*

$$p(\omega) = c(0) + 2\sum_{\tau=1}^{\infty} c(\tau)\cos(\omega\tau), \tag{6.25}$$

where ω *is termed the frequency and has domain* $(0, \pi]$.

Note that the autocovariance function $c(\tau)$ is a discrete function of τ, but its corresponding power spectrum $p(\omega)$ is a continuous function of ω. If the power spectrum decays sufficiently quickly with frequency, then it can be shown that

$$p(\omega) = \lim_{N \to \infty} \mathbb{E}\left[\hat{p}(\omega)\right]. \tag{6.26}$$

Importantly, this expression does not mean that the periodogram $\hat{p}(\omega)$ converges to $p(\omega)$ in the limit $N \to \infty$, since the right-hand side includes an expectation before taking the limit of large N. Procedures for estimating the power spectrum will be discussed in Section 6.8.

Using the orthogonality relation (6.19), it is straightforward to show from (6.26) that

$$c(\tau) = \frac{1}{\pi} \int_{-\pi}^{\pi} p(\omega) \cos(\omega\tau) d\omega. \tag{6.27}$$

Since the power spectrum and autocovariance function can be derived from each other, knowing one function is equivalent to knowing the other. Evaluating (6.27) at $\tau = 0$ gives

$$\text{var}[X_t] = c(0) = \frac{1}{\pi} \int_{-\pi}^{\pi} p(\omega) d\omega. \tag{6.28}$$

Since $c(0)$ is the variance of the process, *the power spectrum quantifies how vari-ance is distributed over frequencies*. In particular, the variance between frequencies ω and $\omega + d\omega$ is $p(\omega)d\omega/\pi$. A few examples should help illustrate this concept.

Example 6.1 (Power Spectrum of White Noise) *Question: What is the power spectrum of a white noise process with variance σ^2? Answer: Recall that the ACF of a white noise process vanishes at all nonzero lags. Substituting this ACF into (6.25) implies that all terms except $\tau = 0$ vanish, giving*

$$p(\omega) = \sigma^2. \tag{6.29}$$

Note that this power spectrum is constant. This is why the process is known as "white noise," because it is a homogeneous mix of all frequencies, just as white light is a homogeneous mix of different frequencies of light.

Example 6.2 (Power Spectrum of an AR(1) Process) *Question: What is the power spectrum of an AR(1) process? Answer: An AR(1) process satisfies $X_t = \phi X_{t-1} + W_t$, where W_t is a white noise process with zero mean and variance σ_W^2, and ϕ is the lag-1 autocorrelation. As shown in Section 5.4, this AR(1) process has ACF $\rho_\tau = \phi^\tau$ and stationary variance $\sigma_W^2/(1-\phi^2)$ (see (5.44)), hence the autocovariance function is $c(\tau) = \phi^\tau \sigma_W^2/(1-\phi^2)$. Substituting this autocovariance function into (6.25) gives*

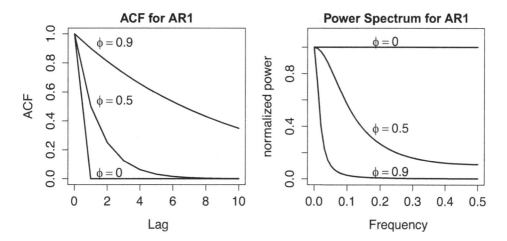

Figure 6.5 The ACF (left) and corresponding normalized power spectra $p(\omega)/p(0)$ (right) for an AR(1) process, for three different values of the AR parameter ϕ. The power spectra are normalized by their maximum value so that each of them can be presented on the same scale.

$$
\begin{aligned}
p(\omega) &= \frac{\sigma_W^2}{(1 - \phi^2)} \left(1 + 2 \sum_{\tau=1}^{\infty} \phi^\tau \cos(\omega\tau) \right) \\
&= \frac{\sigma_W^2}{(1 - \phi^2)} \left(\sum_{\tau=0}^{\infty} \left(\phi^\tau e^{-i\omega\tau} + \phi^\tau e^{i\omega\tau} \right) - 1 \right) \\
&= \frac{\sigma_W^2}{(1 - \phi^2)} \left(\frac{1}{1 - \phi e^{-i\omega}} + \frac{1}{1 - \phi e^{i\omega}} - 1 \right) \\
&= \frac{\sigma_W^2}{1 - 2\phi \cos(\omega) + \phi^2},
\end{aligned}
\tag{6.30}
$$

where (6.12) and standard formulas for geometric series have been used. The normalized power spectra $p(\omega)/p(0)$ for AR(1) processes for different values of ϕ are illustrated in Figure 6.5. At $\phi = 0$, the power spectrum is flat, corresponding to white noise. As ϕ increases, the power at low frequencies increases faster than at high frequencies, giving a peak at zero frequency.

Deriving the power spectrum for a general AR(p) model by this procedure is very cumbersome. A much simpler approach based on linear filtering is discussed in Section 6.10.

Incidentally, the term *red noise* often refers to a process whose spectrum peaks at zero frequency and decays monotonically with increasing frequency. An AR(1) process is a special case of red noise. *Blue noise* refers to a process whose power spectrum peaks at the highest frequency, which can be generated by an AR(1) model with negative AR parameter.

6.6 Periodogram of Gaussian White Noise

To understand how to estimate the power spectrum, it is instructive to consider the periodogram of a Gaussian white noise process:

$$X_t \sim \text{GWN}(0, \sigma_X^2). \tag{6.31}$$

In this case, the Fourier coefficients A_j and B_j defined in (6.8) are normally distributed because they are a linear combination of independent normally distributed random variables. Therefore, we need only compute their mean and variance to fully specify their distributions. The mean of A_j (for $j \neq 0$) is

$$\mathbb{E}[A_j] = \mathbb{E}\left[\frac{2}{N} \sum_{t=1}^{N} X_t \cos(\omega_j t) \right] = \frac{2}{N} \mathbb{E}[X_t] \sum_{t=1}^{N} \cos(\omega_j t) = 0, \tag{6.32}$$

due to identity (6.6). Note that the expectation is an ensemble mean, not a time mean, and hence time t is treated as a fixed parameter. The variance of A_j (for $j \neq 0$ and $j \neq N/2$) is

$$\text{var}[A_j] = \text{var}\left[\frac{2}{N} \sum_{t=1}^{N} X_t \cos(\omega_j t) \right] = \frac{4}{N^2} \text{var}\left[\sum_{t=1}^{N} X_t \cos(\omega_j t) \right]. \tag{6.33}$$

Since X_t is *iid*, the variance of the sum equals the sum of the variances:

$$\text{var}[A_j] = \frac{4}{N^2} \sum_{t=1}^{N} \text{var}[X_t] \cos^2(\omega_j t) = \frac{2 \, \text{var}[X_t]}{N} = \frac{2\sigma_X^2}{N}, \tag{6.34}$$

where (6.3) has been used. We obtain exactly the same mean and variance for B_j. Thus,

$$A_j \sim \mathcal{N}(0, 2\sigma_X^2/N) \quad \text{and} \quad B_j \sim \mathcal{N}(0, 2\sigma_X^2/N) \quad \text{for } j \neq 0 \text{ and } j \neq N/2, \tag{6.35}$$

and

$$A_{N/2} \sim \mathcal{N}\left(0, \sigma_X^2/N\right). \tag{6.36}$$

Moreover, similar calculations for different orthogonality relations yield

$$\text{cov}[A_j, A_k] = 0, \quad \text{cov}[B_j, B_k] = 0, \quad \text{cov}[A_j, B_k] = 0 \quad \text{for} \quad j \neq k. \tag{6.37}$$

These relations imply that the Fourier coefficients of Gaussian white noise are uncorrelated, and therefore independent (because of the Gaussian assumption).

It is convenient to consider the standardized variables

$$Z_A = \frac{A_j}{\left(\frac{\sigma_X}{\sqrt{N}} \right)} \sim \mathcal{N}(0, 1) \quad \text{and} \quad Z_B = \frac{B_j}{\left(\frac{\sigma_X}{\sqrt{N}} \right)} \sim \mathcal{N}(0, 1). \tag{6.38}$$

Because Z_A and Z_B are independent and distributed as $\mathcal{N}(0, 1)$, $Z_A^2 + Z_B^2$ has a chi-squared distribution with two degrees of freedom (see Theorem 1.3). Therefore,

$$\frac{A_j^2 + B_j^2}{\frac{2\sigma_X^2}{N}} = \frac{2\hat{p}(\omega_j)}{\sigma_X^2} \sim \chi_2^2 \quad \text{for } j \neq N/2. \tag{6.39}$$

Similarly,

$$\frac{A_{N/2}^2}{\frac{\sigma_X^2}{N}} = \frac{\hat{p}(\omega_{N/2})}{\sigma_X^2} \sim \chi_1^2. \tag{6.40}$$

This derivation reveals an important property of the periodogram. Using the fact that $\mathbb{E}[\chi_2^2] = 2$, (6.39) implies that

$$\mathbb{E}[\hat{p}(\omega_j)] = \sigma_X^2 \quad \text{for } j \neq N/2. \tag{6.41}$$

Thus, for Gaussian white noise, the expected value of the periodogram ordinate equals the power spectrum. However, (6.39) implies that the *variance* of the periodogram ordinate is *independent of N*. This fact has important implications for estimating power spectra, as will be discussed in Section 6.8.

6.7 Impact of a Deterministic Periodic Component

So far, X_t has been assumed to be stationary. Now instead, consider the process

$$X_t' = X_t + D\cos(\omega_k t), \tag{6.42}$$

where X_t is a stationary process, ω_k is a Fourier frequency, and D is the amplitude of the sinusoid. Although X_t is stationary, X_t' is nonstationary because its expectation equals $D\cos(\omega_k t)$, which depends on time. Because of the orthogonality properties (6.3)–(6.6), the Discrete Fourier Transform of X_t' is exactly the same as that of X_t except at the frequency ω_k, where the corresponding Fourier Coefficient is

$$A_k' = \frac{2}{N}\sum_{t=1}^{N}(X_t + D\cos(\omega_k t))\cos(\omega_k t) = A_k + D, \tag{6.43}$$

where A_k denotes the Fourier Coefficient for X_t. It follows that the corresponding periodogram ordinate computed from (6.23) is

$$\hat{p}'(\omega_k) = \frac{N}{2}\left((A_k + D)^2 + B_k^2\right) = \hat{p}(\omega_k) + \frac{N}{2}(2A_k D + D^2). \tag{6.44}$$

Therefore, the expected periodogram ordinate is

$$\mathbb{E}[\hat{p}'(\omega_j)] = \begin{cases} \mathbb{E}[\hat{p}(\omega_k)] + \frac{N}{2}D^2 & \text{for } j = k \\ \mathbb{E}[\hat{p}(\omega_j)] & \text{for } j \neq k \end{cases}. \tag{6.45}$$

Taking the limit $N \to \infty$, as in (6.26), reveals that the power spectrum *is unbounded* at ω_k, but equals $p(\omega)$ at all other frequencies. In essence, adding deterministic sinusoid components to a stationary process adds *delta functions* to the power spectrum.

6.8 Estimation of the Power Spectrum

From (6.26), one might assume (incorrectly) that the periodogram converges to the power spectrum as $N \to \infty$. However, this is not the case! To illustrate this fact, consider periodograms for Gaussian white noise for different N, as shown in Figure 6.6. The figure shows that as sample size increases, more periodogram ordinates crowd together, but individual ordinates show no tendency to converge toward the true power spectra. On the other hand, the *average* of the periodogram ordinates is nearly the same. Thus, periodogram ordinates seem to cluster about the true power spectrum, but the variances do not decrease with N. The fact that the variances do not decrease with N is a consequence of (6.39), which shows that the distribution of a periodogram ordinate is independent of N.

The fact that the periodogram ordinates seem to cluster about the true power spectrum suggests that an improved estimate can be obtained by *averaging periodogram estimates over a local frequency interval*. The following theorem supports this approach:

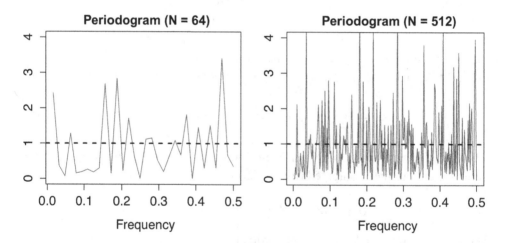

Figure 6.6 Periodogram (thin grey curve) of a realization of Gaussian white noise with zero mean and unit variance of sample size 64 (left) and 512 (right). The horizontal dashed line shows the population power spectrum.

Theorem 6.1 (Asymptotic Distribution of the Periodogram) *Let* $X_1, \ldots, X_N$ *be a stationary Gaussian process with power spectrum* $p(\omega)$*. Then, in the limit* $N \rightarrow \infty$,

$$\frac{2\hat{p}(\omega_j)}{p(\omega_j)} \sim \chi_2^2, \tag{6.46}$$

where $\hat{p}(\omega_j)$ *is defined in (6.22). Also,* $\hat{p}(\omega_j)$ *and* $\hat{p}(\omega_k)$ *are independent for all* $j \neq k$.

This theorem implies that, for large N, a periodogram estimate $\hat{p}(\omega)$ has chi-square distribution clustered around the true power spectrum $p(\omega)$. Thus, the power spectrum may be estimated as follows. First, one specifies a frequency interval over which the periodogram estimates are averaged. This interval is known as the *bandwidth*. Ideally, the bandwidth is chosen such that the power spectrum is approximately constant within that frequency range. The larger the bandwidth, the more terms that are averaged and the smaller the variance. On the other hand, if the power spectrum varies substanatially within the bandwidth, then the averaging procedure will smooth out that structure and produce a biased estimate of the power. Thus, there is a trade-off between bias and variance: the shorter the bandwidth, the more structure that can be resolved, but at the cost of more variance; the larger the bandwidth, the less variance, but at the cost of resolving less structure in the power spectrum.

Averaging the periodogram leads to the discrete spectral average estimator:

Definition 6.3 (Discrete Spectral Average Estimator) *The discrete spectral average estimator of the spectral density* $p(\omega_j)$ *has the form*

$$\hat{p}_M(\omega_j) = \frac{1}{2M+1} \sum_{m=-M}^{M} \hat{p}(\omega_{j+m}) \quad \text{for } M < j < \frac{N}{2} - M, \tag{6.47}$$

where M *is a parameter that determines the* bandwidth.

This type of estimator is a "running average" or a *smoother* of the periodogram. Theorem (6.46) implies that the sum of periodogram ordinates over the bandwidth also has a chi-squared distribution, but with degrees of freedom equal to the sum of the degrees of freedom of the individual ordinates. For instance, if three ordinates are averaged, then

$$2 \left(\frac{\hat{p}(\omega_{j-1}) + \hat{p}(\omega_j) + \hat{p}(\omega_{j+1})}{p(\omega_j)} \right) \sim \chi_6^2, \tag{6.48}$$

where the power spectrum is assumed to be constant over the frequency interval $(\omega_{j-1}, \omega_{j+1})$. This result is generalized to arbitrary stochastic processes in the next theorem.

Theorem 6.2 (Distribution of the Discrete Spectral Average Estimator) *Let $X_1, \ldots, X_N$ be a stochastic process with smooth power spectrum $p(\omega)$. Then in the limit $N \to \infty$, the discrete spectral average estimator defined in (6.47) has distribution*

$$2(2M+1)\frac{\hat{p}_M(\omega_j)}{p(\omega_j)} \sim \chi^2_{2(2M+1)}, \tag{6.49}$$

where $\omega_j = 2\pi j/N$ for $j = M+1, \ldots, N/2 - M - 1$. In addition,

$$\hat{p}_M(\omega_j) \quad \text{and} \quad \hat{p}_M(\omega_k) \tag{6.50}$$

are independent when $|j - k| \geq 2M+1$. The interval $(\omega_{j-M}, \omega_{j+M})$ is known as the bandwidth of the spectral estimator.

An approximate confidence interval for the power spectrum can be derived from the fact that the periodogram ordinates have approximately a chi-squared distribution. For example, a 90% confidence interval is given by (L, U), where

$$L = \frac{2(2M+1)}{\chi^2_{0.05, 2(2M+1)}} \hat{p}_M(\omega_j) \quad \text{and} \quad U = \frac{2(2M+1)}{\chi^2_{0.95, 2(2M+1)}} \hat{p}_M(\omega_j). \tag{6.51}$$

Note that the width of the confidence interval is proportional to the periodogram estimate. Thus, the width of the interval increases as the periodogram ordinate increases. This property is illustrated in the left panel of Figure 6.7 for an arbitrary power spectrum. This dependence makes it awkward to compare power spectra estimates. For instance, the significance of a difference in spectra depends on the magnitude of the power spectrum. A happy coincidence is that these confidence

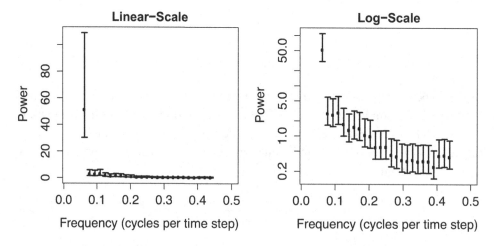

Figure 6.7 A smoothed periodogram of an AR(1) process displayed using a linear (left) and log (right) scale for the y-axis. The error bars show the 90% confidence interval.

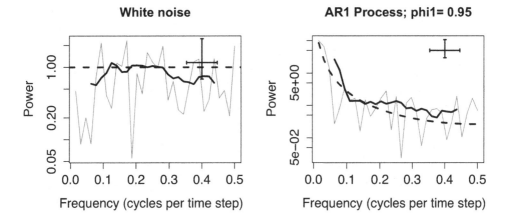

Figure 6.8 Periodogram (thin grey curves) of a realization of a stochastic process for white noise (left) and an AR(1) process (right). The solid curve shows the discrete spectral average estimate using $M = 3$. The dashed curve shows the respective population power spectrum. The error bars in the upper right show the 90% confidence interval and bandwidth of the spectral average estimator. The spectrum on the right is the same as that shown in Figure 6.7.

intervals have constant widths when plotted on a log-scale (or equivalently, when plotting the logarithm of the power). This fact can be seen by taking the difference of the logs of the upper and lower confidence limits:

$$\log U - \log L = \log \frac{U}{L} = \log \frac{\chi^2_{0.05,\,2(2M+1)}}{\chi^2_{0.95,\,2(2M+1)}}, \tag{6.52}$$

which is independent of $\hat{p}_M(\omega_j)$. This property is illustrated in the right panel of Figure 6.7. Because the confidence intervals are constant on a log scale, they can be indicated by showing a *single* error bar, as illustrated in Figure 6.8.

Smoothing estimators become problematic near the ends of the frequency range $\omega = 0$ and $\omega = \pi$, since the bandwidth goes outside the frequency range. The bandwidth can be shortened as the end points are approached, but then the confidence interval no longer applies. One might indicate the correct confidence interval by showing that the interval increases as the end points are approached. Alternatively, one can simply ignore the low or high frequencies for which the smoothing cannot be performed (as done in Figure 6.8).

One reason to estimate a power spectrum is to infer the type of stochastic process. To illustrate this application, consider white noise, which has a constant power spectrum. A white noise power spectrum is shown as the dash curve in the left panel of Figure 6.8. A realization is generated and then used to estimate the periodogram and the discrete spectral average, shown as the thin and thick solid curves, respectively.

In this case, the spectral average fluctuates entirely within the 90% confidence interval, indicating no significant difference in variance between frequency bands. Thus, one would conclude that the spectral average estimate is "flat" and consistent with white noise. Now consider an AR(1) process. The power spectrum of an AR(1) process decays with frequency, as illustrated in the right panel of Figure 6.8 for a particular choice of parameters. In this case, the spectral average estimate at low frequencies (that is, $f < 0.1$) and high frequencies (that is, $f > 0.4$) lie more than two confidence intervals apart (that is, their confidence intervals do not overlap), and lie more than two bandwidths apart, suggesting that the variance at low frequencies differs significantly from the variance at high-frequencies. Thus, the spectral average estimate is inconsistent with white noise but consistent with an AR(1) process. These considerations show how the *shape* of the spectral estimate, in relation to the confidence intervals and bandwidth, can be used to infer the type of stochastic process.

Another reason to estimate a power spectrum is to identify deterministic periodicities. This application was illustrated in Section 6.1. In that example, a key point is that the period of the deterministic signals were postulated unambiguously *before* looking at the data. In particular, they were postulated to be the periods of the Milankovic cycles. If, by contrast, one first estimates a power spectrum and then, *after looking at the estimates*, hypothesizes a signal at the location of a spectral peak, then the confidence interval computed earlier is not the appropriate standard for significance. The problem is that the identification of spectral peaks is a form of *screening*, as discussed in Chapter 11. After all, to identify a spectral peak, one must look at all the estimates and choose the largest one. The distribution of the most extreme ordinate differs from the distribution of a randomly chosen ordinate. In certain cases, this screening process can be taken into account to derive a test for hidden periodicities of unspecified frequency (Brockwell and Davis, 1991, chapter 10).

6.9 Presence of Trends and Jump Discontinuities

Power spectra are often used to infer properties of a stochastic process. In making these inferences, one should be aware that trends and discontinuities in time series can produce strong peaks at low frequencies. To illustrate, consider the straight line in Figure 6.9a. The grey curves show various approximations of the straight line using a total of 1, 3, 7 Fourier harmonics. The amplitudes of the Fourier harmonics decay very rapidly with frequency, as confirmed by the periodogram of the straight-line shown in Figure 6.9b. Thus, in general, a time series containing a strong trend will have a steeply decaying periodogram. The same effect occurs in the case of a jump discontinuity, shown in Figures 6.9c–d: the power spectrum exhibits a strong peak near zero frequency and decays rapidly.

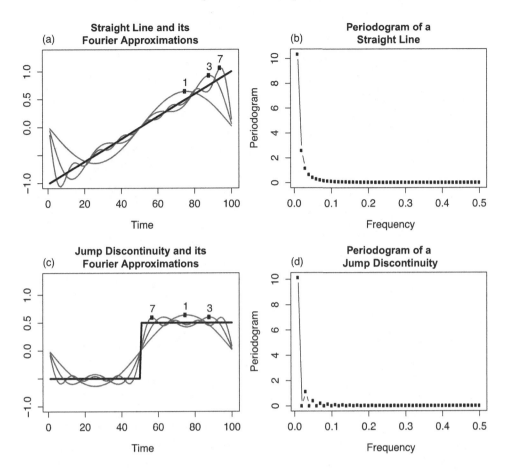

Figure 6.9 (a) A straight line (thick line) and various approximations of the straight line using a total of 1, 3, 7 Fourier harmonics (grey curves). (b) The periodogram of the straight line. (c) A jump discontinuity (thick curve) and its various approximations using 1, 3, 7 Fourier harmonics (grey curves). (d) The periodogram of the jump discontinuity.

To illustrate the impact of trends and jump discontinuities, Figure 6.10a shows three time series labeled A, B, C. These time series were generated by adding a linear function of time to the same realization of white noise. The corresponding smoothed periodograms are shown in Figure 6.10b. The time series marked A has no trend and the corresponding spectrum is indistinguishable from white noise; that is, the spectrum varies within the confidence interval. However, the other time series generate spectra with significant peaks at low frequency. As a second example, Figure 6.10c shows time series generated by adding a jump discontinuity to white noise. The corresponding spectra for B and C, shown in 6.10d, exhibit the same kind of enhanced power at low frequencies. In general, abrupt shifts or long-period oscillations are characterized by enhanced power at low frequencies.

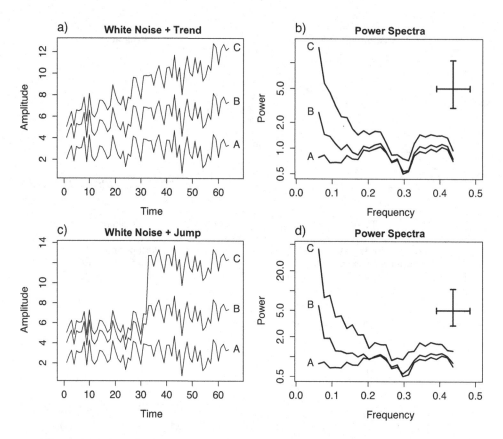

Figure 6.10 Sample time series (left) and corresponding estimated power spectra (right). The top row shows results for three time series comprising white noise plus a straight line (a "trend"), and the bottom row shows results for three time series comprising white noise plus a "jump" (a function that is constant for the first half and a different constant for the second half of the period). The error bars in the top right of the power spectra shows the 90% confidence interval and the bandwidth.

6.10 Linear Filters

The above material completes our review of power spectrum estimation. Before closing this chapter, it is worth discussing *linear filters*. A linear filter takes a time series U_t as "input" and produces an "output" Y_t based on the linear transformation

$$Y_t = \sum_{j=-\infty}^{\infty} a_j U_{t-j}. \qquad (6.53)$$

Such filters are common in climate studies. For instance, a *running mean* is often used to filter out high-frequency variability. As a concrete example, a five-point running mean is

$$Y_t = \frac{U_{t-2} + U_{t-1} + U_t + U_{t+1} + U_{t+2}}{5}, \tag{6.54}$$

which is a special case of the linear filter (6.53), in which the coefficients are

$$a_j = \begin{cases} 1/5 & \text{if } -2 \le j \le 2 \\ 0 & \text{otherwise} \end{cases}. \tag{6.55}$$

How is the power spectrum of the output Y_t related to the power spectrum of the input U_t? The answer is surprisingly simple. The relation is derived most simply by using the complex form of the power spectrum, which is

$$p_Y(\omega) = \sum_{\tau=-\infty}^{\infty} c_Y(\tau) e^{i\omega\tau}. \tag{6.56}$$

To confirm that this definition is in fact equivalent to (6.25), recall that the autocovariance function is even (that is, $c_Y(\tau) = c_Y(-\tau)$), hence the above expression can be written as

$$p_Y(\omega) = c_Y(0) + \sum_{\tau=1}^{\infty} c_Y(\tau) e^{i\omega\tau} + \sum_{\tau=-\infty}^{-1} c_Y(\tau) e^{i\omega\tau} \tag{6.57}$$

$$= c_Y(0) + \sum_{\tau=1}^{\infty} c_Y(\tau) e^{i\omega\tau} + \sum_{\tau=1}^{\infty} c_Y(-\tau) e^{-i\omega\tau} \tag{6.58}$$

$$= c_Y(0) + \sum_{\tau=1}^{\infty} c_Y(\tau) \left(e^{i\omega\tau} + e^{-i\omega\tau} \right), \tag{6.59}$$

which yields (6.25) after substituting (6.12).

Now consider the autocovariance function of Y_t:

$$c_Y(\tau) = \text{cov}[Y_{t+\tau}, Y_t] \tag{6.60}$$

$$= \text{cov}\left[\sum_{j=-\infty}^{\infty} a_j U_{t+\tau-j}, \sum_{k=-\infty}^{\infty} a_k U_{t-k} \right] \tag{6.61}$$

$$= \sum_{j=-\infty}^{\infty} \sum_{k=-\infty}^{\infty} a_j a_k \, \text{cov}\left[U_{t+\tau-j}, U_{t-k} \right] \tag{6.62}$$

$$= \sum_{j=-\infty}^{\infty} \sum_{k=-\infty}^{\infty} a_j a_k c_U(\tau - j + k). \tag{6.63}$$

Multiplying both sides by $\exp(i\omega\tau)$ and rearranging yields

$$c_Y(\tau)e^{i\omega\tau} = \sum_{j=-\infty}^{\infty}\sum_{k=-\infty}^{\infty} a_j a_k c_U(\tau - j + k)e^{i\omega\tau} \tag{6.64}$$

$$= \sum_{j=-\infty}^{\infty}\sum_{k=-\infty}^{\infty} a_j e^{i\omega j} a_k e^{-i\omega k} c_U(\tau - j + k)e^{i\omega(\tau-j+k)} \tag{6.65}$$

Summing over τ to compute the power spectrum yields

$$p_Y(\omega) = \sum_{j=-\infty}^{\infty}\sum_{k=-\infty}^{\infty} a_j e^{i\omega j} a_k e^{-i\omega k} \left(\sum_{\tau=-\infty}^{\infty} c_U(\tau - j + k)e^{i\omega(\tau-j+k)} \right). \tag{6.66}$$

The term in parentheses is just the power spectrum of U_t, namely $p_U(\omega)$. This term is independent of j and k because τ ranges over $(-\infty, \infty)$ and j and k merely shift τ. Thus

$$p_Y(\omega) = \left(\sum_{j=-\infty}^{\infty} a_j e^{i\omega j} \right) \left(\sum_{k=-\infty}^{\infty} a_k e^{-i\omega k} \right) p_U(\omega). \tag{6.67}$$

The first summation is just the complex Fourier Transform of the filter coefficients:

$$\tilde{a}(\omega) = \sum_{j=-\infty}^{\infty} a_j e^{i\omega j}. \tag{6.68}$$

This function is known as the *transfer function* of the linear filter. In this notation, $p_Y(\omega)$ is

$$p_Y(\omega) = |\tilde{a}(\omega)|^2 p_U(\omega), \tag{6.69}$$

where $|z|$ denotes the *modulus* of z. The identity (6.69) states that the power spectrum of a linearly filtered process can be *factored* into a term that depends *only* on filter properties, and another term that depends *only* on the spectrum of the original stochastic process. This factorization has numerous important consequences. A few are illustrated next.

Example 6.3 (Transfer Function for the Five-Point Running Mean) *The transfer function for the five-point running mean (6.54) can be computed by substituting the coefficients (6.55) into transfer function Equation (6.68), which yields*

$$\tilde{a}(\omega) = \sum_{j=-2}^{2} a_j e^{i\omega j} = \frac{1}{5}\left(1 + 2\cos(\omega) + 2\cos(2\omega)\right), \tag{6.70}$$

where (6.12) has been used. The squared modulus of the transfer function is plotted in Figure 6.11. To interpret this result, consider a white noise process U_t. The

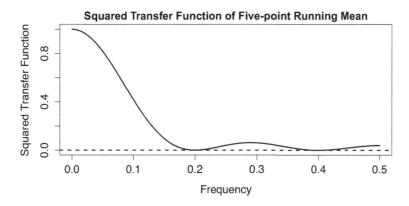

Figure 6.11 The squared transfer function of the five-point running mean filter (6.54).

power spectrum of a white noise process is flat– that is, $p_U(\omega) =$ constant. Therefore, according to (6.69), the power spectrum of the five-point running mean has the same shape as the transfer function. This example illustrates a very important point: To the extent that the power spectrum of a process is constant over a range of frequencies, the power spectrum of a linearly filtered process is determined completely by the linear filter, not by the process itself (see Cane et al., 2017, for further discussion). The squared transfer function shown in Figure 6.11 reveals that, contrary to what one might expect from a running average, the spectrum does not decay monotonically with frequency. Instead, the power spectrum reaches secondary maxima at $f = \cos^{-1}(1/4)/(2\pi) \approx 0.29$ and at $f = 1/2$. The existence of the latter peak can be seen by considering a time series that contains only the frequency $f = 1/2$, which corresponds to a time series that alternates between 1 and -1:

$$U_t = (-1)^t. \tag{6.71}$$

A five-point running mean of this function equals $(-1)^t/5$, which also is an alternating function but with less amplitude. More generally, the power spectrum of any N-point running mean does not decay uniformly to zero, but instead reaches secondary maxima depending on the length of the running mean. It is possible to select coefficients a_j to produce a squared transfer function that decays almost monotonically to zero, but in this case the coefficients a_j decay away from the central point.

Example 6.4 (Power Spectrum of an Arbitrary Autoregressive Model) *Consider the general autoregressive model of order p:*

$$X_t = \phi_1 X_{t-1} + \cdots + \phi_p X_{t-p} + \epsilon_t, \tag{6.72}$$

where ϵ_t is a white noise process. This equation can be written equivalently as

$$\epsilon_t = X_t - \phi_1 X_{t-1} + \cdots + \phi_p X_{t-p}. \tag{6.73}$$

Note that this equation is a linear filter of the form (6.53) *where*

$$
\begin{pmatrix} a_0 \\ a_1 \\ \vdots \\ a_p \end{pmatrix} = \begin{pmatrix} 1 \\ -\phi_1 \\ \vdots \\ -\phi_p \end{pmatrix}. \tag{6.74}
$$

The transfer function is therefore

$$
\tilde{\phi}(\omega) = 1 - \phi_1 e^{i\omega} - \phi_2 e^{i\omega 2} - \cdots - \phi_p e^{i\omega p}. \tag{6.75}
$$

It follows from (6.69) *that*

$$
p_\epsilon(\omega) = \left| \tilde{\phi}(\omega) \right|^2 p_X(\omega). \tag{6.76}
$$

The power spectra of a white noise process ϵ is a constant equal to σ_ϵ^2. Thus, the power spectra of the AR(p)-process is

$$
p_X(\omega) = \frac{\sigma_\epsilon^2}{\left| \tilde{\phi}(\omega) \right|^2}. \tag{6.77}
$$

As a specific example, consider an AR(2) model. The squared modulus of the associated transfer function is

$$
|\tilde{\phi}(\omega)| = \left| 1 - \phi_1 e^{i\omega} - \phi_2 e^{i\omega 2} \right|^2 \tag{6.78}
$$

$$
= \left(1 - \phi_1 e^{i\omega} - \phi_2 e^{i\omega 2} \right) \left(1 - \phi_1 e^{-i\omega} - \phi_2 e^{-i\omega 2} \right) \tag{6.79}
$$

$$
= 1 + \phi_1^2 + \phi_2^2 + 2\phi_1(\phi_2 - 1) \cos(\omega) - 2\phi_2 \cos(2\omega). \tag{6.80}
$$

It follows that the power spectra of an AR(2) process is

$$
p(\omega) = \frac{\sigma_W^2}{1 + \phi_1^2 + \phi_2^2 + 2\phi_1(\phi_2 - 1) \cos(\omega) - 2\phi_2 \cos(2\omega)}. \tag{6.81}
$$

Examples of power spectra for different AR(2) processes are shown in Figure 5.13. The figure shows that AR(2) processes generate spectral peaks that depend on the choice of parameter. In contrast, AR(1) processes generate spectral peaks only at zero frequency.

6.11 Tying Up Loose Ends

Relation between the Periodogram and Autocovariance Function

A rather remarkable fact is that the periodogram and autocovariance function are each other's Fourier Transform not only in a population sense but also in a *sample sense*. Here, we prove (6.18).

Expanding the squares in (6.22) gives

$$\left(\sum_{t=1}^{N} \left(X_t - \overline{X} \right) \cos(\omega t) \right)^2 = \sum_{t=1}^{N} \sum_{s=1}^{N} \left(X_t - \overline{X} \right) \left(X_s - \overline{X} \right) \cos(\omega t) \cos(\omega s)$$

$$\left(\sum_{t=1}^{N} \left(X_t - \overline{X} \right) \sin(\omega t) \right)^2 = \sum_{t=1}^{N} \sum_{s=1}^{N} \left(X_t - \overline{X} \right) \left(X_s - \overline{X} \right) \sin(\omega t) \sin(\omega s).$$

Substituting these identities in (6.22) and using the trigonometric identity

$$\cos(a - b) = \cos(a)\cos(b) + \sin(a)\sin(b), \tag{6.82}$$

gives

$$\hat{p}(\omega) = \frac{1}{N} \sum_{t=1}^{N} \sum_{s=1}^{N} \left(X_t - \overline{X} \right) \left(X_s - \overline{X} \right) \cos(\omega(t - s)). \tag{6.83}$$

We now change variables so that all quantities can be expressed in terms of the difference variable $\tau = t - s$. To reduce the notational burden, define the temporary variable

$$f(t,s) = \frac{1}{N} \left(X_t - \overline{X} \right) \left(X_s - \overline{X} \right) \cos(\omega(t - s)). \tag{6.84}$$

Then, transforming (t,s)-coordinates into (t,τ)-coordinates maps the original square domain into a parallelogram, as indicated in Figure 6.12, with the final summation becoming

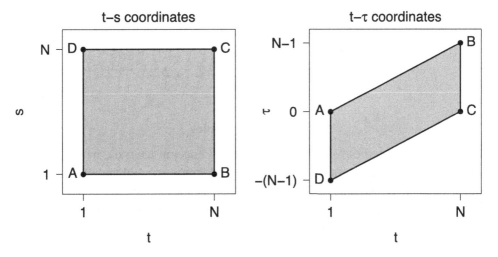

Figure 6.12 Summation domain (shaded) in $t-s$ coordinates and $t-\tau$ coordinates, where $\tau = t - s$. The square ABCD in $t - s$ coordinates is mapped into the parallelogram ABCD in $t - \tau$ coordinates.

$$\sum_{t=1}^{N}\sum_{s=1}^{N} f(t,s) = \sum_{\tau=-(N-1)}^{0}\sum_{t=1}^{\tau+N} f(t,t-\tau) + \sum_{\tau=1}^{N-1}\sum_{t=\tau+1}^{N} f(t,t-\tau)$$

$$= \sum_{\tau=-(N-1)}^{0}\sum_{t=1}^{N-|\tau|} f(t,t+|\tau|) + \sum_{\tau=1}^{N-1}\sum_{t=\tau+1}^{N} f(t,t-\tau)$$

By another change of variable $u = t - \tau$, the last sum can be re-written as

$$\sum_{t=\tau+1}^{N} f(t,t-\tau) = \sum_{u=1}^{N-\tau} f(u+\tau,u).$$

Since u is a dummy variable, we can replace it with t:

$$\sum_{t=1}^{N}\sum_{t=1}^{N} f(t,s) = \sum_{\tau=-(N-1)}^{0}\sum_{t=1}^{N-|\tau|} f(t,t+|\tau|) + \sum_{\tau=1}^{N-1}\sum_{t=1}^{N-\tau} f(t+\tau,t).$$

Note that the function $f(t,s)$ is symmetric with respect to the two arguments (that is, $f(t,s) = f(s,t)$). Therefore,

$$\sum_{t=1}^{N}\sum_{t=1}^{N} f(t,s) = \sum_{\tau=-(N-1)}^{0}\sum_{t=1}^{N-|\tau|} f(t+|\tau|,t) + \sum_{\tau=1}^{N-1}\sum_{t=1}^{N-|\tau|} f(t+|\tau|,t).$$

The innermost sums on the right-hand side are identical. Moreover,

$$\frac{1}{N}\sum_{t=1}^{N-|\tau|} f(t+|\tau|,t) = \hat{c}(\tau)\cos(\omega\tau).$$

Therefore, we have shown that

$$\hat{p}(\omega) = \sum_{\tau=-(N-1)}^{N-1} \hat{c}(\tau)\cos(\omega\tau) = \hat{c}(0) + 2\sum_{\tau=1}^{N-1} \hat{c}(\tau)\cos(\omega\tau).$$

6.12 Further Topics

This chapter covered only the bare essentials of spectral analysis, namely power spectra, aliasing, estimation based on the periodogram, spectral smoothing, bandwidth, and confidence intervals. These concepts lay a foundation for understanding more advanced concepts in spectral analysis. One should recognize that there is no unique Fourier Transform, since any transformed quantity can be multiplied by a scalar as long as the scalar is divided out before the final inverse transform is computed, so different texts may define the transform in different ways (usually involving different factors of 2 or N or π). Therefore, the reader should beware that the equations appearing in this chapter may appear in different forms in different texts because of the different definitions of the Fourier Transform.

These techniques are *nonparameteric* because they make no assumption about the functional form of the spectrum, aside from the fact that it varies smoothly with frequency (so that neighboring frequencies can be pooled). If, however, the power spectra is known to have a specific functional form, then it is often more accurate to estimate the few parameters directly from data and then generate the power spectra by substituting those estimates into the function. For instance, Chapter 5 reviewed the first-order autoregressive model AR(1). If the process is AR(1), then one could estimate the AR(1) parameters ϕ_1 and σ_W^2 and then substitute them into the equation for the power spectra of an AR(1) model. This procedure can be generalized to AR(p) models (and ARMA(p,q) models). This method is known as *rational power spectrum* estimation, because ARMA models have power spectra that can be written as a rational polynomial function of $\exp(i\omega)$. Parametric methods are superior to nonparametric methods when the assumed functional form is close to correct, and can resolve narrow peaks that are too close to resolve by nonparametric methods.

Spectral analysis is especially useful in analyzing time-invariant linear models. This usefulness stems from the fact that the Fourier Transform possesses remarkable properties in such systems. One property is the *convolution theorem*. If $y(t)$ is governed by a linear model, then the response to forcing $f(t)$ can be written as a convolution integral

$$y(t) = \int_{-\infty}^{\infty} h(s)f(t-s)ds, \tag{6.85}$$

where $h(s)$ is the *transfer function*. The linear filter (6.53) is a finite discretization of the convolution integral. The Fourier Transform of the integral convolution is

$$Y(\omega) = H(\omega)F(\omega), \tag{6.86}$$

where $Y(\omega), H(\omega), F(\omega)$ are the Fourier Transforms of $y(t), h(t), f(t)$. The Fourier Transform considerably simplifies the analysis of time-invariant linear models. For instance, the transfer function can be estimated directly from spectra and then inverted to obtain the transfer function in the time domain. In contrast, statistically estimating the transfer function in the time domain using the convolution integral is difficult. Moreover, the solution (6.86) has the form of a linear model, and therefore many concepts in linear regression can be brought to bear. For instance, the transfer function can be interpreted as a frequency-dependent *regression coefficient* that can be estimated for each frequency separately. Furthermore, the degree of relation between response and forcing can be quantified by the *coherency spectrum*, which is analogous to a correlation coefficient in the time domain. The coherency spectrum was the basis for discovering the Madden–Julian Oscillation (MJO) (Madden and

Julian, 1994). Virtually the entire apparatus of linear regression theory can be used in the spectral domain to test hypotheses about spectral relations between variables. Jenkins and Watts (1968) and Priestley (1981) are recommended for additional details.

Estimation of power spectra occupies a major part of spectral analysis. This chapter considered only one method of estimation, namely a version of the *Daniell* estimator. This estimator is essentially a running average of the periodogram, where the weights used in the averaging window are equal. Such a set of weights are called a *rectangular* spectral window. Running averages of the periodogram are effectively discrete convolution integrals, hence the convolution theorem implies that such averaging is equivalent to computing the spectrum as a weighted sum of autocovariances:

$$\hat{p}(\omega_j) = \hat{c}(0) + 2 \sum_{\tau=1}^{N-1} w_\tau \hat{c}(\tau) \cos[\omega_j \tau], \qquad (6.87)$$

where w_τ, the *lag window*, is the Fourier Transform of the spectral window. The use of lag windows to estimate power spectra is often known as the *Blackmon–Tukey method*, but this method is equivalent to smoothing the periodogram owing to the convolution theorem. Unfortunately, the rectangular spectral window is discontinuous at the edges. Discontinuous functions produce Fourier Transforms that oscillate over relatively wide intervals, which is undesirable. For instance, the Fourier Transform of the rectangular window is the *sinc function* $w_\tau = \sin(\tau)/\tau$, which oscillates about zero (and therefore becomes negative). It is natural, then, to consider alternative spectral windows that moderate sudden changes, examples of which include the *triangular, Parzen, Welch, Hamming,* and *Hanning* windows. Unfortunately, these alternative spectral windows have nonuniform weights and so the distribution of the spectral estimates no longer exactly follows a chi-squared distribution. In practice, the spectral estimates have approximate chi-squared distributions; the scaling and degrees of freedom are determined by equating the first two moments of the spectral estimates with those of the chi-squared distribution. Jenkins and Watts (1968) and Priestley (1981) are highly recommended for more details.

Generalizing spectral analysis to multiple variables leads to the concept of *cross spectra*. Unlike power spectra, cross spectra are noneven functions of frequency and hence negative frequencies carry information. Moreover, power spectra are characterized by the *spectral covariance matrix*. Many linear regression techniques extend naturally to the multivariate case. For instance, *multiple coherence* generalizes the concept of multiple correlation to the spectral domain; the transfer function becomes a vector or matrix, depending on the number of outputs; discrimination, cluster analysis, and principal component analysis can also be generalized to the spectral domain (Shumway and Stoffer, 2006).

As mentioned earlier, complex variables offer numerous advantages in spectral analysis, so a serious user of spectral analysis will have to become familiar with complex variables.

6.13 Conceptual Questions

1 What is the relation between power spectra and variance?
2 What is the relation between the power spectrum and autocovariance function?
3 What is the power spectrum of white noise? Is the Gaussian assumption necessary?
4 What characteristics of a time series can you learn from the shape of a power spectrum?
5 What is the Nyquist frequency? Why is it important?
6 What is the periodogram?
7 Why doesn't more data produce a smoother periodogram?
8 Why should you smooth a periodogram?
9 What does the bandwidth refer to in an estimated power spectrum?
10 What is the relation between confidence intervals and bandwidth?
11 What happens to the estimated power spectrum if you superimpose a trend or discontinuity on a time series?
12 If a time series is run through a linear filter, how are the power spectra of the input and output related?

7

Introduction to Multivariate Methods

> Most problems in scientific computing eventually lead to solving a matrix equation.[1]
>
> *Gene Golub*

In climate science, many scientific questions lead to hypotheses about random vectors. For instance, the question of whether global warming has occurred over a geographic region is a question about whether temperature has changed *at each spatial location* within the region. One approach to addressing such a question is to apply a univariate test to each location *separately* and then use the results collectively to make a decision. This approach is known as *multiple testing or multiple comparisons* and is common in genomics for analyzing gene expressions. The disadvantage of this approach is that it does not fully account for correlation between variables. Multivariate techniques provide a framework for hypothesis testing that takes into account correlations between variables. Although multivariate tests are more comprehensive, they require estimating more parameters and therefore have low power when the number of variables is large. This chapter lays a foundation for understanding multivariate statistical analysis. This foundation draws heavily from linear algebra concepts, including vector, matrix, rank, linear independence, and orthogonal transformations. This foundation also includes a generalization of the normal distribution, called the multivariate normal distribution, whose population parameters are the *mean vector* and the *covariance matrix*. A particularly important tool in multivariate statistics is the linear transformation. Linear transformations are used for a number of purposes, including to (1) filter out unwanted noise, (2) remove correlations between variables, or (3) reduce the dimension of the analysis space. Many of these concepts are illustrated by generalizing the t-test to test equality of mean vectors.

---

[1] Reprinted by permission from American Mathematical Society, *777 Mathematical Conversation Starters*. dePillis (2002).

7.1 The Problem

Many scientific questions involve *multivariate* hypotheses; that is, hypotheses about two or more population parameters. As an example, a climate scientist might suspect that temperature over a *geographic region* has warmed over the twentieth century. Temperature in a geographic region is often specified by values at a discrete set of points on a grid. In such cases, the natural null hypothesis H_0 is that the population mean *at each grid point* is constant in time. Thus, if the region contains 100 grid points, then H_0 asserts that 100 population means are constant in time. Suppose the scientist applies the t-test to each grid point, one at a time, to decide if the means over 30-year periods at the beginning and end of the century are equal. If the t-test rejects the hypothesis at one location, should H_0 be rejected? No! The problem is that the *t*-test was performed one-at-a-time, and then the most significant result among the tests was selected. This process is known as *screening* and leads to a false impression of significance if not taken into account (see Chapter 11). In climate applications, there is an additional complication that variables are correlated in space. It is instructive to temporarily assume that samples from different spatial locations are independent. In this case, if H_0 were true, then the probability of at least one rejection at the 5% significance level is

$$P(\text{at least one rejection})$$
$$= 1 - P(\text{all 100 tests accept})$$
$$= 1 - P(\text{test 1 accepts})P(\text{test 2 accepts})\ldots P(\text{test 100 accepts})$$
$$= 1 - P(\text{a test accepts})^{100}$$
$$= 1 - (0.95)^{100} \approx 99.4\%.$$

In other words, at least one of the tests is virtually certain to reject. This probability is known as the family-wise error rate (FWER) and it is discussed further in Chapter 11. This result should be no surprise – the probability of type-I error is 5%, thus we *expect* an average of 5 false rejections out of 100 independent tests.

A multivariate approach to this hypothesis is the following. Let t_i denote the t-statistic at the i'th station. Under H_0 and for large sample size, $t_i \sim \mathcal{N}(0,1)$, which is equivalent to $t_i^2 \sim \chi_1^2$ for each $i = 1, \ldots, 100$ (see Theorem 1.3). If the variables are independent (again, a very unrealistic assumption), then

$$t_1^2 + t_2^2 + \cdots + t_{100}^2 \sim \chi_{100}^2, \quad \text{if } H_0 \text{ is true and the } t_i\text{'s are independent.}$$

This sampling distribution provides a way to test the multivariate hypothesis H_0. Specifically, one can compute the *sum* $t_1^2 + \cdots + t_{100}^2$, and then compare that *single* value to the 5% critical value of the appropriate chi-squared distribution.

Conceivably, the multivariate test could reject H_0 even if none of the 100 *individual* t-tests reject the null hypothesis. After all, the 5% critical value for χ_{100}^2 is 124, hence H_0 would be rejected if, say, each t-squared statistic slightly exceeded 1.24.

In contrast, for an individual station, the t-squared statistic would need to exceed $1.96^2 \approx 3.84$ to reject equality of means. This example illustrates that *separate* tests may not justify rejecting H_0, but several weak results *pointing in the same direction* justify rejecting H_0.

This example assumed that samples from different stations were independent. This assumption is unrealistic – temperature in one city tends to be correlated with temperature in a neighboring city. Such correlations need to be taken into account to obtain meaningful conclusions. After all, if the variables are highly correlated, then a "false positive" at one location would lead to false positives at the neighboring locations, creating a false impression of significance. Multivariate analysis provides a framework for accounting for correlations between variables when testing hypotheses. One approach is to exploit the fact that any set of correlated variables can be transformed into a set of uncorrelated variables. Therefore, if the variables are jointly Gaussian, then such transformation produces a new set of variables that are independent, after which the previously described procedure can be applied to test the multivariate hypothesis H_0. This approach requires knowing the population covariance between every pair of stations. In practice, these covariances are unknown, but this requirement is not as restrictive as it might seem. In particular, the uncertainty in estimating these covariances can be taken into account, which is the basis of *Hotelling's T^2-test* (see Section 7.11).

Merely rejecting a multivariate hypothesis does not tell us which features in the data contributed to this rejection. Often, multivariate tests identify linear combinations of variables that explain how the variables contribute jointly to reject the hypothesis. Such transformations are the focus of multivariate tests in subsequent chapters.

Given the advantages of multivariate hypothesis testing, why not *always* use them? The problem is that multivariate tests require estimating more parameters than univariate tests. Specifically, the dependence between each pair of variables needs to be taken into account, hence the number of population parameters increases dramatically with the number of variables under investigation. As a result, multivariate tests rapidly lose power as the number of variables increases. In climate studies, a standard approach to mitigating this effect is to project the data onto a lower-dimensional space and then perform the test in this space. This strategy is discussed in more detail in the individual chapters in which it is used.

This chapter lays a foundation for multivariate statistical analysis. This foundation relies heavily on linear algebra. There exist numerous excellent texts on linear algebra on which we could hardly improve. Accordingly, the key results that are needed for the rest of the text are merely summarized, and some of the elementary proofs are outlined. This review covers vectors, matrices, linear independence, matrix rank, and orthogonal matrices. At the end of the chapter, these concepts are brought to bear to demonstrate that any set of correlated variables can be transformed into a set of uncorrelated variables. This result is then used to illustrate how univariate tests can be generalized to multivariate hypotheses. Specifically, we generalize the t-test into a multivariate t-test, called Hotelling's T^2-test

Notation

The notation used in previous chapters needs to be modified to be less cumbersome in a multivariate context. Accordingly, a scalar will be denoted by lower case italic, regardless of whether the variable is random or a particular realization. Bold lower case letters denote vectors, and bold upper case letters denote matrices (defined shortly). Thus, x is a scalar, $\mathbf{x}$ is a vector, and $\mathbf{X}$ is a matrix.

7.2 Vectors

Consider M variables $x_1, x_2, \ldots, x_M$. These variables can be collected into a *column vector*, which is an *ordered* set of variables denoted

$$\mathbf{x} = \begin{pmatrix} x_1 \\ x_2 \\ \vdots \\ x_M \end{pmatrix}. \tag{7.1}$$

The number of elements is called the *dimension* of the vector. $\mathbf{x}$ is said to be an M-dimensional vector. The vector concept is very flexible: the vector could describe M different variables, a single variable observed at M separate times (e.g., a time series) or M separate geographic locations (e.g., a spatial field), or any combination thereof.

The elements of an M-dimensional vector can be interpreted as defining the coordinates of a *point* in an M-dimensional space. This perspective has already been used in generating scatter plots, where each pair of values represents a point in 2-dimensional space. However, the vector concept is more flexible and can describe the same data set in different ways. For instance, instead of a scatter plot, all the x-values could be collected into a single M-dimensional vector $\mathbf{x}$. To visualize this vector, we connect a line from the origin to the point $\mathbf{x}$ to create a *geometric vector*. Then, new realizations of $\mathbf{x}$ correspond to new realizations of this geometric vector. One basic attribute of this vector is its length. The *Euclidean length* of the vector $\mathbf{x}$ is denoted $\|\mathbf{x}\|$ and follows from Pythagoras' theorem:

$$\|\mathbf{x}\| = (x_1^2 + x_2^2 + \cdots + x_M^2)^{1/2}. \tag{7.2}$$

The Euclidean length is also known as *the norm* of $\mathbf{x}$ (or the L^2-norm, to distinguish it from other vector norms). A *unit vector* is a vector of unit length. The set of points $\mathbf{x}$ that satisfy $\|\mathbf{x}\| = 1$ also satisfy the equation $x_1^2 + x_2^2 + \cdots + x_M^2 = 1$, which defines a sphere of unit radius. Thus, $\|\mathbf{x}\| = k$ defines a sphere of radius k in M-dimensional space.

Now consider two M-dimensional vectors $\mathbf{x}$ and $\mathbf{y}$. A basic attribute of two vectors is the angle between them. This angle can be written in terms of the inner product.

Inner Product

The *inner product* (also known as the *dot product*) of two vectors $\mathbf{x}$ and $\mathbf{y}$ is defined as $\mathbf{x} \cdot \mathbf{y} = x_1 y_1 + x_2 y_2 + \cdots + x_M y_M$. The length of a vector can be denoted equivalently as $\|\mathbf{x}\| = (\mathbf{x} \cdot \mathbf{x})^{1/2}$.

Angle

The angle between two nonzero vectors $\mathbf{x}$ and $\mathbf{y}$ is given by

$$\cos \theta = \frac{\mathbf{x} \cdot \mathbf{y}}{\|x\| \, \|y\|}. \tag{7.3}$$

Orthogonal

Two vectors $\mathbf{x}$ and $\mathbf{y}$ are *orthogonal* if the cosine of their angle vanishes. Equivalently, two vectors are orthogonal if their inner product vanishes; that is, $\mathbf{x} \cdot \mathbf{y} = 0$. Vectors that are orthogonal *and normalized to unit length* are said to be *orthonormal*.

7.3 The Linear Transformation

The linear transformation is one of the most important tools in multivariate statistics. Linear transformations are often used to highlight something interesting about the data, filter out unwanted noise, or remove correlations between variables. As a simple example, suppose S is a "signal" that we would like to isolate from the observations Z_1 and Z_2, which are related to the signal as

$$Z_1 = S + N \quad \text{and} \quad Z_2 = N, \tag{7.4}$$

where N is unwanted "noise." Then, the linear combination $Z_1 - Z_2 = S$ removes the noise and extracts the signal. Practical filtering problems are more complex, but the concept is the same – combine variables to reduce the noisy components.

Linear transformations are built up from a few simple operations. One such operation is to multiply each vector element by a scalar. This operation is denoted as

$$a\mathbf{x} = \begin{pmatrix} ax_1 \\ ax_2 \\ \vdots \\ ax_M \end{pmatrix}. \tag{7.5}$$

Geometrically, scalar multiplication of a vector corresponds to stretching or contracting the vector without changing the direction of the vector, as illustrated in

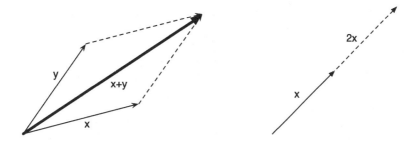

Figure 7.1 Illustration of summation (left) and scalar multiplication (right) of vectors.

Figure 7.1. Another operation is to add two vectors, which is performed element-by-element:

$$\mathbf{x} + \mathbf{y} = \begin{pmatrix} x_1 + y_1 \\ x_2 + y_2 \\ \vdots \\ x_M + y_M \end{pmatrix}. \tag{7.6}$$

Geometrically, addition of vectors follows the familiar rule in physics in which one of the vectors is translated without changing its orientation such that its "tail" rests on the "head" of the other vector. This procedure is illustrated in Figure 7.1.

The transpose of a *column vector* is a *row vector* (or vice versa) and is indicated by

$$\begin{pmatrix} x_1 \\ x_2 \\ \vdots \\ x_M \end{pmatrix}^T = \begin{pmatrix} x_1 & x_2 & \cdots & x_M \end{pmatrix}. \tag{7.7}$$

The product of a row vector and a column vector is the inner product:

$$\mathbf{x}^T \mathbf{y} = \mathbf{x} \cdot \mathbf{y} = x_1 y_1 + \cdots + x_M y_M. \tag{7.8}$$

Notation
Hereafter, inner products will be denoted using the transpose rather than by the dot. For instance, Euclidean length will be denoted $\|\mathbf{x}\| = (\mathbf{x}^T \mathbf{x})^{1/2}$.

Example 7.1 (Mean, Variance, Correlation in Terms of Inner Products) *Many sample estimates can be expressed through the above notation. For instance, the sample mean is*

$$\hat{\mu}_X = \frac{1}{M} \sum_{m=1}^{M} x_m = \frac{1}{M} \begin{pmatrix} 1 & 1 & \cdots & 1 \end{pmatrix} \begin{pmatrix} x_1 \\ x_2 \\ \vdots \\ x_M \end{pmatrix}. \tag{7.9}$$

If we define the vector **j** as

$$\mathbf{j} = \begin{pmatrix} 1 \\ 1 \\ \vdots \\ 1 \end{pmatrix}, \tag{7.10}$$

then the sample mean can be written more concisely as

$$\hat{\mu}_X = \frac{1}{M}\mathbf{j}^T\mathbf{x}. \tag{7.11}$$

The number of elements in **j** is understood from context, otherwise it is explicitly indicated as a subscript; e.g., $\mathbf{j}_M$ is a vector of M ones. A centered variable is a variable whose sample mean has been subtracted; that is, $\tilde{x}_n = x_n - \hat{\mu}_X$ (see Chapter 1). In vector notation, a centered variable can be written as

$$\begin{pmatrix} \tilde{x}_1 \\ \tilde{x}_2 \\ \vdots \\ \tilde{x}_M \end{pmatrix} = \begin{pmatrix} x_1 \\ x_2 \\ \vdots \\ x_M \end{pmatrix} - \begin{pmatrix} \hat{\mu}_X \\ \hat{\mu}_X \\ \vdots \\ \hat{\mu}_X \end{pmatrix}, \tag{7.12}$$

which can be written more concisely as

$$\tilde{\mathbf{x}} = \mathbf{x} - \mathbf{j}\hat{\mu}_X. \tag{7.13}$$

Because $\tilde{\mathbf{x}}$ is centered, the sum of its elements vanishes: $\mathbf{j}^T\tilde{\mathbf{x}} = 0$. Using this notation, the sample variance of **x** is proportional to its squared length,

$$\hat{\sigma}_X^2 = \frac{\|\tilde{\mathbf{x}}\|^2}{M-1}, \tag{7.14}$$

and the sample covariance of **x** and **y** is proportional to their inner product,

$$\hat{c}_{XY} = \frac{\tilde{\mathbf{x}}^T\tilde{\mathbf{y}}}{M-1}. \tag{7.15}$$

These results imply that the correlation coefficient between **x** and **y** is

$$\hat{\rho}_{XY} = \frac{\hat{c}_{XY}}{\sqrt{\hat{\sigma}_X^2\hat{\sigma}_Y^2}} = \frac{\tilde{\mathbf{x}}^T\tilde{\mathbf{y}}}{\|\tilde{\mathbf{x}}\|\,\|\tilde{\mathbf{y}}\|}. \tag{7.16}$$

Comparison with (7.3) reveals that the correlation coefficient is the cosine of the angle between the geometric vectors $\tilde{\mathbf{x}}$ and $\tilde{\mathbf{y}}$.

Given an arbitrary number of vectors $\mathbf{a}_1, \mathbf{a}_2, \ldots, \mathbf{a}_M$, each multiplied by an arbitrary scalar $x_1, x_2, \ldots, x_M$, a general linear transformation is

$$\mathbf{x}' = x_1\mathbf{a}_1 + x_2\mathbf{a}_2 + \cdots + x_M\mathbf{a}_M. \tag{7.17}$$

We will express Equation (7.17) more concisely using *matrix-vector multiplication*

$$\mathbf{x}' = \mathbf{A}\mathbf{x}, \tag{7.18}$$

where **x** is defined in (7.1) and **A** is a *matrix* defined as

$$\mathbf{A} = \begin{bmatrix} \mathbf{a}_1 & \mathbf{a}_2 & \cdots & \mathbf{a}_M \end{bmatrix} = \begin{pmatrix} a_{11} & a_{12} & \cdots & a_{1M} \\ a_{21} & a_{22} & \cdots & a_{2M} \\ \vdots & \vdots & \ddots & \vdots \\ a_{N1} & a_{N2} & \cdots & a_{NM} \end{pmatrix}. \tag{7.19}$$

We call **x** the *input*, **x**' the *output*, and **A** the *transformation matrix*. A matrix is denoted by a bold capital letter, whereas an individual element is denoted by a lower case letter with a subscript indicating the row and column. Matrix-vector multiplication satisfies

$$\mathbf{A}\,(b\mathbf{x} + c\mathbf{y}) = b\mathbf{A}\mathbf{x} + c\mathbf{A}\mathbf{y}, \tag{7.20}$$

where b, c are scalars. This identity is the defining property of a *linear transformation*.

Equation (7.17) highlights that $\mathbf{A}\mathbf{x}$ is defined only when the dimension of **x** equals the number of columns of **A**. Vectors $\mathbf{a}_1, \ldots, \mathbf{a}_M$ are known as the *column vectors* of **A**. Transformation (7.18) transforms an M-dimensional vector **x** to an N-dimensional vector **x**'. The full set of equations in (7.18) is

$$x_1' = a_{11}x_1 + a_{12}x_2 + \cdots + a_{1M}x_M$$
$$x_2' = a_{21}x_1 + a_{22}x_2 + \cdots + a_{2M}x_M$$
$$\vdots$$
$$x_N' = a_{N1}x_1 + a_{N2}x_2 + \cdots + a_{NM}x_M.$$

In order for this set of equations to be consistent with the matrix notation in (7.18), the matrix-vector product $\mathbf{x}' = \mathbf{A}\mathbf{x}$ follows the rule

$$x_n' = \sum_{m=1}^{M} a_{nm}x_m \quad \text{for } n = 1, 2, \ldots, N. \tag{7.21}$$

Matrix **A** (7.19) is said to have dimension $N \times M$. If $M = N$, then **A** is said to be *square*.

7.4 Linear Independence

In climate applications, data often comes on a spatial grid with thousands of points, which can be overwhelming. To highlight important features, the data is often transformed into another vector of smaller dimension which is easier to interpret. As an example, the Niño 3.4 index is the output of a transformation whose input is sea surface temperatures (SSTs) on a globe. When transforming a high-dimensional vector to a low-dimensional one, some information will be lost, which is the price of simplifying the data.

To illustrate this concept, consider the following example:

$$\begin{bmatrix} 1 & 2 & -1 \\ 5 & 2 & 3 \end{bmatrix} \begin{bmatrix} 1 \\ 1 \\ 1 \end{bmatrix} = \begin{bmatrix} 2 \\ 10 \end{bmatrix} \quad \text{and} \quad \begin{bmatrix} 1 & 2 & -1 \\ 5 & 2 & 3 \end{bmatrix} \begin{bmatrix} 2 \\ 0 \\ 0 \end{bmatrix} = \begin{bmatrix} 2 \\ 10 \end{bmatrix}. \tag{7.22}$$

This example shows that $\mathbf{Ax} = \mathbf{Ay}$ even though $\mathbf{x} \neq \mathbf{y}$, where

$$\mathbf{A} = \begin{bmatrix} 1 & 2 & -1 \\ 5 & 2 & 3 \end{bmatrix}. \tag{7.23}$$

Equivalently, $\mathbf{A}(\mathbf{x} - \mathbf{y}) = \mathbf{0}$ for some $\mathbf{x} - \mathbf{y} \neq \mathbf{0}$. It is readily verified that the vector

$$\mathbf{n}_0 = \begin{pmatrix} 1 \\ -1 \\ -1 \end{pmatrix}, \tag{7.24}$$

satisfies $\mathbf{An}_0 = \mathbf{0}$. Such a vector is known as a *null vector*. When a null vector $\mathbf{n}_0$ exists, then

$$\mathbf{A}\,(\mathbf{x} + \alpha\mathbf{n}_0) = \mathbf{Ax} \quad \text{for all } \alpha. \tag{7.25}$$

That is, the product of $\mathbf{A}$ and any vector $\mathbf{x} + \alpha\mathbf{n}_0$ is independent of α, hence knowing the product $\mathbf{Ax}$ does not uniquely specify $\mathbf{x}$. Thus, when $\mathbf{A}$ has a null vector, the transformation $\mathbf{x}' = \mathbf{Ax}$ cannot be inverted to recover $\mathbf{x}$.

Expressing $\mathbf{A}$ in terms of its column vectors, as in (7.19), condition $\mathbf{Ax} = \mathbf{0}$ becomes

$$x_1\mathbf{a}_1 + \cdots + x_M\mathbf{a}_M = 0. \tag{7.26}$$

The vectors $\mathbf{a}_1, \ldots, \mathbf{a}_M$ are said to be *linearly dependent* if this equation holds for some scalars $x_1, \ldots, x_M$, *not all zero*. In this case, one vector can be expressed as a linear combination of the others. In contrast, the vectors $\mathbf{a}_1, \ldots, \mathbf{a}_M$ are said to be *linearly independent* if no one of them can be written as a linear combination of the others. If the column vectors of $\mathbf{A}$ are linearly independent, then $\mathbf{Ax} = \mathbf{0}$ implies $\mathbf{x} = \mathbf{0}$.

Returning to the specific example of $\mathbf{A}$ in (7.23), we see that the column vectors of $\mathbf{A}$ are linearly dependent because

$$\begin{bmatrix} 1 \\ 5 \end{bmatrix} - \begin{bmatrix} 2 \\ 2 \end{bmatrix} - \begin{bmatrix} -1 \\ 3 \end{bmatrix} = \begin{bmatrix} 0 \\ 0 \end{bmatrix}. \tag{7.27}$$

This is just another way of writing $\mathbf{A}\mathbf{n}_0 = \mathbf{0}$. However, if the columns of $\mathbf{A}$ were linearly independent, then $\mathbf{A}\mathbf{x} = \mathbf{A}\mathbf{y}$ *does* imply $\mathbf{x} = \mathbf{y}$. This can be seen by subtracting $\mathbf{A}\mathbf{y}$ from both sides to give $\mathbf{A}(\mathbf{x} - \mathbf{y}) = \mathbf{0}$, in which case linear independence of the columns of $\mathbf{A}$ imply $\mathbf{x} - \mathbf{y} = \mathbf{0}$, and therefore $\mathbf{x} = \mathbf{y}$. When transforming from high to low dimensions, the column vectors of the transformation matrix are linearly dependent.

These concepts have a geometric interpretation. The column vectors $\mathbf{a}_1, \ldots, \mathbf{a}_M$ define a *coordinate system*, and the elements of $\mathbf{x}$ are the associated *coordinates*. The set of all linear combinations of $\mathbf{a}_1, \mathbf{a}_2, \ldots, \mathbf{a}_M$ is termed a *space*. The space is said to be *spanned* by $\mathbf{a}_1, \mathbf{a}_2, \ldots, \mathbf{a}_M$. The dimension of a space is not necessarily the dimension of the vectors $\mathbf{a}_1, \mathbf{a}_2, \ldots, \mathbf{a}_M$. To see this, consider two vectors $\mathbf{a}_1$ and $\mathbf{a}_2$. One might naturally assume that two vectors span a two-dimensional space. However, if the two vectors are parallel to each other, then they lie on a line and span only one dimension. Thus, the dimension of the space spanned by a group of vectors might be *less* than the number of vectors in the group. Vectors that are parallel to each other are known as *collinear*. If $\mathbf{a}_1$ and $\mathbf{a}_2$ are collinear, then one is proportional to the other, that is, $\mathbf{a}_1 = \alpha\mathbf{a}_2$ for some α. Equivalently, the two collinear vectors satisfy $\mathbf{a}_1 - \alpha\mathbf{a}_2 = \mathbf{0}$, which is the condition for linear dependence stated in (7.26). Thus, in two dimensions, *linear dependence corresponds to collinearity*. In contrast, if two vectors are linearly independent, then they lie in a two-dimensional space. In the case of three vectors, the vectors might all be parallel to each other, in which case they span only one dimension, or one of the vectors may span the same plane as the other two, in which case the three vectors span only two dimensions. Vectors that span a single plane are termed *coplanar*. If three vectors are coplanar, then one of them can be expressed as the sum of the other two, which means they are linearly dependent. Thus, in three dimensions, linear dependence corresponds to coplanarity or collinearity. In contrast, if three vectors are linearly independent, then they lie in three dimensions.

These considerations illustrate that the *dimension of a space equals the number of linearly independent vectors that span it*. A set of linearly independent vectors that span a space is known as a *basis*. Basis vectors are not unique, e.g., in two dimensions, each vector could be rotated by the same angle within that space and still constitute a basis for that space. The number of basis vectors that span a space is known as the *dimension* of the space.

Rank

If the vectors $\mathbf{a}_1, \ldots, \mathbf{a}_M$ comprise the column vectors of $\mathbf{A}$, then the dimension of the space spanned by the column vectors of $\mathbf{A}$ is known as the *rank* of $\mathbf{A}$. Equivalently, the rank is the number of linearly independent column vectors. The rank cannot exceed the minimum of the number of rows or columns; that is, the rank of an $N \times M$ matrix cannot exceed $\min(M, N)$. To see this, note that each column vector contains N rows and hence is embedded in an N-dimensional space. Thus, the dimension cannot be larger than N. On the other hand, there are M column vectors, so the dimension also cannot be larger than M. Thus, the smaller of these two numbers is an upper bound on the rank of a matrix.

- If the rank equals $\min(M, N)$, then the matrix is said to be *full rank*.
- If the rank is less than $\min(M, N)$, then the matrix is said to be *rank deficient*
- If $\mathbf{A}$ has rank K, then $\mathbf{A}$ is said to be a *rank-K matrix*.
- A not-obvious fact is that $\mathbf{A}^T$ and $\mathbf{A}$ have the same rank. That is, the number of linearly independent columns of $\mathbf{A}$ equals the number of linearly independent rows of $\mathbf{A}$.

> **Example 7.2** (Rank) *What is the rank of the matrix $\mathbf{A}$ in (7.23)? According to the last bullet point, rank can be decided by examining either the rows or the columns. It is easier to consider the rows since there are only two of them. For two vectors, linear dependence means one vector is a multiple of the other. Clearly, the first row of $\mathbf{A}$ is not a multiple of the second, hence the rows are linearly independent. Thus, the rank is 2; that is, the number of linearly independent columns of $\mathbf{A}$ is 2. It follows that the column vectors of $\mathbf{A}$ are linearly dependent, as was confirmed earlier by explicitly showing $\mathbf{A}\mathbf{n}_0 = \mathbf{0}$, where $\mathbf{n}_0$ is (7.24).*

7.5 Matrix Operations

After transforming a vector as $\mathbf{A}\mathbf{x} = \mathbf{x}'$, suppose the output $\mathbf{x}'$ is used as the input to a *second* transformation $\mathbf{B}\mathbf{x}' = \mathbf{x}''$. Eliminating $\mathbf{x}'$ in the two equations yields

$$\mathbf{x}'' = \mathbf{B}\mathbf{A}\mathbf{x}. \tag{7.28}$$

Two consecutive transformations are equivalent to a single transformation $\mathbf{x}'' = \mathbf{C}\mathbf{x}$, where

$$\mathbf{C} = \mathbf{B}\mathbf{A}, \tag{7.29}$$

is defined through the *matrix multiplication rule*:

$$(\mathbf{C})_{mk} = (\mathbf{B}\mathbf{A})_{mk} = \sum_n b_{mn}a_{nk}. \tag{7.30}$$

This definition requires that the number of columns in $\mathbf{B}$ equal the number of rows in $\mathbf{A}$, in which case the matrices are said to be *conformable*. In the matrix product (7.29), $\mathbf{C}$ has the same number of rows as $\mathbf{B}$, and the same number of columns as $\mathbf{A}$. The columns of $\mathbf{C}$ are found by applying $\mathbf{B}$ to the K columns of $\mathbf{A}$. Matrix multiplication is not generally commutative; that is, $\mathbf{B}\mathbf{A} \neq \mathbf{A}\mathbf{B}$.

Linear transformations satisfy the distributive law $\mathbf{A}\mathbf{x} + \mathbf{B}\mathbf{x} = (\mathbf{A} + \mathbf{B})\mathbf{x}$, which implies that addition of matrices is performed element-by-element:

$$A + B = \begin{pmatrix} a_{11} + b_{11} & a_{12} + b_{12} & \cdots & a_{1N} + b_{1N} \\ a_{21} + b_{21} & a_{22} + b_{22} & \cdots & a_{2N} + b_{2N} \\ \vdots & \vdots & \ddots & \vdots \\ a_{M1} + b_{M1} & a_{M2} + b_{M2} & \cdots & a_{MN} + b_{MN} \end{pmatrix}. \tag{7.31}$$

Similarly, scalar multiplication of matrices is performed element-by-element:

$$cA = \begin{pmatrix} ca_{11} & ca_{12} & \cdots & ca_{1N} \\ ca_{21} & ca_{22} & \cdots & ca_{2N} \\ \vdots & \vdots & \ddots & \vdots \\ ca_{M1} & ca_{M2} & \cdots & ca_{MN} \end{pmatrix}. \tag{7.32}$$

The transpose of a matrix is an operation in which rows and columns are exchanged. Thus, the transpose of the matrix A is

$$A^T = \begin{pmatrix} a_{11} & a_{21} & \cdots & a_{M1} \\ a_{12} & a_{22} & \cdots & a_{M2} \\ \vdots & \vdots & \ddots & \vdots \\ a_{1N} & a_{2N} & \cdots & a_{MN} \end{pmatrix}. \tag{7.33}$$

The definitions of transpose and matrix multiplication are consistent with the prior definitions of the transpose of a column vector and the dot product. If a matrix equals its transpose (that is, $A = A^T$), then the matrix is said to be *symmetric*.

Transpose of a Matrix Product

For any two conformable matrices X and Y, $(XY)^T = Y^T X^T$. Recursive application of this identity shows that the transpose of a product equals the product of the transposes *in reversed order*. For instance,

$$(XYZ)^T = Z^T Y^T X^T. \tag{7.34}$$

Also $(X + Y)^T = X^T + Y^T$.

Diagonal

A is *diagonal* if $a_{mn} = 0$ when $m \neq n$. The i'th *diagonal element* refers to the element a_{ii} of the matrix. A diagonal matrix can be specified by listing its diagonal elements, as in $\text{diag}(a_{11}, a_{22}, \ldots, a_{nn})$. The *off-diagonal elements* refer to elements that are not along the diagonal (that is, elements a_{mn} such that $m \neq n$). If A is a square diagonal matrix, then the transformation Ax corresponds to re-scaling individual elements of x. A diagonal matrix need not be square.

Identity

The *identity matrix* is a square matrix I such that $Ix = x$ for all vectors x. By considering a series of vectors $x_1, x_2, \ldots$, each with a single nonzero element, it is straightforward to prove that the identity matrix is a diagonal matrix of the form

$$\mathbf{I} = \begin{pmatrix} 1 & 0 & \cdots & 0 \\ 0 & 1 & \cdots & 0 \\ \vdots & \vdots & \ddots & \vdots \\ 0 & 0 & \cdots & 1 \end{pmatrix}. \tag{7.35}$$

The identity matrix can also be denoted $\mathrm{diag}(1, 1, \ldots, 1)$. Furthermore, $\mathbf{AI} = \mathbf{IA} = \mathbf{A}$ for any square matrix $\mathbf{A}$. The dimension of an identity matrix is understood from the fact that it is conformable with the matrices or vectors to which it is multiplied. If the dimension is not clear from context, then $\mathbf{I}_M$ is used to indicate an $M \times M$ identity matrix. The identity matrix elements can also be written using the *Kronecker delta* notation

$$(\mathbf{I})_{jk} = \delta_{jk} = \begin{cases} 0 & \text{if } j \neq k \\ 1 & \text{if } j = k \end{cases}. \tag{7.36}$$

Outer product

The outer product of two vectors $\mathbf{x}$ and $\mathbf{y}$ is denoted $\mathbf{xy}^T$ and is a *matrix*. The (i, j) element of this matrix is

$$\left(\mathbf{xy}^T\right)_{ij} = x_i y_j \quad \text{which is consistent with matrix multiplication in 7.30.} \tag{7.37}$$

Note that the outer product combines two vectors to form a matrix. If $\mathbf{x}$ and $\mathbf{y}$ are N and M dimensional vectors, respectively, then the outer product $\mathbf{xy}^T$ is an $N \times M$ matrix. Note that each column of $\mathbf{xy}^T$ is proportional to $\mathbf{x}$, hence the outer product is a rank-1 matrix.

7.6 Invertible Transformations

Statistical analysis often requires transforming back and forth between spaces, which raises questions about when such transformations are reversible. If the dimensions of the input and output spaces differ, then the transformation cannot be perfectly reversible, since some information is lost going from high to low dimensions. However, even if the input and output dimensions are the same, the transformation still may not be perfectly reversible. Conditions for a transformation to be reversible are discussed next.

To reverse the transformation $\mathbf{x}' = \mathbf{Ax}$, there must exist a matrix $\mathbf{L}$ such that

$$\mathbf{x} = \mathbf{Lx}' \quad \text{for all } \mathbf{x}. \tag{7.38}$$

Substituting the original transformation $\mathbf{x}' = \mathbf{Ax}$ into Equation (7.38) yields

$$\mathbf{LAx} = \mathbf{x}. \tag{7.39}$$

For this equation to hold for all $\mathbf{x}$, we must have

$$\mathbf{LA} = \mathbf{I}. \tag{7.40}$$

A matrix **L** with this property is known as a *left inverse* of **A**. A left inverse is a special case of a *pseudoinverse* (which will appear in later chapters). If the columns of **A** are linearly dependent, then **A** does not have left inverse. After all, in this case there exists a null vector $\mathbf{x} = \mathbf{n}_0$ that does not satisfy (7.39). It follows that for the left inverse to exist, the column vectors of **A** must be linearly independent.

An important class of transformations are those for which inputs and outputs have the same dimensions; that is, the transformation matrix is square.

Inverse

The inverse of a square matrix **A** is another square matrix $\mathbf{A}^{-1}$ satisfying

$$\mathbf{A}^{-1}\mathbf{A} = \mathbf{I} \quad \text{and} \quad \mathbf{A}\mathbf{A}^{-1} = \mathbf{I}. \tag{7.41}$$

The first condition is merely (7.40) with $\mathbf{L} = \mathbf{A}^{-1}$. If a matrix possesses a left inverse, it can be shown that the inverse *of a square matrix* automatically satisfies the second condition $\mathbf{A}\mathbf{A}^{-1} = \mathbf{I}$. The inverse of a matrix is unique when it exists. However, not all square matrices have an inverse. If the inverse of **A** exists, then **A** is said to be *invertible* or *nonsingular*. If **A** does not have an inverse, then **A** is said to be *singular*.

Theorem 7.1 *If the columns of **A** are linearly dependent, then **A** is singular.*

This theorem can be proven by contradiction. Suppose the columns of **A** are linearly dependent. Then, by definition, $\mathbf{A}\mathbf{x} = \mathbf{0}$ for some $\mathbf{x} \neq \mathbf{0}$. Now, assuming the inverse of **A** exists, it can be multiplied on the left to give

$$\mathbf{A}^{-1}\mathbf{A}\mathbf{x} = \mathbf{A}^{-1}\mathbf{0}$$

$$\mathbf{x} = \mathbf{0},$$

which contradicts the original assumption that $\mathbf{x} \neq \mathbf{0}$. It follows that a matrix with linearly dependent columns does not possess an inverse. Equivalently, if $\mathbf{A}\mathbf{x} = \mathbf{0}$ for some **x** not equal to the zero vector, then **A** has no inverse.

Theorem 7.1 tells us when **A** is *not* invertible, but it doesn't tell us when it *is* invertible.

Theorem 7.2 *A square matrix is invertible if and only if it has linearly independent columns, or equivalently, it is full rank.*

The proof of this theorem can be found in standard linear algebra books (Horn and Johnson, 1985; Noble and Daniel, 1988; Strang, 2016).

Properties of the inverse of a matrix

- For any two invertible matrices **X** and **Y** of the same dimensions, $(\mathbf{X}\mathbf{Y})^{-1} = \mathbf{Y}^{-1}\mathbf{X}^{-1}$.
- A diagonal matrix with a zero on the diagonal is singular.

- If $\mathbf{Ax} = \mathbf{0}$ for some $\mathbf{x} \neq \mathbf{0}$, then $\mathbf{A}$ is singular.
- If $\mathbf{A}$ is singular, $\mathbf{Ax} = \mathbf{Ay}$ may hold even if $\mathbf{x} \neq \mathbf{y}$.
- If the diagonal elements are nonzero, then the inverse of a diagonal matrix is:

$$\mathbf{A} = \begin{pmatrix} a_1 & 0 & \cdots & 0 \\ 0 & a_2 & \cdots & 0 \\ \vdots & \vdots & \ddots & \vdots \\ 0 & 0 & \cdots & a_M \end{pmatrix} \implies \mathbf{A}^{-1} = \begin{pmatrix} 1/a_1 & 0 & \cdots & 0 \\ 0 & 1/a_2 & \cdots & 0 \\ \vdots & \vdots & \ddots & \vdots \\ 0 & 0 & \cdots & 1/a_M \end{pmatrix} \tag{7.42}$$

Example 7.3 (Invertibility of a Product Does not Imply Invertibility of Individual Terms) *Certain multivariate analyses require inverting matrix products of the form* $\mathbf{AA}^T$, *hence we would like to know when such inversion is possible. As a numerical example, consider the matrix* $\mathbf{A}$ *in (7.23), which gives*

$$\mathbf{AA}^T = \begin{bmatrix} 6 & 6 \\ 6 & 38 \end{bmatrix}. \tag{7.43}$$

This matrix is invertible, even though $\mathbf{A}$ *itself is not. Importantly, one cannot write* $(\mathbf{AA}^T)^{-1} = (\mathbf{A}^{-1})^T \mathbf{A}^{-1}$ *because the inverse of* $\mathbf{A}$ *does not exist. Now consider the alternative product*

$$\mathbf{A}^T \mathbf{A} = \begin{bmatrix} 26 & 12 & 14 \\ 12 & 8 & 4 \\ 14 & 4 & 10 \end{bmatrix}. \tag{7.44}$$

This matrix is singular, since the first column equals the sum of the other two columns. Alternatively, recall that $\mathbf{An}_0 = \mathbf{0}$; *therefore,* $\mathbf{A}^T \mathbf{An}_0 = \mathbf{0}$. *This example illustrates the general fact that if* $\mathbf{A}$ *is a rectangular matrix, the products* $\mathbf{A}^T \mathbf{A}$ *and* $\mathbf{AA}^T$ *cannot both be invertible.*

7.7 Orthogonal Transformations

Sometimes we want a transformation to preserve certain attributes of the original vectors. Two basic attributes of vectors are their lengths, and the angles between them. A transformation from one space to the same space that preserves lengths and angles is known as an *orthogonal transformation*. Orthogonal transformations correspond to rigid rotations, reflections, or a combination thereof.

Suppose $\mathbf{A}$ is an orthogonal transformation. Then, if $\mathbf{x}' = \mathbf{Ax}$, the requirement that $\mathbf{x}$ and $\mathbf{x}'$ have the same length requires

$$\|\mathbf{x}'\|^2 = \|\mathbf{x}\|^2 \implies \mathbf{x}^T \mathbf{A}^T \mathbf{Ax} = \mathbf{x}^T \mathbf{x}. \tag{7.45}$$

If the transformation also preserves angles, then the angle between $\mathbf{x}$ and $\mathbf{y}$ equals the angle between $\mathbf{x}' = \mathbf{Ax}$ and $\mathbf{y}' = \mathbf{Ay}$, for all $\mathbf{x}$ and $\mathbf{y}$. According to (7.3), this requires

$$\frac{\mathbf{x}'^T\mathbf{y}'}{\|\mathbf{x}'\|\|\mathbf{y}'\|} = \frac{\mathbf{x}^T\mathbf{y}}{\|\mathbf{x}\|\|\mathbf{y}\|} \quad \Longrightarrow \quad \mathbf{x}^T\mathbf{A}^T\mathbf{A}\mathbf{y} = \mathbf{x}^T\mathbf{y}. \tag{7.46}$$

The last constraint shows that an orthogonal transformation *preserves the inner product*.

Theorem 7.3 *A matrix* **A** *that preserves the inner product between two vectors has orthonormal column vectors.*

To prove this theorem, note that condition (7.46) holds for all **x** and **y**, so they may be chosen arbitrarily. Accordingly, let **x** be a vector of zeroes except for a 1 at the j'th element, and let **y** be a vector of zeroes except for a 1 at the k'th element. Then, $\mathbf{x}^T\mathbf{A}^T\mathbf{A}\mathbf{y}$ extracts the (j,k) element of $\mathbf{A}^T\mathbf{A}$. Now, $j \neq k$ implies $\mathbf{x}^T\mathbf{y} = 0$, and (7.46) implies that all off-diagonal elements of $\mathbf{A}^T\mathbf{A}$ vanish. On the other hand, $j = k$ implies $\mathbf{x}^T\mathbf{y} = 1$ and (7.46) implies that all diagonal elements of $\mathbf{A}^T\mathbf{A}$ equal one. Consolidating these results,

$$\mathbf{A}^T\mathbf{A} = \mathbf{I}. \tag{7.47}$$

This equation can be written equivalently in terms of the column vectors (7.19) as

$$\mathbf{a}_j^T\mathbf{a}_k = \delta_{jk}. \tag{7.48}$$

Thus, the column vectors of **A** have unit length and are mutually orthogonal.

Theorem 7.4 *If* $\mathbf{a}_1, \mathbf{a}_2, \ldots, \mathbf{a}_M$ *are orthogonal, then they must be linearly independent.*

We prove this by contradiction. First, collect the vectors into the matrix **A**, as in (7.19). Then, assuming the column vectors are linearly dependent, $\mathbf{A}\mathbf{x} = \mathbf{0}$ for some vector $\mathbf{x} \neq \mathbf{0}$. Multiply both sides of this equation by $\mathbf{A}^T$, which gives $\mathbf{A}^T\mathbf{A}\mathbf{x} = \mathbf{0}$. Since the column vectors are orthogonal, $\mathbf{A}^T\mathbf{A} = \mathbf{I}$, which implies $\mathbf{x} = \mathbf{0}$, which contradicts the assumption that $\mathbf{x} \neq \mathbf{0}$. This proves that orthogonal vectors must be linearly independent.

Corollaries: A matrix with M orthogonal column vectors has rank M. The dimension of a space spanned by orthogonal vectors equals the number of vectors in the set.

A square matrix with orthonormal column vectors is known as an *orthogonal matrix*. It follows from Theorems 7.2 and 7.4 that an orthogonal matrix is full rank and invertible. Therefore, from (7.47),

$$\mathbf{A}^{-1} = \mathbf{A}^T \quad \text{if } \mathbf{A} \text{ is an orthogonal matrix.} \tag{7.49}$$

Thus, the inverse of an orthogonal matrix is simply the transpose of the matrix.

These results imply that if **A** is an orthogonal matrix, then the transformation $\mathbf{x}' = \mathbf{A}\mathbf{x}$ may be inverted very simply as

$$\mathbf{x} = \mathbf{A}^T\mathbf{x}'. \tag{7.50}$$

Furthermore, multiplying (7.49) on the left by $\mathbf{A}$ implies

$$\mathbf{I}_M = \mathbf{A}\mathbf{A}^T. \tag{7.51}$$

Expanding (7.51) in terms of column vectors leads to a sum of outer products:

$$\mathbf{A}\mathbf{A}^T = \begin{bmatrix} \mathbf{a}_1 & \cdots & \mathbf{a}_M \end{bmatrix} \begin{bmatrix} \mathbf{a}_1^T \\ \vdots \\ \mathbf{a}_M^T \end{bmatrix} = \mathbf{a}_1\mathbf{a}_1^T + \cdots + \mathbf{a}_M\mathbf{a}_M^T = \mathbf{I}_M. \tag{7.52}$$

A common application of orthogonal basis vectors is to project a vector into a lower-dimensional space. For instance, suppose the M-dimensional vector $\mathbf{x}$ is transformed into the K-dimensional vector $\tilde{\mathbf{x}}$ using $\tilde{\mathbf{x}} = \dot{\mathbf{A}}\mathbf{x}$, where $K < M$ and

$$\dot{\mathbf{A}} = \begin{bmatrix} \mathbf{a}_1 & \ldots & \mathbf{a}_K \end{bmatrix}. \tag{7.53}$$

Because the column vectors are orthogonal, the identity $\dot{\mathbf{A}}^T\dot{\mathbf{A}} = \mathbf{I}_K$ holds, which is analogous to (7.47). However, $\dot{\mathbf{A}}\dot{\mathbf{A}}^T \neq \mathbf{I}_M$, because the sum in (7.52) does not include all M terms. In this case, the column vectors of $\dot{\mathbf{A}}$ are said to be orthogonal but not *complete*.

7.8 Random Vectors

Now consider $\mathbf{x}$ as a vector of *random* variables $x_1, \ldots, x_M$. The distribution of a random vector is described by a *joint* distribution $p(x_1, \ldots x_M)$. The expectation of a random vector is simply the expectation of individual elements:

$$\boldsymbol{\mu}_X = \mathbb{E}[\mathbf{x}] = \begin{pmatrix} \mathbb{E}[x_1] \\ \vdots \\ \mathbb{E}[x_M] \end{pmatrix}. \tag{7.54}$$

Covariance Matrix

The multivariate generalization of covariance is the *covariance matrix*. The covariance matrix of $\mathbf{x}$ specifies the covariance between every pair of elements,

$$(\boldsymbol{\Sigma})_{ij} = \mathbb{E}[(x_i - \mathbb{E}[x_i])(x_j - \mathbb{E}[x_j])], \tag{7.55}$$

where $\boldsymbol{\Sigma}$ is capital sigma. Using the outer product, the covariance matrix can be written as

$$\text{cov}[\mathbf{x}] = \boldsymbol{\Sigma} = \mathbb{E}[(\mathbf{x} - \mathbb{E}[\mathbf{x}])(\mathbf{x} - \mathbb{E}[\mathbf{x}])^T]. \tag{7.56}$$

All covariance matrices are symmetric.

A natural estimate of μ_X is the sample mean

$$\hat{\mu}_X = \frac{1}{N} \sum_{n=1}^{N} \mathbf{x}_n. \tag{7.57}$$

A natural estimate of Σ_X is the *sample covariance matrix*

$$\hat{\Sigma}_X = \frac{1}{N-1} \sum_{n=1}^{N} \left(\mathbf{x}_n - \hat{\mu}_X \right) \left(\mathbf{x}_n - \hat{\mu}_X \right)^T. \tag{7.58}$$

Theorem 7.5 (Mean and Covariance of the Sample Mean) *Let the vectors* $\mathbf{x}_1, \mathbf{x}_2, \ldots, \mathbf{x}_N$ *be independent samples from a multivariate distribution with mean* μ *and covariance matrix* Σ. *Then the sample mean* $\hat{\mu}$ *has the following expectation and covariance matrix:*

$$\mathbb{E}[\hat{\mu}] = \mu \quad \text{and} \quad \text{cov}[\hat{\mu}] = \frac{\Sigma}{N}. \tag{7.59}$$

These are simply the multivariate generalizations of $\mathbb{E}[\hat{\mu}] = \mu$ *and* $\text{var}[\hat{\mu}] = \sigma^2/N$.

Proof

$$\mathbb{E}[\hat{\mu}] = \mathbb{E}\left[\frac{1}{N} \sum_{j=1}^{N} \mathbf{x}_j \right] = \frac{1}{N} \sum_{j=1}^{N} \mathbb{E}[\mathbf{x}_j] = \frac{1}{N} \sum_{j=1}^{N} \mu = \mu \tag{7.60}$$

$$\begin{aligned}
\text{cov}[\hat{\mu}] &= \mathbb{E}\left[(\hat{\mu} - \mu)(\hat{\mu} - \mu)^T \right] \\
&= \mathbb{E}\left[\left(\frac{1}{N} \sum_{j=1}^{N} (\mathbf{x}_j - \mu) \right) \left(\frac{1}{N} \sum_{k=1}^{N} (\mathbf{x}_k - \mu)^T \right) \right] \\
&= \frac{1}{N^2} \sum_{j=1}^{N} \sum_{k=1}^{N} \mathbb{E}\left[(\mathbf{x}_j - \mu)(\mathbf{x}_k - \mu)^T \right]
\end{aligned} \tag{7.61}$$

Because $\mathbf{x}_j$ and $\mathbf{x}_k$ are independent if $j \neq k$, the only nonzero terms in the summation are those for which $j = k$. That is, all the "cross terms" vanish, leaving

$$\text{cov}[\hat{\mu}] = \frac{1}{N^2} \sum_{j=1}^{N} \mathbb{E}\left[(\mathbf{x}_j - \mu)(\mathbf{x}_j - \mu)^T \right] = \frac{\Sigma}{N}. \qquad \square$$

Not all symmetric matrices can be covariance matrices. To see how covariance matrices are constrained, consider the linear combination of variables

$$z = w_1 x_1 + w_2 x_2 + \cdots + w_M x_M = \mathbf{w}^T \mathbf{x}, \tag{7.62}$$

where $\mathbf{w}$ is a constant vector and $\mathbf{x}$ is a random vector with mean μ and covariance Σ. Then,

$$\text{var}[z] = \text{var}\left[\sum_i w_i x_i\right] = \mathbb{E}\left[\left(\sum_i w_i(x_i - \mu_i)\right)^2\right] \tag{7.63}$$

$$= \mathbb{E}\left[\left(\sum_i w_i(x_i - \mu_i)\right)\left(\sum_j w_j(x_j - \mu_j)\right)\right] \tag{7.64}$$

$$= \sum_i \sum_j w_i w_j \mathbb{E}\left[(x_i - \mu_i)(x_j - \mu_j)\right] \tag{7.65}$$

$$= \sum_i \sum_j w_i w_j \left(\boldsymbol{\Sigma}\right)_{ij}. \tag{7.66}$$

Quadratic Form

Any expression of the form

$$\sum_i \sum_j x_i x_j A_{ij} = \mathbf{x}^T \mathbf{A} \mathbf{x} \tag{7.67}$$

is termed a *quadratic form*. Only square matrices can have quadratic forms. Since variance cannot be negative, it follows that

$$\text{var}[z] = \mathbf{w}^T \boldsymbol{\Sigma} \mathbf{w} \geq 0 \quad \text{for all } \mathbf{w}. \tag{7.68}$$

A matrix $\boldsymbol{\Sigma}$ that satisfies this inequality for all $\mathbf{w}$ is said to be *positive semi-definite*. Thus, all covariance matrices are positive semi-definite. If $\mathbf{w}^T \boldsymbol{\Sigma} \mathbf{w}$ is strictly positive for all $\mathbf{w}$, then the matrix $\boldsymbol{\Sigma}$ is said to be *positive definite*.

The diagonal elements of a positive definite matrix must be positive. This can be shown by choosing $\mathbf{x}$ to be the i'th column of the identity matrix, in which case $\mathbf{x}^T \boldsymbol{\Sigma} \mathbf{x} = \Sigma_{ii}^2$. Thus, if $\boldsymbol{\Sigma}$ is positive definite, each diagonal element is positive. Similarly, the diagonal elements of a positive semi-definite matrix must be nonnegative. However, a positive definite or semi-definite matrix *may have negative off-diagonal elements*.

Finally, a positive definite matrix is invertible. This can be proven by contradiction. Specifically, assume a positive definite $\boldsymbol{\Sigma}$ has linearly dependent column vectors. Then, there exists a non-zero vector $\mathbf{x}$ such that $\boldsymbol{\Sigma} \mathbf{x} = \mathbf{0}$. Therefore, $\mathbf{x}^T \boldsymbol{\Sigma} \mathbf{x} = 0$ and hence $\boldsymbol{\Sigma}$ is not positive definite, which contradicts the premise. So, $\boldsymbol{\Sigma}$ has linearly independent column vectors. Since $\boldsymbol{\Sigma}$ also is square, theorem 7.2 implies that $\boldsymbol{\Sigma}$ is invertible.

Theorem 7.6 (Mean and Covariance of Linear Transformations) *Consider the linear transformation*

$$\mathbf{y} = \mathbf{A} \mathbf{x}. \tag{7.69}$$

Then, the expectation of $\mathbf{y}$ is

$$\mathbb{E}[\mathbf{y}] = \mathbf{A} \mathbb{E}[\mathbf{x}]. \tag{7.70}$$

The covariance matrix of **y** *is*

$$\text{cov}[\mathbf{y}] = \mathbb{E}\left[(\mathbf{Ax} - \mathbf{A}\mathbb{E}[\mathbf{x}])(\mathbf{Ax} - \mathbf{A}\mathbb{E}[\mathbf{x}])^T\right] = \mathbf{A}\,\text{cov}[\mathbf{x}]\mathbf{A}^T. \tag{7.71}$$

Equivalently, using a more concise notation,

$$\boldsymbol{\mu}_Y = \mathbf{A}\boldsymbol{\mu}_X \quad and \quad \boldsymbol{\Sigma}_Y = \mathbf{A}\boldsymbol{\Sigma}_X\mathbf{A}^T. \tag{7.72}$$

These identities will be used repeatedly in subsequent chapters. For two linear transformations $\mathbf{x}' = \mathbf{Ax}$ *and* $\mathbf{y}' = \mathbf{By}$, *the covariance matrix between* $\mathbf{x}'$ *and* $\mathbf{y}'$ *is*

$$\boldsymbol{\Sigma}_{Y'X'} = \text{cov}[\mathbf{x}', \mathbf{y}'] = \text{cov}[\mathbf{Ax}, \mathbf{By}] = \mathbf{A}\,\text{cov}[\mathbf{x}, \mathbf{y}]\mathbf{B}^T = \mathbf{A}\boldsymbol{\Sigma}_{XY}\mathbf{B}. \tag{7.73}$$

7.9 Diagonalizing a Covariance Matrix

Variables that are *iid* have attractive properties that facilitate hypothesis testing. Importantly, if multivariate normal variables are not *iid*, then they may be transformed into *iid* variables. To see how this is done, note that if $x_1, \ldots, x_M$ are independent, then the covariance between any pair vanishes, and therefore the corresponding covariance matrix is *diagonal*. Thus, a transformation to uncorrelated variables is tantamount to a transformation to a diagonal covariance matrix. To find such a transformation, let **x** be a random vector with covariance matrix $\boldsymbol{\Sigma}_X$, and consider a transformation matrix $\mathbf{U}^T$ such that

$$\mathbf{x}' = \mathbf{U}^T\mathbf{x}. \tag{7.74}$$

We seek a transformation matrix that renders the covariance matrix of $\mathbf{x}'$ diagonal. If the diagonal elements are denoted $\lambda_1, \ldots, \lambda_M$, then

$$\text{cov}[\mathbf{x}'] = \text{diag}(\lambda_1, \ldots, \lambda_M) = \boldsymbol{\Lambda}, \tag{7.75}$$

where $\boldsymbol{\Lambda}$ is capital λ. By Theorem 7.6,

$$\text{cov}[\mathbf{x}'] = \text{cov}[\mathbf{U}^T\mathbf{x}] = \mathbf{U}^T\boldsymbol{\Sigma}_X\mathbf{U}. \tag{7.76}$$

Setting (7.75) and (7.76) equal to each other yields

$$\mathbf{U}^T\boldsymbol{\Sigma}_X\mathbf{U} = \boldsymbol{\Lambda}. \tag{7.77}$$

We are particularly interested in *orthogonal* transformation matrices, which satisfy $\mathbf{U}^{-1} = \mathbf{U}^T$ (see (7.49)). Orthogonal transformations satisfying (7.77) are said to *diagonalize* the covariance matrix. Assuming **U** is an orthogonal matrix, multiplying both sides of (7.76) on the left by **U** yields

$$\boldsymbol{\Sigma}_X\mathbf{U} = \mathbf{U}\boldsymbol{\Lambda}. \tag{7.78}$$

Let the column vectors of **U** be denoted as

$$\mathbf{U} = \begin{bmatrix} \mathbf{u}_1 & \cdots & \mathbf{u}_M \end{bmatrix}. \tag{7.79}$$

Then, writing (7.78) column by column yields the equations

$$\boldsymbol{\Sigma}_X \mathbf{u}_m = \lambda_m \mathbf{u}_m, \quad \text{for } m = 1, 2, \ldots, M. \tag{7.80}$$

An equation of this form is known as an *eigenvalue problem*. A nonzero vector $\mathbf{u}_m$ that satisfies this equation is known as an *eigenvector*, and the corresponding scalar λ_m is known as an *eigenvalue*. All mathematical packages (e.g., Lapack, Matlab, R, Python) contain functions for solving eigenvalue problems. If the matrix $\boldsymbol{\Sigma}_X$ is $M \times M$, then there can be at most M distinct eigenvalues. To each distinct eigenvalue corresponds a unique eigenvector. An eigenvalue may appear more than once, in which case it is known as a *degenerate eigenvalue*. For simplicity, we assume the population covariance matrix has distinct eigenvalues. The matrix $\mathbf{U}$ is known as the *eigenvector marix*, and the matrix $\boldsymbol{\Lambda}$ is known as the *eigenvalue matrix*.

Strictly speaking, this derivation merely showed that *if* an orthogonal transformation exists for diagonalizing $\boldsymbol{\Sigma}_X$, then it must satisfy the eigenvalue problem (7.80). The next theorem *guarantees* that such a transformation exists.

Theorem 7.7 (Eigenvectors and Eigenvalues of Symmetric Matrices) *If $\mathbf{B}$ is a symmetric $M \times M$ matrix, then $\mathbf{B}$ has exactly M real eigenvalues (not necessarily distinct), and M eigenvectors that form a complete, orthonormal set.*

Because all covariance matrices are symmetric, this theorem guarantees that the eigenvectors of $\boldsymbol{\Sigma}_X$ are orthogonal. However, this theorem does not specify whether the eigenvalues are positive or negative. The sign of the eigenvalues is important because eigenvalues correspond to the *variances* of the transformed variables, as indicated in (7.75), and thus should be nonnegative. The following theorem guarantees that the eigenvalues of a covariance matrix are nonnegative.

Theorem 7.8 (Eigenvalues of Positive Definite Matrices) *If $\mathbf{B}$ is symmetric, then $\mathbf{B}$ is positive definite if and only if all its eigenvalues are positive, and $\mathbf{B}$ is positive semi-definite if and only if all its eigenvalues are nonnegative.*

Since all covariance matrices are positive semi-definite, Theorem 7.8 implies that the eigenvalues of $\boldsymbol{\Sigma}_X$ are nonnegative.

Eigenvectors are defined *up to a multiplicative factor*. For instance, if $\mathbf{u}$ is an eigenvector satisfying (7.80), then so is $\alpha \mathbf{u}$ for any α. Thus, eigenvectors may be normalized to any length without loss of generality. Because we assume an orthogonal transformation, the eigenvectors are normalized to unit length; that is, $\mathbf{u}^T \mathbf{u} = 1$. After this normalization, the eigenvectors form an orthonormal set, and the eigenvector matrix $\mathbf{U}$ is an orthogonal matrix. For future reference, we note that (7.76) implies

$$\boldsymbol{\Sigma}_X = \mathbf{U} \boldsymbol{\Lambda} \mathbf{U}^T, \tag{7.81}$$

which is known as the *eigenvector decomposition of* $\boldsymbol{\Sigma}_X$.

These results guarantee that any set of variables can be transformed into an uncorrelated set of variables by an orthogonal transformation. However, the variances of the transformed variables $\lambda_1, \ldots, \lambda_M$ are not equal, hence they are not identically distributed. To obtain variables that have equal variances, we simply normalize each variable by its standard deviation. This can be done by defining the transformation matrix

$$\mathbf{P} = \mathbf{\Lambda}^{-1/2}\mathbf{U}^T, \tag{7.82}$$

where

$$\mathbf{\Lambda}^{-1/2} = \begin{pmatrix} 1/\sqrt{\lambda_1} & 0 & \cdots & 0 \\ 0 & 1/\sqrt{\lambda_2} & \cdots & 0 \\ \vdots & \vdots & \ddots & \vdots \\ 0 & 0 & \cdots & 1/\sqrt{\lambda_M} \end{pmatrix}. \tag{7.83}$$

The matrix $\mathbf{P}$ is called a *pre-whitening operator* and is defined only for positive eigenvalues $\lambda_1, \ldots, \lambda_M$, which requires $\mathbf{\Sigma}$ to be positive definite. It is readily verified that

$$\mathrm{cov}[\mathbf{Px}] = \mathbf{P}\mathbf{\Sigma}_X\mathbf{P}^T = \left(\mathbf{\Lambda}^{-1/2}\mathbf{U}^T\right)\left(\mathbf{U}\mathbf{\Lambda}\mathbf{U}^T\right)\left(\mathbf{U}\mathbf{\Lambda}^{-1/2}\right) = \mathbf{I}, \tag{7.84}$$

where (7.81) has been used. The term "white" is used because the variances are equal, just as "white noise" refers to a process with equal power across frequencies. Finally, if $\mathbf{x} \sim \mathcal{N}(\boldsymbol{\mu}_X, \mathbf{\Sigma}_X)$, then

$$\mathbf{x}' = \mathbf{P}(\mathbf{x} - \boldsymbol{\mu}_X) \sim \mathcal{N}_M(\mathbf{0}, \mathbf{I}). \tag{7.85}$$

Thus, for multivariate normal distributions, $\mathbf{P}$ transforms the correlated variables $\mathbf{x} - \boldsymbol{\mu}_X$ into *iid* variables $\mathbf{x}'$.

We now prove that an orthogonal transformation preserves the sum of the variances of the random variables, termed the *sum total variance*. In terms of the covariance matrix, the sum total variance is the sum of the diagonal elements, termed the *trace*.

Trace

The trace is the sum of the diagonal elements of a (square) matrix. It is denoted $\mathrm{tr}[X] = \sum_i x_{ii}$. If X and Y^T have the same dimension, then, $\mathrm{tr}[\mathbf{XY}] = \mathrm{tr}[\mathbf{YX}]$. It follows from this that

$$\mathrm{tr}[\mathbf{XYZ}] = \mathrm{tr}[\mathbf{YZX}] = \mathrm{tr}[\mathbf{ZXY}]. \tag{7.86}$$

The matrix products $\mathbf{XYZ}, \mathbf{YZX}, \mathbf{ZXY}$ are said to be *cyclic permutations*. They are cyclic in the sense that they preserve the order if the terms were joined end-to-end.

To prove that orthogonal transformations preserve the sum total variance, start with definition (7.76), then apply cyclic permutations:

$$\mathrm{tr}[\mathbf{\Lambda}] = \mathrm{tr}[\mathbf{U}^T\mathbf{\Sigma}_X\mathbf{U}] = \mathrm{tr}[\mathbf{\Sigma}_X\mathbf{U}\mathbf{U}^T] = \mathrm{tr}[\mathbf{\Sigma}_X]. \tag{7.87}$$

This proves that the sum total variance of variables that have undergone an orthogonal transformation equals the sum total variance of the original variables.

7.10 Multivariate Normal Distribution

The generalization of the normal distribution from a scalar x to a vector $\mathbf{x}$ is

$$p(x_1, x_2, ..., x_M) = p(\mathbf{x}) = \frac{1}{(2\pi)^{M/2}|\mathbf{\Sigma}|^{1/2}} e^{-(\mathbf{x}-\boldsymbol{\mu})^T \mathbf{\Sigma}^{-1}(\mathbf{x}-\boldsymbol{\mu})/2}. \tag{7.88}$$

This distribution will be denoted as $\mathcal{N}_M(\boldsymbol{\mu}, \mathbf{\Sigma})$. For expression (7.88) to be sensible, the covariance matrix $\mathbf{\Sigma}$ must be positive definite to be invertible.

Determinant

Equation (7.88) involves a new term, $|\mathbf{\Sigma}|$, which is known as the *determinant* of $\mathbf{\Sigma}$. The determinant is a scalar defined for square matrices. The precise definition of the determinant is not needed in this book and so will not be presented, but it can be found in any matrix algebra book. Some properties of determinants include the following:

- the determinant of a singular matrix is zero.
- the determinant of a diagonal matrix equals the product of its diagonal elements.
- for any two square matrices $\mathbf{A}$ and $\mathbf{B}$ that are conformable, $|\mathbf{AB}| = |\mathbf{A}||\mathbf{B}|$.
- $|\mathbf{A}| = |\mathbf{A}^T|$
- $|\mathbf{A}^{-1}| = |\mathbf{A}|^{-1}$.
- the determinant of a matrix $|\mathbf{A}|$ equals the product of the eigenvalues of $\mathbf{A}$.

The following expressions show that if $\mathbf{x} \sim \mathcal{N}_M(\boldsymbol{\mu}, \mathbf{\Sigma})$, then the density function integrates to one, μ_i is the mean of X_i, and Σ_{ij} is the covariance between X_i and X_j:

$$\iiint_{-\infty}^{\infty} p(\mathbf{x}) dx_1 dx_2 ... dx_M = 1, \tag{7.89}$$

$$\iiint_{-\infty}^{\infty} x_i \, p(\mathbf{x}) dx_1 dx_2 ... dx_M = \mu_i, \tag{7.90}$$

$$\iiint_{-\infty}^{\infty} (x_i - \mu_i)(x_j - \mu_j) p(\mathbf{x}) dx_1 dx_2 ... dx_M = \Sigma_{ij}. \tag{7.91}$$

Theorem 7.9 (Distribution of the Sample Mean) *If $\mathbf{x}_1, \ldots, \mathbf{x}_N$ are independent vectors from the distribution $\mathcal{N}_M(\boldsymbol{\mu}_X, \mathbf{\Sigma}_X)$, then the sample mean is distributed as*

$$\hat{\boldsymbol{\mu}}_X \sim \mathcal{N}_M \left(\boldsymbol{\mu}_X, \frac{1}{N}\mathbf{\Sigma}_X \right). \tag{7.92}$$

Theorem 7.10 (Distribution of Linear Transformations) *If $\mathbf{x}$ is distributed as $\mathcal{N}_M(\boldsymbol{\mu}, \mathbf{\Sigma})$ and $\mathbf{A}$ is a constant $N \times M$ matrix, then the distribution of $\mathbf{Ax}$ is*

$$\mathbf{Ax} \sim \mathcal{N}_N \left(\mathbf{A}\boldsymbol{\mu}, \mathbf{A}\mathbf{\Sigma}\mathbf{A}^T \right). \tag{7.93}$$

We already knew the mean and variance of the linear combination from Theorem 7.6. What's new in this theorem is that we now know its *distribution*.

Theorem 7.11 (The Multivariate Central Limit Theorem) *Let the M-dimensional vectors* $\mathbf{x}_1, \mathbf{x}_2, \ldots, \mathbf{x}_N$ *be independent samples from any population with mean* $\boldsymbol{\mu}$ *and covariance* $\boldsymbol{\Sigma}$. *Then, for large* $N - M$ $(N \gg M)$,

$$\hat{\boldsymbol{\mu}} \sim \mathcal{N}_M\left(\boldsymbol{\mu}, \frac{1}{N}\boldsymbol{\Sigma}\right) \tag{7.94}$$

Note: this theorem differs from Theorem 7.9 in that the population is not necessarily Gaussian, and the distribution holds only for sufficiently large N. The similarity of Theorem 7.11 with the univariate version of the Central Limit Theorem should be obvious.

7.11 Hotelling's T-squared Test

We now discuss an example that brings together many of the concepts discussed in this chapter. The example is a test for the hypothesis that two populations have the same *vector means*, that is, $\boldsymbol{\mu}_X = \boldsymbol{\mu}_Y$. To simplify the example, the population covariances are assumed to be known. Obviously, such an assumption is unrealistic, but is made here to remove complications resulting from estimating the covariance. As it turns out, this is no real limitation, and an exact test with estimated covariance matrices is given at the end of this section.

First, let us briefly derive a *univariate* difference-in-means test for when the population variance is known to be σ^2. Let N_X and N_Y be the size of the samples drawn independently from the same population. Then, the sample means have the distributions

$$\hat{\mu}_X \sim \mathcal{N}\left(\mu, \frac{\sigma^2}{N_X}\right) \quad \text{and} \quad \hat{\mu}_Y \sim \mathcal{N}\left(\mu, \frac{\sigma^2}{N_Y}\right), \tag{7.95}$$

and their difference has the distribution

$$\hat{\mu}_X - \hat{\mu}_Y \sim \mathcal{N}\left(0, \left(\frac{1}{N_X} + \frac{1}{N_Y}\right)\sigma^2\right). \tag{7.96}$$

The standardized version of this distribution is

$$\frac{\text{observed} - \text{hypothesized}}{\text{standard error}} = \frac{1}{\sqrt{\frac{1}{N_X} + \frac{1}{N_Y}}} \frac{\hat{\mu}_X - \hat{\mu}_Y}{\sigma} \sim \mathcal{N}(0, 1). \tag{7.97}$$

Assuming no prior knowledge of the sign of the difference, a two-sided test will be used. An equivalent one-sided test is obtained by squaring the statistic and using Theorem 1.3:

$$Z^2 = \frac{1}{\frac{1}{N_X} + \frac{1}{N_Y}} \frac{(\hat{\mu}_X - \hat{\mu}_Y)^2}{\sigma^2} \sim \chi_1^2. \tag{7.98}$$

The criterion for rejecting H_0 is then of the (one-sided) form $Z^2 > \chi_{1,\alpha}^2$.

Now consider the hypothesis of equal *vector* means,

$$H_0 : \boldsymbol{\mu}_X = \boldsymbol{\mu}_Y, \tag{7.99}$$

where $\boldsymbol{\mu}_X$ and $\boldsymbol{\mu}_Y$ are M-dimensional vectors. This hypothesis states that *each* element of $\boldsymbol{\mu}_X$ equals the corresponding element of $\boldsymbol{\mu}_Y$, comprising M distinct hypotheses. The alternative hypothesis is that $\boldsymbol{\mu}_X \neq \boldsymbol{\mu}_Y$; that is, at least one element of $\boldsymbol{\mu}_X$ differs from the corresponding element of $\boldsymbol{\mu}_Y$.

For simplicity, let us first consider the case in which the covariance matrix of each random vector is $\sigma^2 \mathbf{I}$. Then, the elements of a vector are *independent*, hence Z^2 (7.98) can be evaluated for *each* of the M variables *separately*, giving $Z_1^2, \ldots, Z_M^2$. Then, owing to the additivity property of the chi-squared distribution,

$$Z_1^2 + Z_2^2 + \cdots + Z_M^2 \sim \chi_M^2 \quad \text{if } H_0 \text{ is true.} \tag{7.100}$$

Note that this statistic is the sum square difference in sample means:

$$Z_1^2 + \cdots + Z_M^2 = \frac{\left(\hat{\mu}_X - \hat{\mu}_Y\right)_1^2 + \cdots + \left(\hat{\mu}_X - \hat{\mu}_Y\right)_M^2}{\sigma^2 \left(\frac{1}{N_X} + \frac{1}{N_Y}\right)}. \tag{7.101}$$

Naturally, this statistic tends to be small if H_0 is true and tends to be large if H_0 is false. The hypothesis H_0 is rejected if $\sum_m Z_m^2$ exceeds the critical value for a chi-squared distribution with M degrees of freedom. Note, a *single* statistic is used to test *multiple* hypotheses.

Now suppose the variables are not independent. Therefore, the covariance matrix for $\mathbf{x}$ and $\mathbf{y}$ is not diagonal. However, we assume the covariance matrix $\boldsymbol{\Sigma}$ is the same for $\mathbf{x}$ and $\mathbf{y}$, just as the t-test assumes the variance of the two populations is the same. Then, by Theorem (7.9), the sample means have distributions

$$\hat{\mu}_X \sim \mathcal{N}_M \left(\boldsymbol{\mu}_X, \frac{1}{N_X}\boldsymbol{\Sigma}\right) \quad \text{and} \quad \hat{\mu}_Y \sim \mathcal{N}_M \left(\boldsymbol{\mu}_Y, \frac{1}{N_Y}\boldsymbol{\Sigma}\right). \tag{7.102}$$

As discussed in Section 7.9, we may define a pre-whitening transformation (7.82). Applying this transformation yields

$$\mathbf{P}\hat{\mu}_X \sim \mathcal{N}_M \left(\mathbf{P}\boldsymbol{\mu}_X, \frac{1}{N_X}\mathbf{I}\right) \quad \text{and} \quad \mathbf{P}\hat{\mu}_Y \sim \mathcal{N}_M \left(\mathbf{P}\boldsymbol{\mu}_Y, \frac{1}{N_Y}\mathbf{I}\right), \tag{7.103}$$

and therefore

$$\left(\frac{1}{N_X} + \frac{1}{N_Y}\right)^{-1/2} \left(\mathbf{P}\hat{\mu}_X - \mathbf{P}\hat{\mu}_Y\right) \sim \mathcal{N}_M \left(\mathbf{0}, \mathbf{I}\right). \tag{7.104}$$

This result shows that the elements of the vector are *iid*. As a result, each element may be squared and then added together to derive a chi-squared statistic, as in (7.100):

$$\chi_M^2 \sim \left(\frac{1}{N_X} + \frac{1}{N_Y} \right)^{-1} \left(\mathbf{P} \hat{\boldsymbol{\mu}}_X - \mathbf{P} \hat{\boldsymbol{\mu}}_Y \right)^T \left(\mathbf{P} \hat{\boldsymbol{\mu}}_X - \mathbf{P} \hat{\boldsymbol{\mu}}_Y \right) \tag{7.105}$$

$$\sim \left(\frac{1}{N_X} + \frac{1}{N_Y} \right)^{-1} \left(\hat{\boldsymbol{\mu}}_X - \hat{\boldsymbol{\mu}}_Y \right)^T \mathbf{P}^T \mathbf{P} \left(\hat{\boldsymbol{\mu}}_X - \hat{\boldsymbol{\mu}}_Y \right) \tag{7.106}$$

$$\sim \left(\frac{1}{N_X} + \frac{1}{N_Y} \right)^{-1} \left(\hat{\boldsymbol{\mu}}_X - \hat{\boldsymbol{\mu}}_Y \right)^T \boldsymbol{\Sigma}^{-1} \left(\hat{\boldsymbol{\mu}}_X - \hat{\boldsymbol{\mu}}_Y \right). \tag{7.107}$$

Note that the last expression can be evaluated without explicitly constructing the orthogonal transformation. In other words, we have derived a *statistic* that involves only known quantities (remember $\boldsymbol{\Sigma}$ is assumed known here). Note that (7.107) reduces to the statistic (7.100) if

$$\boldsymbol{\Sigma} = \sigma^2 \mathbf{I}. \tag{7.108}$$

The natural extension of this test to the case of *unknown* covariance matrix $\boldsymbol{\Sigma}$ is to substitute the *pooled* sample covariance matrix for $\boldsymbol{\Sigma}$, which is

$$\hat{\boldsymbol{\Sigma}}_{\text{pooled}} = \frac{(N_X - 1)\hat{\boldsymbol{\Sigma}}_X + (N_Y - 1)\hat{\boldsymbol{\Sigma}}_Y}{N_X + N_Y - 2}. \tag{7.109}$$

The resulting statistic is known as *Hotelling's T-squared statistic* and is given by

$$T^2 = \left(\frac{1}{N_X} + \frac{1}{N_Y} \right)^{-1} \left(\hat{\boldsymbol{\mu}}_X - \hat{\boldsymbol{\mu}}_Y \right)^T \hat{\boldsymbol{\Sigma}}_{\text{pooled}}^{-1} \left(\hat{\boldsymbol{\mu}}_X - \hat{\boldsymbol{\mu}}_Y \right). \tag{7.110}$$

The distribution of T^2 under H_0 is related to the familiar F-distribution as

$$\left(\frac{N_X + N_Y - M - 1}{(N_X + N_Y - 2)M} \right) T^2 \sim F_{M, N_X + N_Y - M - 1}. \tag{7.111}$$

The proof of this distribution requires more advanced methods (see Anderson, 1984).

7.12 Multivariate Acceptance and Rejection Regions

The criterion for rejecting H_0 based on Hotelling's T-squared statistic (7.110) is

$$\mathbf{w}^T \boldsymbol{\Sigma}^{-1} \mathbf{w} > T_C^2, \tag{7.112}$$

where T_C is the appropriate critical value from the F distribution and

$$\mathbf{w} = \hat{\boldsymbol{\mu}}_X - \hat{\boldsymbol{\mu}}_Y \quad \text{and} \quad \boldsymbol{\Sigma} = \left(\frac{1}{N_X} + \frac{1}{N_Y} \right) \hat{\boldsymbol{\Sigma}}_{\text{pooled}}. \tag{7.113}$$

What do the acceptance and rejection regions look like for this test? Expanding the term $\mathbf{w}^T \mathbf{\Sigma}^{-1} \mathbf{w}$ yields a linear combination of second-order products of the elements of $\mathbf{w}$, hence setting it to a constant defines a conic section in $\mathbf{w}$-space. That is, the critical surface must be an ellipse, hyperboloid, or cone. To figure out which one, we consider an orthogonal transformation based on the eigenvectors of $\mathbf{\Sigma}$. An orthogonal transformation preserves lengths and angles and therefore *preserves the shape* of the conic section (see Section 7.7).

Given the eigenvector decomposition (7.81), it follows that the inverse is

$$\mathbf{\Sigma}^{-1} = \mathbf{U}\mathbf{\Lambda}^{-1}\mathbf{U}^T . \tag{7.114}$$

Therefore,

$$\mathbf{w}^T \mathbf{\Sigma}^{-1} \mathbf{w} = \mathbf{w}^T \mathbf{U}\mathbf{\Lambda}^{-1}\mathbf{U}^T \mathbf{w} = \mathbf{y}^T \mathbf{\Lambda}^{-1} \mathbf{y}, \tag{7.115}$$

where we have defined $\mathbf{y} = \mathbf{U}^T \mathbf{w}$. Thus, $\mathbf{y}$ and $\mathbf{w}$ are related by an orthogonal transformation. Since $\mathbf{\Lambda}$ is diagonal, this expression can be written equivalently as

$$\tilde{Z} = \frac{y_1^2}{\lambda_1} + \frac{y_2^2}{\lambda_2} + \ldots + \frac{y_M^2}{\lambda_M}. \tag{7.116}$$

Expression (7.116) is a sum of squares with positive coefficients. Therefore, $\mathbf{w}^T \mathbf{\Sigma}^{-1} \mathbf{w} = T_C^2$ defines an *ellipsoid* in $\mathbf{y}$-space with major axes oriented along the coordinate axes. In $\mathbf{w}$-space, the equality defines an ellipsoid in $\mathbf{w}$-space that is centered at the origin, and has major axes oriented along the eigenvectors of $\mathbf{u}_1, \ldots, \mathbf{u}_M$ with lengths proportional to $\sqrt{\lambda_1}, \ldots, \sqrt{\lambda_M}$. The decision rule is to reject H_0 when $\mathbf{w} = \hat{\boldsymbol{\mu}}_X - \hat{\boldsymbol{\mu}}_Y$ lies outside the ellipsoid. An illustration of a possible rejection region is shown in Figure 7.2.

The shape of the ellipsoid is determined by the covariance matrix $\mathbf{\Sigma}$. The closer the ellipsoid is to a circle (or equivalently, a hypersphere), the closer the variables are to being *iid*. If the major axes are along the coordinate axes, then the variables are uncorrelated. Major axes tilted with respect to the coordinate axes indicate the variables are correlated. Different lengths of the major axes indicate that the variables have different variances.

7.13 Further Topics

An introduction to linear algebra with a strong emphasis on intuitive understanding is Strang (1988) (and its later editions). Noble and Daniel (1988) is also recommended for accessible proofs and good coverage. Other accessible introductions to matrix analysis include Johnson and Wichern (2002) (from a statistical point of view) and Horn and Johnson (1985) (from a mathematical point of view). The "bible" for numerical matrix analysis is Golub and Van Loan (1996). Frequent

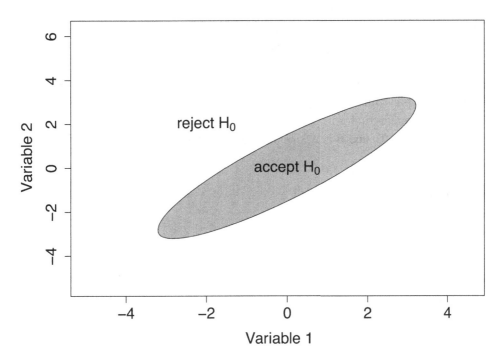

Figure 7.2 Illustration of the acceptance and rejection regions for a two-variable difference-in-means test.

users of matrix packages should be familiar with LAPACK (Anderson et al., 1992). Comprehensive compilations of linear algebra and matrix theorems also exist, such as Bernstein (2009).

7.14 Conceptual Questions

1 When testing multiple hypotheses, what is the advantage of multivariate tests, compared to performing univariate tests separately and then combining them?
2 When testing multiple hypotheses, why not always use multivariate tests?
3 If N samples of the random variable are interpreted as the vector $\mathbf{x}$, what is the geometric interpretation of the sample variance? Also, if N samples of Y are interpreted as a vector $\mathbf{y}$, what is the geometric interpretation of the sample correlation?
4 What is the transpose of a vector? Of a matrix?
5 What is an invertible transformation? When is a linear transformation invertible?
6 If $\mathbf{x}_1, \ldots, \mathbf{x}_M$ are linearly dependent, what property do the vectors satisfy?
7 What is the definition of the rank of a matrix? What does "full rank" mean?
8 What is the relation between rank and invertibility?

9 What is an orthogonal transformation?

10 What is a covariance matrix? What does it tell you?

11 What does a diagonal covariance matrix signify?

12 How do you diagonalize a covariance matrix?

13 What is a pre-whitening operator?

14 What is a quadratic form? When does it arise?

15 If $\mathbf{x} \sim \mathcal{N}_M(\boldsymbol{\mu}, \boldsymbol{\Sigma})$ and $\mathbf{B}$ is a constant matrix, what is the distribution of $\mathbf{Bx}$?

8

Linear Regression
Least Squares Estimation

> In most investigations where the object is to reduce the most accurate
> possible results from observational measurements, we are led to a system
> of equations of the form
>
> $$E = a + bx + cy + fz + \&c.,$$
>
> in which $a, b, c, f, \&c.$ are known coefficients, varying from one equa-
> tion to the other, and $x, y, z \&c.$ are unknown quantities, to be determined
> by the condition that each value of E is reduced either to zero, or to a
> very small quantity... Of all the principles that can be proposed for this
> purpose, I think there is none more general, more exact, or easier to
> apply, than that which we have used in this work; it consists of making
> the sum of the squares of the errors a *minimum*. By this method, a kind
> of equilibrium is established among the errors which, since it prevents
> the extremes from dominating, is appropriate for revealing the state of
> the system which most nearly approaches the truth.[1]
>
> *Legendre, 1805 (note "&c" is an old form of "Et cetera")*

Some variables can be modeled by an equation in which one variable equals a linear
combination of other random variables, plus random noise. Such models are used to
quantify the relation between variables, to make predictions, and to test hypotheses
about the relation between variables. After identifying the variables to include in
a model, the next step is to *estimate* the coefficients that multiply them, which
are known as the regression parameters. This chapter discusses the *least squares
method* for estimating regression parameters. The least squares method estimates
the parameters by minimizing the sum of squared differences between the fitted
model and the data. This chapter also describes measures for the goodness of fit and
an illuminating geometric interpretation of least squares fitting. The least squares
method is illustrated on various routine calculations in weather and climate analysis
(e.g., fitting a trend). Procedures for testing hypotheses about linear models are
discussed in the next chapter.

---

[1] Translation reprinted by permission from THE HISTORY OF STATISTICS: THE MEASUREMENT OF
UNCERTAINTY BEFORE 1900, by Stephen M. Stigler, Cambridge, MA: The Belknap Press of Harvard
University Press, Copyright ©1986 by the President and Fellows of Harvard College.

8.1 The Problem

Some physical laws take the form $Y = \beta X$, where X and Y are observed quantities and β is a coefficient that is characteristic of a substance. For example:

- adding heat Q to a substance increases its temperature by $\Delta T = \beta Q$.
- incident R_I and transmitted R_T radiation through a substance are related as $R_T = \beta R_I$.
- the volume V of a dry gas at constant pressure is related to its temperature as $T = \beta V$.

Typically, β is unknown and must be measured experimentally. Suppose experiments to measure β are performed and yield the results shown in Figure 8.1. As can be seen, the experimental results do not lie exactly on a line. Deviations from the line may be caused by experimental errors, observational uncertainty, or model inadequacy. Nevertheless, if the errors are random with equal chances of positive and negative values, it is reasonable to suppose that a line "passing through the data" may produce a useful estimate of β. How should the line be determined? There is no unique answer. By far the most common approach is the *method of least squares*, which finds the coefficient β that minimizes the sum square difference between Y and βX. Mathematically, this method finds β by minimizing

$$\sum_{n=1}^{N} (Y_n - \beta X_n)^2, \qquad (8.1)$$

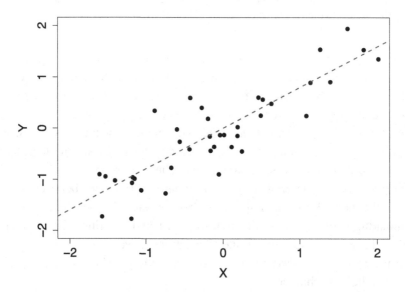

Figure 8.1 An example of experimental data to test the relation $Y = kX$ and the corresponding line derived from (8.2).

where X_n and Y_n are the n'th experimental values of X and Y, respectively, and N is the total number of experimental values. The minimum value can be found by setting the derivative of this expression to zero and solving for β, which yields

$$\hat{\beta} = \frac{\sum\limits_{n=1}^{N} Y_n X_n}{\sum\limits_{n=1}^{N} X_n^2}, \tag{8.2}$$

where the caret ˆ indicates a sample estimate. The resulting line $Y = \hat{\beta} X$ is shown in Figure 8.1 and gives a reasonable fit to the data. In this chapter, we consider the more general problem of estimating the coefficients $\beta_1, \ldots, \beta_M$ in a model of the form

$$Y = X_1 \beta_1 + X_2 \beta_2 + \cdots + X_M \beta_M + \epsilon, \tag{8.3}$$

where the terms in the model have the following characteristics:

- Y is an observed variable known as the *dependent* or *response* variable in the context of linear modeling, or the *predictand* in the context of prediction.
- $X_1, \ldots, X_M$ are observed variables, known as the *independent* or *explanatory* variables in the context of linear modeling, or the *predictors* in the context of prediction.
- $\beta_1, \ldots, \beta_M$ are unobserved population quantities known as the *regression parameters*.
- ϵ is unobserved *random error* to account for random deviations from the linear relation.

Model (8.3) is known as a *univariate regression model*, or a *univariate multiple regression model*, where "multiple" indicates that more than one (non-constant) predictor occurs on the right-hand side, and "univariate" indicates only one Y-variable is modeled.

A property of model (8.3) that is essential to the methods developed in this chapter is that it is a *linear function* of the regression parameters. Thus, a model of the form

$$Y = \beta_1 + \beta_2 X + \beta_3 X^2 + \beta_4 X^3 + \epsilon, \tag{8.4}$$

is also classified as linear because the model depends linearly *on the regression parameters* $\beta_1, \ldots, \beta_4$. The fact that the predictors may be nonlinear functions of X is irrelevant. Some nonlinear models can be transformed into a linear model. For instance, the model

$$W = a e^{bX + \epsilon} \tag{8.5}$$

can be transformed into a linear model by taking the log of both sides,

$$\log(W) = \log(a) + bX + \epsilon, \tag{8.6}$$

where $\log(a)$ and b are the regression parameters and ϵ is the error.

This chapter reviews the *least squares method* for estimating the coefficients of a linear model. Chapter 9 discusses methods for testing hypotheses about the coefficients.

8.2 Method of Least Squares

Consider the problem of estimating the regression parameters $\beta_1, \ldots, \beta_M$ in the model (8.3). The data for this problem consists of N realizations of the vector $(Y, X_1, \ldots, X_M)$. The n'th realization of this vector comes from a model denoted

$$Y_n = X_{n1}\beta_1 + X_{n2}\beta_2 + \cdots + X_{nM}\beta_M + \epsilon_n. \tag{8.7}$$

The equations for $n = 1, \ldots, N$ can be written concisely in matrix-vector form as

$$\begin{array}{ccccccc} \mathbf{y} & = & \mathbf{X} & \boldsymbol{\beta} & + & \boldsymbol{\epsilon}. \\ N \times 1 & & N \times M & M \times 1 & & N \times 1 \end{array} \tag{8.8}$$

The least squares method finds the $\beta_1, \ldots, \beta_M$ that minimizes

$$\mathrm{SSE} = \sum_{n=1}^{N} (Y_n - \beta_1 X_{n1} - \beta_2 X_{n2} - \cdots - \beta_M X_{nM})^2, \tag{8.9}$$

where SSE stands for *sum square error*. This expression can be written equivalently as

$$\mathrm{SSE} = \|\mathbf{y} - \mathbf{X}\boldsymbol{\beta}\|^2, \tag{8.10}$$

where $\| \cdot \|$ denotes the length of a vector (7.2). Geometrically, the problem is illustrated in Figure 8.2 for the special case of the line $Y = aX + b$. The line in the figure corresponds to a specific value of (a, b). One can think of adjusting the parameters (a, b) until the differences $\epsilon_n = Y_n - aX_n - b$ have the minimum possible sum square.

Three types of problems should be distinguished. If $N < M$, then the number of equations in (8.7) is less than the number of regression parameters. In this case, an infinitely many sets of regression parameters can be found that reduce SSE to zero. In the context of fitting a line as in Figure 8.2, $M = 2$ and $N < M$ means only one data point exists, through which an infinite number of lines may pass. Such problems are said to be *under-determined*. If $N = M$, then $\mathbf{X}$ is square. If $\mathbf{X}$ is also invertible, then the equations can be solved exactly as $\boldsymbol{\beta} = \mathbf{X}^{-1}\mathbf{y}$ with $\epsilon = 0$.

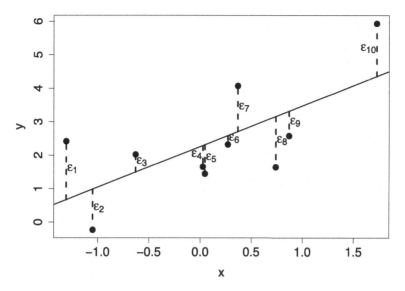

Figure 8.2 Scatter diagram of (X, Y) data (dots) and the line $y = ax + b$ corresponding to a specific choice of (a, b) (solid line). The vertical dashed lines show the residuals between the data and line.

In the context of fitting a line, this corresponds to fitting a line through two points, which can be done so that the line passes exactly through the two points, leaving no error. If $N > M$, then the number of equations exceeds the number of parameters being estimated. Such problems are said to be *over-determined*. In the context of fitting a line, this corresponds to a situation like that illustrated in Figure 8.2, where no single line can fit all the points simultaneously. This chapter considers only the over-determined case $N > M$.

To find the minimum of (8.9), we compute the derivative of SSE with respect to each of the regression parameters, set the result to zero, and solve for the β's. To do this, define

$$\mathbf{e} = \mathbf{y} - \mathbf{X}\boldsymbol{\beta}, \tag{8.11}$$

which we interpret as a function of $\boldsymbol{\beta}$. Then, the sum square error is

$$\text{SSE} = \sum_{n=1}^{N} e_n^2, \tag{8.12}$$

and the derivative with respect to β_k, where k is an integer between 1 and M, is,

$$\frac{\partial \text{SSE}}{\partial \beta_k} = \sum_{n=1}^{N} \left(2e_n \frac{\partial e_n}{\partial \beta_k} \right). \tag{8.13}$$

The partial derivative on the right-hand side is

$$\frac{\partial e_n}{\partial \beta_k} = \frac{\partial}{\partial \beta_k}\left(y_n - \sum_m X_{nm}\beta_m\right) = -\sum_m X_{nm}\frac{\partial \beta_m}{\partial \beta_k}. \tag{8.14}$$

Using the fact that

$$\frac{\partial \beta_m}{\partial \beta_k} = \begin{cases} 1 & \text{if } k = m \\ 0 & \text{if } k \neq m, \end{cases} \tag{8.15}$$

we obtain

$$\frac{\partial e_n}{\partial \beta_k} = -X_{nk}. \tag{8.16}$$

Substituting this into (8.13) gives

$$\frac{\partial \text{SSE}}{\partial \beta_k} = -2\sum_{n=1}^{N}(e_n X_{nk}) = -2\left(\mathbf{X}^T\mathbf{e}\right)_k. \tag{8.17}$$

Setting the derivative to zero for each k leads to the system of equations

$$\mathbf{X}^T\mathbf{e} = \mathbf{0}. \tag{8.18}$$

Substituting (8.11) to eliminate **e** yields

$$-2\mathbf{X}^T\left(\mathbf{y} - \mathbf{X}\hat{\boldsymbol{\beta}}\right) = 0, \tag{8.19}$$

where $\hat{\boldsymbol{\beta}}$ indicates the value of $\boldsymbol{\beta}$ that satisfies this equation, which will be termed the *least squares estimate of* $\boldsymbol{\beta}$. Distributing the terms and manipulating gives

$$\left(\mathbf{X}^T\mathbf{X}\right)\hat{\boldsymbol{\beta}} = \mathbf{X}^T\mathbf{y}, \tag{8.20}$$

which are known as the *normal equations*. Solving for $\hat{\boldsymbol{\beta}}$ gives

$$\hat{\boldsymbol{\beta}} = \left(\mathbf{X}^T\mathbf{X}\right)^{-1}\mathbf{X}^T\mathbf{y}, \tag{8.21}$$

assuming an inverse exists. The condition for an inverse is given in the following theorem.

Theorem 8.1 $\mathbf{X}^T\mathbf{X}$ *is invertible if and only if* $\mathbf{X}$ *has linearly independent columns.*

Proof

We prove this by proving the contrapositive; that is, by proving: $X^T X$ is singular if and only if X has linearly dependent columns. It is easy to see that if X has linearly dependent columns, then $X^T X$ is singular. After all, if the columns of X were linearly dependent, then $Xw = 0$ for some vector w not equal to zero, and multiplying both sides on the left by X^T gives $X^T Xw = 0$, which shows that the matrix is singular. To prove the converse, suppose $X^T X$ is singular. Then $X^T Xw = 0$ for some vector w not equal to zero. Multiplying both sides by w^T gives

$$w^T X^T Xw = \|Xw\|^2 = 0. \tag{8.22}$$

It may help to think about a vector z, and to recall the definition

$$\|z\|^2 = z_1^2 + z_2^2 + \cdots + z_M^2. \tag{8.23}$$

Since each term is nonnegative, the only way that the sum can vanish is if *each and every* z_i vanishes. Thus, $\|z\|^2 = 0$ implies $z = 0$. Therefore, (8.22) implies

$$Xw = 0, \tag{8.24}$$

which in turn implies X has linearly dependent columns. It follows logically that $X^T X$ has an inverse if and only if the columns of X are linearly independent.

If the columns of X are linearly dependent, then there exist solutions to (8.20), but they are not unique. To understand this situation, consider the regression model

$$y = \beta_1 x_1 + \beta_2 x_2 + \epsilon. \tag{8.25}$$

If x_1 and x_2 are linearly dependent, then $x_1 = \alpha x_2$ for some α, and therefore

$$y = \beta_1 (\alpha x_2) + \beta_2 x_2 + \epsilon = (\alpha \beta_1 + \beta_2) x_2 + \epsilon. \tag{8.26}$$

This model is equivalent to the model

$$y = \beta_2^* x_2 + \epsilon \tag{8.27}$$

with

$$\beta_2^* = \alpha \beta_1 + \beta_2. \tag{8.28}$$

Thus, the model with two regression parameters can be *re-parameterized* into a model with only one regression parameter. While β_2^* is unique, β_1 and β_2 are not and therefore cannot be estimated. More generally, if the columns of X are linearly dependent, then at least one of the predictors can be represented as a linear combination of the others, and hence the regression model can be re-parameterized into another model with fewer regression parameters. Accordingly, when the inverse in (8.21) does not exist, one approach to fixing that situation is to identify the linear dependency and then eliminate that dependency by re-parameterizing the original model into another model with fewer regression parameters.

In practice, predictors may be *almost* linearly dependent, in the sense that one predictor is nearly (but not perfectly) a linear combination of the other predictors. This situation is known as *multicollinearity*. When multicollinearity occurs, parameter estimates can change dramatically for small changes in data, and confidence intervals on individual parameters can be unreasonably large.

To show that this solution *minimizes* sum square error, consider $\|\mathbf{y} - \mathbf{X}\boldsymbol{\beta}^*\|^2$, where $\boldsymbol{\beta}^*$ is an arbitrary value. For any $\mathbf{a}$ and $\mathbf{b}$, $\|\mathbf{a} + \mathbf{b}\|^2 = \|\mathbf{a}\|^2 + \|\mathbf{b}\|^2 + 2\mathbf{a}^T\mathbf{b}$. Using this fact, add and subtract $\mathbf{X}\hat{\boldsymbol{\beta}}$ to the residual to write:

$$\|\mathbf{y} - \mathbf{X}\boldsymbol{\beta}^*\|^2 = \|\mathbf{y} - \mathbf{X}\hat{\boldsymbol{\beta}} + \mathbf{X}\hat{\boldsymbol{\beta}} - \mathbf{X}\boldsymbol{\beta}^*\|^2$$
$$= \|\mathbf{y} - \mathbf{X}\hat{\boldsymbol{\beta}}\|^2 + \|\mathbf{X}(\hat{\boldsymbol{\beta}} - \boldsymbol{\beta}^*)\|^2 + 2(\hat{\boldsymbol{\beta}} - \boldsymbol{\beta}^*)^T \left[\mathbf{X}^T(\mathbf{y} - \mathbf{X}\hat{\boldsymbol{\beta}})\right].$$

$$(8.29)$$

The last term on the right vanishes because of (8.19). Therefore,

$$\|\mathbf{y} - \mathbf{X}\boldsymbol{\beta}^*\|^2 = \|\mathbf{y} - \mathbf{X}\hat{\boldsymbol{\beta}}\|^2 + \|\mathbf{X}(\hat{\boldsymbol{\beta}} - \boldsymbol{\beta}^*)\|^2. \qquad (8.30)$$

The first term on the right-hand side does not depend on $\boldsymbol{\beta}^*$. The second term is non-negative, and therefore is minimized when $\boldsymbol{\beta}^* = \hat{\boldsymbol{\beta}}$, demonstrating that the least squares estimate minimizes the sum square error.

In practice, numerical calculation of $\hat{\boldsymbol{\beta}}$ is usually based on the SVD or the QR-algorithm, rather than on (8.21), primarily because these matrix decompositions give more accurate solutions and provide a natural framework for finding least squares solutions when $\mathbf{X}$ has linearly dependent columns (see Press et al., 2007, for a description of this procedure).

8.3 Properties of the Least Squares Solution

The *least squares prediction* of $\mathbf{y}$ is defined as

$$\hat{\mathbf{y}} = \mathbf{X}\hat{\boldsymbol{\beta}}, \qquad (8.31)$$

and the *least squares residuals* are defined as

$$\hat{\boldsymbol{\epsilon}} = \mathbf{y} - \mathbf{X}\hat{\boldsymbol{\beta}}. \qquad (8.32)$$

These quantities have important properties that follow directly from the solution.

In what follows, only models (8.3) that include $X_1 = 1$ as a predictor are considered. The associated regression parameter β_1 is known as the *intercept*.

The Residuals Are Orthogonal to the Predictors

According to (8.19), the least squares residuals satisfy the constraint

$$\mathbf{X}^T\hat{\boldsymbol{\epsilon}} = \mathbf{0}, \qquad (8.33)$$

which states that the residuals are *orthogonal to the predictors*. If the column vectors of $\mathbf{X}$ are denoted $\mathbf{x}_1, \ldots, \mathbf{x}_M$, then Equation (8.33) gives the M distinct constraints:

$$\mathbf{x}_1^T \hat{\boldsymbol{\epsilon}} = 0, \quad \mathbf{x}_2^T \hat{\boldsymbol{\epsilon}} = 0, \quad \ldots, \quad \mathbf{x}_M^T \hat{\boldsymbol{\epsilon}} = 0. \tag{8.34}$$

This property implies that the *residuals are uncorrelated with each predictor*. Because of this property, the residuals $\hat{\boldsymbol{\epsilon}}$ are sometimes described as representing $\mathbf{y}$ after $\mathbf{X}$ has been "removed," or after $\mathbf{X}$ has been "regressed out."

The Least Squares Prediction Is Invariant to Invertible Linear Transformations on X Suppose a new set of predictors $\mathbf{X}^*$ are obtained from the old predictors $\mathbf{X}$ by

$$\mathbf{X}^* = \mathbf{X}\mathbf{A}, \tag{8.35}$$

where $\mathbf{A}$ is an invertible square matrix. A special case of this transformation is simple re-scaling of the predictors. For instance, if X_1 is rescaled as $X_1^* = X_1/a_1$, and X_2 is rescaled as $X_2^* = X_2/a_2$, and so on, then

$$\mathbf{A} = \begin{pmatrix} 1/a_1 & 0 & \cdots & 0 \\ 0 & 1/a_2 & \cdots & 0 \\ \vdots & \vdots & \ddots & \vdots \\ 0 & 0 & \cdots & 1/a_M \end{pmatrix}. \tag{8.36}$$

The least squares estimate of $\boldsymbol{\beta}^*$ in the model $\mathbf{y} = \mathbf{X}^* \boldsymbol{\beta}^* + \boldsymbol{\epsilon}$ is found by substituting $\mathbf{X}^*$ into (8.21), which yields

$$\hat{\boldsymbol{\beta}}^* = \left(\mathbf{X}^{*T}\mathbf{X}^*\right)^{-1}\mathbf{X}^{*T}\mathbf{y} = \left(\mathbf{A}^T\mathbf{X}^T\mathbf{X}\mathbf{A}\right)^{-1}\mathbf{A}^T\mathbf{X}^T\mathbf{y}$$
$$\mathbf{A}^{-1}\left(\mathbf{X}^T\mathbf{X}\right)^{-1}\mathbf{A}^{-1T}\mathbf{A}^T\mathbf{X}^T\mathbf{y} = \mathbf{A}^{-1}\hat{\boldsymbol{\beta}}. \tag{8.37}$$

Therefore, the least squares prediction based on the new predictors is

$$\hat{\mathbf{y}}^* = \mathbf{X}^*\hat{\boldsymbol{\beta}}^* = \mathbf{X}\mathbf{A}\mathbf{A}^{-1}\hat{\boldsymbol{\beta}} = \mathbf{X}\hat{\boldsymbol{\beta}} = \hat{\mathbf{y}}, \tag{8.38}$$

which shows that the least squares prediction is the same for the old and new variables. In other words, the least squares prediction is *invariant to invertible linear transformations of the predictors*. One might intuitively expect this invariance because the new predictors are merely linear combinations of the old predictors, and therefore contain the same *information* as the old predictors. Note that $\hat{\mathbf{y}}$ is invariant, but the coefficients $\hat{\boldsymbol{\beta}}$ change.

One can also show that since the model includes an intercept term, then the least squares prediction $\hat{\mathbf{y}}$ is invariant to an arbitrary shift of each predictor by an additive constant.

An important consequence of these invariance properties is that the least squares prediction does not depend on the physical units in which the predictors are measured. For example, if two models are considered, one with $\mathbf{x}$ measured in Celcius, another with $\mathbf{x}'$ measured in Fahrenheit, then the resulting regression coefficients β and β' differ, but the least squares predictions $\hat{\mathbf{y}}$ are the same. More generally, if the units of $X_1, \ldots, X_M$ are changed separately – that is, each predictor is rescaled and shifted by a constant separately – the least squares prediction $\hat{\mathbf{y}}$ remains the same (provided the model also contains an intercept).

Least Squares Fit to a Constant
Consider the intercept-only model

$$y_n = \beta_1 + \epsilon_n, \tag{8.39}$$

where β_1 is an *unknown* constant. This equation is of the form (8.8) with $\mathbf{X} = \mathbf{j}$, where $\mathbf{j}$ is a vector of ones. The least squares estimate of β_1 is obtained from (8.21) using $\mathbf{X} = \mathbf{j}$:

$$\hat{\beta}_1 = \left(\mathbf{j}^T \mathbf{j}\right)^{-1} \mathbf{j}^T \mathbf{y} = \frac{1}{N} \sum_{n=1}^{N} y_n = \hat{\mu}_Y. \tag{8.40}$$

This result shows that the sample mean $\hat{\mu}_Y$ satisfies an optimization property: it is the constant that minimizes the sum square difference to the sample. Because the residuals of this model are orthogonal to $\mathbf{j}$, the residuals have zero mean:

$$0 = \mathbf{j}^T \hat{\boldsymbol{\epsilon}} = \sum_{n=1}^{N} \hat{\epsilon}_n. \tag{8.41}$$

In other words, the residuals $\hat{\boldsymbol{\epsilon}}$ are *centered*. The sum square error of this model is known as the *total sum of squares* and defined as

$$\text{total sum of squares} = \text{SST} = \sum_{n=1}^{N} \left(y_n - \hat{\mu}_Y\right)^2 = \|\mathbf{y} - \hat{\mu}_Y \mathbf{j}\|^2. \tag{8.42}$$

Coefficient of Determination
A measure of how well a model fits data is known as a *goodness of fit* measure. A natural measure of goodness of fit is the sum square error SSE

$$\text{SSE} = \sum_{n=1}^{N} \left(\hat{y}_n - y_n\right)^2 = \|\hat{\mathbf{y}} - \mathbf{y}\|^2. \tag{8.43}$$

SSE has units of Y-square and tends to increase with N, which makes it awkward for comparing models with different sample sizes, or models whose different

Y-variables have different units. Accordingly, SSE is often normalized by the total sum of squares, which leads to the most popular goodness of fit measure,

$$R^2 = 1 - \frac{\text{SSE}}{\text{SST}}. \tag{8.44}$$

The parameter R^2 is known as the *coefficient of determination*, or *R-squared*. The intercept-only model (8.39) is a special case of (8.3) obtained by setting $\beta_2 = \cdots = \beta_M = 0$. Since this choice is always available regardless of other predictors, SSE $\leq$ SST. Thus, R^2 ranges between 0 and 1, with 0 indicating a poor fit and 1 indicating a perfect fit.

Other properties of R^2 include the following (which are discussed in the exercises):

- The positive square root of R^2 is known as the *multiple correlation R*.
- If the regression model contains only one predictor $\mathbf{x}$ and an intercept, then R^2 equals the squared correlation between $\mathbf{y}$ and $\mathbf{x}$.
- R^2 can be interpreted equivalently as the fraction of variance "explained" by the predictors. More precisely, it can be shown that

$$R^2 = \frac{\sum_n (\hat{y}_n - \hat{\mu}_Y)^2}{\sum_n (y_n - \hat{\mu}_Y)^2} = \frac{\|\hat{\mathbf{y}} - \hat{\boldsymbol{\mu}}\|^2}{\text{SST}}, \tag{8.45}$$

- R is the correlation between the prediction $\hat{y}$ and y:

$$R = \text{cor}[\hat{y}, y]. \tag{8.46}$$

- R is the maximum correlation between $\mathbf{y}$ and any linear combination of $\mathbf{X}$:

$$R = \max_{\boldsymbol{\beta}} \text{cor}[\mathbf{y}, \mathbf{X}\boldsymbol{\beta}]. \tag{8.47}$$

In this sense, R is a generalization of $\text{cor}[Y, X]$ to more than one X.
- R is invariant to invertible transformations of the form

$$\mathbf{y}' = a\mathbf{y} + b \quad \text{and} \quad \mathbf{X}' = \mathbf{X}\mathbf{A} + \mathbf{j}\mathbf{c}^T, \tag{8.48}$$

where $a \neq 0$, $\mathbf{A}$ is nonsingular, and b and $\mathbf{c}$ are arbitrary. This invariance implies that R is independent of the units used to measure the predictors and predictands, and that any set of predictors in the same space as the columns of $\mathbf{X}$ will have the same R.
- The coefficient of determination can be written equivalently as

$$R^2 = \frac{\mathbf{y}^T \mathbf{X} \left(\mathbf{X}^T \mathbf{X}\right)^{-1} \mathbf{X}^T \mathbf{y}}{\text{SST}}. \tag{8.49}$$

8.4 Geometric Interpretation of Least Squares Solutions

The algebraic properties of the least squares solution may seem curious and unconnected with each other. However, a fascinating geometrical interpretation exists in which these relations emerge naturally and are connected to each other through a picture.

Consider the linear regression model

$$\mathbf{y} = \mathbf{X}\boldsymbol{\beta} + \boldsymbol{\epsilon}. \tag{8.50}$$

To interpret this model geometrically, note that the columns of $\mathbf{X}$ are vectors in an N-dimensional space. Let the columns of $\mathbf{X}$ be denoted

$$\mathbf{X} = \begin{bmatrix} \mathbf{x}_1 & \mathbf{x}_2 & \cdots & \mathbf{x}_M \end{bmatrix}. \tag{8.51}$$

Then, the term $\mathbf{X}\boldsymbol{\beta}$ is nothing more than a linear combination of the column vectors of $\mathbf{X}$:

$$\mathbf{X}\boldsymbol{\beta} = \begin{bmatrix} \mathbf{x}_1 & \mathbf{x}_2 & \cdots & \mathbf{x}_M \end{bmatrix} \begin{pmatrix} \beta_1 \\ \beta_2 \\ \vdots \\ \beta_M \end{pmatrix} = \beta_1 \mathbf{x}_1 + \beta_2 \mathbf{x}_2 + \dots + \beta_M \mathbf{x}_M. \tag{8.52}$$

The columns of $\mathbf{X}$ span a space known as the *range* of $\mathbf{X}$. Every vector in the range of $\mathbf{X}$ is a linear combination of $\mathbf{x}_1, \mathbf{x}_2, \dots, \mathbf{x}_M$. This range is indicated by the shading in Figure 8.3.

If $\mathbf{y}$ lies in the range of $\mathbf{X}$, then $\mathbf{y} = \mathbf{X}\boldsymbol{\beta}$ for some $\boldsymbol{\beta}$. However, we usually consider the case in which $\mathbf{y}$ does not lie in the range of $\mathbf{X}$. This situation is illustrated by the point $\mathbf{y}$ in Figure 8.3. What is the shortest line between the point $\mathbf{y}$ and the range of $\mathbf{X}$? It is obvious geometrically that the shortest line segment is *perpendicular* to the range of $\mathbf{X}$. More precisely, the shortest distance is the vector $\hat{\boldsymbol{\epsilon}}$ that is *orthogonal* to every column vector in $\mathbf{X}$. The fact that $\hat{\boldsymbol{\epsilon}}$ should be orthogonal to $\mathbf{X}$ can be expressed as

$$\mathbf{X}^T \hat{\boldsymbol{\epsilon}} = \mathbf{X}^T \left(\mathbf{y} - \mathbf{X}\hat{\boldsymbol{\beta}} \right) = \mathbf{0}, \tag{8.53}$$

which is identical to (8.19). This result shows that the residuals are orthogonal to the predictors, as discussed in Section 8.3. The interesting part here is that the orthogonality constraint (8.53) *emerges naturally* in this geometrical interpretation. Moreover, solving this equation leads to the least squares estimate of $\boldsymbol{\beta}$:

$$\hat{\boldsymbol{\beta}} = \left(\mathbf{X}^T \mathbf{X} \right)^{-1} \mathbf{X}^T \mathbf{y}, \tag{8.54}$$

which shows that the least squares estimate can be *derived* by geometric reasoning alone.

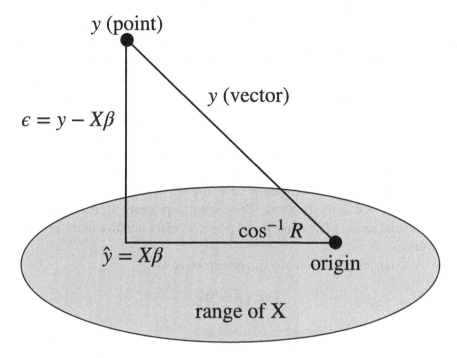

Figure 8.3 Schematic of the least squares solution. The shaded oval indicates the range of $\mathbf{X}$ as a two-dimensional surface, and the point $\mathbf{y}$ lies outside this range in the third dimension. The linear combination $\mathbf{X}\boldsymbol{\beta}$, where $\boldsymbol{\beta}$ varies over all values, spans the range of $\mathbf{X}$. The difference $\mathbf{y} - \mathbf{X}\boldsymbol{\beta}$ is a line segment between $\mathbf{y}$ and a point in the range of $\mathbf{X}$. The minimum length of this line segment occurs when this line is orthogonal to every column vector of $\mathbf{X}$, which is the least squares solution. The figure assumes $\mathbf{y}$ is centered, in which case the cosine of the angle between $\mathbf{y}$ and $\mathbf{X}\boldsymbol{\beta}$ is R.

Once it is known that the residual vector $\hat{\boldsymbol{\epsilon}}$ is orthogonal to the predictors $\mathbf{x}_1, \ldots, \mathbf{x}_M$, a right triangle can be set up as in Figure 8.3 to derive the identity

$$\|\mathbf{y}\|^2 = \|\hat{\boldsymbol{\epsilon}}\|^2 + \|\mathbf{X}\hat{\boldsymbol{\beta}}\|^2. \tag{8.55}$$

This identity is obvious geometrically, which illustrates the value of geometric reasoning. Obviously, if all vectors were translated in space by the same amount, the triangle would be preserved. Furthermore, if $\mathbf{j}$ is included in $\mathbf{X}$, then $\hat{\boldsymbol{\epsilon}}$ is orthogonal to $\mathbf{j}$ and so shifting vectors by a constant has no impact on $\hat{\boldsymbol{\epsilon}}$. Therefore, shifting vectors by $\boldsymbol{\mu}_Y = \mathbf{j}\hat{\mu}_Y$ gives

$$\|\mathbf{y} - \hat{\boldsymbol{\mu}}_Y\|^2 = \|\hat{\boldsymbol{\epsilon}}\|^2 + \|\mathbf{X}\hat{\boldsymbol{\beta}} - \hat{\boldsymbol{\mu}}_Y\|^2, \tag{8.56}$$

which can be written equivalently as SST = SSE + $\|\mathbf{X}\hat{\boldsymbol{\beta}} - \hat{\boldsymbol{\mu}}_Y\|^2$. Recall that the coefficient of determination is defined as $R^2 = 1 - \text{SSE}/\text{SST}$. This identity can used to derive the alternative form of R^2 given in (8.45). Moreover, it is clear from the figure that R is the cosine of the angle between vectors $\mathbf{y}$ and $\mathbf{X}\hat{\boldsymbol{\beta}}$ (provided $\mathbf{y}$ has been centered). When $R = 1$, $\mathbf{y}$ is in the range of $\mathbf{X}$ – the variability of Y can be explained completely as a linear combination of X-variables.

Another property of least squares mentioned earlier is that the prediction $\hat{\mathbf{y}} = \mathbf{X}\hat{\boldsymbol{\beta}}$ is invariant to transformations of the form $\mathbf{X} \rightarrow \mathbf{XA}$, where $\mathbf{A}$ is a nonsingular square matrix. Since the transformed predictors are linearly independent combinations of the original predictors, the range is preserved and hence the least squares prediction and residuals must be the same. This transformation merely changes the basis set of the range of $\mathbf{X}$ without changing the range itself, so the geometry of least squares is exactly the same. Thus, again, a property of least squares that requires several steps of linear algebra to prove becomes intuitive in the geometric framework.

The least squares prediction of $\mathbf{y}$ can be expressed as

$$\hat{\mathbf{y}} = \mathbf{X}\hat{\boldsymbol{\beta}} = \mathbf{Hy}, \tag{8.57}$$

where $\mathbf{H}$ is known as *the hat matrix*, defined as

$$\mathbf{H} = \mathbf{X}\left(\mathbf{X}^T\mathbf{X}\right)^{-1}\mathbf{X}^T. \tag{8.58}$$

The hat matrix $\mathbf{H}$ is a *projection operator* that projects an arbitrary vector onto the range of $\mathbf{X}$. One can think of projection as shining a light straight down onto the range of $\mathbf{X}$ and observing the shadow $\hat{\mathbf{y}} = \mathbf{X}\hat{\boldsymbol{\beta}}$ cast by the *vector* $\mathbf{y}$. This interpretation explains why the hat matrix is idempotent; that is, why $\mathbf{H}^2 = \mathbf{H}$. Specifically, $\mathbf{H}$ simply projects a vector onto a certain space, and projecting that vector twice onto the same space yields the same result as projecting the vector once. Also, this interpretation implies $\mathbf{HX} = \mathbf{X}$, since projecting $\mathbf{X}$ onto its own range simply recovers $\mathbf{X}$. This identity can be verified algebraically as well:

$$\mathbf{HX} = \mathbf{X}\left(\mathbf{X}^T\mathbf{X}\right)^{-1}\mathbf{X}^T\mathbf{X} = \mathbf{X}. \tag{8.59}$$

The residual errors can also be written in terms of the hat matrix as

$$\hat{\boldsymbol{\epsilon}} = \mathbf{y} - \hat{\mathbf{y}} = \mathbf{y} - \mathbf{Hy} = (\mathbf{I} - \mathbf{H})\,\mathbf{y}. \tag{8.60}$$

Since $\mathbf{I} - \mathbf{H}$ projects vectors on the space *orthogonal* to the range of $\mathbf{X}$, one can anticipate the identity $\mathbf{H}(\mathbf{I} - \mathbf{H}) = \mathbf{0}$. This can be verified directly as

$$\mathbf{H}\,(\mathbf{I} - \mathbf{H}) = \mathbf{H} - \mathbf{H}^2 = \mathbf{H} - \mathbf{H} = \mathbf{0}. \tag{8.61}$$

It is fun to be able to predict such mathematical identities geometrically.

8.5 Illustration Using Atmospheric CO$_2$ Concentration

To illustrate the least squares method, consider Figure 8.4, which shows the monthly mean concentration of atmospheric carbon dioxide (CO_2) at a measuring station. Two distinctive features of this time series are clearly noticeable: (1) the general increase in concentration with year and (2) the nearly periodic cycle of local peaks and valleys. These variations are caused by different mechanisms. Specifically, the periodic variations have a period of twelve months and are caused primarily by plants taking in CO_2 during their growth phase and releasing CO_2 during their decay phase (the precise annual cycle is somewhat complicated because the growth and decay cycles of vegetation are out of phase between northern and southern hemispheres, and most of the land mass is in the northern hemisphere). In contrast, the overall increase with year is caused by human activities, primarily fossil fuel emissions and land use changes (Archer, 2010).

To quantify changes in CO_2 concentration, a typical approach is to assume that the observed variable Y_t is generated by the model

$$Y_t = \mu_t + \epsilon_t, \tag{8.62}$$

where μ_t is a deterministic function of time and ϵ_t is a random variable whose statistical properties are independent of time. In contrast to the problem discussed at the beginning of this chapter, the model in this example *is not known*. Selecting

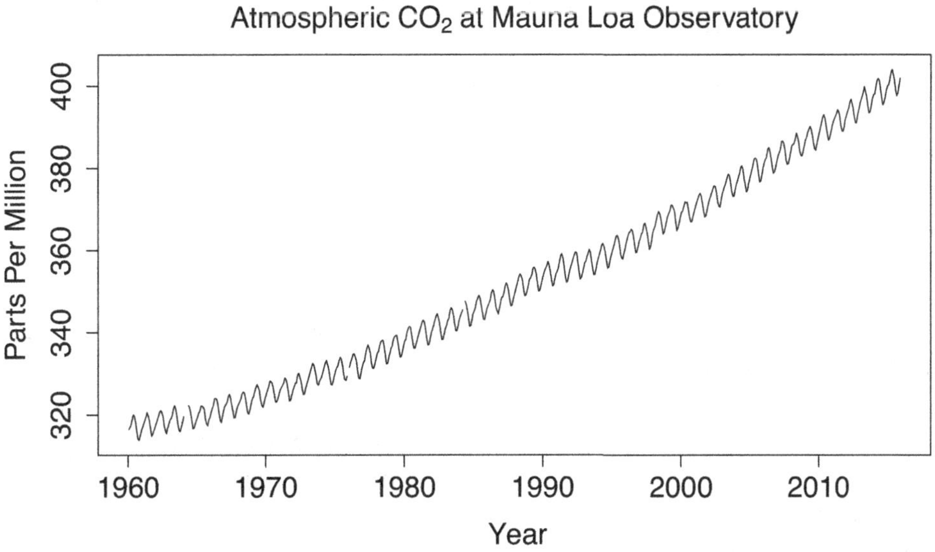

Figure 8.4 Concentration of atmospheric carbon dioxide measured at Mauna Loa Observatory. Gaps in the curve on February–April 1964, December 1975, and April 1984 correspond to missing data.

the model is part of the problem. Here, we select a model with two components in the deterministic term: the annual cycle and the general increase in time. We denote this decomposition as

$$\mu_t = \mu_{\text{seasonal}}(t) + \mu_{\text{secular}}(t). \tag{8.63}$$

The annual cycle is a periodic function of time and therefore can be represented by a suitable sum of sines and cosines, known as a *Fourier Series*. Defining t to be the number of months from an arbitrary reference, the seasonal component can be represented by

$$\mu_{\text{seasonal}}(t) = \sum_{k=1}^{K} \left(a_k \cos \left(\frac{2\pi kt}{12} \right) + b_k \sin \left(\frac{2\pi kt}{12} \right) \right), \tag{8.64}$$

where a_k and b_k are coefficients to be *estimated from data*. K is the number of harmonics.

The time series also generally increases in time. This component is known as the *secular component*. "Secular" is a term used in astronomy to refer to changes in the motion of the sun or planets that are much slower than the motion associated with the obvious periodic orbits. This term is particularly apt here since we are identifying slow changes relative to the annual cycle, which is caused by the earth's orbit around the sun. A common approach for modeling the secular changes is to use a low-order polynomial in time,

$$\mu_{\text{secular}}(t) = \sum_{l=0}^{L} c_l t^l, \tag{8.65}$$

where $c_0, \ldots, c_L$ are coefficients to be *estimated from data*. L is the polynomial order.

Choosing K and L is a *model selection problem*, which will be discussed in Chapter 10. Here, we simply assert that reasonable choices are $K = 2$ and $L = 2$. In this case, the model for μ_t has $M = 7$ parameters: $a_1, b_1, a_2, b_2, c_0, c_1, c_2$.

The typical next step is to estimate these parameters from data. The time series shown in Figure 8.4 has $N = 672$ time steps. These values cannot be fitted exactly to a seven-parameter model. We use the method of least squares to estimate the seven parameters.

Incidentally, one may wonder why Fourier Transform techniques are not used to determine the coefficients of the seasonal model (8.64). In this example, the time series shown in Figure 8.4 has missing data, hence Fourier Transform methods cannot be applied.

In climate studies, the process of estimating annual cycles and long-term trends is so common that it is worthwhile to discuss these two estimation problems separately. After doing that, we discuss the estimation of these two components simultaneously.

8.5.1 The Trend

We first apply the least squares method to estimate a linear growth rate. This is called a *trend analysis*. Accordingly, consider the equation

$$y_n = \beta_1 + \beta_2 t_n + \epsilon, \qquad (8.66)$$

where t_n is time. The parameter β_1 is known as the *intercept* and β_2 is known as the *slope*. The data is monthly, so $n = 1$ corresponds to January 1960 and subsequent values of n count the number of months after January 1960. It is convenient to define the time parameter

$$t_n = (n - 1)/12, \qquad (8.67)$$

so that t_n has units of years. To perform trend analysis, the following variables are defined

$$\mathbf{y} = \begin{pmatrix} y_1 \\ y_2 \\ \vdots \\ y_N \end{pmatrix} \quad \text{and} \quad \mathbf{X} = \begin{pmatrix} 1 & t_1 \\ 1 & t_2 \\ \vdots & \vdots \\ 1 & t_N \end{pmatrix} \quad \text{and} \quad \boldsymbol{\beta} = \begin{pmatrix} \beta_1 \\ \beta_2 \end{pmatrix}. \qquad (8.68)$$

Note that the first column of $\mathbf{X}$ is just ones, reflecting the fact that β_1 is the same for each month. Next, we substitute the CO$_2$ time series in $\mathbf{y}$, and the time index t_n in $\mathbf{X}$, to specify $\mathbf{y}$ and $\mathbf{X}$ numerically. These are then substituted into (8.21) to obtain the least squares estimate of $\boldsymbol{\beta}$. Numerical calculations give $\hat{\boldsymbol{\beta}} = \begin{bmatrix} 309 & 1.5 \end{bmatrix}^T$, which corresponds to

$$\hat{\mu}_{\text{trend}}(t) = 309 + 1.5t. \qquad (8.69)$$

Because the predictor t has units of years, the slope parameter 1.5 has units of parts per million (ppm) per year. Thus, according to this model, atmospheric CO$_2$ concentration is increasing at the average rate of 1.5 ppm per year.

Recall that the residuals are orthogonal to the predictors, as discussed in Section 8.3. Thus, the residuals of this model have vanishing linear trend and zero mean. As a result, the residuals of a line fit are often said to be the *detrended* time series.

Is the line fit reasonable? The coefficient of determination is $R^2 = 0.98$, which means that the linear model captures 98% of the variability in the CO$_2$ concentration. Despite the large R^2, it is worth looking at the results more closely.

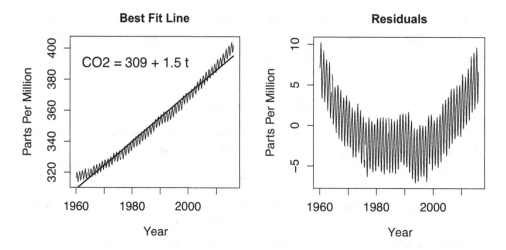

Figure 8.5 Left panel: the atmospheric CO_2 concentration (thin curve) and best fit line to $t = $ year $- 1960$ (thick straight line). Right panel: the residuals (that is, difference between atmospheric CO_2 concentration and the best fit line).

The best fit line is shown in the left panel of Figure 8.5. Although the line does capture quite a lot of the variability, it fails to capture the annual oscillations (as expected), and it systematically underestimates CO_2 concentration at the beginning and end of the record, suggesting the existence of other secular variations. An important step in judging a model fit is to examine the residuals. The residual between CO_2 and the best fit line is shown in the right panel of Figure 8.5. These residuals are known as the *detrended* atmospheric CO_2 concentration. The residual plot clearly shows a secular variation that would be well fit by a *quadratic* function of time (in addition to the periodic seasonal variations). A quadratic polynomial will be considered shortly.

8.5.2 The Annual Cycle

In this section we apply the least squares method to "remove the annual cycle," which is a very common application in climate studies. Consider the periodic model (8.64). To represent this model in matrix notation, define $\omega_k = 2\pi k/12$ and let

$$
c_k = \begin{pmatrix} \cos(\omega_k) \\ \cos(2\omega_k) \\ \vdots \\ \cos(N\omega_k) \end{pmatrix} \quad \text{and} \quad s_k = \begin{pmatrix} \sin(\omega_k) \\ \sin(2\omega_k) \\ \vdots \\ \sin(N\omega_k). \end{pmatrix} \tag{8.70}
$$

As discussed earlier, we use $K = 2$, hence the predictor matrix is

$$\mathbf{X} = \begin{pmatrix} \mathbf{j} & \mathbf{c}_1 & \mathbf{s}_1 & \mathbf{c}_2 & \mathbf{s}_2 \end{pmatrix}. \tag{8.71}$$

Applying the method of least squares yields the best fit regression model

$$\hat{\boldsymbol{\mu}}_{\text{seasonal}} = 352\mathbf{j} - 1.49\mathbf{c}_1 + 2.10\mathbf{s}_1 + 0.77\mathbf{c}_2 - 0.29\mathbf{s}_2. \tag{8.72}$$

In some studies, especially prediction studies, we are interested in variations *about* the seasonal component. Such variations can be examined by subtracting the best fit model from the original time series, yielding the *anomaly* time series $\mathbf{y}'$:

$$\mathbf{y}' = \mathbf{y} - \hat{\boldsymbol{\mu}}_{\text{seasonal}}. \tag{8.73}$$

The resulting $\mathbf{y}'$ is a monthly time series in which the annual cycle has been subtracted from each month. The anomaly time series for atmospheric CO$_2$ concentration is shown in Figure 8.6 (the intercept term 352 has been added so that it can be shown graphically on the same scale as the original time series). The figure shows that the anomaly time series has removed the annual cycle variations and thereby clarifies the secular variations.

One might wonder about the fact that there are two ways to fit a sum of sines and cosines to data: one based on least squares, and another based on the Fourier Transform (discussed in Chapter 6). These two methods are identical when there are no missing data and the time span is an integral multiple of 12 months. Neither condition holds in the present example.

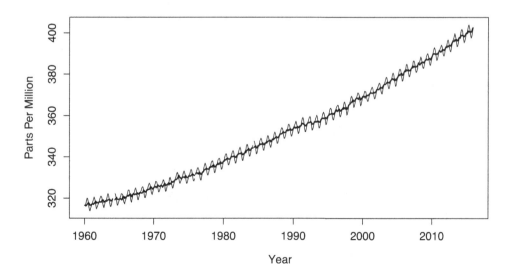

Figure 8.6 Atmospheric CO$_2$ concentration (thin curve) and the anomaly value in which the annual cycle has been regressed out (thick curve).

8.5.3 *Estimating Secular and Seasonal Components Simultaneously*

Although the annual cycle and trend have been computed separately in the previous sections, they cannot be added together to obtain the deterministic term. The problem is that the annual cycle and trend are not perfectly orthogonal, hence adding the two models together would "double count" the variations that are common between the two models. The correct way to estimate the deterministic term is to include both secular and periodic models in the deterministic model and then estimate all regression parameters simultaneously. To do this, the predictor matrix is defined as

$$\mathbf{X} = \begin{pmatrix} \mathbf{j} & \mathbf{c}_1 & \mathbf{s}_1 & \mathbf{c}_2 & \mathbf{s}_2 & \mathbf{t} & \mathbf{t}_2 \end{pmatrix}, \tag{8.74}$$

where the predictor matrix (8.71) is augmented by two polynomials in time:

$$(\mathbf{t})_n = t_n \quad \text{and} \quad (\mathbf{t}_2)_n = t_n^2. \tag{8.75}$$

The least squares prediction is found to be

$$\hat{\mu} = 316\mathbf{j} - 1.67\mathbf{c}_1 + 2.29\mathbf{s}_1 + 0.79\mathbf{c}_2 - 0.04\mathbf{s}_2 + 0.86\mathbf{t} + 0.01\mathbf{t}_2. \tag{8.76}$$

A plot of this function is indistinguishable from the actual data (not shown). Consistent with this, $R^2 > 0.9999$, which implies the model explains practically all the variability. Usually, simple models do not explain this much variability in climate observations. The residuals of this model are shown in Figure 8.7. The residuals are

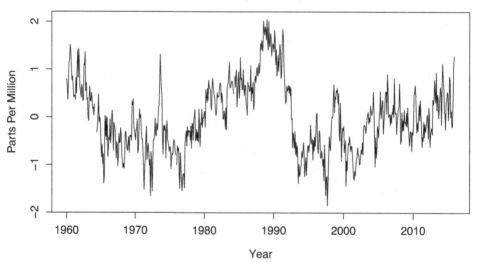

Figure 8.7 Atmospheric CO_2 concentration after removing the first two harmonics of the annual cycle and a second order polynomial in time.

on the order of 1 ppm and exhibit slow variations with a time scale around 30 years. One could attempt to remove these variations by introducing Fourier harmonics on 30-year time scales, but this would be recommended only if the cause of these variations were understood to be a deterministic function of time. However, there is no obvious *natural* mechanism for the 30-year oscillation. Therefore, this residual time series becomes the object of further investigation. This example illustrates how regression can be used to remove variations as a result of well-understood mechanisms to isolate variations that remain to be explained.

Incidentally, the growth rate of CO_2 concentration is the derivative of the secular terms in (8.76), which is

$$\left(\frac{\partial CO_2}{\partial t}\right)_n = 0.86 + 0.02t_n. \tag{8.77}$$

According to this model, the growth rate of CO_2 concentration was about 0.86 ppm/year in 1960 and about 2.2 ppm/year in 2015. The 1.5 ppm/year obtained in (8.69) represents a best fit *constant* rate over the period and is intermediate between the two rates.

8.6 The Line Fit

The line fit is sufficiently simple that the coefficients can be written explicitly in terms of inner products. This analytical simplicity is sometimes illuminating. A line fit assumes that two variables X and Y are related to each other through the model

$$Y = \beta_1 + \beta_2 X + \epsilon, \tag{8.78}$$

where β_1 is the *intercept*, β_2 is the *slope*, and ϵ is a random variable that is independent of X. The data is in the form of N pairs of data points (x_n, y_n), $n = 1, 2, \ldots, N$, which satisfy a model of the form (8.8) under the identifications

$$\mathbf{y} = \begin{pmatrix} y_1 \\ y_2 \\ \vdots \\ y_n \end{pmatrix} \quad \text{and} \quad \mathbf{X} = \begin{pmatrix} 1 & x_1 \\ 1 & x_2 \\ \vdots & \vdots \\ 1 & x_N \end{pmatrix} \quad \text{and} \quad \boldsymbol{\beta} = \begin{pmatrix} \beta_1 \\ \beta_2 \end{pmatrix}. \tag{8.79}$$

The least squares estimate is (8.21). Substituting $\mathbf{X}$ and $\mathbf{y}$ gives

$$\mathbf{X}^T \mathbf{X} = \begin{pmatrix} N & \sum_n x_n \\ \sum_n x_n & \sum_n x_n^2 \end{pmatrix} \quad \text{and} \quad \mathbf{X}^T \mathbf{y} = \begin{pmatrix} \sum_n y_n \\ \sum_n x_n y_n \end{pmatrix}. \tag{8.80}$$

These quantities can be re-written equivalently as

$$\mathbf{X}^T \mathbf{X} = N \begin{pmatrix} 1 & \hat{\mu}_X \\ \hat{\mu}_X & \hat{\sigma}_X^2 + \hat{\mu}_X^2 \end{pmatrix} \qquad \mathbf{X}^T \mathbf{y} = N \begin{pmatrix} \hat{\mu}_Y \\ \hat{c}_{YX} + \hat{\mu}_X \hat{\mu}_Y \end{pmatrix}, \tag{8.81}$$

where the sample mean, variance, and covariance have been defined as

$$\hat{\mu}_X = \frac{1}{N}\sum_{n=1}^{N} x_n, \quad \hat{\sigma}_X^2 = \frac{1}{N}\sum_{n=1}^{N} x_n^2 - \hat{\mu}_X^2, \quad \hat{c}_{yx} = \frac{1}{N}\sum_{n=1}^{N} y_n x_n - \hat{\mu}_y\hat{\mu}_x. \quad (8.82)$$

Using the analytic inverse of a 2×2 matrix (A.5) gives

$$(\mathbf{X}^T\mathbf{X})^{-1} = \frac{1}{N\hat{\sigma}_X^2}\begin{pmatrix} \hat{\sigma}_X^2 + \hat{\mu}_X^2 & -\hat{\mu}_X \\ -\hat{\mu}_X & 1 \end{pmatrix}. \quad (8.83)$$

Substituting this expression into (8.21) gives the results

$$\hat{\beta}_1 = \hat{\mu}_Y - \left(\frac{\hat{c}_{YX}}{\hat{\sigma}_X^2}\right)\hat{\mu}_X \quad \text{and} \quad \hat{\beta}_2 = \frac{\hat{c}_{YX}}{\hat{\sigma}_X^2}. \quad (8.84)$$

This solution shows that least squares estimates of β_1 and β_2 depend only on the sample means, sample variances, and the sample covariance between x_n and y_n.

8.7 Always Include the Intercept Term

In Section 8.5, a linear growth rate was estimated by fitting an equation of the form (8.66), which contained two parameters, the slope and the intercept. Since only the slope was of interest, could the same result have been obtained by fitting an equation without the intercept term? No! To see why, let us fit the data to the equation $y_n = \beta t_n + \epsilon_n$, which does not contain an intercept term. The result is shown in Figure 8.8. Clearly, this particular line fit is a poor model of atmospheric CO_2 concentration. Moreover, the resulting slope is 10 ppm per year, which is several times greater than the slope obtained from including the intercept. What is going on? Omitting the intercept term is tantamount to constraining the line to *pass through the origin*. This implicit constraint is very unrealistic: atmospheric CO_2 concentration does not vanish at January 1960, the origin of the model. Note that the line fit without an intercept term is still "best" in the sense that it minimizes the sum square residuals – it is the best line that fits the data while constrained to pass through the origin.

There is another good reason to include the intercept term in a regression equation. When considering regression models of the form (8.3), we often implicitly assume that the errors ϵ have zero mean. Thus, it is desirable for the residuals to have zero mean. Requiring that the residuals have zero mean is equivalent to requiring

$$\mathbf{j}^T\hat{\epsilon} = \mathbf{0}, \quad (8.85)$$

where $\mathbf{j}$ is a vector of ones. Recall from Section 8.3 that the residuals are orthogonal to the predictors. Thus, constraint (8.85) is satisfied automatically by including $\mathbf{j}$ as one of the columns of $\mathbf{X}$, that is, by including an intercept term in the model.

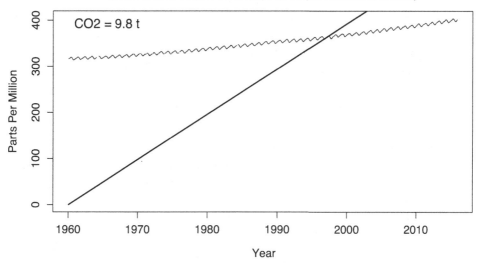

Figure 8.8 Best fit line to atmospheric CO_2 concentration without including the intercept term. Not recommended!

Equivalently, including an intercept in the model ensures that the residuals of the least squares fit have zero mean.

One exception to the rule "always include the intercept term" is when all the data have been centered, because in this case the intercept term will be zero and therefore need not be included in the model. The fact that the intercept term will vanish if the data are centered follows from the orthogonalizing procedure for multiple regression, which is discussed in Section 9.7.

8.8 Further Topics

One problem with solving the least squares problem is that it requires inverting a matrix. Although a matrix might not be singular, it may be *ill-conditioned* – that is, the inverse of the matrix will be sensitive to small errors in the matrix elements. This sensitivity usually occurs when predictors are "close" to being linearly dependent. Such situations can be diagnosed by matrix decompositions based on orthogonal transformations, including the Singular Value Decomposition and the QR-algorithm (Golub and Van Loan, 1996). In practice, algorithms based on matrix decompositions are recommended for solving least squares problems because they are more stable and comprehensively diagnose ill-conditioning.

The model discussed in this chapter assumed that the predictor matrix **X** had no random errors. The resulting solution method is known as *Ordinary Least Squares*.

Accounting for random errors in the predictors leads to the *Total Least Squares Problem* or *Errors-in-Variables Model* (see the end of Section 12.2). To understand the difference, recall that in two dimensions, the least squares method corresponds to measuring errors in the *vertical* direction, as indicated in Figure 8.2. In contrast, the Total Least Squares method corresponds to measuring errors along the *normal* to the line, which is a special case of the Errors-in-Variables method. A nice property of the Total Least Squares method is that the solution is invariant to switching the X and Y labels, whereas this invariance does not hold for the least squares method.

The results of this chapter hold regardless of the distribution of the random term ϵ. That is, the least squares solution always minimizes the sum square errors regardless of the distribution of the errors. A remarkable result, known as the Gauss–Markov theorem, states that the least squares estimate gives the *best linear unbiased estimator (BLUE)* for any linear regression model in which the errors have equal variances, zero expectation, and are uncorrelated. Here "best" means that the estimated parameters have the minimum variance. This theorem is remarkable because it does not depend on the distribution of the errors. However, there exist situations in which these assumptions are not appropriate. For instance, if the predictand is always positive, then the errors cannot have constant variance. For instance, if X is small, then the error must be sufficiently small to avoid negative values of Y; if X is large, the error can have larger variance. To deal with such problems, one could modify the predictand to be the logarithm of Y. An alternative approach is to specify the distribution of the errors and then use the *Maximum Likelihood Method* to derive parameter estimates. This latter approach is the basis of the *Generalized Linear Model* and provides a unifying framework for linear regression, logistic regression, and Poisson regression.

The Gauss–Markov theorem seems to imply that it would be hard to improve on the least squares solution when the errors are *iid* with zero mean. Surprisingly, it turns out that *biased* parameter estimates can give smaller mean square error *in independent data* than unbiased estimates. Indeed, least squares estimates often give smaller mean square error in independent data by *shrinking them toward zero*. Methods for performing such shrinkage include *ridge regression* and *Bayesian regression*. The ridge regression estimate is

$$\hat{\beta}_{\text{ridge}} = \left(\mathbf{X}^T\mathbf{X} + \lambda\mathbf{I}\right)^{-1}\mathbf{X}^T\mathbf{y}, \tag{8.86}$$

where λ is an adjustable parameter. In contrast to ordinary least squares, ridge regression estimates can be computed even if the matrix $\mathbf{X}^T\mathbf{X}$ is singular! Unlike ordinary least squares, ridge regression estimate depends on the units used to measure predictors.

Many common estimation problems are *under-determined*. Some of these problems will be discussed in later chapters in the context of multivariate analysis.

The key to solving these problems is to introduce *constraints* in the estimation problem. This approach is known as *regularized regression*. Ridge regression is an example of regularized regression that emerges when the norm of the coefficients $\|\boldsymbol{\beta}\|$ is constrained.

This chapter assumed that all variables are *continuous*. It is of interest to develop regression models for *discrete* variables too. If the variable is *binary* – that is, takes on only two values, such as rain/no-rain or present/absent – then *logistic regression* is a common method for estimating regression models for binary variables.

Generalizing regression to the case of multiple predictands is called *multivariate regression*. Multivariate regression is central to many other techniques discussed in this book, including canonical correlation analysis, predictable component analysis, and discriminant analysis. Overfitting is also an important concept that will be discussed in Chapter 10.

A fascinating history of the method of least squares is presented in Stigler (1986).

8.9 Conceptual Questions

1 What is the method of least squares used for?
2 After using the method of least squares to fit a model, in what sense is the fit "best."
3 After fitting a linear model to data using the method of least squares, what is the relation between the residuals and the predictors?
4 If you change the units of $\mathbf{X}$, what changes in the least squares model? What does not change?
5 What is the constant that minimizes the sum square difference between it and the data?
6 What is the coefficient of determination? What are some alternative interpretations of it?
7 In what sense is R^2 a generalization of the correlation coefficient?
8 How do you remove the annual cycle using the method of least squares?
9 How do you use the method of least squares to estimate a trend?
10 Why is it recommended to always include an intercept in your model?

9

Linear Regression
Inference

> There are many hypotheses we can test about regression lines, but the
> most common and important one is that the slope of the line of means
> is zero. This hypothesis is equivalent to estimating the chance that we
> would observe a linear trend as strong or stronger than the data show
> when there is actually no relationship between the dependent and inde-
> pendent variables.[1]
>
> *Glantz and Slinker (2016)*

The previous chapter discussed the method of least squares for estimating parameters in a linear model. This method will fit *any* model to a set of data points. The
question arises as to whether the resulting model is a "good" one. One criterion is
that the model should fit the data significantly better than a simpler model in which
certain predictors are omitted. After all, if the fit is not significantly better, then
the predictors in question can just be omitted to produce a model that is almost as
good. In a linear model framework, this approach is equivalent to testing if selected
regression parameters vanish. This chapter discusses procedures for testing such
hypotheses. Deciding whether a predictor can be omitted from a model depends
in part on the other predictors in the model. If a regression parameter is deemed
significant, then the associated predictor is said to influence the predictand while
"controlling" for the other predictors. This concept is illustrated in two examples. In
the first, a variable is shown to have a significant trend after controlling for ENSO.
In the second, the relation between two variables is shown to depend on whether
the regression model controls for the seasonal cycle. In the second example, the
seasonal cycle plays the role of a confounding variable. The equivalence between
controlling for variables and regressing them out is discussed. This chapter also
discusses *detection and attribution of climate change*, which can be framed in a
regression model framework.

---

[1] Republished with permission of McGraw Hill LLC, from *Primer of Applied Regression & Analysis of
Variance*, Glantz and Slinker (2016), p59; permission conveyed through the Copyright Clearance Center, Inc.

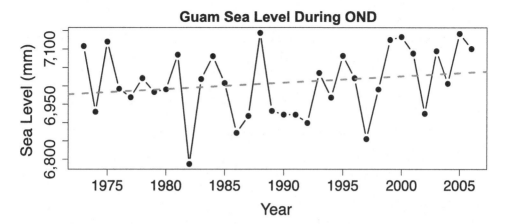

Figure 9.1 October–December mean sea level at Guam (solid lines with dots). Also shown is the least-squares linear trend (dashed grey line).

9.1 The Problem

Is sea level rising? The earth is warming and this warming causes water to expand and glaciers to melt, contributing to *global* sea level rise. On the other hand, *local* sea level is influenced by many factors, including ocean currents, atmospheric pressure, and tectonic processes. As an example, Figure 9.1 shows the observed sea level at a station in Guam averaged over October to December.[2] To quantify sea level rise, recall that global warming is a gradual process, hence it can be modeled by a low-order polynomial in time:

$$\text{sea level} = \beta_1 + \beta_2 * \text{year} + \epsilon_Y, \tag{9.1}$$

where β_1 and β_2 are the intercept and slope parameters, and ϵ_Y is random. Applying the method of least squares (see Chapter 8) yields estimates for β_1 and β_2, where the latter is

$$\hat{\beta}_2 \approx 1.7 \text{ mm / year.} \tag{9.2}$$

The resulting least squares prediction of sea level is shown as the dashed line in Figure 9.1. The estimated rate is positive, consistent with expectation. Does this result confirm sea level rise at Guam? Not necessarily! After all, year-to-year changes often exceed 100 mm, which are large compared to the trend. That is, the estimated rise might reflect random sampling variability. These considerations lead us to consider the hypothesis $\beta_2 = 0$. Rejecting this hypothesis would support the hypothesis of sea level rise in Guam. This chapter discusses such hypothesis tests. We will find

---

[2] The data is obtained from the Permanent Service for Mean Sea Level (www.psmsl.org/data/obtaining/stations/540.php).

that this rate is not statistically significant. However, this conclusion is not the final word: A more complete analysis reveals that sea level rise is indeed statistically significant, *after controlling for ENSO.*

9.2 The Model

Following Chapter 8, we consider the linear model

$$
\underset{N \times 1}{\mathbf{y}} \;=\; \underset{N \times M}{\mathbf{X}} \quad \underset{M \times 1}{\boldsymbol{\beta}} \;+\; \underset{N \times 1}{\boldsymbol{\epsilon}} \,, \tag{9.3}
$$

where $\mathbf{X}$ specifies the predictors, $\boldsymbol{\beta}$ are regression parameters, and $\boldsymbol{\epsilon}$ a random vector whose elements are independently and identically distributed. The least squares method, discussed in Chapter 8, can be used to estimate the regression parameters $\boldsymbol{\beta}$. The result is

$$
\hat{\boldsymbol{\beta}} = \left(\mathbf{X}^T\mathbf{X}\right)^{-1}\mathbf{X}^T\mathbf{y}, \tag{9.4}
$$

where the caret $\hat{}$ denotes a sample estimate. The least squares method applies regardless of the distribution of the errors $\boldsymbol{\epsilon}$. However, to test hypotheses and derive confidence intervals the distribution of the errors needs to be known. Here we assume the errors are independent and normally distributed as $\mathcal{N}(0, \sigma_\epsilon^2)$, which can be expressed concisely as

$$
\boldsymbol{\epsilon} \sim \mathcal{N}_N(\mathbf{0}, \sigma_\epsilon^2 \mathbf{I}). \tag{9.5}
$$

9.3 Distribution of the Residuals

The source of randomness in (9.3) is the term $\boldsymbol{\epsilon}$. A natural estimate of the variance of $\boldsymbol{\epsilon}$ is based on the sum square residuals

$$
\mathrm{SSE} = \hat{\boldsymbol{\epsilon}}^T \hat{\boldsymbol{\epsilon}}, \tag{9.6}
$$

where $\hat{\boldsymbol{\epsilon}}$ denotes the residuals

$$
\hat{\boldsymbol{\epsilon}} = \mathbf{y} - \mathbf{X}\hat{\boldsymbol{\beta}}. \tag{9.7}
$$

Substituting (9.4) into (9.7) gives

$$
\hat{\boldsymbol{\epsilon}} = (\mathbf{I} - \mathbf{H})\,\mathbf{y}. \tag{9.8}
$$

where $\mathbf{H}$, termed the *projection matrix* (or the "hat matrix"; see Chapter 8), is defined as

$$
\mathbf{H} = \mathbf{X}\left(\mathbf{X}^T\mathbf{X}\right)^{-1}\mathbf{X}^T. \tag{9.9}
$$

It is easily verified that $\mathbf{H}^2 = \mathbf{H}$ and $\mathbf{HX} = \mathbf{X}$. It follows from these properties that $(\mathbf{I} - \mathbf{H})^2 = (\mathbf{I} - \mathbf{H})$. Given these properties, substituting (9.3) into (9.8) yields

$$\hat{\boldsymbol{\epsilon}} = (\mathbf{I} - \mathbf{H})(\mathbf{X}\boldsymbol{\beta} + \boldsymbol{\epsilon}) = (\mathbf{I} - \mathbf{H})\boldsymbol{\epsilon}, \tag{9.10}$$

where we have used $(\mathbf{I} - \mathbf{H})\mathbf{X}\boldsymbol{\beta} = \mathbf{0}$. In Section 9.13, we prove that

$$\frac{\text{SSE}}{\sigma_\epsilon^2} \sim \chi^2_{N-M}. \tag{9.11}$$

Note that the degrees of freedom is $N - M$ while $\hat{\boldsymbol{\epsilon}}$ is an N-dimensional vector. The degrees of freedom follows from the fact that the residual errors satisfy the constraints

$$\mathbf{X}^T \hat{\boldsymbol{\epsilon}} = \mathbf{0}, \tag{9.12}$$

(see (8.19)). This equation represents M linearly independent constraints, thereby rendering the number of degrees of freedom $N - M$. Since $\mathbb{E}[\chi_v^2] = v$, taking the mean of both sides of (9.11) and solving for σ_ϵ^2 shows that an unbiased estimate of the variance σ_ϵ^2 is

$$\hat{\sigma}_\epsilon^2 = \frac{\text{SSE}}{N - M}. \tag{9.13}$$

We consider only over-determined problems in which $N > M$.

9.4 Distribution of the Least Squares Estimates

The least squares estimate $\hat{\boldsymbol{\beta}}$ obtained from (9.4) is a linear function of the random variable $\mathbf{y}$, hence $\hat{\boldsymbol{\beta}}$ is itself a random variable. To determine the distribution of the least squares estimate, substitute (9.3) into (9.4), which gives

$$\begin{aligned}
\hat{\boldsymbol{\beta}} &= \left(\mathbf{X}^T\mathbf{X}\right)^{-1}\mathbf{X}^T\mathbf{y} \\
&= \left(\mathbf{X}^T\mathbf{X}\right)^{-1}\mathbf{X}^T(\mathbf{X}\boldsymbol{\beta} + \boldsymbol{\epsilon}) \\
&= \left(\mathbf{X}^T\mathbf{X}\right)^{-1}\left((\mathbf{X}^T\mathbf{X})\boldsymbol{\beta} + \mathbf{X}^T\boldsymbol{\epsilon}\right) \\
&= \boldsymbol{\beta} + \left(\mathbf{X}^T\mathbf{X}\right)^{-1}\mathbf{X}^T\boldsymbol{\epsilon}.
\end{aligned} \tag{9.14}$$

This equation shows that $\hat{\boldsymbol{\beta}}$ is a constant, $\boldsymbol{\beta}$, plus a linear combination of independent, normally distributed random variables, hence $\hat{\boldsymbol{\beta}}$ is multivariate Gaussian. The distribution is thus specified completely by its mean and covariance matrix. The mean of $\hat{\boldsymbol{\beta}}$ is

$$\mathbb{E}\left[\hat{\boldsymbol{\beta}}\right] = \mathbb{E}\left[\boldsymbol{\beta} + \left(\mathbf{X}^T\mathbf{X}\right)^{-1}\mathbf{X}^T\boldsymbol{\epsilon}\right] = \boldsymbol{\beta} + \left(\mathbf{X}^T\mathbf{X}\right)^{-1}\mathbf{X}^T\mathbb{E}[\boldsymbol{\epsilon}] = \boldsymbol{\beta}, \tag{9.15}$$

where we have used the linear property of the expectation operator and the fact that $\mathbb{E}[\epsilon] = \mathbf{0}$. This result shows that $\hat{\beta}$ is an unbiased estimate of β. The covariance matrix of $\hat{\beta}$ is

$$
\begin{aligned}
\operatorname{cov}\left[\hat{\beta}\right] &= \mathbb{E}\left[\left(\hat{\beta} - \mathbb{E}[\hat{\beta}]\right)\left(\hat{\beta} - \mathbb{E}[\hat{\beta}]\right)^T\right] \\
&= \mathbb{E}\left[\left(\mathbf{X}^T\mathbf{X}\right)^{-1}\mathbf{X}^T\epsilon\epsilon^T\mathbf{X}\left(\mathbf{X}^T\mathbf{X}\right)^{-1}\right] \\
&= \left(\mathbf{X}^T\mathbf{X}\right)^{-1}\mathbf{X}^T\mathbb{E}\left[\epsilon\epsilon^T\right]\mathbf{X}\left(\mathbf{X}^T\mathbf{X}\right)^{-1} \\
&= \sigma_\epsilon^2\left(\mathbf{X}^T\mathbf{X}\right)^{-1},
\end{aligned}
\tag{9.16}
$$

where we used the fact that (9.5) implies $\mathbb{E}\left[\epsilon\epsilon^T\right] = \sigma_\epsilon^2\mathbf{I}$. It follows from the two previous results that $\hat{\beta}$ is distributed as

$$
\hat{\beta} \sim \mathcal{N}_M\left(\beta, \sigma_\epsilon^2\left(\mathbf{X}^T\mathbf{X}\right)^{-1}\right).
\tag{9.17}
$$

The regression parameters are also independent of the residuals. To show this, first note that their covariance matrix vanishes,

$$
\begin{aligned}
\operatorname{cov}[\hat{\beta}, \hat{\epsilon}] &= \mathbb{E}\left[\left(\hat{\beta} - \mathbb{E}[\hat{\beta}]\right)\left(\hat{\epsilon} - \mathbb{E}[\hat{\epsilon}]\right)^T\right] \\
&= \mathbb{E}\left[\left(\mathbf{X}^T\mathbf{X}\right)^{-1}\mathbf{X}^T\epsilon\epsilon^T\left(\mathbf{I} - \mathbf{H}\right)\right] \\
&= \left(\mathbf{X}^T\mathbf{X}\right)^{-1}\mathbf{X}^T\mathbb{E}[\epsilon\epsilon^T]\left(\mathbf{I} - \mathbf{H}\right) \\
&= \sigma_\epsilon^2\left(\mathbf{X}^T\mathbf{X}\right)^{-1}\mathbf{X}^T\left(\mathbf{I} - \mathbf{H}\right) \\
&= \mathbf{0},
\end{aligned}
\tag{9.18}
$$

where we use the fact $\mathbf{X}^T\left(\mathbf{I} - \mathbf{H}\right) = \mathbf{0}$. Since both the residuals $\hat{\epsilon}$ and regression coefficients $\hat{\beta}$ are derived from multivariate Gaussian random variables, the fact that their covariance vanishes implies that they are independent. This independence is not obvious because $\hat{\beta}$ and $\hat{\epsilon}$ are derived from the same data. The reason these quantities are nevertheless independent is because they depend on orthogonal combinations of the sample.

The distribution of any *single* regression coefficient can be obtained from (9.17) using Theorem 7.10. Specifically, the distribution of the i'th regression coefficient is

$$
\hat{\beta}_i \sim \mathcal{N}\left(\beta_i, \sigma_\epsilon^2 d_i\right).
\tag{9.19}
$$

where d_i is the i'th diagonal element of $\left(\mathbf{X}^T\mathbf{X}\right)^{-1}$:

$$
d_i = \left(\left(\mathbf{X}^T\mathbf{X}\right)^{-1}\right)_{ii}
\tag{9.20}
$$

Thus, the standard error of $\hat{\boldsymbol{\beta}}_i$ is $\sigma_\epsilon \sqrt{d_i}$, and

$$z = \frac{\text{observed} - \text{hypothesized}}{\text{standard error}} = \frac{\hat{\beta}_i - \beta_i}{\sigma_\epsilon \sqrt{d_i}} \sim \mathcal{N}(0, 1). \qquad (9.21)$$

Unfortunately, this expression involves the unknown population parameter σ_ϵ. However, (9.13) and (9.18) can be invoked with Theorem 2.1 to derive the t-statistic

$$t = \frac{\hat{\beta}_i - \beta_i}{\hat{\sigma}_\epsilon \sqrt{d_i}} \sim t_{N-M}. \qquad (9.22)$$

9.5 Inferences about Individual Regression Parameters

The distribution (9.22) provides the basis for testing hypotheses about individual regression parameters. For instance, to test the null hypothesis $\beta_i = 0$, we use the test statistic

$$t = \frac{\hat{\beta}_i}{\hat{\sigma}_\epsilon \sqrt{d_i}}. \qquad (9.23)$$

Generally, the alternative hypothesis is $\beta_i \neq 0$, regardless of the *sign* of the regression parameter, implying a two-tailed test. Thus, the decision rule for $\alpha 100\%$ significance is

$$\begin{array}{llll} \text{reject } \beta_j = 0 & \text{if} & |t| > t_{\alpha/2, N-M} \\ \text{accept } \beta_j = 0 & \text{if} & |t| < t_{\alpha/2, N-M} \end{array}. \qquad (9.24)$$

A confidence interval for the regression coefficient is then

$$(1 - \alpha)100\% \text{ confidence interval} = \hat{\beta}_j \pm t_{\alpha/2, N-M} \hat{\sigma}_\epsilon \sqrt{d_j}. \qquad (9.25)$$

Example 9.1 *To illustrate this test, consider the sea level example of Section 9.1. Model (9.1) is effectively the model*

$$Y = \beta_1 + \beta_2 X + \epsilon, \qquad (9.26)$$

where X is "year." The hypothesis of no sea level rise corresponds to the hypothesis $\beta_2 = 0$. The statistic for testing this hypothesis is

$$t = \frac{\hat{\beta}_2}{\hat{\sigma}_\epsilon \sqrt{d_2}}, \quad \text{where} \quad d_2 = \left(\left(\mathbf{X}^T \mathbf{X} \right)^{-1} \right)_{22}. \qquad (9.27)$$

The matrix $\mathbf{X}^T \mathbf{X}$ is 2×2, but, following (9.20), the second diagonal element is extracted because we are testing a hypothesis about β_2, that is, the second *regression parameter. Because $\mathbf{X}^T \mathbf{X}$ is 2×2, its inverse can be computed analytically. As shown* *in Section 8.6, this inversion yields $d_2 = 1/(N\hat{\sigma}_X^2)$. Substituting this value gives the statistic*

$$t = \sqrt{N}\hat{\beta}_2 \frac{\hat{\sigma}_X}{\hat{\sigma}_\epsilon}, \tag{9.28}$$

which has a central t distribution with $N - 2$ degrees of freedom under the null hypothesis $\beta_2 = 0$. The $(1 - \alpha)100\%$ confidence interval for the slope parameter is

$$(1 - \alpha)100\% \text{ confidence interval} = \hat{\beta}_2 \pm \frac{1}{\sqrt{N}} \frac{\hat{\sigma}_\epsilon}{\hat{\sigma}_X} t_{\alpha/2, N-2}. \tag{9.29}$$

The 95% confidence interval derived from (9.29) gives $(-1.5, 5.0)$, which includes zero, implying that the trend is not significant.

9.6 Controlling for the Influence of Other Variables

Despite finding no statistically significant trend in Guam sea level, the question of whether sea level is rising at Guam is not settled. In particular, sea level along the western part of the Pacific basin is affected by more than global warming. One major source of variability is ENSO: during El Niño, the trade winds along the equatorial Pacific weaken, causing below-average sea levels in the western Pacific; conversely, during La Niña, surface trade winds strengthen, causing above-average sea levels in the western Pacific. This year-to-year variability associated with ENSO might be obscuring long-term sea level rise. To investigate this, one can examine sea level separately for El Niño, La Niña, and neutral ENSO years. An analysis of this type is shown in Figure 9.2. Here, instead of using lines to join data between consecutive years, the lines are used to join data during La Niñas ("C" for "cold"), El Niños ("W" for "Warm"), and Neutral years *separately*. The figure shows that sea level is consistently higher during La Niña than during El Niño events, indicating a systematic relation with ENSO. Also, long-term sea level rise is more evident in the neutral and warm ENSO states separately. One might be tempted to test hypotheses about sea level rise for each ENSO state separately, but this approach has several shortcomings: (1) it significantly reduces the sample size and thus increases uncertainty, (2) it involves testing a hypothesis three times for three different ENSO states, which leads to a *multiple testing problem* (see Chapter 11), and (3) the conclusions may be sensitive to the arbitrary separation into El Niño/La Niña/neutral categories.

   An alternative approach is to consider a regression model that explicitly includes a dependence on ENSO. For instance, using Niño 3.4 as an index of ENSO, consider

$$\text{sea level} = \beta_1' + \beta_2' * \text{year} + \beta_3' * \text{Niño 3.4} + \epsilon_Y'. \tag{9.30}$$

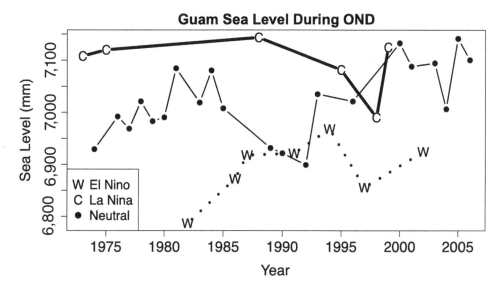

Figure 9.2 October to December mean sea level at Guam, as in Figure 9.1, except that the years categorized as El Niño ("W" for "warm"), La Niña ("C" for "cold"), and neutral are each distinguished by separate symbols, and lines or small dots are used to join data in the same category. Warm and cold ENSO years are defined based on when the Niño 3.4 anomaly is greater or less than 1K, respectively.

The interpretation of β_2' in (9.30) differs fundamentally from the interpretation of β_2 in (9.1), despite the fact that both coefficients multiply the same predictor. Specifically, β_2' quantifies how much sea level would change in one year *if Niño 3.4 were held constant.*

To clarify this interpretation, consider the general model

$$Y = \beta_1 X_1 + \beta_2 X_2 + \cdots + \beta_M X_M + \epsilon_n. \tag{9.31}$$

What is the change in Y resulting from one unit change in X_1? The answer is not β_1! The reason is that when X_1 changes, the other X's may change systematically with X_1. For instance, suppose every time X_1 goes up 1 unit, X_2 goes down by 1 unit, while the other X's are independent. Then, the change in Y resulting from a change in X_1 is $\beta_1 - \beta_2$, because the change β_1 resulting from one unit increase in X_1 is compensated by the change β_2 resulting from one unit decrease in X_2. Thus, β_1 measures the average change in Y per unit change in X_1 *while holding all other predictors in the regression equation constant.* If the noise term ϵ were not present, then β_m is the *partial derivative* of Y with respect to X_m; that is, it is the derivative of Y with respect to X_m while holding the other X's constant:

$$\frac{\partial Y}{\partial X_m} = \beta_m. \tag{9.32}$$

(As a result, β_m is sometimes called a *partial regression coefficient*). Testing the hypothesis $\beta_1 = 0$ in the multiple regression model (9.31) is commonly referred to as testing for a relation between Y and X_1 while *controlling* for $X_2, \ldots, X_M$.

Another way to understand this concept is through *conditional independence*. Y and X_2 are said to be conditionally independent, given X_1, if

$$p(y \mid x_1, x_2) = p(y \mid x_1). \tag{9.33}$$

This identity states that relative to the conditional distribution $p(y \mid x_1)$, adding X_2 as a conditioning variable does not change the distribution of Y. In terms of the regression model $Y = \beta_1 X_1 + \beta_2 X_2 + \epsilon$, this means $\beta_2 = 0$. For Gaussian distributions, hypotheses about partial regression coefficients are hypotheses about conditional distributions.

Thus, to test for a linear growth in sea level rise while controlling for ENSO, we assume the data were generated by model (9.30) and test the hypothesis $\beta_2 = 0$. This test is based on the t-statistic (9.23), which differs from that of the simple regression model (9.1) because $\hat{\beta}_i$ and d_i from (9.30) include ENSO as a predictor. Using the least squares method to estimate the parameters in (9.30) yields estimates for $\beta_1, \beta_2, \beta_3$. The estimate for β_2 is

$$\hat{\beta}_2' = 2.6 \text{ mm / year}, \tag{9.34}$$

which is *larger* than when not controlling for ENSO. Moreover, the corresponding 95% confidence interval is (0.5, 4.8), which does NOT include zero, and hence is statistically significant. These results suggest that Guam sea level is indeed increasing with time, but this increase is obscured by random ENSO variability that is superimposed on the gradual sea level rise.

As a by-product of this analysis, the least squares estimate of β_3' in (9.30) is

$$\hat{\beta}_3' = -63 \text{ mm / Niño 3.4 index}. \tag{9.35}$$

Moreover, the corresponding 95% confidence interval is $(-83, -44)$, which does not include zero, and thus is statistically significant. This result implies that, in any given year, Guam sea-level drops 63 mm for each unit increase in Niño 3.4 index. As anticipated, this year-to-year change is very large compared to the change in sea-level resulting from the long-term trend.

9.7 Equivalence to "Regressing Out" Predictors

There exists yet another way to interpret these results. Specifically, the dependence between two variables, while controlling for a third set of variables, can also be quantified by regressing out the third set of variables, and then analyzing the depen-

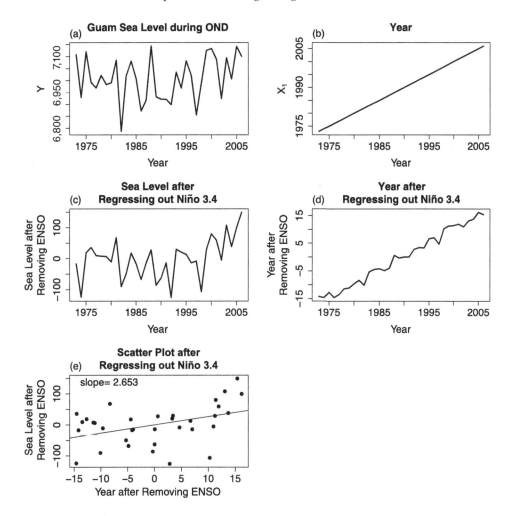

Figure 9.3 Example of regressing out predictors to quantify dependencies. The Y-variable is Guam sea level during October to December (a), and the predictor X_1 whose dependence is being investigated is the year (b). The Niño 3.4 index, plus a constant, were regressed out of Y (c) and X_1 (d). A scatter plot of the two variables after Niño 3.4 has been regressed out is shown in (e), along with the least squares line fit and slope.

dence between the resulting residuals. We first show a particular example, then show it generally.

Let us consider again whether Guam sea level has a trend. The sea level time series is shown in Figure 9.3a, while the trend component X_1 is shown in Figure 9.3b. In contrast to the previous section, we *regress out* ENSO by separately fitting the following model:

$$\text{sea level} = \beta_1' + \beta_3' * \text{Niño 3.4} + \epsilon', \tag{9.36}$$

and then compute the residuals

$$Y^* = \text{sea level} - \hat{\beta}'_1 - \hat{\beta}'_3 * \text{Niño 3.4.} \tag{9.37}$$

By construction, Y^* is centered and uncorrelated with the Niño 3.4 index. The variable Y^* is said to be the sea level after regressing out ENSO, or sea level after "removing" ENSO. This variable is shown in Figue 9.3c. The trend in sea level is much more apparent after ENSO has been regressed out (e.g., compare Y^* in Figure 9.3c versus Y in Figure 9.3a). However, for consistency, ENSO must be regressed out of *all* other time series, including the trend term. Therefore, we consider the model

$$\text{year} = \beta''_1 + \beta''_3 * \text{Niño 3.4} + \epsilon'', \tag{9.38}$$

and corresponding residuals

$$X^*_1 = \text{year} - \hat{\beta}''_1 - \hat{\beta}''_3 * \text{Niño 3.4.} \tag{9.39}$$

Although the year is a deterministic variable, ENSO must be regressed out of *every* variable for consistency. The year after regressing out ENSO is shown in Figure 9.3d and is still mostly a linear function of year. Having computed the two residuals X^*_1 and Y^*, we then plot them on a scatter plot in Figures 9.3e and fit a line, which yields a slope of 2.6 mm/year, precisely the same slope as obtained from the full model (9.30) (see (9.34)).

This numerical example illustrates a general property of regression. To state this property precisely, let $\mathbf{X}_1$ identify the M_1 variables whose relation with $\mathbf{y}$ is of interest, while $\mathbf{X}_2$ identifies the M_2 variables to be controlled. The regression equation becomes

$$\mathbf{y} = \mathbf{X}_1\boldsymbol{\beta}_1 + \mathbf{X}_2\boldsymbol{\beta}_2 + \boldsymbol{\epsilon}. \tag{9.40}$$

We want to regress out variability associated with $\mathbf{X}_2$. To do this, we construct the projection matrix that projects a vector onto the range of $\mathbf{X}_2$ (see Section 8.4):

$$\mathbf{H}_2 = \mathbf{X}_2 \left(\mathbf{X}_2^T\mathbf{X}_2\right)^{-1} \mathbf{X}_2^T. \tag{9.41}$$

This matrix has the property $\mathbf{H}_2\mathbf{X}_2 = \mathbf{X}_2$, as one would expect of a projection matrix. Thus, regressing out $\mathbf{X}_2$ from $\mathbf{y}$ is done by computing the *difference* $\mathbf{y}' = \mathbf{y} - \mathbf{H}_2\mathbf{y} = (\mathbf{I} - \mathbf{H}_2)\mathbf{y}$. The matrix $\mathbf{I} - \mathbf{H}_2$ projects vectors to the space *orthogonal* to $\mathbf{X}_2$. Thus,

$$\mathbf{y}_{|2} = (\mathbf{I} - \mathbf{H}_2)\,\mathbf{y}, \tag{9.42}$$

gives $\mathbf{y}$ after $\mathbf{X}_2$ has been regressed out. We use the subscript "$|2$" to indicate that $\mathbf{X}_2$ is regressed out. Similarly,

$$\mathbf{X}_{1|2} = (\mathbf{I} - \mathbf{H}_2)\,\mathbf{X}_1. \tag{9.43}$$

gives $\mathbf{X}_1$ after $\mathbf{X}_2$ has been regressed out. Multiplying both sides of (9.40) by $\mathbf{I} - \mathbf{H}_2$ gives

$$\mathbf{y}_{|2} = \mathbf{X}_{1|2}\boldsymbol{\beta}_1 + (\mathbf{I} - \mathbf{H}_2)\,\boldsymbol{\epsilon}. \tag{9.44}$$

Note that the regression parameters $\boldsymbol{\beta}_2$ have been removed by this procedure. Therefore, the regression parameter $\boldsymbol{\beta}_1$ can be estimated by the method of least squares:

$$\hat{\boldsymbol{\beta}}_1 = \left(\mathbf{X}_{1|2}^T \mathbf{X}_{1|2}\right)^{-1} \mathbf{X}_{1|2}^T \mathbf{y}_{|2}. \tag{9.45}$$

It can be shown that the solution (9.45) is identical to the estimates for $\hat{\boldsymbol{\beta}}_1$ obtained from the full regression model

$$\mathbf{y} = \mathbf{X}_1\boldsymbol{\beta}_1 + \mathbf{X}_2\boldsymbol{\beta}_2 + \boldsymbol{\epsilon} = \begin{bmatrix} \mathbf{X}_1 & \mathbf{X}_2 \end{bmatrix} \begin{pmatrix} \boldsymbol{\beta}_1 \\ \boldsymbol{\beta}_2 \end{pmatrix} + \boldsymbol{\epsilon} = \mathbf{X}\boldsymbol{\beta} + \boldsymbol{\epsilon}. \tag{9.46}$$

In the latter model, the least squares estimates of $\boldsymbol{\beta}_1$ and $\boldsymbol{\beta}_2$ are known immediately:

$$\hat{\boldsymbol{\beta}} = \begin{pmatrix} \hat{\boldsymbol{\beta}}_1 \\ \hat{\boldsymbol{\beta}}_2 \end{pmatrix} = \left(\mathbf{X}^T\mathbf{X}\right)^{-1} \mathbf{X}^T\mathbf{y}. \tag{9.47}$$

Substituting $\mathbf{X} = \begin{bmatrix} \mathbf{X}_1 & \mathbf{X}_2 \end{bmatrix}$ leads to *block patterned* matrices that can be inverted in block form, as given in (A.9). The calculation is tedious, so we give the final result:

$$\hat{\boldsymbol{\beta}}_1 = \left(\mathbf{X}_1^T\,(\mathbf{I} - \mathbf{H}_2)\,\mathbf{X}_1\right)^{-1} \left(\mathbf{X}_1^T\,(\mathbf{I} - \mathbf{H}_2)\,\mathbf{y}\right), \tag{9.48}$$

which is identical to (9.45). Incidentally, the t-statistic (9.23) is the same in the two calculations, hence hypothesis tests of $\boldsymbol{\beta}_1$ also are equivalent, provided that one accounts properly for the degrees of freedom. Specifically, the degrees of freedom in the noise is not $N - M_1$, as one might guess (incorrectly) by counting the number of predictors in (9.44), but $N - M_1 - M_2 = N - M$, because an additional M_2 degrees of freedom were lost through the process of projecting onto the space orthogonal to the range of $\mathbf{X}_2$.

The impact of regressing out variables can also be quantified through correlations. The correlation between sea level and year is 0.19 and is not significant, consistent with our conclusion in Section 9.5 (this is no coincidence; we show in Section 9.9 that these two hypothesis tests are always consistent). However, the correlation between these variables after ENSO has been regressed out is 0.41 (just to be clear, this is the correlation between X_1^* and Y^* shown in Figure 9.3e). This correlation is known as the *partial correlation*. In this example, the partial correlation is larger than the standard correlation, which illustrates the usefulness of regressing out variables. Since β_2' in (9.30) was found to be significant, so too is the correlation between X_1^* and Y^*. In the special case in which only a single variable Z is regressed out, the partial correlation between X and Y after Z has been regressed out can be derived equivalently as

$$\rho_{XY|Z} = \frac{\rho_{XY} - \rho_{XZ}\rho_{YZ}}{\sqrt{1 - \rho_{XZ}^2}\sqrt{1 - \rho_{YZ}^2}}. \tag{9.49}$$

This equivalence justifies a certain practice in climate studies. Specifically, a standard first step in many climate studies is to subtract out the mean and annual cycle from each time series, yielding so-called anomaly time series. Then, other regression analyses, such as trend analysis or correlation analysis, are performed on the anomalies. The equivalence implies that this two-step regression procedure (i.e., first compute anomalies, then analyze the anomalies) yields the same conclusions as a single-step procedure applied to the original data using a regression model that includes the mean and annual cycle.

9.8 Seasonality as a Confounding Variable

In the previous example, a third variable (ENSO) had the effect of *hiding* a relation between two other variables (sea level and time). The opposite effect can happen too: a third variable may produce an apparent relation between two other variables that otherwise are not related, or are related in a different way. In such cases, the third variable is said to be a *confounding variable*. A confounding variable is a variable Z that is correlated with both X and Y. When the confounding variable is held fixed, then the relation between Y and X may disappear or even reverse sign!

A confounding variable that every climate scientist should be aware of is the annual cycle. To illustrate, consider the question of whether daily temperature in Seattle is related to daily temperature in Tallahassee. Weather disturbances typically do not span the entire American continent, and therefore are unlikely to cause a relation between these cities. To investigate a relation nonetheless, one might follow the (now familiar) procedure of testing the hypothesis of zero correlation. A time series of maximum daily temperature at these two cities is shown in Figure 9.4. A scatter plot of the two time series, shown in the right panel of Figure 9.4, does indeed suggest that the two temperatures are related: when one city has a relatively high temperature, the other city tends to have a high temperature too, and similarly for cool temperatures. Quantitatively, the sample correlation between these time series is 0.65, which presumably is significant given the large sample size, daily data for 1950–2000 exceeds 18,000 numbers (we say "presumably" because serial correlation has not been taken into account). However, the most obvious relation between the two time series is that the peaks and valleys occur at nearly the same time of year. In other words, the relation merely reflects the fact that winters are colder than summers! However, this is likely not the question we intended to ask.

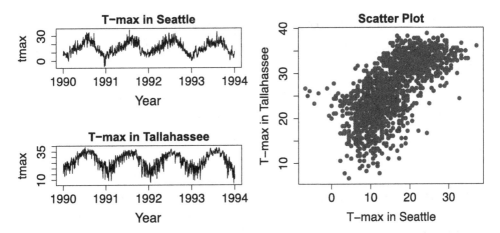

Figure 9.4 The daily maximum temperature for a weather station in Seattle, WA (top left) and Tallahassee, FL (bottom left), and their scatter plot (right panel).

Instead, we were attempting to ask if a relation between daily temperatures at the two cities exists *after controlling for the annual cycle*.

To investigate whether temperatures in Seattle and Tallahassee are related after controlling for the annual cycle, we consider the model

$$S_n = \mu + \beta_T T_n + \sum_{k-1}^{2} \left(a_k c_{k,n} + b_k s_{k,n} \right) + \epsilon_n, \tag{9.50}$$

where S_n and T_n denote the temperature in Seattle and Tallahassee, respectively, n specifies time in days, a_k and b_k are parameters defining the annual cycle, and

$$c_{k,n} = \cos\left(\frac{2\pi kn}{365.25} \right), \quad \text{and} \quad s_{k,n} = \sin\left(\frac{2\pi kn}{365.25} \right). \tag{9.51}$$

The fundamental period of the periodic components is 365.25 days to account for leap years. The decision to select two annual harmonics is a *model selection problem*, which will discussed in Chapter 10. The decision procedure is not important for this example, so we simply assert that two harmonics is sufficient. The question of whether Seattle and Tallahassee temperatures are related after controlling for the annual cycle is tantamount to the question of whether β_T is non-zero. The 95% confidence interval for β_T is found to be $(-0.15, -0.12)$, which does not include zero. Moreover, β_T is *negative*. Therefore, after holding the phase of the annual cycle constant, daily temperatures in Seattle and Tallahassee are *anti-correlated*, contrary to our initial analysis that did not account for the annual cycle. These calculations have ignored serial correlations in the data. After taking serial correlation into account, the confidence interval may widen and hence β_T may not be significant.

9.9 Equivalence between the Correlation Test and Slope Test

Two procedures for testing a linear relation between X and Y have been encountered:

1 test the hypothesis that the correlation between two variables vanishes.
2 test the hypothesis that the slope between two variables vanishes.

These two hypotheses appear to be equivalent because the slope and the correlation coefficient both measure the linear relation between two variables. On the other hand, the corresponding hypothesis test procedures appear very different. However, appearances are deceiving: given the same data, the decision rule for rejecting the hypothesis of vanishing correlation is identical to the decision rule for rejecting the hypothesis that the slope vanishes. To show this equivalence, first note that from the solutions (8.84):

$$\hat{\beta}_2 = \frac{\hat{c}_{XY}}{\hat{\sigma}_X^2} = \hat{\rho}_{XY}\frac{\hat{\sigma}_Y}{\hat{\sigma}_X}. \tag{9.52}$$

Also, the sample correlation is related to the sum square error in (8.44):

$$\hat{\rho}_{XY}^2 = 1 - \frac{\text{SSE}}{\text{SST}} = 1 - \frac{\text{SSE}}{N\hat{\sigma}_Y^2}, \tag{9.53}$$

which implies

$$\text{SSE} = N\hat{\sigma}_Y^2\left(1 - \hat{\rho}_{XY}^2\right). \tag{9.54}$$

Substituting these expressions and $\hat{\sigma}_\epsilon^2 = \text{SSE}/(N-2)$ in the t statistic (9.28) gives

$$t = \sqrt{N}\hat{\beta}_2\frac{\hat{\sigma}_X}{\hat{\sigma}_\epsilon}$$

$$= \sqrt{N}\hat{\rho}_{XY}\frac{\hat{\sigma}_Y}{\hat{\sigma}_X}\hat{\sigma}_X\sqrt{\frac{N-2}{N\hat{\sigma}_Y^2\left(1 - \hat{\rho}_{XY}^2\right)}}$$

$$= \frac{\hat{\rho}_{XY}\sqrt{N-2}}{\sqrt{1 - \hat{\rho}_{XY}^2}}. \tag{9.55}$$

This relation is identical to (2.40), which was used for the correlation test. It follows that the decision rule for testing the significance of a correlation is identical to that for testing significance of a slope, hence the two tests of independence are equivalent.

9.10 Generalized Least Squares

Up to now, the elements of ϵ have been assumed to be *iid*. We now discuss *generalized least squares*, which accounts for arbitrary correlations in ϵ. Recall that if ϵ is multivariate normal, then it is described completely by its mean μ_ϵ and covariance matrix cov[ϵ]. Without loss of generality, ϵ can be assumed to have zero mean, provided an intercept term is included in the predictor matrix $\mathbf{X}$. Thus,

$$\epsilon \sim \mathcal{N}\left(\mathbf{0}, \sigma_\epsilon^2 \mathbf{\Sigma}_\epsilon\right), \tag{9.56}$$

where $\mathbf{\Sigma}_\epsilon$ is known. Note that assuming $\mathbf{\Sigma}_\epsilon$ is known is a strong assumption that never holds in realistic applications.

Recall from Chapter 7 that any set of correlated variables can be transformed into *iid* variables. This transformation can be obtained from the eigenvector decomposition

$$\mathbf{\Sigma}_\epsilon = \mathbf{U}\mathbf{\Lambda}\mathbf{U}^T, \tag{9.57}$$

where $\mathbf{U}$ is an orthogonal matrix and $\mathbf{\Lambda}$ is a diagonal matrix with positive diagonal elements. Then, the *pre-whitening operator* is

$$\mathbf{P} = \mathbf{\Lambda}^{-1/2}\mathbf{U}^T. \tag{9.58}$$

Applying this transformation operator to (9.3) yields

$$\mathbf{y}^* = \mathbf{X}^* \boldsymbol{\beta} + \boldsymbol{\epsilon}^*, \tag{9.59}$$

where

$$\mathbf{y}^* = \mathbf{P}\mathbf{y}, \quad \mathbf{X}^* = \mathbf{P}\mathbf{X}, \quad \text{and} \quad \boldsymbol{\epsilon}^* = \mathbf{P}\boldsymbol{\epsilon}. \tag{9.60}$$

This equation is of the form (9.3), hence the least squares estimate of $\boldsymbol{\beta}$ from (9.4) is

$$\hat{\boldsymbol{\beta}} = \left((\mathbf{X}^*)^T\mathbf{X}^*\right)^{-1}(\mathbf{X}^*)^T\mathbf{y}^*. \tag{9.61}$$

By construction, $\boldsymbol{\epsilon}^*$ is *iid*, hence the distribution from (9.17) is

$$\hat{\boldsymbol{\beta}} \sim \mathcal{N}\left(\boldsymbol{\beta}, \sigma_\epsilon^2 \left((\mathbf{X}^*)^T\mathbf{X}^*\right)^{-1}\right). \tag{9.62}$$

Returning to the original variables by substituting (9.60) into these equations yields

$$\hat{\boldsymbol{\beta}} = \left(\mathbf{X}^T\mathbf{\Sigma}_\epsilon^{-1}\mathbf{X}\right)^{-1}\mathbf{X}^T\mathbf{\Sigma}_\epsilon^{-1}\mathbf{y}, \tag{9.63}$$

and

$$\hat{\boldsymbol{\beta}} \sim \mathcal{N}\left(\boldsymbol{\beta}, \sigma_\epsilon^2 \left(\mathbf{X}^T \boldsymbol{\Sigma}_\epsilon^{-1} \mathbf{X}\right)^{-1}\right), \tag{9.64}$$

where we have used

$$\mathbf{P}^T \mathbf{P} = \boldsymbol{\Sigma}_\epsilon^{-1}. \tag{9.65}$$

Similarly, the least squares estimate of σ_ϵ^2 is

$$\hat{\sigma}_\epsilon^2 = \frac{\left(\mathbf{y} - \mathbf{X}\hat{\boldsymbol{\beta}}\right)^T \boldsymbol{\Sigma}_\epsilon^{-1} \left(\mathbf{y} - \mathbf{X}\hat{\boldsymbol{\beta}}\right)}{N - M}. \tag{9.66}$$

Note that (9.63) and (9.64) differ from (9.4) and (9.17) merely by insertion of $\boldsymbol{\Sigma}_\epsilon^{-1}$ in suitable places.

9.11 Detection and Attribution of Climate Change

We now discuss the use of generalized least squares to detect climate change. This framework separates climate variability into two additive parts: forced and unforced variability. *Forced* variability refers to the response of the climate system to external mechanisms, such as changes in solar insolation, atmospheric composition as a result of volcanic activity and human-caused emissions, land-use practices, etc. *Unforced* variability, also known as *internal variability*, refers to variability that occurs naturally from fluid instabilities in the atmosphere and from atmosphere-ocean-ice-land interactions. El Niño and weather variability are examples of internal variability. The goal of detection studies is to quantify how much an observed climate change results from forced and unforced variability, and to partition the forced variability into separate components, such as human-caused versus natural.

Let the observed climate change be denoted by $\mathbf{o}$. This vector may be a single time series, such as the global mean temperature, or a set of time series consolidated into a single vector, such as temperatures from an observation network. Then, we assume that

$$\mathbf{o} = \mathbf{f} + \boldsymbol{\epsilon}, \tag{9.67}$$

where $\mathbf{f}$ and $\boldsymbol{\epsilon}$ are the forced and unforced components, respectively. The forced and unforced components come from physical or statistical models. For instance, the forced component may come from an energy balance model while the unforced component may come from a stochastic model. In any case, these models constitute hypotheses about forced and unforced variability, and the goal is to test these hypotheses based on data.

Historically, the earliest detection and attribution studies focused only on man-made global warming resulting from increasing greenhouse gas concentrations. Later, it was realized that atmospheric aerosols play a major role in forced variability. Atmospheric aerosols are solid particles or liquid droplets suspended in the atmosphere. Without going into detailed mechanisms, certain aerosols tend to cool the planet by reflecting sunlight directly and increasing the reflectivity of clouds, while other aerosols absorb solar radiation and warm the local air mass. The net effect is difficult to estimate, but most studies point to a net cooling effect. Estimates of the magnitude of this cooling have important implications for predicting future climate change. For instance, if we knew how much of the observed climate change could be attributed to just greenhouse gases, then we could use past observations to estimate the sensitivity of the climate system to changing greenhouse gas concentrations.

Accordingly, we consider the model

$$\mathbf{o} = \mathbf{f}_{\text{GHG}} + \mathbf{f}_{\text{AER}} + \mathbf{f}_{\text{NAT}} + \boldsymbol{\epsilon}, \tag{9.68}$$

where $\mathbf{f}_{\text{NAT}}$ denotes the forced response from natural mechanisms, primarily volcanic activity and solar insolation; $\mathbf{f}_{\text{AER}}$ denotes the forced response from anthropogenic atmospheric aerosols; $\mathbf{f}_{\text{GHG}}$ refers to the forced response of everything else (ozone, land-use change, etc), which is dominated by changes in greenhouse gas composition.

As mentioned earlier, the terms $\mathbf{f}_{\text{GHG}}, \mathbf{f}_{\text{AER}}, \mathbf{f}_{\text{NAT}}, \boldsymbol{\epsilon}$ come from physical or statistical models. Of course, these models are not perfect. A standard approach is to allow that the *scale* or amplitude of the response may be wrong. Accordingly, we introduce *scaling factors* $\beta_{\text{GHG}}, \beta_{\text{AER}}, \beta_{\text{NAT}}$ to account for such errors, which leads to the model

$$\mathbf{o} = \beta_{\text{GHG}}\, \mathbf{f}_{\text{GHG}} + \beta_{\text{AER}}\, \mathbf{f}_{\text{AER}} + \beta_{\text{NAT}}\, \mathbf{f}_{\text{NAT}} + \boldsymbol{\epsilon}. \tag{9.69}$$

The value of β_i for the i'th forced response informs us about the quality of the hypothesized forced response. For instance, $\beta_i = 1$ indicates that the hypothesized forced response has the right magnitude, whereas $\beta_i > 1$ or $\beta_i < 1$ indicates that the physical model underestimated or overestimated the forced response, respectively. On the other hand, $\beta_i = 0$ indicates that the observed variability can be explained without invoking the i'th forced response. These considerations lead us to consider hypotheses about β_i. If $\beta_i = 0$ is rejected, then the i'th forced response is said to have been *detected*. If $\beta_i = 1$ cannot be rejected, then the observed variability is said to be *consistent* with, or has been *attributed* to, the i'th forced response. If $\beta_i > 1$, then the hypothesized forcing $\mathbf{f}_i$ is said to *underestimate* the magnitude of the i'th forced response. After all, $\beta_i > 1$ implies that the forced response needs to be inflated to match observations. Similarly, $\beta_i < 1$ implies the hypothesized forcing

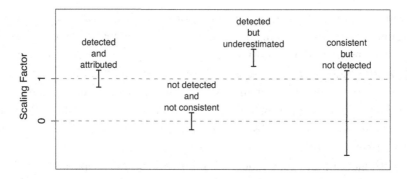

Figure 9.5 Illustration of confidence intervals for the scaling factor and their associated interpretation in terms of detection and attribution.

$\mathbf{f}_i$ *overestimates* the magnitude of the i'th forced response. These hypothesis tests can be assessed at once by plotting confidence intervals, as illustrated in Figure 9.5.

Model (9.69) can be written in matrix notation as

$$\mathbf{o} = \mathbf{F}\boldsymbol{\beta} + \boldsymbol{\epsilon}, \tag{9.70}$$

where

$$\mathbf{F} = \begin{bmatrix} \mathbf{f}_{\text{GHG}} & \mathbf{f}_{\text{AER}} & \mathbf{f}_{\text{NAT}} \end{bmatrix} \quad \text{and} \quad \boldsymbol{\beta} = \begin{pmatrix} \beta_{\text{GHG}} \\ \beta_{\text{AER}} \\ \beta_{\text{NAT}} \end{pmatrix}. \tag{9.71}$$

Without loss of generality, $\boldsymbol{\epsilon}$ can be assumed to have zero mean provided $\mathbf{o}, \mathbf{f}_{\text{GHG}},$ $\mathbf{f}_{\text{AER}}, \mathbf{f}_{\text{NAT}}$ are centered. The resulting estimation problem for $\boldsymbol{\beta}$ *looks* like ordinary least squares, but there are important differences when applied to realistic data. First, ordinary least squares assumes the elements of $\boldsymbol{\epsilon}$ are *iid*, whereas in fact realistic climate data exhibit correlations in time and in space. These correlations can be taken into account through *generalized least squares*. Second, ordinary least squares assumes the predictors $\mathbf{f}_{\text{GHG}}, \mathbf{f}_{\text{AER}}, \mathbf{f}_{\text{NAT}}$ are known *exactly* up to a multiplicative constant, whereas in some cases they are estimated from climate models by ensemble averaging, which leaves some uncertainty. Accounting for such uncertainties leads to an *Error-in-Variables* model and the *total least squares problem* (which is discussed in more detail in chapter 12). This latter approach is more advanced and will not be discussed here (see Van Huffel and Vandewalle, 1991; Allen and Stott, 2003; DelSole et al., 2018).

To account for correlations in $\boldsymbol{\epsilon}$, we use generalized least squares as discussed in Section 9.10, which yields

$$\hat{\boldsymbol{\beta}} = \left(\mathbf{F}^T \boldsymbol{\Sigma}_{\epsilon}^{-1} \mathbf{F}\right)^{-1} \mathbf{F}^T \boldsymbol{\Sigma}_{\epsilon}^{-1} \mathbf{o}, \tag{9.72}$$

and

$$\hat{\boldsymbol{\beta}} \sim \mathcal{N}\left(\boldsymbol{\beta}, \sigma_\epsilon^2 \left(\mathbf{F}^T \boldsymbol{\Sigma}_\epsilon^{-1} \mathbf{F}\right)^{-1}\right). \tag{9.73}$$

The question arises as to how to specify $\boldsymbol{\Sigma}_\epsilon$. In many studies, this covariance matrix is estimated from a climate model that is run without year-to-year variations in external forcing. Such runs are known as *control runs*. Unfortunately, the dimension of $\boldsymbol{\Sigma}_\epsilon$ is the same as that of **o**, which means that extremely long simulations are needed to estimate this matrix. For instance, if **o** denotes annual means over a hundred year period, then $\boldsymbol{\Sigma}_\epsilon$ is estimated by sampling *one-hundred-year segments* of the simulation, which obviously requires many such segments, which requires large computational resources. Often, the covariance matrix is estimated by *overlapping* one-hundred-year segments and *pooling* control runs from different models, but the remaining sampling errors are still too large. There are two standard approaches to dealing with this problem. One approach is to reduce the dimension of the problem by projecting the data onto the leading principal components of the control simulation (Allen and Tett, 1999). The second approach is to use a *regularized* estimate of the covariance matrix (Ribes et al., 2009). Each approach has advantages and disadvantages and involves advanced statistical techniques that are beyond the scope of this book.

For the purpose of illustration, we follow a simpler approach. First, only a single time series for **o** is investigated. Second, internal variability is assumed to be stationary, hence covariances depend only on time lag. As a result, the covariance matrix for $\boldsymbol{\epsilon}$ has the form

$$\boldsymbol{\Sigma}_\epsilon = \text{cov}\begin{bmatrix} \epsilon_t \\ \epsilon_{t+1} \\ \epsilon_{t+2} \\ \vdots \\ \epsilon_{t+N} \end{bmatrix} = \sigma_\epsilon^2 \begin{pmatrix} 1 & \rho_1 & \rho_2 & \cdots & \rho_{N-1} \\ \rho_1 & 1 & \rho_1 & \cdots & \rho_{N-2} \\ \rho_2 & \rho_1 & 1 & \cdots & \rho_{N-3} \\ \vdots & \vdots & \vdots & \ddots & \vdots \\ \rho_{N-1} & \rho_{N-2} & \rho_{N-3} & \cdots & 1 \end{pmatrix}, \tag{9.74}$$

where ρ_τ is the autocorrelation function of the process ϵ_t. We further assume the process is an autoregressive process. The parameters of the autoregressive model are estimated from a control simulation. These parameter estimates can then be used to specify the autocorrelation function for all lags, thereby specifying the covariance matrix (9.74).[3]

To illustrate this approach, we use simulations from a climate model known as the Canadian Earth System model (CanESM-LES; Kirchmeier-Young et al., 2016).

---

[3] This approach neglects uncertainty in the autoregressive parameters, but this uncertainty is small if the control simulation is long. Incidentally, the inverse of the covariance matrix can be computed analytically given the autoregressive model, but this is a technical trick that is not important for this example.

This model has 50 ensemble members for each forcing scenario and a 996-year control run. Each ensemble member is initialized from a slightly different initial condition. A realization of **o** is taken from a single ensemble member of the all-forcings run. This realization is known as an *observation proxy*, and its use ensures that all components of the regression model (9.70) come from the same population. In contrast, if real observations were used to specify **o**, then the consistency between model and observation would require further assessment. Only a single time series is considered for **o**, namely 2m-temperature averaged over an observational grid taken from the HadCRUT4 data set (Morice et al., 2012). This grid is illustrated in Figure 9.6a, and the resulting time series is shown as the line with dots in Figure 9.6b. To be clear, we are using only the observational grid from HadCRUT4, we are not using any of the data from HadCRUT4. The forced response is estimated by averaging over the remaining 49 ensemble members of the particular forcing scenario, which is sufficiently large that uncertainty in the forced responses can be ignored. The estimated response to all forcings, aerosols from human emissions, and natural forcings, are shown in Figures 9.6b–d.

Finally, the autocorrelation function of internal variability estimated from the control run is shown in Figure 9.6e. An AR(1) fit tends to overestimate the autocorrelation at long lags (not shown), so we use an AR(2) fit, which is shown in Figure 9.6e and gives an excellent fit (the two curves are nearly indistinguishable). The specific model is

$$\epsilon_t = 0.45\epsilon_{t-1} - 0.13\epsilon_{t-2} + w_t, \tag{9.75}$$

where w_t is a Gaussian white noise process with a variance chosen so that the variance of ϵ_t matches the control variance, which is 0.011.

Note that the forcing scenarios shown in Figure 9.6 do not match those in (9.69). In particular, Figure 9.6 shows the response to all-forcings, whereas model (9.69) separates this into three forcing scenarios, one of which is denoted by GHG, which is not a forcing scenario simulated with CanESM-LES. This situation is common: the particular forcing scenario used in model simulations may not match the forcing scenario one wants to detect. However, because the all-forcings scenario is a linear combination of all the forcings, it is possible to estimate the scaling factor β_{GHG} by linear transformation of other scaling factors. To do this, consider a linear model based on the forcing scenarios that do exist, namely

$$\mathbf{o} = \beta_1 \mathbf{f}_{\text{ALL}} + \beta_2 \mathbf{f}_{\text{AER}} + \beta_3 \mathbf{f}_{\text{NAT}} + \epsilon. \tag{9.76}$$

It may seem strange to consider a model in which the ALL predictor contains the other two predictors AER and NAT, since the model seems to count AER and NAT twice. However, recall that $\beta_1, \beta_2, \beta_3$ are *partial* coefficients that quantify the contribution of a given forcing *while holding the others constant*. Thus, β_1 is not the contribution of ALL, but rather the contribution of ALL while holding AER and

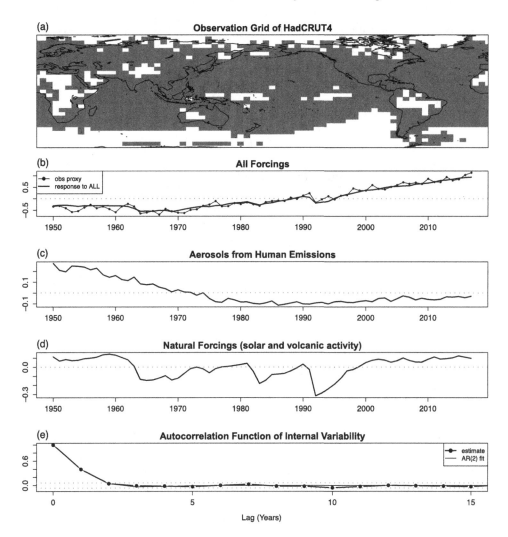

Figure 9.6 Results from the large-ensemble Canadian Earth System Model (CanESM-LES) for 1950–2017. Global averages are computed over an observation grid taken from the HadCRUT4 data set (a). The time series are the annual-mean, ensemble-mean (49-members) 2m-temperature of simulations with all forcings (b), aerosols from human emissions (c), and natural forcings (d) (note the different y-axes). The bottom figure (e) shows the autocorrelation function estimated from a 996-year control simulation (solid with dots), the least squares AR(2) fit (solid), and $\pm\sqrt{2/68}$ (dotted lines). The line with dots in (b) is a single realization of the all-forcings runs, which is used as a proxy for observations, whereas the other 49 members are used to estimate the response to all forcings.

NAT constant. To make proper identifications, substitute $\mathbf{f}_{\text{ALL}} = \mathbf{f}_{\text{GHG}} + \mathbf{f}_{\text{AER}} + \mathbf{f}_{\text{NAT}}$ into Equation (9.76) and re-group as

$$\mathbf{0} = \beta_1 \mathbf{f}_{\text{GHG}} + (\beta_1 + \beta_2)\mathbf{f}_{\text{AER}} + (\beta_1 + \beta_3)\mathbf{f}_{\text{NAT}} + \boldsymbol{\epsilon}. \tag{9.77}$$

Comparing this model to (9.69) indicates

$$\beta_{\text{GHG}} = \beta_1, \quad \beta_{\text{AER}} = \beta_1 + \beta_2, \quad \text{and} \quad \beta_{\text{NAT}} = \beta_1 + \beta_3. \tag{9.78}$$

These identities can be written equivalently as

$$\begin{pmatrix} \beta_{\text{GHG}} \\ \beta_{\text{AER}} \\ \beta_{\text{NAT}} \end{pmatrix} = \mathbf{C} \begin{pmatrix} \beta_1 \\ \beta_2 \\ \beta_3 \end{pmatrix}, \tag{9.79}$$

where

$$\mathbf{C} = \begin{pmatrix} 1 & 0 & 0 \\ 1 & 1 & 0 \\ 1 & 0 & 1 \end{pmatrix}. \tag{9.80}$$

From (9.73) and Theorem 7.10, it follows that

$$\begin{pmatrix} \beta_{\text{GHG}} \\ \beta_{\text{AER}} \\ \beta_{\text{NAT}} \end{pmatrix} \sim \mathcal{N}\left(\mathbf{C}\boldsymbol{\beta}_{123}, \sigma_\epsilon^2 \mathbf{C}\left(\mathbf{F}^T \boldsymbol{\Sigma}_\epsilon^{-1}\mathbf{F}\right)^{-1}\mathbf{C}^T \right), \tag{9.81}$$

where

$$\boldsymbol{\beta}_{123} = \begin{pmatrix} \beta_1 \\ \beta_2 \\ \beta_3 \end{pmatrix}. \tag{9.82}$$

This result can be used to construct confidence intervals for the scaling factors. For instance, the 95% confidence interval for β_{AER} is

$$\hat{\beta}_1 + \hat{\beta}_2 \pm 1.96 \,\hat{\sigma}_\epsilon^2 \sqrt{\left(\mathbf{C}\left(\mathbf{F}^T \boldsymbol{\Sigma}_\epsilon^{-1}\mathbf{F}\right)^{-1}\mathbf{C}^T\right)_{22}}, \tag{9.83}$$

where $\hat{\sigma}_\epsilon^2$ is estimated from (9.66). The resulting confidence intervals are shown in Figure 9.7. The confidence intervals include 1, as expected from this "perfect model" setup. Also, each forcing has been detected because the intervals exclude zero. We emphasize that this result has ignored uncertainties as a result of using finite sample sizes for estimating the forced response and for estimating the autoregressive model for internal variability.

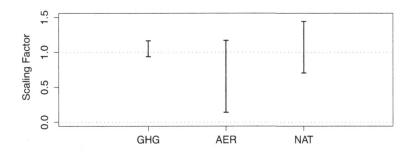

Figure 9.7 95% confidence intervals for the scaling factors for $f_{GHG}, f_{AER}, f_{NAT}$ derived from the CanESM-LES, using one ensemble member of an all-forcings run as an observational proxy.

9.12 The General Linear Hypothesis

All hypothesis tests discussed in this chapter are special cases of a framework known as the *general linear hypothesis*. This framework considers hypotheses that can be expressed as

$$\underset{Q \times M}{\mathbf{C}} \quad \underset{M \times 1}{\beta} \quad = \quad \underset{Q \times 1'}{\mathbf{c}} \tag{9.84}$$

where Q is the number of constraints. For instance, hypothesis $\beta_2 = 0$ in (9.1) can be expressed in the form of (9.84)

$$\mathbf{C} = \begin{pmatrix} 0 & 1 \end{pmatrix}, \quad \beta = \begin{pmatrix} \beta_1 & \beta_2 \end{pmatrix}^T, \quad \text{and} \quad \mathbf{c} = \begin{pmatrix} 0 \end{pmatrix}, \tag{9.85}$$

while hypothesis $\beta_2 = 0$ in (9.30) can be expressed as

$$\mathbf{C} = \begin{pmatrix} 0 & 1 & 0 \end{pmatrix}, \quad \beta = \begin{pmatrix} \beta_1 & \beta_2 & \beta_3 \end{pmatrix}^T, \quad \text{and} \quad \mathbf{c} = \begin{pmatrix} 0 \end{pmatrix}. \tag{9.86}$$

However, the general linear model allows many other hypotheses to be tested, such as the hypothesis that β_2 and β_3 vanish *simultaneously* in (9.30), or the hypothesis $\beta_1 + \beta_2 = 0$, which is needed to detect β_{AER} in the previous example (see (9.78)). This approach to hypothesis testing is based on the following theorem.

Theorem 9.1 (The General Linear Hypothesis) *Consider the model*

$$\underset{N \times 1}{\mathbf{y}} \quad = \quad \underset{N \times M}{\mathbf{X}} \quad \underset{M \times 1}{\beta} \quad + \quad \underset{N \times 1}{\epsilon} \tag{9.87}$$

where $\epsilon \sim N(\mathbf{0}, \sigma_\epsilon^2 \mathbf{I})$. Consider a constraint on the population parameters of the form

$$\underset{Q \times M}{\mathbf{C}} \quad \underset{M \times 1}{\beta} \quad = \quad \underset{Q \times 1'}{\mathbf{c}} \tag{9.88}$$

where Q gives the number of independent constraints. Let SSE_c and SSE_{uc} denote the sum square errors of the constrained and unconstrained regression models, respectively. If the general linear hypothesis (9.88) is true, then the statistic

$$F = \frac{SSE_c - SSE_{uc}}{SSE_{uc}} \frac{N - M}{Q} \tag{9.89}$$

has an F distribution with Q and $N - M$ degrees of freedom.

Theorem 9.1 compares the sum square error of two regression models, one of which is *nested* within the other. By nested, we mean that one model reduces to the other after imposing a linear constraint of the form (9.88). Intuitively, if the constraint is true, then removing the constraint should not improve the fit much, so the change in sum square errors should be small, leading to a small F. Conversely, if the constraint is false, then removing the constraint should lead to a big improvement in the fit, and hence a large F.

All of the hypothesis tests considered so far for normal distributions can be framed as special cases of testing the general linear hypothesis. These include testing the significance of an individual regression parameter, testing the significance of a correlation, and testing the difference between two means.

Another important application of the general linear hypothesis is to test the hypothesis that all the regression parameters (besides the intercept) vanish simultaneously. If β_1 is the intercept, then this hypothesis is

$$\beta_2 = 0, \quad \beta_3 = 0, \quad \text{and} \quad \beta_M = 0, \tag{9.90}$$

which constitutes $Q = M - 1$ hypotheses. The SSE of the model under this hypothesis is just SST, defined in (8.42). The F-statistic (9.89) then becomes

$$F = \left(\frac{SST - SSE}{SSE} \right) \frac{N - M}{M - 1} = \left(\frac{R^2}{1 - R^2} \right) \frac{N - M}{M - 1}. \tag{9.91}$$

We can think of this test as testing the hypothesis $R^2 = 0$, which in turn is the generalization of the hypothesis $\rho = 0$ to more than one predictor.

9.13 Tying Up Loose Ends

Distribution of the Sum Square Error

Here we prove that the sum square error has a chi-squared distribution as in (9.11). The sum square error can be written as

$$SSE = \epsilon^T (I - H) \epsilon. \tag{9.92}$$

Because H is symmetric, it permits an eigenvalue decomposition of the form

$$H = U\Lambda U^T, \tag{9.93}$$

where $\mathbf{U}$ is an orthogonal matrix and $\boldsymbol{\Lambda}$ is a diagonal matrix whose diagonal elements equal the eigenvalues. Therefore, the sum square error can be written equivalently as

$$\text{SSE} = \boldsymbol{\epsilon}^T \mathbf{U} \, (\mathbf{I} - \boldsymbol{\Lambda}) \, \mathbf{U}^T \boldsymbol{\epsilon}. \tag{9.94}$$

Define the new random variable $\mathbf{z} = \mathbf{U}^T \boldsymbol{\epsilon}$. Then the sum square error can be written as

$$\text{SSE} = \mathbf{z}^T \, (\mathbf{I} - \boldsymbol{\Lambda}) \, \mathbf{z}.$$

Since $\mathbf{I} - \boldsymbol{\Lambda}$ is diagonal, the sum square error can be written equivalently as

$$\text{SSE} = (1 - \lambda_1) \, z_1^2 + (1 - \lambda_2) \, z_2^2 + \cdots + (1 - \lambda_N) \, z_N^2. \tag{9.95}$$

Because the matrix $\mathbf{H}$ satisfies $\mathbf{H}^2 = \mathbf{H}$, it is idempotent and therefore its eigenvalues take on only two values, 0 and 1 (this is a standard exercise that readers may want to prove for themselves). The number of eigenvalues that equal 1 equals the trace of $\mathbf{H}$, which is

$$\text{tr}[\mathbf{H}] = \text{tr}[\mathbf{X} \left(\mathbf{X}^T \mathbf{X}\right)^{-1} \mathbf{X}^T] = \text{tr}[(\mathbf{X}^T \mathbf{X})^{-1} \mathbf{X}^T \mathbf{X}] = \text{tr}[\mathbf{I}_M] = M,$$

where we have used the property $\text{tr}[\mathbf{AB}] = \text{tr}[\mathbf{BA}]$. Thus, $\mathbf{H}$ has M eigenvalues equal to 1, and $N - M$ eigenvalues equal to 0. Looking at the coefficients in (9.95), note that if $\lambda_i = 0$, then $1 - \lambda_i = 1$, and if $\lambda_i = 1$, then $1 - \lambda_i = 0$; that is, the 1s and 0s are "flipped." Therefore, the sum square error in (9.95) is a sum of $N - M$ nonzero terms.

It remains to determine the distribution of $\mathbf{z}$. By definition, $\mathbf{z}$ is a linear combination of independent, normally distributed random variables, and therefore is itself multivariate Gaussian. Thus, only its mean and covariance matrix are needed. The mean is

$$\mathbb{E}[\mathbf{z}] = \mathbb{E}[\mathbf{U}^T \boldsymbol{\epsilon}] = \mathbf{U}^T \mathbb{E}[\boldsymbol{\epsilon}] = \mathbf{0}. \tag{9.96}$$

The covariance matrix of $\mathbf{z}$ is

$$\text{cov}[\mathbf{z}, \mathbf{z}] = \mathbb{E}[\mathbf{z}\mathbf{z}^T] = \mathbb{E}[\mathbf{U}^T \boldsymbol{\epsilon}\boldsymbol{\epsilon}^T \mathbf{U}] = \mathbf{U}^T \mathbb{E}[\boldsymbol{\epsilon}\boldsymbol{\epsilon}^T] \mathbf{U} = \sigma_\epsilon^2 \mathbf{U}^T \mathbf{U} = \sigma_\epsilon^2 \mathbf{I}.$$

Since the covariance matrix of $\mathbf{z}$ is proportional to the identity matrix, each element of $\mathbf{z}$ is independently and identically distributed as a normal distribution with zero mean and variance σ_ϵ^2. It follows that the sum square error in (9.95) has a chi-squared distribution. Furthermore, $N - M$ of the coefficients equal 1 and the others equal 0, implying

$$\frac{\text{SSE}}{\sigma_\epsilon^2} \sim \chi_{N-M}^2.$$

9.14 Conceptual Questions

1 How do you test the hypothesis that X and Y are linearly related using a linear regression framework? How does this test differ from the correlation test?

2 How do you test the hypothesis that X and Y are linearly related after controlling for Z?

3 Consider the two regression models

$$Y = \beta_X X + \epsilon_1 \tag{9.97}$$

$$Y = \beta_X X + \epsilon_2 + \beta_Z Z. \tag{9.98}$$

How does the interpretation of the hypothesis $\beta_X = 0$ differ for the two models?

4 How is it possible that X and Y may be positively related, but then be negatively related after holding Z fixed?

5 Suppose you wanted to test the hypothesis $\beta_X = 0$ in the model

$$Y = \beta_X X + \epsilon_2 + \beta_Z Z. \tag{9.99}$$

Describe how you could test this hypothesis by "regressing out variables."

6 Why should the annual cycle be taken into account when quantifying the relation between variables?

7 What is generalized least squares? Why not always use generalized least squares?

8 In a linear regression framework, how do you detect climate change? How do you attribute climate change to human activities?

10

Model Selection

You know, the most amazing thing happened to me tonight. I was coming here, on the way to the lecture, and I came in through the parking lot. And you won't believe what happened. I saw a car with the license plate ARW 357. Can you imagine? Of all the millions of license plates in the state, what was the chance that I would see that particular one tonight? Amazing! [1]

Richard Feynman

This chapter discusses the problem of selecting predictors in a linear regression model. This problem arose several times in previous chapters. For instance, when modeling the annual cycle, the number of annual harmonics needed to be chosen; when modeling secular changes, the order of the polynomial needed to be chosen; when modeling serial correlation, the order of the autoregressive model needed to be chosen. Each of these choices is an example of *model selection*. One might think that the best model is the one with the most predictors. However, each predictor is associated with a parameter that must be estimated, and errors in the estimation add uncertainty to the final prediction. Thus, when deciding whether to include certain predictors or not, the associated gain in prediction skill should exceed the loss due to estimation error. Model selection is not easily addressed using a hypothesis testing framework because *multiple testing* is involved. Instead, the standard approach is to define a criterion for preferring one model over another. Sometimes, this criterion is framed as choosing the "true" model among a set of models, but this framing is artificial since it is unlikely that any model invented by humans is the true model. A more realistic goal is to select the model that gives the best predictions *of independent data*. By independent data, we mean data that is generated independently of the sample that was used to inform the model building process. Criteria for identifying the model that gives the best predictions in independent data include Mallows' C_p, Akaike's Information Criterion, Bayesian Information Criterion, and cross-validated error.

---

[1] From *Feynman Lectures on Physics* by Richard P. Feynman, copyright ©2000. Reprinted by permission of Basic Books, an imprint of Hachette Book Group, Inc.

10.1 The Problem

Often, a climate time series is modeled as the sum of two terms: an annual cycle and random variability. Accordingly, let a weather variable y_t come from the model

$$y_t = a_t + \eta_t, \tag{10.1}$$

where $\eta_t \overset{iid}{\sim} N(0, \sigma_\eta^2)$ models randomness and a_t is the true annual cycle, defined as

$$a_t = \cos(\omega_1 t + b \sin^2(\omega_1 t)), \tag{10.2}$$

where t denotes time, in days, and $\omega_1 = 2\pi/365.25$ days. Let $\sigma_\eta^2 = 0.5$, $b = 1.4$, and sample t every 10 days for two consecutive years. A plot of a_t is shown in Figure 10.1a. In practice, we do not know (10.2), or even the functional form of its dependence on t. Accordingly, we have to *guess* a model for the annual cycle. Since the annual cycle is periodic, a natural model is a Fourier Series, which has the form

$$\sum_{k=0}^{K} (A_k \cos(\omega_k t) + B_k \sin(\omega_k t)), \tag{10.3}$$

where A_k and B_k are unknown coefficients that will be estimated from data, and ω_k is the frequency of the k'th annual harmonic, defined as

$$\omega_k = \frac{2\pi k}{365.25 \text{ days}}. \tag{10.4}$$

B_0 is ignored because $\sin(\omega_0 t) = 0$. The problem here is to choose the upper limit K. As a point of reference, we show in Figure 10.1a the best-fit Fourier Series to a_t

$$\hat{a}_t = \sum_{k=0}^{K} \left(\hat{A}_k \cos(\omega_k t) + \hat{B}_k \sin(\omega_k t) \right), \tag{10.5}$$

using $K = 1$ and $K = 10$, where $\hat{A}_k$ and $\hat{B}_k$ are least squares estimates of A_k and B_k (see Chapter 8). The fit using $K = 1$ has obvious discrepancies, whereas the fit using $K = 10$ lies on top of the true annual cycle and therefore is perfect for all practical purposes.

However, in practice, we never observe the annual cycle directly, rather, we observe y_t, which has superposed random perturbations η_t. A realization of y_t is shown as dots in Figure 10.1b, along with the best-fit Fourier Series using $K = 10$.

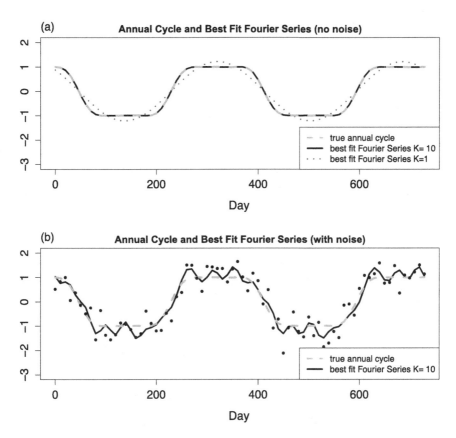

Figure 10.1 (a) The annual cycle function (10.2) (grey dashed) and the best-fit Fourier Series using $K = 1$ (dotted) and $K = 10$ harmonics (solid). The annual cycle and best-fit Fourier Series for $K = 10$ lie on top of each other. (b) The annual cycle function (10.2) (grey dashed), a realization of y_t from (10.1) (dots), and the Fourier Series using $K = 10$ fitted to the data points (solid). The best fit Fourier Series does not appear exactly periodic in the figure because it is sampled every 10 days.

In this case, the estimated annual cycle $\hat{a}_t$ contains "extra" variability that is not in a_t. Bear in mind that, in practice, we do not know a_t, so we do not know if the extra variability is "wrong." To generalize beyond one sample, we generate 100 realizations of y_t and derive $\hat{a}_t$ from each realization. The mean square difference between $\hat{a}_t$ and a_t as a function of K is shown in Figure 10.2. As can be seen, the mean square difference decreases from $K = 0$ to $K = 3$, then *increases* for $K \geq 4$. That is, the estimated model *gets worse* beyond a certain number of predictors, even though the model can fit the true annual cycle nearly perfectly in the absence of noise (see Figure 10.1a). Why does the model get worse as more predictors are

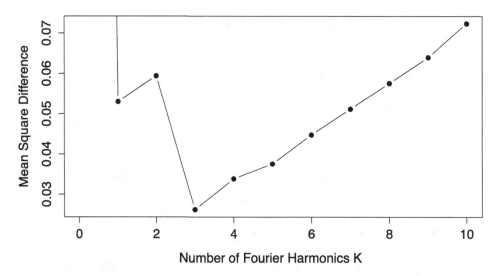

Figure 10.2 Mean square difference between predicted and actual annual cycle for $K = 0, 1, \ldots, 10$, estimated from 100 realizations of y_t.

added? How do we choose K when we don't know the truth? These questions will be addressed in this chapter.

10.2 Bias–Variance Trade off

To gain insight into why the regression model deviates further from the true model as the number of predictors increases, it is helpful to compare the predicted and actual annual cycles at a fixed time. The mean and standard deviation of the predicted annual cycle $\hat{a}_t$ at $t = 150$ for multiple realizations of y_t is shown in Figure 10.3. At $K = 1$, the prediction *systematically underestimates* the true value, as would be anticipated from Figure 10.1. As the number of harmonics increases, the *mean* of the prediction approaches the true value, but the *variance* of the predictions *increases*. Thus, for small K, the predictions are *biased* but have relatively little variance, while for large K the predictions have less bias but have relatively large variance. This *bias–variance tradeoff* is seen at all values of t, not just at $t = 150$ chosen for this example.

Since we know the true model, we should be able to express the previously described behavior mathematically. Note that estimating model (10.3) is equivalent to considering the model

$$\underset{N \times 1}{\mathbf{y}} = \underset{N \times M}{\mathbf{X}} \quad \underset{M \times 1}{\boldsymbol{\beta}} + \underset{N \times 1}{\boldsymbol{\epsilon}}, \tag{10.6}$$

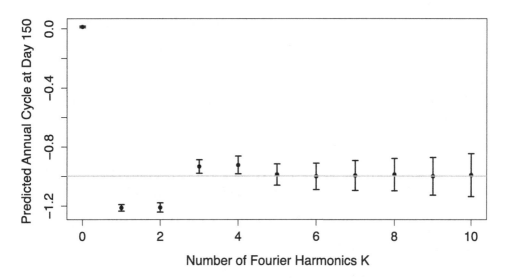

Figure 10.3 The predicted annual cycle at day 150, $\hat{a}_{150}$, for $K = 0, 1, \ldots, 10$. The dot shows the mean and the error bars show two standard deviations over 1,000 realizations of y_t. The horizontal grey line shows the true value.

where the columns of $\mathbf{X}$ are sines and cosines, $\boldsymbol{\beta}$ contains the A_k and B_k coefficients, and $M = 2K + 1$ (see Section 8.5.2). As discussed in Chapter 8, the least squares estimate of $\boldsymbol{\beta}$ is

$$\hat{\boldsymbol{\beta}} = \left(\mathbf{X}^T\mathbf{X}\right)^{-1}\mathbf{X}^T\mathbf{y}, \tag{10.7}$$

the prediction model is $\hat{\mathbf{a}} = \mathbf{X}\hat{\boldsymbol{\beta}}$, and the residual errors are

$$\hat{\boldsymbol{\epsilon}} = \mathbf{y} - \mathbf{X}\hat{\boldsymbol{\beta}}. \tag{10.8}$$

Let the data generated from the model (10.1) be written as $\mathbf{y} = \mathbf{a} + \boldsymbol{\eta}$. Then the difference between predicted and true annual cycle, which we name the *model difference* $\mathbf{d}$, is

$$\mathbf{d} = \hat{\mathbf{a}} - \mathbf{a} = \mathbf{X}\hat{\boldsymbol{\beta}} - \mathbf{a} = \mathbf{H}\mathbf{y} - \mathbf{a} = \mathbf{H}(\mathbf{a} + \boldsymbol{\eta}) - \mathbf{a}$$
$$= \mathbf{H}\boldsymbol{\eta} - (\mathbf{I} - \mathbf{H})\mathbf{a}, \tag{10.9}$$

where $\mathbf{H}$ is the hat matrix

$$\mathbf{H} = \mathbf{X}\left(\mathbf{X}^T\mathbf{X}\right)^{-1}\mathbf{X}^T. \tag{10.10}$$

As discussed in Chapter 8, the hat matrix is a *projection matrix* that projects quantities onto the space spanned by the columns of $\mathbf{X}$, which here are the sines and cosines. Thus, $\mathbf{H}\boldsymbol{\eta}$ quantifies how random variations from $\boldsymbol{\eta}$ "infect" $\hat{\boldsymbol{\beta}}$ and ultimately project onto the range of $\mathbf{X}$ to produce randomness in $\hat{\mathbf{a}}$. Thus, $\mathbf{H}\boldsymbol{\eta}$

contributes *variance* to $\hat{\mathbf{a}}$. On the other hand, $\mathbf{I} - \mathbf{H}$ projects quantities onto the space orthogonal to the range of $\mathbf{X}$. It follows that $(\mathbf{I} - \mathbf{H})\,\mathbf{a}$ is the variability in $\mathbf{a}$ that cannot be represented by the column vectors of $\mathbf{X}$. This term contributes *bias* to $\hat{\mathbf{a}}$.

Because $\mathbf{H}$ is a projection matrix, it is idempotent; that is, $\mathbf{H}^2 = \mathbf{H}$, and $(\mathbf{I} - \mathbf{H})^2 = (\mathbf{I} - \mathbf{H})$. Thus, the expected sum squared difference $\mathbb{SSD}$ is

$$\mathbb{SSD} = \mathbb{E}\left[\mathbf{d}^T\mathbf{d}\right] = \mathbb{E}\left[\boldsymbol{\eta}^T\mathbf{H}\boldsymbol{\eta}\right] + \mathbf{a}^T\left(\mathbf{I} - \mathbf{H}\right)\mathbf{a}, \tag{10.11}$$

where we have used the fact that $\mathbf{a}$ is deterministic and therefore $\mathbb{E}[\boldsymbol{\eta}\mathbf{a}^T] = \mathbb{E}[\boldsymbol{\eta}]\mathbf{a}^T = \mathbf{0}$. The first term on the right-hand side can be simplified as

$$\mathbb{E}\left[\boldsymbol{\eta}^T\mathbf{H}\boldsymbol{\eta}\right] = \text{tr}\left[\mathbf{H}\mathbb{E}\left[\boldsymbol{\eta}\boldsymbol{\eta}^T\right]\right] = \sigma_\eta^2\,\text{tr}\,[\mathbf{H}] = M\sigma_\eta^2, \tag{10.12}$$

where we have used $\mathbb{E}[\boldsymbol{\eta}\boldsymbol{\eta}^T] = \sigma_\eta^2\mathbf{I}$ and

$$\text{tr}\,[\mathbf{H}] = \text{tr}\left[\mathbf{X}\left(\mathbf{X}^T\mathbf{X}\right)^{-1}\mathbf{X}^T\right] = \text{tr}\left[\left(\mathbf{X}^T\mathbf{X}\right)^{-1}\left(\mathbf{X}^T\mathbf{X}\right)\right] = \text{tr}\,[\mathbf{I}_M] = M. \tag{10.13}$$

Thus, the expected sum square difference is

$$\mathbb{SSD} = M\sigma_\eta^2 + \mathbf{a}^T\left(\mathbf{I} - \mathbf{H}\right)\mathbf{a}, \tag{10.14}$$

which expresses the bias–variance tradeoff mathematically. The result of evaluating this equation for our example, shown in Figure 10.4, confirms that this population

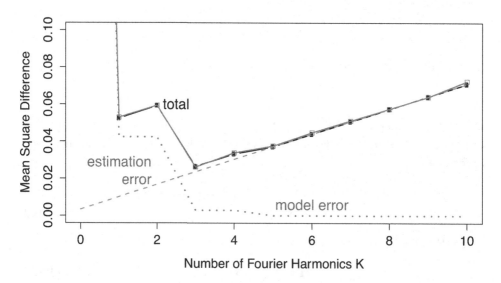

Figure 10.4 Mean square difference between the predicted and actual annual cycle for the realizations shown in Figure 10.1 (solid grey curve with open squares; reproduced from Figure 10.2), the theoretical expectation (10.14) (solid black curve with closed circles), the model error (grey dotted curve), and the estimation error (grey dashed line).

equation explains the results from the 100 realizations very well. The individual terms in (10.14) are also shown in the figure. The squared bias will be known as *Model Error* ($\mathbb{ME}$) and defined as

$$\mathbb{ME} = \mathbf{a}^T \left(\mathbf{I} - \mathbf{H} \right) \mathbf{a}. \tag{10.15}$$

$\mathbb{ME}$ measures the variability in $\mathbf{a}$ that cannot be represented by the predictors. If $\mathbf{a}$ lies in the range of $\mathbf{X}$, then $\mathbb{ME}$ vanishes. $\mathbb{ME}$ is a nonincreasing function of K and vanishes if $M = N$. On the other hand, $M\sigma_\eta^2$ arises from uncertainty in $\hat{\boldsymbol{\beta}}$ and quantifies the *error due to estimating model parameters*. This variance will be known as *Estimation Error* ($\mathbb{EE}$),

$$\mathbb{EE} = M\sigma_\eta^2. \tag{10.16}$$

These results reveal a fundamental principle in model building, namely that estimation error increases with each parameter that is estimated. Accordingly, the predictors should be chosen judiciously, since each one adds to the estimation error. On the other hand, using too few predictors or a poor set of predictors results in large model error because the predictors cannot capture the true variability. The model that minimizes $\mathbb{SSD}$ strikes a balance between these conflicting sources of error. If K is much larger than this optimal value, then estimation error dominates and the model is said to be *overfitted*.

10.3 Out-of-Sample Errors

If $\mathbb{SSD}$ could be computed, it would be a simple matter to select the model that minimizes $\mathbb{SSD}$. However, computing $\mathbb{SSD}$ requires knowing the true model $\mathbf{a}$, which is unknown. We need a way to quantify model performance based on the data in hand, namely $(\mathbf{y}, \mathbf{X})$.

It is tempting to try using the residual errors $\hat{\boldsymbol{\epsilon}}$ to measure model performance. This approach has shortcomings. To see why, note that the residual errors are related to $\mathbf{a}$ as

$$\hat{\boldsymbol{\epsilon}} = \mathbf{y} - \mathbf{X}\hat{\boldsymbol{\beta}} = \left(\mathbf{I} - \mathbf{H} \right) \left(\mathbf{a} + \boldsymbol{\eta} \right). \tag{10.17}$$

Let the sum square errors be denoted as

$$\mathrm{SSE}_{\mathrm{IN}} = \hat{\boldsymbol{\epsilon}}^T \hat{\boldsymbol{\epsilon}}, \tag{10.18}$$

(the reason for the subscript IN will be explained shortly). Following the same steps as those used to derive (10.14), the expectation of $\mathrm{SSE}_{\mathrm{IN}}$ is

$$\mathbb{E} \left[\mathrm{SSE}_{\mathrm{IN}} \right] = \sigma_\eta^2 \left(N - M \right) + \mathbf{a}^T \left(\mathbf{I} - \mathbf{H} \right) \mathbf{a}. \tag{10.19}$$

This result shows that the expected sum square errors *decreases* with M, which is a very different behavior from SSD. In fact, SSE_{IN} vanishes for $M = N$, because in this case the model can fit the data *exactly*. Thus, if our selection criterion were to choose the model that minimizes SSE_{IN}, then the inescapable conclusion would be to always choose $M = N$.

Let us think more carefully about the purpose of our model. Often, the purpose is not to *fit* the data, but to make *predictions*. By prediction, we do not necessarily mean predicting the *future*, but rather, predicting data *that was not available during the model building process*. Therefore, quantifying prediction performance requires distinguishing between *in-sample* and *out-of-sample* error. In-sample errors are the regression residuals $\hat{\epsilon}$. These errors are "in-sample" in the sense that they come from the same sample used to estimate β. Out-of-sample error is the error of predicting a realization of y that was generated independently of the sample used to estimate β. To compute out-of-sample error, let (y_0, X_0) be another *independent sample* from the same population from which (y, X) was drawn. We assume the dimensions of the two sets are the same, which simplifies some expressions derived later. The sample (y, X) is the "training sample" used to estimate β in the regression model (10.6), which yields the in-sample errors (10.8). The sample (y_0, X_0) is known as the *validation sample* and is used to estimate out-of-sample prediction error

$$\epsilon_0 = \text{predicted} - \text{actual} = X_0\hat{\beta} - y_0 = X_0 \left(X^T X\right) X^T y - y_0. \tag{10.20}$$

Since the annual cycle repeats, $X_0 = X$ and $a_0 = a$, hence $y = a + \eta$ and $y_0 = a + \eta_0$, where η and η_0 are independent. Substituting these expressions into (10.20) gives

$$\epsilon_0 = Hy - y_0 = H(a + \eta) - (a + \eta_0) = (H\eta - \eta_0) - (I - H) a. \tag{10.21}$$

Following the same steps as those used to derive (10.14), the expected $SSE_0 = \epsilon_0^T \epsilon_0$ is

$$\mathbb{E}[SSE_0] = \mathbb{E}\left[\epsilon_0^T \epsilon_0\right] = (N + M)\sigma_\eta^2 + a^T (I - H) a, \tag{10.22}$$

where we used the fact that η_0 and η are independent. This expression has a very different dependence on M than the in-sample error (10.19); specifically, it increases with M, consistent with (10.14). In fact, comparison with (10.14) shows

$$\mathbb{E}[SSE_0] = SSD + N\sigma_\eta^2. \tag{10.23}$$

This expression shows that $\mathbb{E}[SSE_0]$ differs from SSD by the additive term $N\sigma_\eta^2$, which merely shifts quantities upward without changing the location of the minimum. The term $N\sigma_\eta^2$ comes from η, which is the *irreducible* randomness of the population model (10.1) that is unpredictable even if a were perfectly known. Thus, (10.23) implies that minimizing $\mathbb{E}[SSE_0]$ is equivalent to minimizing SSD. This equivalence is important because $\mathbb{E}[SSE_0]$ is capable of being estimated (e.g., by predicting independent data), whereas $\mathbb{SSD}$ is not *directly* computable because the true **a** is unknown.

This equivalence allows us to re-frame model selection as follows: *the goal of model selection is to select the model that minimizes the out-of-sample prediction error.* There are four basic strategies for doing this:

1 Estimate the prediction error from in-sample quantities.
2 Estimate the prediction error using cross-validation techniques.
3 Estimate parameters using methods that yield better out-of-sample predictions (e.g., ridge regression or LASSO, although these require a model selection criterion too).
4 Combine predictions from different models using Bayesian methods.

This chapter reviews the first two strategies. The third strategy is discussed in more detail in Hastie et al. (2009) and the fourth strategy is discussed in Hoeting et al. (1999).

10.4 Model Selection Criteria

This section reviews some common criteria for selecting regression models. In each case, the procedure is to evaluate the criterion for each of the candidate models and then choose the model that minimizes the criterion. There are no good guidelines for deciding which criterion to use, which is unfortunate since it is tempting to cherry-pick the criterion that gives the most desirable result. We first review each criterion and then show an application.

10.4.1 Mallows' C_p

Since our goal is to minimize out-of-prediction error, it is natural to estimate $\mathbb{E}[\text{SSE}_0]$. Unfortunately, $\mathbb{E}[\text{SSE}_0]$ cannot be estimated directly from (10.22) because **a** is unknown. However, comparing (10.19) with (10.22) shows that

$$\mathbb{E}[\text{SSE}_0] = \mathbb{E}[\text{SSE}_{\text{IN}}] + 2M\sigma_\eta^2. \tag{10.24}$$

This equation shows that in-sample errors tend to *underestimate* out-of-sample errors. More importantly, model error has cancelled out, hence this equation suggests a way to estimate $\mathbb{E}[\text{SSE}_0]$. Specifically, the first term on the right can be estimated by SSE_{IN}. Then, if an estimate of σ_η^2 can be found, it could be substituted in (10.24) to derive a selection criterion. According to (10.19), $\text{SSE}_{\text{IN}}/(N - M)$ is an unbiased estimate of σ_η^2 *if the model error is small*. Recall from Figure 10.4 that model error vanishes after five or more harmonics are included. Thus, σ_η^2 may be estimated from the residual errors provided $K \geq 5$. In more realistic situations, we do not know how many predictors should be included to reduce the model error to zero. Hopefully, the candidate predictors have been chosen such that model error is negligible when all candidate predictors are included. Thus, one strategy for estimating σ_η^2 is to include *all the candidate predictors* and then estimate the noise from the residuals. In general, including all predictors would give a poor prediction model, but this is of no concern since the model will not be used for predictions, it is used only to estimate σ_η^2. Let us denote this estimate as $\hat{\sigma}_{kmax}^2$. Then, an unbiased estimate of $\mathbb{E}[SSE_0]$ is

$$\widehat{SSE}_0 = SSE_{IN} + 2M\hat{\sigma}_{kmax}^2. \tag{10.25}$$

Note that this quantity depends on the units of y_t. Mallows (1973) proposed a re-scaling that has some attractive properties. Specifically, Mallows suggested the criterion

$$C_p = \frac{SSE_{IN}}{\hat{\sigma}_{kmax}^2} - N + 2M, \tag{10.26}$$

which is called *Mallows' C_p*. Note that $\widehat{SSE}_0$ and C_p are minimized at the same model, since C_p is derived from $\widehat{SSE}_0$ by dividing by $\hat{\sigma}_{kmax}^2$ and subtracting N, operations that do not alter the location of the minimum. A nice property of Mallows' C_p is that if model error is small, then $C_p \approx M$.

10.4.2 Akaike's Information Criteria

Our focus so far has been on the *mean* error. However, prediction error is described by a *probability distribution*. A more general criterion is to select the model that minimizes some measure of the difference between the prediction and true distribution. A framework for quantifying differences in probability distributions is *information theory*, but this involves concepts more advanced than assumed here. We refer the reader to Chapter 14 for more details. One of the most widely used criteria derived from this framework is *Akaike's Information Criterion* (AIC). For linear regression models with Gaussian errors, this criterion is

$$AIC = N \log \left(\frac{SSE_{IN}}{N} \right) + N \log 2\pi + 2(M + 1). \tag{10.27}$$

This criterion has a form analogous to Mallows' C_p (10.26), namely, the first term is a goodness-of-fit measure that decreases with M, while the last term acts as a penalty term that increases with M. The $N \log 2\pi$ term is a constant and does not affect model selection, but is included for consistency with later chapters. Choosing the model that minimizes AIC strikes a balance between fitting the data and minimizing the number of parameters.

10.4.3 Corrected AIC (AICc)

AIC is derived from asymptotic analysis, and hence is most appropriate for large N. Hurvich and Tsai (1989) derived a bias correction to AIC that is appropriate for small sample sizes and Gaussian distributions. The bias corrected criterion is

$$\text{AICc} = N \log \left(\frac{\text{SSE}_{\text{IN}}}{N} \right) + N \log 2\pi + \frac{2(M+1)N}{N-M-2}. \tag{10.28}$$

AICc differs from AIC only in the penalty. The penalty term in AICc grows more strongly with M than in AIC, hence AICc tends to select models with fewer predictors than AIC. A subtle assumption in the derivation of AICc is that it assumes that predictors are the same for training and validation samples (DelSole and Tippett, 2021b; Tian et al., 2020). This assumption is valid for our example because sines and cosines are indeed the same in the two samples. The generalization of AICc to random predictors is discussed in Chapter 14.

10.4.4 Bayesian Information Criterion (BIC)

AIC is an *inconsistent estimator*, which means that it does not select the true model in the limit of large N even if the true model is among the models under consideration (Hannan and Quinn, 1979). Other criteria have been derived from different arguments that outperform AIC in these contexts. Perhaps the most commonly used of these is the *Bayesian information criterion (BIC)* (also known as Schwarz information criterion),

$$\text{BIC} = N \log \left(\frac{\text{SSE}_{\text{IN}}}{N} \right) + N \log 2\pi + (M+1) \log(N). \tag{10.29}$$

BIC and AIC differ only in the penalty term. The penalty term in BIC is such that it tends to select models with fewer predictors than AIC.

10.4.5 K-Fold Cross Validation

A major pillar of model selection is *cross validation*. The basic idea is to *withhold* some samples from the fitting process, fit the model to the remaining sample, then use the resulting model to *predict* the withheld sample. The part of the data that is used to fit the model is known as the *training sample*, while the part used to measure model performance is known as the *validation sample*. A strict rule in this procedure is *never use any information from the validation sample to inform the training phase of model building.*

If this procedure were applied to only a single partition of training and validation samples, then the result might be sensitive to the choice of partition. A standard approach that avoids this problem is *K-fold cross validation*. In K-fold cross validation, the data is split into K roughly equal-sized parts, then the k'th part is withheld and the remaining $K - 1$ parts are used to fit the regression model. The resulting regression model is then used to predict the withheld k'th part. Repeating this procedure for $k = 1, \ldots, K$ yields a prediction for each of the N samples. Letting the prediction error for the n'th sample be e_n, the mean square error is computed as

$$\text{MSE}_{\text{CV}} = \frac{1}{N} \sum_{n=1}^{N} e_n^2. \tag{10.30}$$

Note that the errors $e_1, \ldots, e_N$ do not come from the same model, since different prediction models are derived for each k. The case $K = N$ is called *leave-one-out* cross validation.

Sometimes, the mean square error derived from cross-validation is a noisy function of the number of predictors. As a result, the minimum MSE_{CV} may differ from other values by a small amount that may be a result of sampling errors. A more robust criterion is the following. Recall from Chapter 1 that if $Z_1, \ldots, Z_N$ are independent and identically distributed random variables, then the sample mean $\hat{\mu}_Z$ satisfies

$$\text{var}\left[\hat{\mu}_Z \right] = \frac{\text{var}[Z]}{N}. \tag{10.31}$$

Applying this theorem to $e_1^2, \ldots, e_N^2$ leads to the estimated *standard error* of MSE_{CV}

$$\hat{\sigma}_{\text{CVSSE}} = \sqrt{\left(\frac{1}{N-1} \sum_{n=1}^{N} \left(\epsilon_n^2 - \text{MSE} \right)^2 \right) / N}. \tag{10.32}$$

The criterion then is to *select the simplest model within one standard error of the best* (Hastie et al., 2009). This selection criterion is illustrated in Figure 10.5.

It should be recognized that errors from cross validation are not independent, so (10.32) is not strictly valid. However, statistical dependencies tend to inflate variance (see Section 5.3), so $\hat{\sigma}_{\text{CVSSE}}$ will tend to *underestimate* the standard error. Accordingly, $\hat{\sigma}_{\text{CVSSE}}$ represents a lower bound on the standard error.

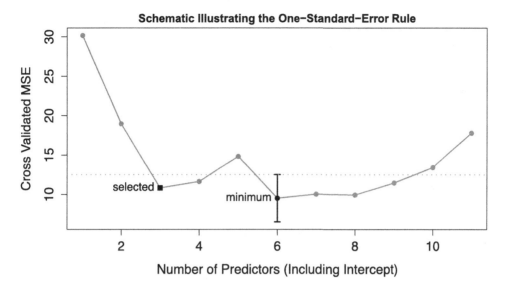

Figure 10.5 Schematic illustrating the one-standard-error rule. The rule is to identify the model with the smallest number of predictors whose cross-validated mean square error lies within one standard error of the minimum mean square error. The minimum MSE is identified by a black circle; the error bars show one standard error above and below the minimum MSE; the horizontal dashed line shows the minimum MSE plus one standard error; the black square identifies the model with the fewest predictors whose MSE lies within a standard error of the minimum MSE.

10.4.6 Comparison of Different Criteria

The result of applying each of these criteria to the sample shown in Figure 10.1 is shown in Figure 10.6. Most criteria agree on selecting $K = 3$, consistent with the exact result in Figure 10.4. The exceptions are BIC and cross validation, which select $K = 1$. Two-fold cross validation was chosen because there were only two years of data and we chose to validate an entire annual cycle; other choices could be defended and the results would be similar.

10.5 Pitfalls

The bias–variance tradeoff suggests that a predictor should be added to a model only if the reduction in model error exceeds the increase in estimation error. Given this constraint, it is tempting to pre-screen the data to find a "good" set of predictors, then use the resulting set as candidate predictors in model selection. Unfortunately, if *the same data* is used to identify candidate predictors and to select the model, then the resulting model often performs poorly in independent samples. To illustrate, suppose 200 random vectors $\mathbf{x}_1, \ldots, \mathbf{x}_{200}$ are considered as predictors for modeling

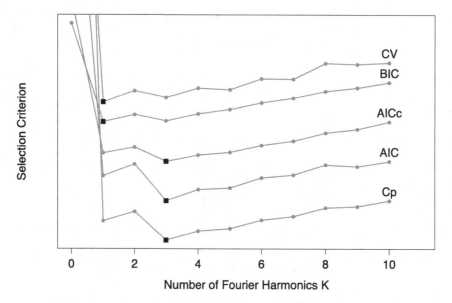

Figure 10.6 Application of different model selection criteria to the annual cycle model (10.3) based on the sample shown in Figure 10.1 (that is, the dots). Each criterion is offset and re-scaled to facilitate comparisons. The minimum value of each criterion is indicated by a black square.

$\mathbf{y} \sim \mathcal{N}_N(0, \mathbf{I})$. By construction, none of the predictors can help predict $\mathbf{y}$ because they are independent of $\mathbf{y}$, but this fact is never known with observed data. Instead, an investigator may have a hunch that a few good predictors are among the 200 predictors, but some sleuthing is required to find them. Accordingly, the investigator computes the correlation between $\mathbf{y}$ and each predictor, and then re-orders the predictors by their absolute correlation from largest to smallest. The resulting candidate predictors are then chosen successively as predictors for $\mathbf{y}$. The result of applying 10-fold cross validation to these re-ordered predictors is shown in Figure 10.7. Results for three realizations of data are shown. As the figure shows, cross validation suggests MSE is less than one and that about a dozen of these predictors should be selected. Yet, by construction, the predictors are independent of $\mathbf{y}$, hence the out-of-sample mean square error of the prediction model is one. This example illustrates the fact that using *all* data to prioritize the predictors, and then using the same data to perform cross-validation, leads to underestimation of the out-of-sample prediction error. Other criteria (AIC, AICc, BIC, and C_p) give similar conclusions (not shown).

Variations of this idea can be found in the literature. Perhaps the most extreme is *all-possible-subsets* regression, which involves evaluating a criterion for each possible subset of predictors and then choosing the subset that optimizes the criterion.

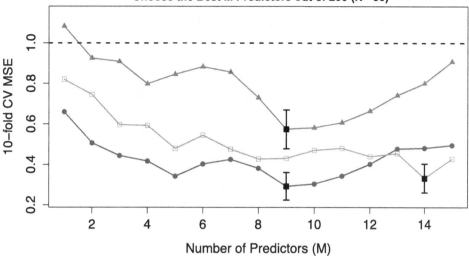

Figure 10.7 Result of 10-fold cross validation on *random data*, but with predictors ordered by their correlation with Y. 200 random time series each of length $N = 30$ were generated, and then re-ordered by their absolute sample correlation relative to a randomly generated Y. Then these variables were used successively as predictors in a regression model. Results for three different realizations are shown. The square shows the minimum value and the error bars show one standard error above and below the minimum value. The horizontal dashed line shows the variance of Y, which in this case is the true minimum out-of-sample MSE because the predictors are independent of Y.

This approach suffers from even more extreme prediction biases than the previous method because of the greater number of comparisons. Another approach to model selection is *forward selection*. In this method, one starts with the intercept-only model and tests each predictor one-at-a-time to find the *single* predictor that reduces SSE_{IN} the most. Depending on a stopping criterion, no predictor may be selected, because the reduction in SSE_{IN} is too small, or the best single predictor is selected, yielding a two-predictor model (the intercept and the predictor). Next, starting with the two-predictor model, the remaining predictors are tested one-at-a-time to find the single predictor that reduces the SSE_{IN} the most. Again, the predictor is included if the reduction in error is large enough, otherwise the procedure stops. Repeating this procedure builds up the model one predictor at a time until the stopping criterion is satisfied.

Because the same data is used to select the predictor *and* to select the model, we may anticipate that forward selection leads to strongly biased prediction error estimates, and often will select models with poor out-of-sample prediction performance. Since forward selection is recommended in some textbooks and software packages, it is worth demonstrating these problems explicitly. As an example, let the number of candidate predictors be 20 and the sample size be $N = 30$. Accordingly, we generate 30 numbers for Y, and 30 * 20 numbers for the candidate predictors. Then, we perform stepwise regression: first, we select the predictor that gives the best R-square; next, we select the predictor that improved the R-square the most, relative to the model previously selected; then we repeat this procedure, yielding a sequence of models with $1, 2, \ldots, 10$ predictors. No significance threshold is considered in this illustration. Then, this entire procedure is repeated 1,000 times to estimate quantiles of R-square when forward selection is applied to independent data. None of these models can give good out-of-sample predictions because the predictors are independent of Y *by construction*. The 5th and 95th percentiles of the R-square values for each $M = 1, 2, \ldots, 10$ are shown in Figure 10.8. For comparison, the figure also shows the 95th percentile of R-square when the predictors are generated independently. As can be seen, the R-square

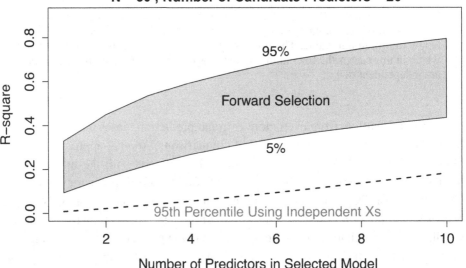

Figure 10.8 Result of applying forward selection to random predictors generated from a Monte Carlo simulation. Random numbers are used to specify **y** and **X**, where $N = 30$ and the number of candidate predictors is 20, then forward selection was used to select predictors for $M = 1, \ldots, 10$. The 5th and 95th percentiles of the R-square for each M are shown as the upper and lower boundaries of the shaded region. The 95th percentile of R-square for independent Xs (with no selection) is shown as the grey dashed curve.

values from forward selection are much larger than those from independent Xs. Indeed, even models at the bottom 5th percentile exceed the critical threshold for independent Xs. One might try to compensate for this selection bias by using more stringent significance thresholds in forward selection, but no accepted way of doing this exists.

Consistent with these considerations, numerous papers in different disciplines call for banning stepwise regression procedures (Henderson and Denison, 1989; Thompson, 1995; Whittingham et al., 2006; Mundry and Nunn, 2009). There is a considerable literature showing that goodness-of-fit measures and p-values are severely biased because of the multiple testing that occurs in the selection procedure (Rencher and Pun, 1980; Harrell, 2001; Miller, 2002). Other variations such as *backward elimination*, where predictors are removed successively from a regression model, and *stepwise regression*, which combines forward selection and backward elimination, suffer from the same shortcomings. Harrell (2001) (page 56) gives a comprehensive list of problems with stepwise regression. We generally discourage stepwise procedures because they violate the fundamental rule that *no validation data should be used to inform the model building process.*

10.6 Further Topics

An alternative approach to model selection is based on *Bayesian linear regression*. In this approach, all predictors under consideration are included, but the associated parameters are estimated while taking into account *prior information* about the parameters. For instance, we may believe that the coefficients are "clustered" near certain values (perhaps zero). Bayesian linear regression provides a framework for incorporating this kind of prior information into the estimation process. Incorporating prior information effectively imposes constraints on the coefficients, which is said to *regularize* the problem. In many cases the estimation procedure is equivalent to minimizing a cost function of the form

$$\text{cost function} = \|\mathbf{y} - \mathbf{X}\boldsymbol{\beta}\|^2 + \text{penalty}, \tag{10.33}$$

where the penalty term depends on the prior information. The procedure reduces to ordinary least squares in the absence of the penalty function. If the prior information is expressed as a normal distribution centered about zero and independent and identically distributed deviations, then the cost function is of the form

$$\text{cost function} = \|\mathbf{y} - \mathbf{X}\boldsymbol{\beta}\|^2 + \lambda\|\boldsymbol{\beta}\|^2, \tag{10.34}$$

where λ is a *regularization parameter*, which is often chosen to minimize prediction error in independent samples. If λ is known, then the $\boldsymbol{\beta}$ that minimizes the cost function is

$$\beta_{\text{ridge}} = \left(\mathbf{X}^T\mathbf{X} + \lambda\mathbf{I}\right)^{-1}\mathbf{X}^T\mathbf{y}. \tag{10.35}$$

This procedure is known as *ridge regression*. An alternative method, known as *LASSO (Least Absolute Shrinkage and Selection Operator)*, is based on the cost function

$$\text{cost function} = \|\mathbf{y} - \mathbf{X}\boldsymbol{\beta}\|^2 + \lambda\|\boldsymbol{\beta}\|_1, \tag{10.36}$$

where $\|\boldsymbol{\beta}\|_1$ denotes the L^1-norm, which is the sum of the absolute values of the regression parameters. Interestingly, this approach tends to set some coefficients *exactly to zero*, which is attractive for interpretation. This approach turns out to be equivalent to Bayesian regression when the prior information is expressed as a zero-mean Laplacian distribution. A modern introduction to these types of methods is Hastie et al. (2009).

Another approach to model building is *Bayesian Model Averaging (BMA)*. In this approach, the models are averaged together using weights determined from Bayesian reasoning. Practical implementation of BMA can be difficult – it involves evaluating implicit integrals and specifying prior distributions of competing models – but this approach can produce forecasts that are superior to that of any single model in the group. Hoeting et al. (1999) give an introduction to this approach with an insightful summary of the literature.

A fascinating approach to model selection is based on *minimum description length (MDL)*. If the variable being modeled is viewed as a sequence of numbers, then any regularities in the sequence can be used to describe it using fewer symbols than a description that literally repeats the sequence. The "best" model is the description that leads to the best compression of data. This approach differs from the usual statistical framework in which the sample is interpreted as a random draw from some population. In practice, this approach leads to model selection criteria of the same form as that of AIC or BIC, but with different penalty terms. Hansen and Yu (2001) give an introduction to this approach.

10.7 Conceptual Questions

1. Why isn't the model with the most predictors generally the best model for making predictions?
2. What is the bias–variance trade-off?
3. Sometimes, omitting some predictors can produce a better model, even though those predictors are known to be part of the true model. Why?
4. Explain in your own words: what is Mallows' C_p? What is AIC?
5. What is K-fold cross-validation?
6. What is the selection criterion based on K-fold cross validation.
7. Why is stepwise regression not a recommended procedure?

11

Screening
A Pitfall in Statistics

It ain't what you don't know that gets you into trouble. It's what you know for sure that just ain't so.

Mark Twain

Scientists often propose hypotheses based on patterns seen in data. However, if a scientist tests a hypothesis using the same data that suggested the hypothesis, then that scientist has violated a rule of science. The rule is: test hypotheses with *independent* data. This rule may sound so obvious as to be hardly worth mentioning. In fact, this mistake occurs frequently, especially when analyzing large data sets. Among the many pitfalls in statistics, *screening* is particularly serious. Screening is the process of evaluating a property for a large number of samples and then selecting samples in which that property is extreme. After a sample has been selected through screening, classical hypothesis tests exhibit *selection bias*, often by a surprisingly large amount. Screening is related to *data fishing, data dredging, or data snooping*. These procedures have been implicated in the so-called *replication crisis* in which certain classic scientific experiments cannot be replicated on subsequent investigation (Ioannidis, 2005; Baker, 2016). In some cases, the effect of screening can be quantified through Monte Carlo techniques or simple probability arguments. These calculations often reveal that selection bias is substantially larger than one might guess. This chapter explains the concept of screening and illustrates it through examples from selecting predictors, interpreting correlation maps, and identifying change points. This chapter assumes familiarity with hypothesis testing concepts discussed in Chapter 2.

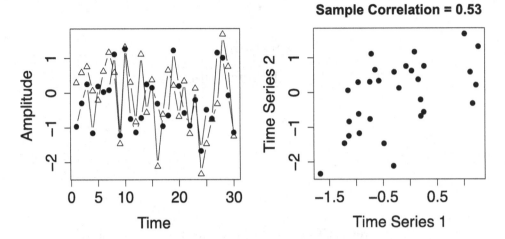

Figure 11.1 Two time series (left) and the corresponding scatter plot (right).

11.1 The Problem

Consider the two time series in Figure 11.1. Are these variables independent? A standard approach to answering this question is to perform a correlation test (see Chapter 2). For this data, the sample correlation is 0.53 while the 5% significance threshold is 0.37. Accordingly, we reject the hypothesis of vanishing correlation. Is there a reason to question this conclusion? The correlation test assumes that samples are independent draws from a bivariate normal distribution. If the data shows no significant serial correlation and no significant departure from normality, then is there any reason to question this conclusion?

A question that might not occur to you is this: How were the variables *selected*? Suppose you learned that the time series were selected as follows. A scientist decides to empirically predict rainfall near his home. To find a suitable predictor, he collects 1,000 weather indices and computes the correlation between each index and observed rainfall. The time series having the largest correlation among the 1,000 cases is the one selected and shown in Figure 11.1. Does this new knowledge alter your interpretation of the correlation? It should!

Recall that a significance level is the probability of rejecting the null hypothesis when it is true. For a standard 5% significance level, 5% of the correlations are *expected* to exceed the threshold by random chance. Therefore, 50 out of 1,000 sample correlations are expected to exceed the threshold. That is, the probability that such a selected variable exceeds the significance threshold is not 5%, but near certainty! The problem here is that the correlation test assumes that X is drawn *independently* of Y, whereas in fact X was selected after an extensive search based on a criterion *that depended on Y*. This is another way of saying X and Y were not drawn from the population assumed in the test.

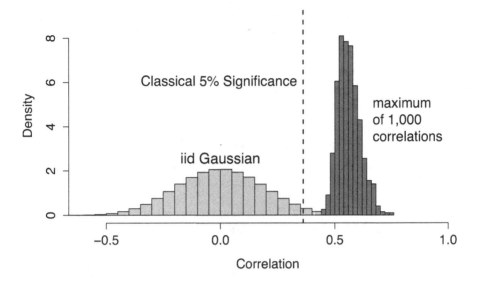

Figure 11.2 Histogram of correlations computed from samples drawn randomly from a normal distribution (light grey shading), and of the maximum correlation out of a 1,000 computed from samples drawn randomly from a normal distribution (dark grey shading). Also shown is the 5% significance threshold corresponding to classical hypothesis testing (dashed line). All correlations are computed from 30 pairs of random numbers.

To appreciate the problem with searching for variables that are correlated, let us *simulate* the procedure using a computer. This is known as a *Monte Carlo* simulation. To illustrate, we first do a Monte Carlo simulation in which the population matches the null hypothesis in the correlation test. The length of the time series in Figure 11.1 is 30, hence we generate 30 random numbers as a surrogate for time series 1, and 30 more random numbers for time series 2. Because they are random, the population correlation of the two time series vanishes exactly. Nevertheless, we repeat this Monte Carlo simulation 1,000 times, yielding 1,000 sample correlations. A histogram of the resulting correlations is shown as light grey shading in Figure 11.2. This procedure simply reproduces the sampling distribution of the correlation coefficient under the null hypothesis that samples are drawn from two independent normal distributions. Specifically, the sampling distribution is approximately Gaussian with zero mean and a 2.5 percentile at 0.37.

Now suppose we simulate the procedure of the forecaster who chooses the time series with the maximum correlation from among 1,000 candidates. This procedure would correspond to selecting the correlation at the far right tail of the light-shaded histogram shown in Figure 11.2. What is the sampling distribution of the forecaster's result under the null hypothesis? The desired distribution can also be estimated by Monte Carlo methods. Specifically, generate 1,000 correlations and select the maximum, then generate another 1,000 correlations and select the maximum, and so on until 1,000 maximum correlations have been generated (a total of one million correlations). The resulting distribution of $\max[\rho_1, \ldots, \rho_{1000}]$ is shown as the dark grey histogram in Figure 11.2. As expected, the latter histogram is concentrated in the tail of the former histogram. The reason for this should be clear: selecting the maximum out of 1,000 sample correlations will tend to select correlations from the extreme positive tail of the null distribution.

Procedures similar to this that involve searching large data sets are often known as *data dredging, data fishing, or data snooping*. Here, we simply use the term *screening*:

> **Definition 11.1** (Screening) *Screening is the process of evaluating a property for a large number of samples and then selecting samples for which that property is most extreme.*

The previous example illustrates a simple point: If a variable has been selected through screening, then the distribution of the associated statistic is *biased toward extreme values* relative to the parent population. This fact has important implications for hypothesis testing. Specifically, rejection regions are usually located in the tails of the parent distribution, hence screening *inflates* the type-I error rate. Moreover, the type-I error rate grows with the amount of screening. This fact does not mean that the hypothesis test is "wrong." On the contrary, the hypothesis test correctly states that the samples were unlikely to have been drawn from the assumed population. A more accurate statement is that the hypothesis test is *uninformative*. After all, if samples were obtained by screening, then we already know that they were not drawn from the distribution assumed under the null hypothesis. The hypothesis test merely confirms what we already knew, hence it is not informative.

Screening breaks a golden rule in statistics: never use the same data *twice*, once to formulate a hypothesis, and again to test the hypothesis. Screening breaks this rule because the same data is used to *select the variables* (that is, formulate a hypothesis) and to test significance. Standard significance tests were not designed to test hypotheses on samples *that were used to formulate the hypothesis*.

Importantly, if variables have been selected through screening, but this fact is hidden from an investigator, then ascertaining that such screening has been performed *using only the given sample* is often difficult (as illustrated by our initial reaction to Figure 11.1). Instead, detecting such screening often requires testing the purported model with *independent* data (that is, data that was not used to infer the original relation).[1]

---

[1] An interesting exception occurs when the population is known theoretically, in which case it is possible to test whether the sample is "too good to be true." An example of this is the analysis of Fisher (1936), who analyzed

11.2 Screening *iid* Test Statistics

The impact of screening can be quantified in certain cases. The first case we consider is a test in which the hypothesis H_0 is rejected when the statistic Z exceeds the threshold z_α. Let the significance level of the test be α, which by definition means

$$P(Z > z_\alpha) = \alpha \quad \text{if } H_0 \text{ is true.} \tag{11.1}$$

Thus, if H_0 is rejected when $Z > z_\alpha$, then the type-I error rate is α. Next, suppose Z is computed M times from independent samples, yielding $Z_1, \ldots, Z_M$, and then the maximum of these values is selected. If we reject H_0 when *this maximum* exceeds z_α, what is the corresponding type-I error rate? This rate is known as the *familywise error rate* and defined

$$\text{FWER} = P(\max\{Z_1, \ldots, Z_M\} > z_\alpha) \quad \text{if } H_0 \text{ is true.} \tag{11.2}$$

The condition $\max\{Z_1, \ldots, Z_M\} > z_\alpha$ is satisfied if at least one of the values $Z_1, \ldots, Z_M$ exceeds z_α. Thus, FWER is the probability of *at least* one type-I error. Now, the probability of any event is one minus the probability of the nonevent:

$$\text{FWER} = 1 - P(\max\{Z_1, \ldots, Z_M\} \leq z_\alpha). \tag{11.3}$$

The condition $\max\{Z_1, \ldots, Z_M\} \leq z_\alpha$ is equivalent to the condition that *each* $Z_m \leq z_\alpha$:

$$\text{FWER} = 1 - P(Z_1 \leq z_\alpha \text{ and } \ldots \text{ and } Z_M \leq z_\alpha).$$

Because $Z_1, \ldots, Z_M$ are independent, the joint distribution on the right can be factored into a product of marginals:

$$\text{FWER} = 1 - P(Z_1 \leq z_\alpha) \ldots P(Z_M \leq z_\alpha). \tag{11.4}$$

Also, because the statistics are identically distributed, the marginals are equal, hence

$$\text{FWER} = 1 - P(Z \leq z_\alpha)^M = 1 - (1 - \alpha)^M. \tag{11.5}$$

The probability α is the type-I error rate on an *individual* test. The FWER as a function of M for $\alpha = 5\%$ is shown in Figure 11.3. As can be seen, the FWER increases rapidly with M. For instance, after screening 10 Z's, the FWER is 40%.

Gregor Mendel's celebrated data on plant fertilization and found that the fit to Mendel's theoretical model was too good to be true, leading to controversies about whether Mendel's data had been adjusted or censored. This "forensic" procedure is rarely useful in climate applications because typical theoretical models in climate science rarely specify the population parameters completely, in contrast to genetic models. For instance, Mendel's model could predict that certain offspring are drawn from a population in which certain traits occur *exactly* in a 3:1 ratio.

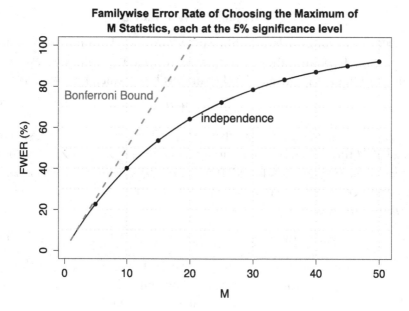

Figure 11.3 The familywise error rate of selecting the maximum of M test statistics, each at the 5% significance level, when the null hypothesis is true. Solid curve with dots shows FWER assuming independence (11.5), and dashed straight line shows the Bonferroni upper bound (11.14).

A natural idea is to *adjust* α to ensure that the FWER is at some prescribed level, e.g., 5%. How small should α be chosen? The answer can be found by inverting (11.5):

$$\alpha_{ADJ} = 1 - (1 - \text{FWER})^{1/M}. \tag{11.6}$$

A plot of the adjusted α for ensuring FWER = 5% is shown in Figure 11.4. The result is shown on a log-log scale and reveals the simple relation

$$\alpha_{ADJ} \sim \text{FWER}/M. \tag{11.7}$$

This relation can be derived from a Taylor expansion of (11.6). Thus, if 10 Z's are screened, then the α of individual tests should be adjusted to 0.5% to ensure FWER = 5%.

Example 11.1 *Question: A scientist investigates whether the mean precipitation over a country differs between El Niño and La Niña. Accordingly, the scientist partitions the country into 10 boxes and performs a difference-in-means test for each box separately (see Chapter 2 for the t-test). These calculations yield 10 t-values $t_1, \ldots, t_{10}$, each computed from 15 El Niños and 15 La Niñas. For just the i'th box, the decision rule would be to reject the hypothesis of equal means if $|t_i| > t_C$, where*

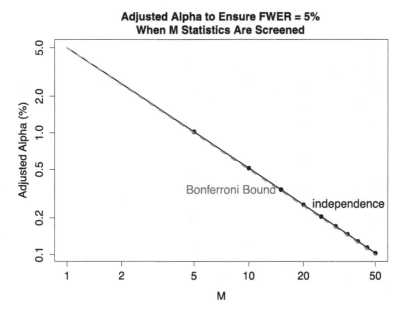

Figure 11.4 The adjusted α to ensure FWER = 5%, when selecting the maximum of M test statistics and the null hypothesis is true. Solid curve with dots shows FWER assuming independence (11.5), and dashed curve shows the Bonferroni bound (11.14). The figure uses a log-log scale.

t_C *is the critical value for a 2-sided t-test with* $15 + 15 - 2 = 28$ *degrees of freedom. Assuming the t-values* $t_1, \ldots, t_{10}$ *are independent, what is the familywise error rate of the decision rule*

$$\text{reject } H_0 \text{ if } \max[|t_1|, \ldots, |t_{10}|] > t_C ? \tag{11.8}$$

Suppose we want to "correct" for the screening. What decision rule of the form

$$\text{reject } H_0 \text{ if } |t_i| > t_{FWER} \text{ for any} \tag{11.9}$$

would ensure that the FWER is 5%?

Answer: *The type-I error rate for a single t-test is 5%. Therefore, the probability of at least one type-I error after choosing the maximum of* 10 *independent values of* $|t|$ *is*

$$FWER = 1 - (0.95)^{10} = 40\%. \tag{11.10}$$

Note how much greater this is compared to 5%. To achieve FWER = 5%, we evaluate (11.6) *for FWER = 5%, which yields* $\alpha = 1 - 0.95^{1/10} \sim 0.5\%$. *Thus, the individual type-I error rate should be 0.5% to achieve a familywise error rate of 5%. Halving this, the upper 0.25% percentile of a t statistic with 28 degrees of freedom is 3.04, thus* $t_{FWER} = 3.04$. *More generally, when multiple testing is involved, the individual type-I error rate needs to be reduced below 5% in order to keep the FWER equal to 5%.*

11.3 The Bonferroni Procedure

The familywise error rate (11.5) assumes independent outcomes. This assumption is rarely realistic. For instance, it is common to test hypotheses across different geographic locations, but most climate variables are correlated with neighboring regions. To derive a general upper bound on the familywise error rate, one may use *Bonferroni's Inequality*. The Bonferroni inequality states that if $A_1, \ldots, A_M$ are a countable set of events, then

$$P(A_1 \text{ or } A_2 \text{ or } \ldots \text{ or } A_M) \leq P(A_1) + P(A_2) + \cdots + P(A_M). \qquad (11.11)$$

This inequality holds for arbitrary distributions. Since the event $\max\{Z_1, Z_2, \ldots, Z_M\} > z_\alpha$ is logically equivalent to the event

$$Z_1 > z_\alpha \text{ or } Z_2 > z_\alpha \text{ or } \ldots \text{ or } Z_M > z_\alpha, \qquad (11.12)$$

Bonferroni's inequality immediately implies

$$P(\max\{Z_1, Z_2, \ldots, Z_M\} > z_\alpha) \leq M P(Z > z_\alpha), \qquad (11.13)$$

or equivalently,

$$\text{FWER} \leq M\alpha. \qquad (11.14)$$

This inequality has two important implications. First, it states that if you perform M tests at the α significance level, then FWER cannot be greater that $M\alpha$. Second, if you adjust α to ensure that the FWER is no greater than 5%, then this inequality implies choosing $\alpha_{\text{ADJ}} = 5\%/M$, because then FWER $\leq M\alpha_{\text{ADJ}} = 5\%$. This adjustment is known as the *Bonferroni procedure* and is the same as the adjustment (11.7) for independent samples. This result shows that independence is not a restrictive assumption when defining α_{ADJ}.

11.4 Screening Based on Correlation Maps

A common example of screening is the use of correlation maps to identify predictors. To illustrate, consider the problem of predicting the seasonal mean of a variable Y. At first sight, predicting seasonal means would seem impossible because day-to-day atmospheric variability is predictable on time scales less than a month (Lorenz, 1963; Shukla, 1981). However, slowly varying components of the climate system, such as the sea surface temperature, are predictable owing to their longer persistence. These slower components might impart predictability on other variables. Unfortunately, this argument is *qualitative* and gives little clue as to what *precise* aspects of SST may be useful for predicting Y. Thus, to explore the relation between Y and SST, a common approach is to construct a correlation map like that shown

Correlation Map between Y and SST (1970–2000)

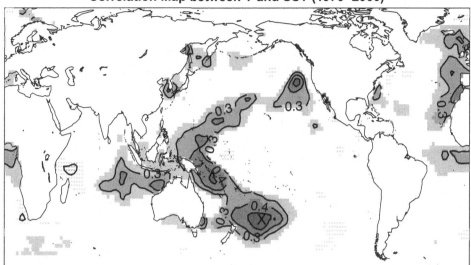

Figure 11.5 Correlation between a variable Y and the June to September sea surface temperature during the period 1970–2000. Correlations that are insignificant at the 5% level, as determined by classical hypothesis testing, have been masked out. The grid point with maximum correlation is marked by "X" (it is located to the east of Australia).

in Figure 11.5, which shows the correlation between Y and the June to September SST during the period 1970–2000. Moreover, values that are insignificant at the 5% level, *locally*, have been masked out. At face value, the figure suggests that Y is related to a "horseshoe" pattern of SSTs centered in the western equatorial Pacific, and to SSTs in the eastern North Atlantic. The grid point with maximum correlation is marked by "X." This grid point is selected as a predictor. A scatter plot between this predictor and Y is shown in Figure 11.6 and confirms the positive correlation. Therefore, it seems that this procedure has helped find a useful predictor of Y.

Surprise! We have played a trick: the variable Y was produced by a random number generator. *The correlation map in Figure 11.5 is purely the result of random noise.* The particular map depends on the realization of Y. For a different realization of Y, a different correlation map would be obtained, but *some* region would still exhibit relatively high correlation.

This behavior is easy to explain. A correlation map displays the correlation between a quantity and a field of predictors. The field is often represented on a grid involving thousands of points. The act of looking at a correlation map and selecting areas with large correlations is tantamount to screening thousands of sample correlations. In the example, there are 8,318 ocean grid points, so 5% of

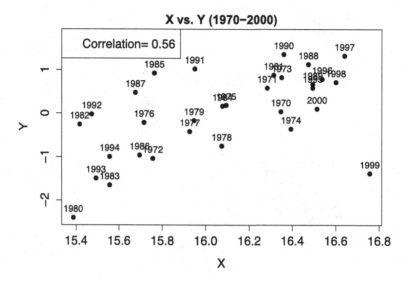

Figure 11.6 Scatter plot between Y and the grid point marked X in Figure 11.5.

these, or about 416 grid points, are expected to exceed the 5% significance level (assuming independence).

One might be misled into believing that SST is related to Y because of the *distinctive pattern* seen in Figure 11.5. After all, if each grid point were independent, then grid points with large correlation would be *randomly* distributed in space. Instead, high correlations are distributed *smoothly* along distinctive shapes. The reason large correlations are distributed smoothly along preferred areas is because the SST is *spatially dependent*: neighboring grid points tend to be correlated with each other. Because of spatial dependence, if a particular grid point happens to be correlated with Y, then neighboring points also tend to be correlated with Y. This spatial correlation reflects some physical connection between grid points, and it is this physical mechanism that one would be explaining in interpreting Figure 11.5. If a new realization of Y were chosen, then the new Y would be highly correlated with some *other* random grid point in the grid point field, and spatial dependence in the SST field would produce a different pattern of high correlations.

What value of α should be used to ensure the familywise error rate is 5%? According to the Bonferroni procedure discussed in Example 11.1, $\alpha = 5\%/8318 \sim 6 * 10^{-6}$. The corresponding t-value with 29 degrees of freedom is $t_c \approx 5.3$. Using (2.42), the corresponding critical correlation should be

$$\rho_{\text{FWER}} = \frac{t_c}{\sqrt{N - 2 + t_c^2}} \sim 0.7. \tag{11.15}$$

As can be seen in Figure 11.5, no point has a correlation exceeding 0.7. On the other hand, climate model simulations suggest that *real* seasonal relations often have correlations below 0.7, so the Bonferroni procedure is so stringent that even real dependencies cannot be detected. *This result is typical*: for a large number of comparisons, the Bonferroni procedure often requires such a high standard of evidence that even real effects cannot be detected.

Procedures for testing hypotheses about population parameters distributed in geographic space are called *field significance* tests. Field significance tests that can detect violations at levels much less stringent than Bonferroni procedures are discussed in Chapter 13.

11.5 Can You Trust Relations Inferred from Correlation Maps?

These considerations suggest that using correlation maps to select predictors is very risky. On the other hand, sometimes the correlations in a correlation map reflect real dependencies. How can you tell the difference between a real relation and an artificial relation? The answer is the same as for any scientific hypothesis: formulate your hypothesis with the data in hand, then test it with *independent* data. That is, after data has been used to construct a correlation map and formulate a testable hypothesis about a relation, then discard that data and use *other* data to test the relation. Such independent data may include (1) data from independent time periods, (2) data from climate model simulations, or (3) future data that was unavailable when you formulated the hypothetical model.

An important requirement in testing hypotheses is that the hypothesis must be *precise*. A *vague* theory cannot be proven wrong. For example, earlier it was argued that Pacific SSTs influence Y. This suspicion constitutes a hypothesis, but it is *too vague to test*. *Which* part of the Pacific? How *early* can the influence be exerted? What is the *directionality*? How *large* is the effect? If the answers to these equations are imprecise, then, with a little talent, *any* observational result can be interpreted as fitting the hypothesis. Instead, a testable hypothesis needs to state, ahead of time, what nature will do within well-defined bounds, so that one can decide if the observations fall outside the bounds implied by the hypothesis. The most precise hypothesis would be a *mathematical model* of the relation between Y and the Pacific SSTs, with well-defined uncertainty ranges for the model parameters. Whatever data is used to formulate this model would need to be excluded in the testing phase.

11.6 Screening Based on Change Points

Another form of screening concerns *change points*. A change point is a point in time at which the distribution of a stochastic process changes. Change points occur naturally in an instrument record when the instrument is moved from one location

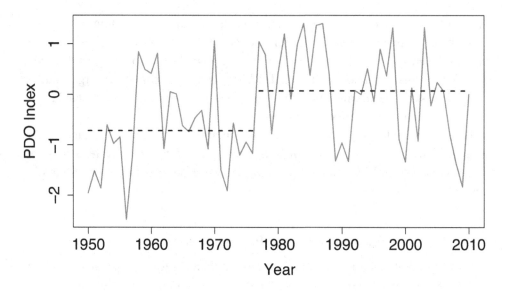

Figure 11.7 The January to March mean Pacific Decadal Oscillation (PDO) index. Horizontal dashed lines show the mean for the periods 1950–1976 and 1977–2010.

to another. Change points also may occur as a result of natural events, such as volcanic eruptions. In these examples, the timing of the change point is known, hence hypotheses about changes across the change point can be investigated using standard tests. However, if the change point is unknown, then selection of the change point involves screening that must be taken into account.

To illustrate, consider an index known as the Pacific Decadal Oscillation (PDO) index. This index measures the state of a particular pattern of sea surface temperatures in the Pacific Ocean poleward of 20N (the precise definition is not important for now). The January to March mean value of this index is plotted in Figure 11.7. After looking at this time series, one might hypothesize that the population means changed around 1976, perhaps reflecting a "climate shift." To investigate this possibility, one performs a t-test using 1950–1976 and 1977–2010 data, which yields an absolute t-statistic of 3.29. The p-value of this statistic is about 0.0017, implying that the probability of finding such a large value under the assumption of equal population means is less than 1 in 500. Is this a valid interpretation?

Hopefully, after reading the previous sections, the flaw in the procedure is evident: The change point 1976 was identified by looking at the time series, and then tested using the same time series. Why was the year 1976 chosen? In essence, because it *maximized the difference that we want to test*, which is screening. The impact of this screening can be quantified using Monte Carlo techniques, as follows. Generate N random numbers from a Gaussian distribution. Let the first i values define group 1, and let the remaining values define group 2. Evaluate the t-statistic (2.25) for $i = 10, 11, \ldots, N - 10$, which ensures that the two groups have at least

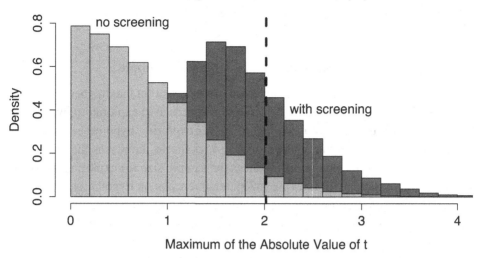

Figure 11.8 Result of Monte Carlo experiments in which 61 random numbers are drawn from a Gaussian distribution and then split into two segments to compute the t-statistic for testing a difference in means. The t-statistic is computed for all possible "change points" in which the minimum sample size is 10. A histogram of the absolute t-values is shown in light grey (labeled "no screening"), and the corresponding 95th percentile is indicated by the vertical dashed line. Then, the maximum absolute value is selected, and this whole procedure of generating 61 random numbers, computing t for the possible change points, and then selecting the maximum, is repeated 10,000 times to produce the histogram in dark shading (labeled "with screening").

ten samples. A histogram of the resulting t-values is shown as the light grey shading in Figure 11.8, and the associated 95th percentile is indicated by the vertical dashed line. Among the t-values derived from different change points, the maximum absolute t value is selected. Repeating this whole procedure – selecting the maximum t-value among all change points – 10,000 times yields the histogram shown in the dark grey in Figure 11.8. Comparison of the two histograms shows that screening significantly biases the distribution of $max(|t|)$ toward larger values relative to the case without screening. Thus, selecting the change point from data can create a dramatic bias in the t values.

Relative to the distribution of $max(|t|)$ shown in Figure 11.8, the observed value $t = 3.29$ is the 99.78th percentile, which corresponds to a p-value of 2.2%. This result suggests that the observed change is significant even after accounting for selection of the change point. However, the Monte Carlo technique assumed independent samples (i.e., no serial correlation). Also, the PDO index was derived from Principal Component Analysis, which maximizes variance (see Chapter 12), which might have contributed to the extreme t-statistic. Thus, the t-statistic might not be significant after these factors are taken into account. There exists an extensive literature on testing change points. See Reeves et al. (2007) for a review.

11.7 Screening with a Validation Sample

Another type of screening involves repeated use of a data set that has been set aside for testing hypotheses. For instance, a common approach in exploratory hypothesis testing is to split the data into two parts: one for exploring hypotheses, known as the *training* sample, and the other for testing hypotheses, known as the *validation* sample. Although splitting the data reduces the sample size for exploring hypothesis, it has the advantage of allowing independent checks on hypotheses. However, if the validation sample gives a disappointing result, it is very tempting to adjust the hypothesis and then *check with the same validation sample*, and then repeat until a successful result emerges. This approach effectively subsumes the validation sample in the screening process, in which case inferences that do not account for screening are easily misinterpreted.

11.8 The Screening Game: Can You Find the Statistical Flaw?

Screening can be quite subtle. Next we present a few more examples and invite you to spot the flaw and explain how to quantify its effect before reading our answer.

11.8.1 Does North American Temperature Depend on the MJO?

Procedure

The Madden–Julian Oscillation (MJO) is a type of organized convection that propagates eastward over the tropical Indian and Pacific Oceans and recurs approximately every 30–60 days. An investigator wants to know if the temperature in his home country is extreme when the MJO is active. To investigate this question, the investigator computes the mean temperature in the region conditioned on the "phase" of the MJO. Since the MJO is a propagating phenomenon, it cannot be characterized by a single index, so two indices often are used. Two standard indices are the RMM1 and RMM2 indices of Wheeler and Hendon (2004). As the MJO evolves, it traces out a curve in the (RMM1, RMM2) plane, as illustrated in Figure 11.9. To define the "phase" of the MJO, the (RMM1, RMM2) plane is partitioned into eight equal "pie slices" labeled $1, 2, \ldots, 8$. The eight sections define eight phases of the MJO. Then, the mean temperature in the region is computed conditioned on each of the eight phases of the MJO. The investigator discovers that during phase 1, the mean temperature is 2.5 standard deviations above normal. Is this result significant?

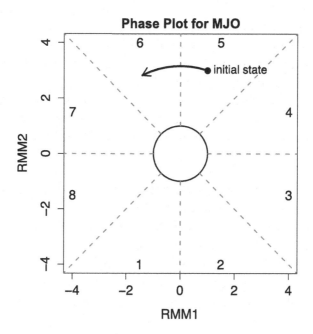

Figure 11.9 A "phase diagram" of the MJO. The RMM1 and RMM2 indices are used as the X- and Y-coordinates. As the MJO evolves a point in the (RMM1, RMM2) plane traces out a curve, as indicated in the figure. The phase diagram is divided into eight equal sections and labeled $1, 2, \ldots, 8$. The unit circle is also indicated.

Answer

In this example, the investigator tested a hypothesis eight times and selected the most extreme result, which is screening. The probability that a random variable from a Gaussian distribution exceeds 2.5 standard deviations is 1.2%, which would be considered significant if that were the only test performed. On the other hand, the probability that at least one of the eight cases would exceed 2.5 standard deviations is

$$\text{FWER} = 1 - (1 - 1.2\%)^8 \approx 9.5\%, \tag{11.16}$$

which is greater than 5% and thus implies that the finding is not statistically significant.

11.8.2 Does the Best Subset of a Multi-Model Forecast Have Skill?

Procedure

A forecaster is interested in predicting a particular event. For each such event in a historical record, the forecaster has nine separate forecasts. The forecaster would like to combine these nine forecasts into a single, superior forecast. The forecaster decides to select the four forecasts that are most correlated with observations from a 20-year historical record, and then average those forecasts to make a prediction. When the four best forecasts are averaged, the correlation between the mean four-member forecast and observations is 0.55, which exceeds the 5% critical value $2/\sqrt{20} \approx 0.45$. Since the actual correlation exceeds the threshold, the forecaster claims that the four-member mean forecast has a statistically significant correlation.

Answer

The forecaster selected the best four out of nine forecasts, which is screening. The impact of screening can be estimated by Monte Carlo methods. To do this, we generate random numbers to simulate 20 years of nine forecasts, plus 20 years of observations. The correlation between each of the nine "forecasts" and "observations" are computed, the four forecasts with largest correlation are averaged to construct a single forecast, and the correlation between this four-member mean and "observations" is computed. This procedure is repeated 10,000 times to produce an empirical distribution for the correlation skill. A histogram of the resulting correlations of the four-member mean is shown in Figure 11.10. First, we see that the correlations are skewed above zero, even though none of the forecasts have skill. Second, the 95th percentile of this distribution, indicated by the vertical dashed line, is 0.61, implying that the correlation would have to exceed 0.61 in order to reject the null hypothesis of zero correlation at the 5% level. Since the actual correlation is less than this, the skill is not significant. In fact, the familywise p-value for this

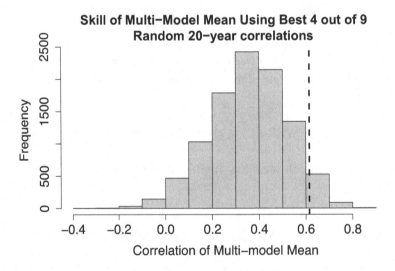

Figure 11.10 Histogram of correlation coefficients of a multi-model forecast derived from averaging the four most correlated "no-skill" forecasts out of nine no-skill forecasts. The vertical dashed line indicates the 95th percentile.

procedure is 11.7% (that is, greater than 5%). Therefore, the correlation skill of 0.55 is not significant when screening is taken into account.

11.8.3 Does October Weather Augur Next Winter?

Procedure

An investigator hypothesizes that a particular index, known as Z, is useful for predicting winter temperature. The 50-year correlation between winter temperature and the December and November values of Z are 0.2 and 0.1, respectively, but 0.3 in October. The 5% critical value for a correlation is $2/\sqrt{50} \approx 0.28$. Accordingly, the investigator claims that the October value of Z is a good predictor of JFM temperature. Is the predictor significant after taking screening into account?

Answer

The investigator computed three correlations and chose the largest, which is screening. The probability that *at least one* of three correlations exceeds the 5% level is

$$\text{FWER} = 1 - (1 - 0.05)^3 \approx 14\%. \tag{11.17}$$

This calculation assumes independence, which might not be correct since the three correlations were computed from the same observations. A more rigorous bound can be computed from the Bonferroni bound, which implies $\text{FWER} \leq M\alpha = 15\%$. Both calculations suggest that the familywise error rate could be as high as 15%, so statistical significance is in doubt. Also, one normally expects a predictor to become less useful as the lead time increases. However, in this example, the correlation is strong in October and weak in the months in closer proximity (November and December). Since this behavior contradicts usual expectations, the investigator is obliged to give a physical explanation of how the October value is a better predictor than the November or December values.

11.9 Screening Always Exists in Some Form

No matter how hard one tries to quantify screening, it is difficult to account for *all* sources of screening. For instance, suppose one day you notice an observation that is "surprising." What does "surprising" mean? Essentially, something is surprising when it is unlikely relative to your personal reference distribution derived from past experience. This process resembles a significance test, except in this case your brain performed the test effortlessly. Note, however, that the observation was singled out from countless observations from your past, which is a form of screening. Thus, if you select something because it seemed "surprising" *after looking at the data*, then screening is often involved.

Now, many scientific hypotheses are formulated based on an observation that seemed "surprising." However, formulating a scientific hypothesis is only part of the scientific method. Another important part is *testing* the hypothesis, and this requires *independent data*. The cleanest tests are those based on *controlled experiments* in which all factors that may influence the outcomes are taken into account.

In statistical forecasting, the choice of selectors often involves numerous choices, which opens opportunities for screening. For instance, the physical variable must be selected. Next, the geographic location and vertical level needs to be chosen. Often, area averaging is performed, requiring the boundaries of the area to be defined. Also, climate variables are typically filtered in time, requiring choices for the time-filtering parameters. If, after making these choices, the resulting predictors turn out to be useless, it is very tempting to adjust decisions to produce a more favorable result. If a sufficiently large number of adjustments are explored, then, eventually, some set of adjustments will produce a desired result. Standard significance tests do not account for the process, and therefore will not correctly characterize the significance of the final result.

Other choices are even more subtle. Scientists are not robots. Experienced scientists know a lot about the data they are analyzing. Such background knowledge often leads scientists to make subconscious choices in defining their analyses to give a favorable outcome. These choices are difficult to quantify in a hypothesis test. This is why *prediction* is such an important component of hypothesis testing – future information is inaccessible and therefore cannot possibly be a source of bias in a scientist's choices. Predicting the future remains one of the most valuable tools for testing scientific understanding.

11.10 Conceptual Questions

1 Explain why the results of a correlation test cannot always be taken at face value.
2 What is screening? What is so important about it?
3 What is the Familywise Error Rate (FWER)?
4 What is the Bonferroni upper bound on the FWER?
5 If you perform $M = 20$ hypothesis tests at the 5% significance level and reject the hypothesis if *any* one test gives a significant result, what is the FWER of this procedure?
6 What is the danger is using correlation maps to select predictors?
7 If you want to test if a relation found from a correlation map is useful, what can you do?
8 What is the danger in identifying "climate shifts" from data?
9 When testing hypotheses, explain why there is no substitute for predicting the future.

12

Principal Component Analysis

... in choosing among the infinity of possible modes of resolution of our variables into components, we begin with a component γ_1 whose contributions to the variances of X have as great a total as possible; that we next take a component γ_2, independent of γ_1, whose contribution to the residual variance is as great as possible; and that we proceed in this way to determine the components ... perhaps neglecting those whose contributions to the total variance are small. This we shall call the *method of principal components*.

Hotelling (1933)

Large data sets are difficult to grasp. To make progress, we often seek a few quantities that capture as much of the information in the data as possible. Typically, the informative components are not known beforehand and therefore must be derived from the data. There are many criteria for defining informative components, such as components that maximize predictability (which leads to Predictable Component Analysis, Chapter 18) or components that maximize correlation (which leads to Canonical Correlation Analysis, Chapter 15). In this chapter, we discuss a procedure known as *Principal Component Analysis (PCA)*, also known as *Empirical Orthogonal Function (EOF) analysis*. PCA finds the components that minimize the sum square difference between the components and the data. The components are ordered such that the first approximates the data the best (in a least squares sense), the second approximates the data the best among all components orthogonal to the first, and so on. In typical climate applications, a principal component consists of two parts: (1) a fixed spatial structure, known as an Empirical Orthogonal Function (EOF) and (2) its time-dependent amplitude, known as a PC time series. The EOFs are orthogonal and the PC time series are uncorrelated. Principal components are often used as input to other analyses, such as linear regression, canonical correlation analysis, predictable components analysis, or discriminant analysis. PCA depends on the norm used to measure the difference between the component and the original data set. The procedure for performing area-weighted PCA is discussed in detail in this chapter.

12.1 The Problem

One of the challenges in climate science is analyzing large data sets. For example, surface temperature is often described by discrete values on a grid. A $2.5° \times 2.5°$ grid on a globe, for instance, has more than $10,000$ points. Furthermore, temperature evolves in time. Therefore, after 100 years, say, there are 100 annual mean surface temperature fields involving over $100 * 10,000 =$ one million numbers. How can a million numbers be grasped?

The problem of understanding large data sets arises in many fields and the way forward is often the same: identify patterns in the data, and then use these patterns to simplify the data description. For example, monthly-mean surface temperature tends to be nearly constant over distances on the order of 100 km (Hansen and Lebedeff, 1987). Thus, if measurements were taken 1-km apart in a linear, 100 km-wide region, monthly averages would essentially consist of the same value repeated 100 times. Storing 100 *redundant* values is wasteful. A more efficient description is to specify a single value for the 100 km-wide linear region.

In general, the more redundancy in a data set, the more it can be compressed while maintaining most or all of the original information. To illustrate, consider the extreme case of a variable Y that is perfectly correlated with itself at every other point. For instance, suppose the variable were simply a fixed pattern with a time-varying amplitude; that is,

$$Y(t,s) = f(t)e(s), \tag{12.1}$$

where s and t are space and time indices, respectively, $f(t)$ is a stationary process, and $e(s)$ is any function of s. Ordinarily, describing the data set $Y(t,s)$ would require NM numbers, where N is the total number of time steps and M is the total number of spatial locations. In contrast, the description (12.1) requires only N numbers for $f(t)$ and M numbers for $e(s)$, which is a total of $N + M$ numbers. This total may be *much* less than recording all MN numbers for $Y(t,s)$. For instance, in the previous example, $N + M = 10,100$, whereas $MN = 1,000,000$, representing a 99% reduction in data size! This dramatic compression reflects an extreme statistical redundancy: at every single instant, $Y(t,s)$ is exactly proportional to $e(s)$. Also, the description (12.1) is more easily visualized because it requires showing only one spatial map $e(s)$, and one amplitude time series $f(t)$, so the entire data set (one million numbers) can be displayed with two figures!

Correlations in a data set are also useful for gaining insight into physical laws. For instance, atmospheric and oceanic fluids generate waves in response to per-turbations. Such waves can transmit information from one point to many other points and thereby impose correlations in space and time. Conversely, correlations provide information about the underlying physical laws, and therefore compression algorithms, which identify correlations in data set, may be used to extract dynamical knowledge or insights from a data set.

Real data is more complicated than (12.1) and requires more than one spatial pattern and time series for its description. This chapter reviews an algorithm known as *Principal Component Analysis (PCA)* for defining the sequence of functions $f_i(t)$ and $e_i(s)$ that best approximate a data set. The product $f_i(t)e_i(s)$ is known as a *component*. A natural measure of "best" is the sum square difference between data and components. For example, the *first principal component* $f_1(t)e_1(s)$ minimizes

$$\|Y(t,s) - f_1(t)e_1(s)\|_F^2 = \sum_t \sum_s (Y(t,s) - f_1(t)e_1(s))^2 . \tag{12.2}$$

The measure $\|\cdot\|_F$ is known as the *Frobenius norm*. More general measures involving weighted sums are considered in Section 12.5. If Y were of the form (12.1), then the Frobenius norm in (12.2) would vanish for suitable choice of $f_1(t)e_1(s)$. Of course, real data is more complicated than (12.1), so the residual $Y(t,s) - f_1(t)e_1(s)$ is typically nonzero. Accordingly, the procedure is repeated *on the residual*, which yields a new component $f_2(t)e_2(s)$ that best approximates the residual $Y(t,s) - f_1(t)e_1(s)$. Again, the *updated residual* $Y(t,s) - f_1(t)e_1(s) - f_2(t)e_2(s)$ is nonzero, hence we repeat the procedure to find a third component. Repeating this procedure K times yields the approximation

$$Y(t,s) \approx Y_K(t,s) = f_1(t)e_1(s) + f_2(t)e_2(s) + \cdots + f_K(t)e_K(s). \tag{12.3}$$

If variables are centered about their sample means, then $\|Y\|_F^2/(N-1)$ is the *sum total variance*. Principal components satisfy certain orthogonality properties, hence the total variance of individual components adds up to the sum total variance. The sum total variance of $Y_K(t,s)$ is known as the *variance explained by the first K principal components*. As K increases, Y_K explains more variance of Y. Replacing Y by Y_K, where K is smaller than either M or N, is a form of *dimension reduction*.

The choice of the parameter K depends on application: one could choose K to (1) approximate a data set at a prescribed level of accuracy, (2) limit the number of bits for storing $e_1, \ldots, e_K$ and $f_1, \ldots, f_K$ (e.g., limit the file size to 100 Mb), or (3) avoid splitting effectively degenerate components (discussed in Section 12.4.2). Often, components beyond K are disregarded, and the first K components become the objects of study. The loss of information resulting from discarding components is often more than compensated by the gain in simplicity as a result of reducing the dimension of the space being investigated.[1]

---

[1] To derive reduced-order dynamical models, PCA is not the best approach. The best approach is a compromise between structures with large variance and structures that optimally excite variance. In realistic fluid dynamical systems, the latter structures often have little variance, and hence would missed by PCA. See Farrell and Ioannou (2001) for further discussion.

Although we have interpreted s and t to mean space and time, PCA does not require any such interpretation. For instance, social scientists frequently use PCA to analyze socio-economic data, such as education, health, and employment. However, in climate science, PCA is used most commonly to analyze a variable that is a *function of space and time*. Accordingly, this technique will be explained in this context (as it was previously). Nevertheless, it should be kept in mind that the technique can be used to analyze a rich variety of data structures quite different from spatio-temporal data.

12.2 Examples

Before describing computational details, we first show a few illustrative examples of PCA. In our first example, we apply PCA to December-mean sea surface temperature (SST). We use the data set of Huang et al. (2015), which represents SST on a $2° \times 2°$ longitude–latitude grid, corresponding to 8,320 ocean grid points. The 56-year period 1960–2015 is analyzed, so the whole data set comprises nearly 500,000 numbers. PCA decomposes this data set into the form (12.3), where $e_i(s)$ is the i'th *Empirical Orthogonal Function (EOF)* and describes the spatial part of the component, while $f_i(t)$ is the i'th *PC time series* and describes the temporal part of the component. The time series at each grid point is centered before applying PCA, hence PC time series describe *anomalies* with respect to the mean. Also, PCA determines only the *product* $f_i(t)e_i(t)$; it does not constrain the magnitudes of $f_i(t)$ and $e_i(t)$ individually. Here, the PC time series are normalized to unit variance.

Principal components are ordered by the degree to which they approximate the data, or equivalently, by the amount of variance they explain. The first component explains 29% of the total variance. The leading EOF $e_1(s)$ is shown in Figure 12.1a and has maximum amplitudes in the equatorial east Pacific. The maximum value is just under 1.2 K. This spatial pattern is often identified with El Niño. The corresponding PC time series $f_1(t)$ is shown in Figure 12.1b. Multiplying the EOF by the PC time series yields a component with fixed spatial pattern and a time-varying amplitude. For instance, during December 1997, the PC time series has a value of 2.6, hence the EOF pattern would be multiplied by 2.6, giving a warming anomaly around 3.1K in the equatorial central Pacific. In contrast, during December 1973, the EOF pattern would be multiplied by -1.9, giving a *cooling* anomaly around 2.3K in that region for that year.

The PC time series are uncorrelated in time and the EOFs are orthogonal in space. These orthogonality properties imply that the total variance is the sum of the variances of individual components. The variances explained by the first 15 components are shown in Figure 12.2 (error bars show 68% confidence intervals).

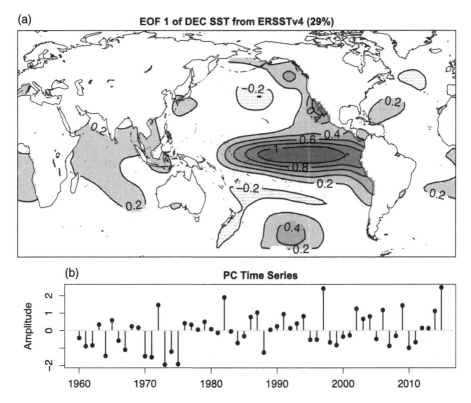

Figure 12.1 (a) Leading EOF of the December SST from the Extended Reconstructed SST of Huang et al. (2015). The percent variance explained by the leading
component is indicated in the title. The value at each grid point gives the departure
from climatology in units of degrees Kelvin per unit PC time series. (b) The
corresponding leading PC time series.

The first component explains 29% of the total variability in SST and is specified by
only 8,320 values in space and 56 values in time, or 8,376 total numbers, which is
less than 2% of the total sample size. Because the components are orthogonal, the
first two components explain $29 + 14 = 43\%$ of the total variability, and no other
component of the form $e_1(s)f_1(t) + e_2(s)f_2(t)$ can approximate the data better than
the first two principal components. Similarly, the variance explained by the first K
components is merely the sum of the variances of the individual components, and
no other set of K components better approximates the data. This property is often
described by saying that the first K components "explain the most variance."

Because principal components are orthogonal in time, they decompose variability
at individual points in space too. For instance, if we focus on spatial point s_0, then
the amount of variance at that point explained by the k'th principal component is the
variance of $f_k(t)e_k(s_0)$, which is $e_k^2(s_0)$, since $f_k(t)$ has unit variance. The fraction

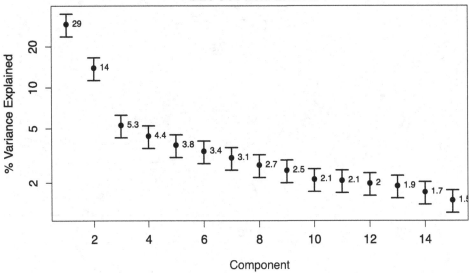

Figure 12.2 Percent variance explained by the principal components of December SST from the Extended Reconstructed of Huang et al. (2015). The error bars show the 68% confidence interval.

of variance explained by the k'th component *at that point* is $e_k^2(s_0)/(e_1^2(s_0) + \cdots + e_R^2(s_0))$, where R is the maximum number of components.

How many principal components K should be chosen? The answer depends on the application. If the goal is to compress data, then K is determined by the size of the data one wants to store. If the goal is to use PC time series as predictors in a regression model, then K is chosen by a model selection criterion. If the goal is to avoid sensitivity to sample (as discussed in Section 12.4.2), then a rule of thumb by North et al. (1982) is to select K such that the confidence interval of the K'th component does not overlap with that of the $(K+1)$'th component. The only choices satisfying this criterion are $K = 1$ and $K = 2$.

Another instructive illustration of PCA is provided by the propagating wave

$$Y(t,s) = A\cos(s - t), \tag{12.4}$$

where s is spatial location and t is time. To create a data set, $Y(t,s)$ is sampled at equally spaced points in space and time that span an integral number of wavelengths and periods. The resulting data set is illustrated in Figure 12.3. A standard trigonometric identity gives

$$\cos(s - t) = \sin(t)\sin(s) + \cos(t)\cos(s). \tag{12.5}$$

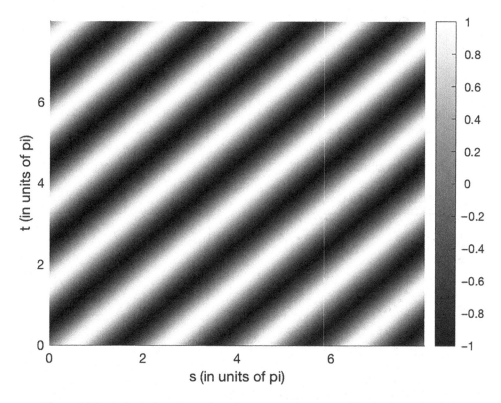

Figure 12.3 A shaded contour plot of a propagating wave. The horizontal axis is space, and the vertical axis is time.

For our discrete sampling, the sines and cosines are orthogonal, hence (12.5) is a decomposition of the form (12.3) with only $K = 2$ components. Despite knowing the analytic solution to PCA in this example, we proceed to apply PCA to $Y(t, s)$ anyway. The results are shown in Figure 12.4. Figure 12.4a shows that only two components have nonzero variance, indicating that the propagating wave can be decomposed *exactly* into two components, each explaining 50% of the variance. This result is precisely to be expected from the decomposition (12.5). The leading EOFs are shown in Figure 12.4b and display the expected cosine and sine wave. Similarly, the leading PC time series are shown in Figure 12.4d and have the form expected from (12.5). Importantly, PCA deduced this decomposition *from the data* $Y(t, s)$, without "knowing" trigonometric identities or propagating waves.

In this example, 100% of the variance can be reconstructed from the first two principal components, hence these two components must encode the propagating wave. In general, *any* propagating structure can be decomposed into a weighted sum of fixed structures. The distinctive characteristic of propagation is that the time series for the individual structures are correlated *after a suitable time shift*. This

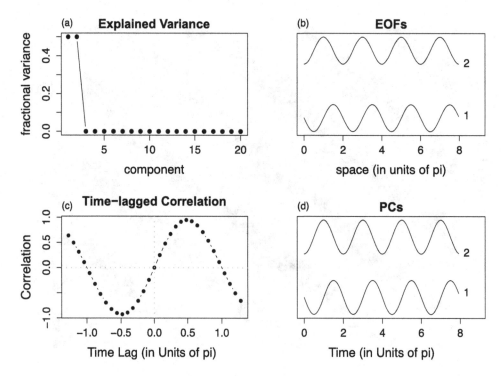

Figure 12.4 Results of applying Principal Component Analysis to a propagating wave. The fraction of variance explained by the PCs is shown in (a). The EOFs and PC time series are shown in (b) and (d), respectively. For clarity, the second EOF and PC time series are offset vertically from the first by a constant. The integers on the far right in (b) and (d) indicate the index of the principal component. The time-lagged correlation between the first two PC time series is shown in (c). X-axes are in units of π (e.g., "2" means "2π").

fact suggests that propagating structures can be identified from PCA by identifying *significant time-lagged correlations between PC time series*. To illustrate, the sample time-lagged correlation between the first two PC time series is shown in Figure 12.4c. In this example, a full period is 2π time units. The time-lagged correlation peaks at a quarter period, a configuration known as *in quadrature*. (More precisely, two oscillations with a phase separation of a quarter period are said to be in quadrature.) Note that the zero-lag correlation vanishes, reflecting the fact that the PC time series are uncorrelated. The use of two principal components to describe propagating structures is common in climate science. For instance, a standard index of the Madden–Julian Oscillation is based on two principal components (Wheeler and Hendon, 2004). More complicated propagating structures may involve more than two time series and more complicated time-lagged relations. In such cases, PCA may not be an efficient method for identifying propagating

structures. Alternative PCA-like methods that are more efficient at identifying propagating features are discussed in Jolliffe (2002).

Examination of Figure 12.4 reveals that the first EOF and PC time series are
proportional to $-\sin(s)$ and $-\sin(t)$, and the second are $-\cos(s)$ and $-\cos(t)$. This
negative sign has no impact on the component since the component depends only
on the *product*. Also, because the explained variances are identical, it is ambiguous
which term in (12.5) should be identified as the first and second. In fact, an arbitrary
phase ϕ may be added to both s and t to obtain

$$\cos(s - t) = \sin(t + \phi)\sin(s + \phi) + \cos(t + \phi)\cos(s + \phi). \qquad (12.6)$$

Since this identity holds for arbitrary ϕ, it follows that the decomposition is *not
unique*. This nonuniqueness occurs whenever the singular values are *equal*, a situation known as *degeneracy*. However, the *space* spanned by the singular vectors of
degenerate singular values is unique. In general, EOFs associated with degenerate
singular values can be transformed into another set through a *rotation and reflection*,
provided the PC time series are similarly rotated or reflected. Indeed, ϕ parameter-
izes such a rotation in (12.6). This flexibility to rotate EOFs is sometimes exploited
to enhance interpretability. Procedures for performing rotations to facilitate physical
interpretation are the subject of *rotated EOFs*. In practice, rotated EOFs are explored
even when the singular values are not degenerate. Such rotations preserve the total
variance of the *subset* of principal components that are rotated, but will alter the
explained variances of individual components.

Because singular values are derived from the sample, they are random. For sufficiently large random variability, the singular values may switch their order and
the singular vectors may "mix," in which case the individual principal components
will be sensitive to the sample. This situation is known as *effective degeneracy*.
North et al.'s rule of thumb for selecting K is intended to avoid splitting effectively
degenerate components.

As a last example, consider two points in space s_1 and s_2, with corresponding
random variables $y_1 = Y(s_1, t)$ and $y_2 = Y(s_2, t)$. Suppose y_1 and y_2 are related as

$$y_2 = \rho y_1 + \epsilon\sqrt{1 - \rho^2}, \qquad (12.7)$$

where y_1 and ϵ are independent random variables from a standardized normal distribution. As shown in Example 1.7, the correlation between y_1 and y_2 is ρ. A random
sample of 200 points drawn from this population model for $\rho = 0.9$ is shown in
Figure 12.5. The first EOF has the largest variance when the points are projected
onto it. The second EOF is orthogonal to the first, by construction. Because the
second EOF is also the last in this example, it has the property that it yields the
smallest variance when the points are projected onto it.

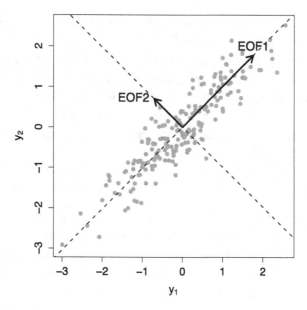

Figure 12.5 Scatter plot of 200 random draws from the population model (12.7) (dots). The first and second EOFs are shown as vectors.

These features suggest that PCA may be used to "fit lines through points." This is indeed true and the associated procedure is known as *orthogonal regression* or *Total Least Squares (TLS)* regression. TLS is not as common as *ordinary least squares regression* (OLS; see Chapter 8). The difference between methods is illustrated in Figure 12.6: TLS minimizes the sum squared *perpendicular* distances between data points and regression line, while OLS minimizes the sum squared *vertical* distances. OLS is appropriate when

$$y_2 = ay_1 + b + \epsilon, \tag{12.8}$$

where a and b are constants, ϵ is *iid*, and y_1 is *observed with little or no error*. In some studies, the error in observing y_1 is not negligible and motivates the alternative model

$$y_1 = y_1^* + \epsilon_1 \tag{12.9}$$
$$y_2 = ay_1^* + \epsilon_2, \tag{12.10}$$

where y_1 and y_2 are observed, but y_1^* is not. ϵ_1 is an observation error with known variance. This model is termed an *Errors-in-Variables* model. TLS reduces to OLS when var$[\epsilon_1] = 0$. Hypothesis testing in TLS is inexact and more difficult than in OLS.

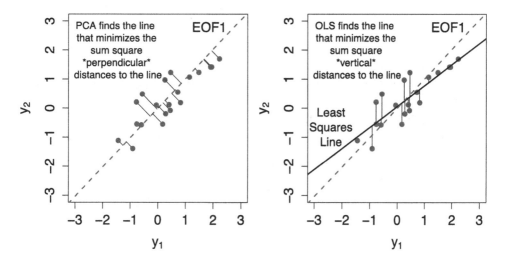

Figure 12.6 Scatter plot of 20 random draws from the model (12.7) (dots), and the corresponding line fit derived from orthogonal regression (left, dashed) and least squares regression (right, solid). The orthogonal regression line is indicated as the dashed grey line in both panels. The dashed lines and data points are identical in the two panels. Orthogonal regression is equivalent to PCA in the sense that the associated regression line is parallel to the first EOF.

12.3 Solution by Singular Value Decomposition

We now give a brief description of PCA. A more detailed description that includes modifications that are common in climate studies is given in Section 12.5. Consider a data set $Y(t, s)$. In practice, this data set is discretized. Let the spatial locations be labeled $s = 1, \ldots, M$ and the time steps be labeled $t = 1, \ldots, N$. Then, the data set is represented by the $N \times M$ matrix $\mathbf{Y}$. PCA is based on the following theorem.

Theorem 12.1 (The Singular Value Decomposition) *Any $N \times M$ matrix $\mathbf{Y}$ can be decomposed in the form*

$$
\begin{array}{ccccc}
\mathbf{Y} & = & \mathbf{U} & \mathbf{S} & \mathbf{V}^T \\
{[N \times M]} & & {[N \times N]} & {[N \times M]} & {[M \times M]}
\end{array},
\tag{12.11}
$$

where $\mathbf{U}$ and $\mathbf{V}$ are orthogonal matrices, $\mathbf{S}$ is a diagonal matrix with non-negative diagonal values. Decomposition (12.11) is called the Singular Value Decomposition.

The SVD can be computed using standard numerical packages (e.g, Lapack, R, Matlab). Further details about the SVD can be found in Strang (2016), Press et al. (2007), and Golub and Van Loan (1996). The columns of $\mathbf{U}$ are denoted

$$
\mathbf{U} = \begin{bmatrix} \mathbf{u}_1 & \mathbf{u}_2 & \cdots & \mathbf{u}_N \end{bmatrix},
\tag{12.12}
$$

and known as the *left singular vectors*. The columns of $\mathbf{V}$ are denoted

$$
\mathbf{V} = \begin{bmatrix} \mathbf{v}_1 & \mathbf{v}_2 & \cdots & \mathbf{v}_M \end{bmatrix},
\tag{12.13}
$$

and known as the *right singular vectors*. Because $\mathbf{U}$ and $\mathbf{V}$ are orthogonal matrices,

$$\mathbf{u}_i^T \mathbf{u}_j = \delta_{ij} \quad \text{and} \quad \mathbf{v}_i^T \mathbf{v}_j = \delta_{ij}. \tag{12.14}$$

The matrix $\mathbf{S}$ is diagonal and the diagonal elements are called *singular values*. The number of nonzero singular values equals the rank of $\mathbf{Y}$, denoted R. It is convention to order singular values from largest to smallest and denote them as $s_1, \ldots, s_R$.

The SVD (12.11) can be represented equivalently as a sum of outer products

$$\mathbf{Y} = s_1 \mathbf{u}_1 \mathbf{v}_1^T + s_2 \mathbf{u}_2 \mathbf{v}_2^T + \cdots + s_R \mathbf{u}_R \mathbf{v}_R^T. \tag{12.15}$$

The product $s_i \mathbf{u}_i \mathbf{v}_i^T$ is the i'th *Principal Component*. By identifying $\mathbf{f}_i = \alpha_i s_i \mathbf{u}_i$ and $\mathbf{e}_i = \mathbf{v}_i / \alpha_i$, where α_i is *any* nonzero scalar, (12.15) may be written equivalently as

$$\mathbf{Y} = \mathbf{f}_1 \mathbf{e}_1^T + \mathbf{f}_2 \mathbf{e}_2^T + \cdots + \mathbf{f}_R \mathbf{e}_R^T, \tag{12.16}$$

which is the discrete version of (12.3). The vector $\mathbf{f}_i$ describes the time variability of the component and is the i'th *PC time series* (to within a multiplicative factor). There is no standard convention for α_i. Our recommended convention is to select α_i so that $\mathbf{f}_i$ has unit variance (see Section 12.5). The vector $\mathbf{e}_i$ describes the spatial variation of the component and is the i'th *Empirical Orthogonal Function (EOF)* (to within a multiplicative factor).

The facts that EOFs are orthogonal and that PCs are uncorrelated follow directly from the orthogonality of right and left singular vectors.

The sample covariance matrix of the data, assuming the data has been centered, is

$$\hat{\boldsymbol{\Sigma}}_Y = \frac{1}{N-1} \mathbf{Y}^T \mathbf{Y} = \frac{1}{N-1} \mathbf{V} \mathbf{S}^T \mathbf{S} \mathbf{V}^T. \tag{12.17}$$

The expression on the far right is the *eigenvector decomposition* of $\hat{\boldsymbol{\Sigma}}_Y$, since $\mathbf{V}$ is an orthogonal matrix and $\mathbf{S}^T \mathbf{S}$ is diagonal. Thus, (12.17) shows that *EOFs are the eigenvectors of the sample covariance matrix*. Many statistics textbooks define PCA in this way.

The sense in which the SVD solves our problem is the following. We want to choose $\mathbf{f}_1$ and $\mathbf{e}_1$ such that their product $\mathbf{f}_1 \mathbf{e}_1^T$ comes as close to $\mathbf{Y}$ as possible. PCA measures closeness by the sum square difference between $\mathbf{Y}$ and $\mathbf{f}_1 \mathbf{e}_1^T$, which is the Frobenius norm

$$\|\mathbf{Y} - \mathbf{f}_1 \mathbf{e}_1^T\|_F^2 = \sum_n \sum_m (Y_{nm} - (\mathbf{f}_1 \mathbf{e}_1^T)_{nm})^2. \tag{12.18}$$

The matrix $\mathbf{f}_1 \mathbf{e}_1^T$ is rank-1 because each column is proportional to $\mathbf{f}_1$. Thus, an alternative statement of our problem is to find the rank-1 matrix that minimizes the Frobenius norm of the difference. The solution to this and related problems is given by the following theorem.

Theorem 12.2 (Eckart–Young–Mirsky Theorem) *Let* $\mathbf{Y}$ *be an* $N \times M$ *matrix. Among all* $N \times M$ *matrices of rank* K, *the matrix* $\mathbf{Y}_K$ *that minimizes the Frobenius norm*

$$\|\mathbf{Y} - \mathbf{Y}_K\|_F^2, \tag{12.19}$$

is

$$\mathbf{Y}_K = s_1 \mathbf{u}_1 \mathbf{v}_1^T + \cdots + s_K \mathbf{u}_K \mathbf{v}_K^T, \tag{12.20}$$

where s_i, $\mathbf{u}_i$, *and* $\mathbf{v}_i$ *are the singular values and left and right singular vectors of* $\mathbf{Y}$.

Theorem 12.2 tells us that the rank-1 matrix that best approximates $\mathbf{Y}$ is

$$\mathbf{Y}_1 = s_1 \mathbf{u}_1 \mathbf{v}_1^T = \mathbf{f}_1 \mathbf{e}_1^T. \tag{12.21}$$

Similarly, the rank-K matrix that best approximates $\mathbf{Y}$ is $\mathbf{Y}_K$, defined in (12.20). The Frobenius norm of the i'th principal component is

$$\|\mathbf{f}_i \mathbf{e}_i^T\|_F^2 = s_i^2, \tag{12.22}$$

where we have used (12.14) and the fact that for any matrix $\mathbf{A}$,

$$\|\mathbf{A}\|_F^2 = \mathrm{tr}\left[\mathbf{A}^T \mathbf{A}\right]. \tag{12.23}$$

Because singular vectors are orthogonal, the Frobenius norm of $\mathbf{Y}$ equals the sum of Frobenius norms of individual components:

$$\|\mathbf{Y}\|_F^2 = \|\mathbf{f}_1 \mathbf{e}_1^T\|^2 + \cdots + \|\mathbf{f}_R \mathbf{e}_R^T\|^2 = s_1^2 + s_2^2 + \cdots + s_R^2. \tag{12.24}$$

The term $\|\mathbf{Y}\|_F^2/(N-1)$ is the variance summed over all spatial locations and hence called the *sum total variance*. The *variance explained* by the first K principal components is $\|\mathbf{Y}_K\|_F^2/(N-1)$. The fraction of variance explained by the first K PCs is

$$\begin{pmatrix} \text{fraction of variance explained by} \\ K \text{ principal components} \end{pmatrix} = \frac{s_1^2 + \cdots + s_K^2}{s_1^2 + s_2^2 + \cdots + s_R^2}. \tag{12.25}$$

12.4 Relation between PCA and the Population

What does PCA describe about the population? If the samples come from a multivariate Gaussian distribution, then the only relevant population parameter is the covariance matrix, since the sample mean is removed prior to performing PCA. The sample covariance matrix is defined in (12.17). Since $\mathbf{V}$ is orthogonal, multiplying both sides on the right by $\mathbf{V}$ yields

$$\hat{\boldsymbol{\Sigma}}_Y \mathbf{V} = \frac{1}{N-1} \mathbf{V} \mathbf{S}^T \mathbf{S}. \tag{12.26}$$

Focusing on the i'th column of this equation yields

$$\hat{\boldsymbol{\Sigma}}_Y \mathbf{v}_i = \hat{\lambda}_i \mathbf{v}_i, \tag{12.27}$$

where $\hat{\lambda}_i = s_i^2/(N-1)$. This equation shows that *EOFs are eigenvectors of the sample covariance matrix*. Thus, PCA gives sample estimates of the eigenvectors and eigenvalues of the population covariance matrix. The corresponding PC time series can be obtained by projection methods, as described in Section 12.5.

The sampling distribution of the eigenvectors/eigenvalues is complicated, but some useful results are known if the sample size is large, as discussed in Theorem 12.3.

> **Theorem 12.3** (Sampling Distribution of EOFs) *Let $\mathbf{y}_1, \mathbf{y}_2, \ldots, \mathbf{y}_N$ be N independent samples from the M-dimensional normal population $\mathcal{N}_M(\boldsymbol{\mu}, \boldsymbol{\Sigma}_Y)$, and let $\hat{\boldsymbol{\Sigma}}_Y$ be a sample estimate of $\boldsymbol{\Sigma}_Y$. Assume that the eigenvalues of $\boldsymbol{\Sigma}_Y$ are distinct and positive. Then, the M eigenvalues $\hat{\lambda}_1, \ldots, \hat{\lambda}_M$ and eigenvectors $\hat{\mathbf{v}}_1, \ldots, \hat{\mathbf{v}}_M$ of the sample covariance matrix $\hat{\boldsymbol{\Sigma}}_Y$ have the following properties for asymptotically large N:*
>
> $$\hat{\lambda}_i \sim \mathcal{N}\left(\lambda_i, \frac{2\lambda_i^2}{N}\right) \tag{12.28}$$
>
> *and*
>
> $$\hat{\mathbf{v}}_i \sim \mathcal{N}_M\left(\mathbf{v}_i, \frac{1}{N}\mathbf{L}_i\right) \quad where \quad \mathbf{L}_i = \lambda_i \sum_{k \neq i} \frac{\lambda_k}{(\lambda_k - \lambda_i)^2} \mathbf{v}_k \mathbf{v}_k^T. \tag{12.29}$$

12.4.1 Confidence Intervals for the Eigenvalues

Theorem 12.3 implies that the $(1-\alpha)100\%$ confidence interval for the eigenvalues is

$$1 - \alpha = p\left(\lambda_i - z_{\alpha/2}\lambda_i\sqrt{\tfrac{2}{N}} < \hat{\lambda}_i \text{ and } \hat{\lambda}_i < \lambda_i + z_{\alpha/2}\lambda_i\sqrt{\tfrac{2}{N}}\right)$$

$$= p\left(\lambda_i < \frac{\hat{\lambda}_i}{1 - z_{\alpha/2}\sqrt{\tfrac{2}{N}}} \text{ and } \frac{\hat{\lambda}_i}{1 + z_{\alpha/2}\sqrt{\tfrac{2}{N}}} < \lambda_i\right) \tag{12.30}$$

$$= p\left(\frac{\hat{\lambda}_i}{1 + z_{\alpha/2}\sqrt{\tfrac{2}{N}}} < \lambda_i < \frac{\hat{\lambda}_i}{1 - z_{\alpha/2}\sqrt{\tfrac{2}{N}}}\right).$$

Since the theorem holds only for large N, $\epsilon = z_{\alpha/2}\sqrt{2/N}$ is small and we may use the approximation $1/(1+\epsilon) \approx 1 - \epsilon$ to approximate the confidence interval as

$$\left(\hat{\lambda}_i - z_{\alpha/2}\hat{\lambda}_i\sqrt{\frac{2}{N}}, \hat{\lambda}_i + z_{\alpha/2}\hat{\lambda}_i\sqrt{\frac{2}{N}}\right). \tag{12.31}$$

Note that the length of the confidence interval is proportional to $\hat{\lambda}_i$, implying that the interval grows with $\hat{\lambda}_i$. Accordingly, plotting the eigenvalues on a *logarithmic scale* produces confidence intervals of constant width. This format was used in Figure 12.2.

Lawley (1956) showed that the expected value of the k'th eigenvalue is

$$\mathbb{E}[\hat{\lambda}_k] = \lambda_k + \frac{\lambda_k}{N}\sum_i \frac{\lambda_i}{\lambda_k - \lambda_i} + O\left(\frac{1}{N^2}\right), \tag{12.32}$$

where $O(1/N^2)$ indicates that the remainder terms decay as $1/N^2$ or faster as $N \to \infty$. This equation implies $\mathbb{E}[\hat{\lambda}_1] \geq \lambda_1$, hence the leading eigenvalue is biased upward. This bias reflects a natural consequence of optimization: The object being optimized tends to be more extreme in the training sample than in independent data. As a result, the first EOF typically explains more variance in the sample from which it was derived than in an independent sample. This bias is important in some studies (e.g., climate detection and attribution).

12.4.2 Effective Degeneracy and North et al.'s Rule of Thumb

One issue we have avoided up to now is that two or more eigenvalues of a matrix may be equal. In such cases, the associated eigenvalues are said to be *degenerate*, and the corresponding eigenvectors are not unique. Instead, the *vector space* is unique, and *any* linearly independent set of vectors that span that space may serve as eigenvectors. In practice, sample eigenvalues are rarely exactly equal to each other, but they can be close. Note that the sum in (12.29) depends inversely on the *difference* in eigenvalues. Thus, if eigenvalues λ_i and λ_{i+1} are sufficiently close, then the sum in (12.29) will be dominated by a single term, in which case the associated eigenvectors have the approximate distributions

$$\hat{\mathbf{v}}_i \sim \mathcal{N}_M\left(\mathbf{v}_i, \frac{1}{N}\frac{\lambda_i\lambda_{i+1}}{(\lambda_i - \lambda_{i+1})^2}\mathbf{v}_{i+1}\mathbf{v}_{i+1}^T\right)$$

$$\hat{\mathbf{v}}_{i+1} \sim \mathcal{N}_M\left(\mathbf{v}_{i+1}, \frac{1}{N}\frac{\lambda_i\lambda_{i+1}}{(\lambda_i - \lambda_{i+1})^2}\mathbf{v}_i\mathbf{v}_i^T\right).$$

It is instructive to recognize that these distributions would be obtained from the models

$$\hat{\mathbf{v}}_i = \mathbf{v}_i + \epsilon_i\mathbf{v}_{i+1} \tag{12.33}$$

$$\hat{\mathbf{v}}_{i+1} = \mathbf{v}_{i+1} + \epsilon_{i+1}\mathbf{v}_i, \tag{12.34}$$

where

$$\epsilon_i \text{ and } \epsilon_{i+1} \stackrel{iid}{\sim} \mathcal{N}\left(0, \frac{1}{N} \frac{\lambda_i \lambda_{i+1}}{(\lambda_i - \lambda_{i+1})^2}\right). \tag{12.35}$$

Model (12.33) shows that $\hat{\mathbf{v}}_i$ equals $\mathbf{v}_i$ plus random errors in the direction of $\mathbf{v}_{i+1}$. Loosely speaking, eigenvector i "mixes" with eigenvector $i + 1$. Similarly, eigenvector $i + 1$ "mixes" with eigenvector i. Importantly, the variance of ϵ_i and ϵ_{i+1} grow *inversely with the difference between eigenvalues*. Thus, when the difference between eigenvalues is comparable to their sampling errors, the eigenvalues are *effectively degenerate*. North et al. (1982) proposed a "rule of thumb" in which an EOF is deemed to be sensitive to sampling errors when the standard error of the associated eigenvalue, namely $\lambda_i \sqrt{2/N}$, "is comparable to or larger than" the spacing between neighboring eigenvalues. In terms of the confidence interval (12.31), a standard error corresponds to $z_{\alpha/2} = 1$, or $\alpha = 0.32\%$. Thus, an equivalent graphical criterion is to declare an EOF to be sensitive to sampling errors when its eigenvalue has a 68% confidence interval that *overlaps* with those of other eigenvalues.[2] Conversely, eigenvalues are said to be *resolved* if their 68% confidence intervals do not overlap. Applying this rule of thumb to the results in Figure 12.2 implies that eigenvectors 1 and 2 are individually resolved, because their intervals do not overlap with other intervals. None of the other eigenvectors are resolved because their corresponding intervals overlap.

This rule of thumb is often invoked to justify the *truncation point K* in PCA. However, this rule of thumb is not always relevant. For instance, if the EOFs are used strictly as a *basis set*, then the EOFs can be held fixed in future applications, in which case their individual uncertainties may not be relevant. The rule of thumb may be useful for identifying propagating structures, since, as seen in the example in Section 12.2, a propagative wave is described by principal components with equal variances and therefore may yield effectively degenerate eigenvalues. Also, the vector space formed by subsets of near-degenerate eigenvalues may sometimes be meaningfully interpreted after certain rotations (Section 11.1 of Jolliffe, 2002). The rule of thumb is quite relevant when robustness of an individual EOF is important. For instance, if one wants to *interpret* a given EOF physically, such an interpretation is meaningful only if that particular EOF is not sensitive to the sample. Conversely, it is wise to be suspicious of attempts to physically interpret a high-order EOF (e.g., the tenth EOF), because these EOFs usually lie in the "continuous spectrum"

---

[2] The precise criterion proposed by North et al. (1982) is ambiguous because of the word "comparable." They discuss an example in their fig. 4, in which two eigenvalues differed by more than a standard error, yet the eigenvalues were considered "comparable." The criterion stated in the text gives the same conclusions as those made by North et al. (1982) when discussing their numerical examples, particularly their figure 4.

where differences between eigenvalues are small and effective degeneracy is unavoidable.

12.5 Special Considerations for Climate Data

As discussed in sec 12.4, PCA finds components that best approximate the original data set, where "best" means "minimizes the Frobenius norm." For climate applications, the Frobenius norm is awkward because it weights each spatial element equally, regardless of area. In this section, we generalize PCA to arbitrary norms and give a detailed description of its properties.

Before generalizing PCA, it is worth explaining some basic issues regarding the data. Typically, $Y(t,s)$ is a single physical variable, such as near-surface temperature, wind speed at a certain height, or some three-dimensional field in the atmosphere or ocean. Variables with different physical units are not additive and therefore should not be combined into a single PCA analysis without some kind of normalization. The field is sampled at discrete points in space and time. The values over all spatial locations can be collected into a single vector, even if the field itself is two- or three-dimensional. PCA does not require regular spatial grids or temporal sampling: the geographic locations may be distributed inhomogeneously in space, and the field may be sampled inhomogeneously in time. Let the spatial locations be labeled $s = 1, \ldots, M$ and the time steps be labeled $t = 1, \ldots, N$. Typically, we are interested in describing variability with respect to some reference, usually the mean. Accordingly, we subtract out the sample mean from each spatial location. The resulting values can be collected in the $N \times M$ matrix $\mathbf{Y}$. The centering implies that the sum of each column vanishes; that is, $\mathbf{Y}^T \mathbf{j} = \mathbf{0}$, where $\mathbf{j}$ is a vector of ones.

As mentioned previously, the SVD of the data matrix $\mathbf{Y}$ has an awkward dependence on the resolution and size of the spatiotemporal dimensions. For instance, $\mathbf{u}_i$ is normalized to unit length. As N increases, the number of terms in the sum increases. To preserve the sum, the average size of each element must shrink to zero as N increases. Similarly, requiring $\mathbf{v}_i$ to be a unit vector makes its amplitude depend on spatial dimension M, and thus on spatial resolution. Importantly, decomposition (12.20) depends only on the product $s_i \mathbf{u}_i \mathbf{v}_i^T$, so as done in sec. 12.4 we may introduce an arbitrary scalar α_i and write this product as $\mathbf{f}_i \mathbf{e}_e^T$, where $\mathbf{f}_i = \alpha_i s_i \mathbf{u}_i$ and $\mathbf{e}_i = \mathbf{v}_i / \alpha_i$, which preserves the identity $s_i \mathbf{u}_i \mathbf{v}_i^T = \mathbf{f}_i \mathbf{e}_e^T$. The scalar α_i may be chosen to define a more meaningful normalization than one that depends on resolution and size of the data. This sort of re-normalization is common in PCA (e.g., Section 3.5 of Jolliffe, 2002).

A second issue is that PCA is based on the Frobenius norm, which weights each spatial location equally. However, on a uniform lat/lon grid, points near the pole are close together and should be weighted differently from points near the equator. Accordingly, standard practice in climate science is to define a norm using an area weighted scheme:

$$\|\mathbf{Y}\|_W^2 = \sum_{n=1}^{N}\sum_{m=1}^{M} w_m Y_{nm}^2, \tag{12.36}$$

where w_m are the weights. For a uniform lat/lon grid, w_m is approximately the cosine of the latitude of spatial location m. The problem now becomes one of finding the rank-K matrix $\mathbf{Y}_K$ that minimizes the *generalized* norm $\|\mathbf{Y} - \mathbf{Y}_K\|_W^2$. Happily, a clever trick allows us to find the solution using the standard SVD. The trick is to define the transformed variable

$$\mathbf{Y}^* = \mathbf{Y}\mathbf{W}^{1/2}, \tag{12.37}$$

where $\mathbf{W}$ is a diagonal matrix with *positive* diagonal elements $w_1, \ldots, w_M$ (which ensures $\mathbf{W}$ is invertible). Then, the generalized norm of Y (12.36) *is* the Frobenius norm of $\mathbf{Y}^*$:

$$\|\mathbf{Y}^*\|_F^2 = \|\mathbf{Y}\|_W^2. \tag{12.38}$$

PCA then can be applied to the transformed variable $\mathbf{Y}^*$, and the transformation can be inverted before plotting. The resulting EOFs will be orthogonal with respect to an area-weighted inner product (this will be shown explicitly shortly).

After selecting $\mathbf{W}$, the next step is to compute the SVD of the transformed matrix $\mathbf{Y}^*$:

$$\text{SVD of } \mathbf{Y}\mathbf{W}^{1/2} = \mathbf{U}\mathbf{S}\mathbf{V}^T. \tag{12.39}$$

Note that the $\mathbf{U}, \mathbf{S}, \mathbf{V}$ in Equation (12.39) differ from those appearing in previous sections because here the SVD is applied to the *transformed* data matrix $\mathbf{Y}^*$. Since $\mathbf{W}$ is positive definite, its inverse square root exists and can be multiplied on both sides to obtain

$$\mathbf{Y} = \mathbf{U}\mathbf{S}\mathbf{V}^T\mathbf{W}^{-1/2}. \tag{12.40}$$

Let the rank of $\mathbf{Y}$ be R. The rank defines the maximum number of terms in a PCA decomposition. Only R singular values will be nonzero. To recover $\mathbf{Y}$, only those singular vectors associated with nonzero singular values need to be kept. Accordingly, let $\dot{\mathbf{U}}, \dot{\mathbf{V}}$ denote the matrices that include only the first R columns of $\mathbf{U}, \mathbf{V}$, respectively, and let $\dot{\mathbf{S}}$ be the upper $R \times R$ block of $\mathbf{S}$ containing positive diagonal elements.

Let the k'th column vector of $\mathbf{U}$ and $\mathbf{V}$ be $\mathbf{u}_k$ and $\mathbf{v}_k$, respectively. The vector $\mathbf{u}_k$ corresponds to a time series and the vector $\mathbf{v}_k$ corresponds to a spatial pattern. In this book, we advocate constraining the time series to have *unit variance*. For instance, this normalization allows PC time series computed from different sample sizes to be plotted on the same axes. Accordingly, we define the *PC time series*[3] as

$$\mathbf{F} = \dot{\mathbf{U}}\sqrt{N-1}. \tag{12.41}$$

The SVD (12.40) can be written equivalently in terms of $\mathbf{F}$ as

$$\begin{array}{ccc} \mathbf{Y} & = & \mathbf{F} & \mathbf{E}^T \\ N \times S & & N \times R & R \times S, \end{array} \tag{12.42}$$

where $\mathbf{F}$ is defined in (12.41) and the *EOFs* are defined as

$$\mathbf{E} = \frac{1}{\sqrt{N-1}}\mathbf{W}^{-1/2}\dot{\mathbf{V}}\dot{\mathbf{S}}^T. \tag{12.43}$$

A summary of the properties of this PCA is given in Table 12.1. For completeness, the derivation of each expression is given next.

The data are the sum of the principal components
The matrix decomposition (12.42) can be written equivalently as

$$\mathbf{Y} = \mathbf{f}_1\mathbf{e}_1^T + \mathbf{f}_2\mathbf{e}_2^T + \cdots + \mathbf{f}_R\mathbf{e}_R^T, \tag{12.44}$$

where $\mathbf{f}_i$ and $\mathbf{e}_i$ denote the i'th column of $\mathbf{F}$ and $\mathbf{E}$, respectively; that is,

$$\mathbf{F} = \begin{bmatrix} \mathbf{f}_1 & \cdots & \mathbf{f}_R \end{bmatrix} \quad \text{and} \quad \mathbf{E} = \begin{bmatrix} \mathbf{e}_1 & \cdots & \mathbf{e}_R \end{bmatrix}. \tag{12.45}$$

$\mathbf{f}_i$ is the i'th *standardized PC time series* and the vector $\mathbf{e}_i$ is the i'th *scaled EOF*.

Standardized PC time series are uncorrelated
The standardized PC time series satisfy

$$\frac{1}{N-1}\mathbf{F}^T\mathbf{F} = \mathbf{I}. \quad \text{or equivalently} \quad \frac{1}{N-1}\mathbf{f}_i^T\mathbf{f}_j = \begin{cases} 0 & \text{if } i \neq j \\ 1 & \text{if } i = j \end{cases} \tag{12.46}$$

This identity states two important properties of the standardized PC time series $\mathbf{f}_k$: they are uncorrelated with each other and they each have unit variance

---

[3] The reader should be aware that these definitions of PC time series and EOFs are re-scaled versions of those described in multivariate statistics books (e.g., Mardia et al., 1979).

Table 12.1. *Summary of equations for Principal Component Analysis for climate data.* $\mathbf{Y}$ *is the centered data matrix, stored in an* $N \times M$ *matrix.* M = *spatial dimension,* N = *time dimension,* R = *rank of* $\mathbf{Y}$. $\mathbf{W}$ *is a positive definite matrix for defining the norm.*

$\mathbf{YW}^{1/2} = \dot{\mathbf{U}}\dot{\mathbf{S}}\dot{\mathbf{V}}^T$	SVD of transformed matrix
$s_i^2/(N-1)$	Variance Explained by i'th EOF/PC
$\dfrac{s_i^2}{s_1^2+s_2^2+\cdots+s_R^2}$	Fraction of Variance Explained by i'th PC
$\mathbf{F} = \sqrt{N-1}\dot{\mathbf{U}}$	standardized PC time series
$\mathbf{E} = \dfrac{1}{\sqrt{N-1}}\mathbf{W}^{-1/2}\dot{\mathbf{V}}\dot{\mathbf{S}}$	Scaled EOFs
$\mathbf{Y} = \mathbf{F}\mathbf{E}^T$	PCA decomposition
$\dfrac{1}{N-1}\mathbf{F}^T\mathbf{F} = \mathbf{I}$	standardized PC time series are uncorrelated and have unit variance
$\mathbf{E}^T\mathbf{W}\mathbf{E} = \dfrac{1}{N-1}\dot{\mathbf{S}}^2$	EOFs are orthogonal; their sum square equals variance
$\dfrac{1}{N-1}\|\mathbf{YW}^{1/2}\|_F^2 = \dfrac{1}{N-1}\left(s_1^2+\cdots+s_R^2\right)$	Total Variance with Respect to Norm W
$\mathbf{E}^i = \sqrt{N-1}\mathbf{W}^{1/2}\dot{\mathbf{V}}\dot{\mathbf{S}}^{-1}$	Generalized inverse of the EOFs
$\dfrac{1}{N-1}\mathbf{Y}^T\mathbf{F} = \mathbf{E}$	EOFs are linear combinations of spatial patterns from $\mathbf{Y}$
$\mathbf{YE}^i = \mathbf{F}$	PC time series are linear combinations of time series from $\mathbf{Y}$

Total variance is the sum of the variances of the individual principal components

The weighted sum of variances over all variables can be expressed as

$$\sum_m w_m \begin{pmatrix} \text{sample variance of} \\ \text{the m'th column of } \mathbf{Y} \end{pmatrix} = \sum_m w_m \left(\frac{1}{N-1}\sum_n (\mathbf{Y})_{nm}^2\right)$$

$$= \frac{1}{N-1}\,\mathrm{tr}\left[\mathbf{YWY}^T\right]$$

$$= \frac{1}{N-1}\,\mathrm{tr}\left[\left(\mathbf{YW}^{1/2}\right)\left(\mathbf{YW}^{1/2}\right)^T\right]$$

$$= \frac{1}{N-1}\,\mathrm{tr}\left[\left(\mathbf{USV}^T\right)\left(\mathbf{USV}^T\right)^T\right]$$

$$= \frac{1}{N-1}\sum_k s_k^2. \tag{12.47}$$

Thus, the sum square singular values divided by $N - 1$ equals the weighted sum of the variances of all variables. Thus, we may define

variance explained by the k'th principal component $= s_k^2/(N - 1)$. (12.48)

Moreover, the ratio

$$\frac{s_k^2}{s_1^2 + s_2^2 + \cdots + s_R^2},$$ (12.49)

is the *fraction of variance explained by the k'th principal component*.

Scaled EOFs are orthogonal

The scaled EOFs satisfy the identity

$$\mathbf{E}^T \mathbf{W} \mathbf{E} = \frac{1}{N-1} \dot{\mathbf{S}} \dot{\mathbf{V}}^T \dot{\mathbf{V}} \dot{\mathbf{S}}^T = \frac{1}{N-1} \dot{\mathbf{S}}^2.$$ (12.50)

Since $\dot{\mathbf{S}}$ is diagonal, the right-hand side is diagonal and the identity can be written as

$$\mathbf{e}_i^T \mathbf{W} \mathbf{e}_j = \begin{cases} \frac{s_i^2}{N-1} & \text{if } i = j \\ 0 & \text{if } i \neq j \end{cases}.$$ (12.51)

The fact that $\mathbf{e}_i^T \mathbf{W} \mathbf{e}_j$ vanishes when $i \neq j$ implies that the EOFs are *orthogonal with respect to an area weighted inner product*. When $i = j$ the identity becomes

$$\sum_m w_m E_{mi}^2 = \frac{s_i^2}{N-1}.$$ (12.52)

Comparison with (12.48) shows that the weighted sum square of a scaled EOF equals the *explained variance* of that component.

PCs are linear combinations of the data

Suppose you have two data sets, $\mathbf{Y}$ and $\mathbf{Y}'$, and an EOF is computed only from $\mathbf{Y}$. A natural question is whether the time variability of the EOF is consistent between data sets. To answer this question, the time series of the EOF needs to be computed for $\mathbf{Y}$ and $\mathbf{Y}'$. For $\mathbf{Y}$, the time series is merely the PC time series described previously. For $\mathbf{Y}'$, the time series can be obtained by *projection*. If the projection coefficients are denoted $\mathbf{E}^i$, then the time series would be computed from $\mathbf{Y}'\mathbf{E}^i$. To find the coefficients, note that they should give the PC time series when applied to $\mathbf{Y}$, hence

$$\mathbf{F} = \mathbf{Y}\mathbf{E}^i.$$ (12.53)

Substituting the SVD (12.40), one can readily verify that

$$\mathbf{E}^i = \sqrt{N-1} \mathbf{W}^{1/2} \dot{\mathbf{V}} \dot{\mathbf{S}}^{-1}.$$ (12.54)

Thus, (12.53) shows that PC time series can be derived as a *linear combination of the* columns of $\mathbf{Y}$. The projection operator satisfies the identity

$$E^T E^i = I, \tag{12.55}$$

which motivates calling E^i the *pseudoinverse* of E. This identity states that the columns of E and E^i form a *bi-orthogonal set*. The time series for the EOFs in Y' is given by

$$F' = Y'E^i. \tag{12.56}$$

This calculation can be described as *projecting Y' onto the EOFs*. Note that E^i involves the reciprocal of singular values, implying that components with the least variance contribute the most. As a result, the generalized inverse tends to be "noisy" and difficult to interpret.

PC time series are centered

The fact that PC time series are linear combinations of the data implies that if each column of Y has zero mean, then any linear combination of columns must also have zero mean, hence PC time series derived from Y have zero mean.

EOFs are linear combinations of the data

EOFs can be obtained by projecting data onto specified *time series*. To find the projection operator, multiply both sides of (12.42) on the left by F^T and invoke the orthogonality relation (12.46):

$$F^T Y = F^T F E^T \quad \Rightarrow \quad E = Y^T F/(N-1). \tag{12.57}$$

This expression shows that each EOF e_k is a linear combination of the rows of Y. If a second data set Y' exists, then the patterns associated with the PC time series are

$$E' = Y'^T F/(N-1). \tag{12.58}$$

This calculation may be described as *projecting Y' onto the PC time series*.

EOFs are regression coefficients

The EOFs can be interpreted as *regression coefficients* e_{mi} in the model

$$Y_{nm} = e_{mi} (f_i)_n + \epsilon_n \quad \text{where } i \text{ and } m \text{ are fixed and } n = 1, \ldots, N. \tag{12.59}$$

The least squares estimate of e_{mi} can be obtained as discussed in Section 8.6. This solution can be evaluated for each m and written concisely as the M-dimensional vector

$$\hat{e}_i = Y^T f_i \left(f_i^T f_i \right)^{-1}. \tag{12.60}$$

Substituting the decomposition (12.42) and invoking orthogonality of PC time series gives

$$\hat{e}_i = Y^T f_i \left(f_i^T f_i \right)^{-1} = E F^T f_i \left(f_i^T f_i \right)^{-1} = e_i \left(f_i^T f_i \right) \left(f_i^T f_i \right)^{-1} = e_i. \tag{12.61}$$

This equation shows that each element of the i'th EOF is the regression coefficient between the i'th PC time series and the corresponding spatial location of $\mathbf{Y}$. The term "explained variance" comes from regression analysis (see Chapter 8) and equals the product of the squared regression coefficient and the variance of the predictor time series. Since the predictor time series $\mathbf{f}_i$ in the model is normalized to unit variance, the explained variance is merely the square of the EOF coefficient. This interpretation rationalizes our use of the term "explained variance" to refer to the sum total variance of the principal component.

The units of $\mathbf{e}_i$ and $\mathbf{f}_i$

Because the PC time series are normalized to unit variance, $\mathbf{f}_i$ is interpreted as a *nondimensional index*. It then follows that each element of the EOF vector $\mathbf{e}_i$ has units of "$\mathbf{y}$ per unit PC index." Because the PC time series have been standardized to unit variance, it follows that the absolute value of each element of an EOF vector gives the *standard deviation* of the contribution of that EOF to the variability of $\mathbf{y}$.

Alternative Choices for W

It is worth mentioning that $\mathbf{W}$ need not be chosen to give an area-weighted norm. For instance, some studies apply PCA to *standardized* variables (that is, variables normalized by their standard deviation). Standardization is common if variables with different physical units, such as winds, temperature, and pressure, are combined together. This version of PCA corresponds to choosing a diagonal $\mathbf{W}$ with diagonal elements equal to the inverse variance. Alternatively, some variables have disparate amplitudes in different geographic domains. For instance, the variance of surface temperature differs considerably between tropics and midlatitudes and between ocean and land. As a result, PCA of surface temperature may highlight certain geographic areas simply because of their relatively large variances. In such cases, it would not be unreasonable to choose a weighting matrix $\mathbf{W}$ to account for domain dependence. Sometimes PCA is used to analyze a streamfunction. However, only gradients of streamfunction are physically important. For gridded data, spatial gradients require defining a nondiagonal $\mathbf{W}$. Fortunately, the expressions here still apply provided $\mathbf{W}^{1/2}$ is interpreted as a *square root matrix* (see Chapter 21). One might even choose $\mathbf{W}$ to maximize the detection of a signal. This idea is the basis of generalized discriminant analysis, as discussed in Chapter 16. Whatever the norm, $\mathbf{W}$ must be positive definite. For diagonal $\mathbf{W}$, this means that all weights must be positive. If some weights vanish, then the associated elements contribute nothing to the norm and should be dropped.

12.6 Further Topics

PCA is discussed in all books on multivariate statistics. A comprehensive and accessible text on PCA is Jolliffe (2002). PCA is attractive when variability is characterized by *fixed* spatial structures that vary in time. However, if *propagating* structures are dominant, then PCA has limitations, as discussed in Section 12.2. An alternative decomposition that may clarify propagating structures is *Extended EOF (EEOF) Analysis* (Weare and Nasstrom, 1982), which is closely related to *multi-channel Singular Spectrum Analysis (MCSSA)*. In this approach, the random vector $\mathbf{y}(t)$ is concatenated with L lagged copies of itself,

$$\mathcal{Y}(t) = \begin{pmatrix} \mathbf{y}(t) \\ \mathbf{y}(t-1) \\ \vdots \\ \mathbf{y}(t-L+1) \end{pmatrix}, \tag{12.62}$$

then PCA is applied to the augmented vector $\mathcal{Y}(t)$. In this decomposition, a principal component is characterized by L spatial structures and a time series. Propagating structures often correspond to closely spaced eigenvalues. The parameter L is known as the window length, delay parameter, or embedding dimension. The univariate version of this analysis, known as *Singular Spectrum Analysis (SSA)*, was proposed by Broomhead and King (1986) in the context of extracting qualitative information about nonlinear dynamical systems from time series generated from those systems. This approach is reviewed by Ghil et al. (2002).

In climate applications, the leading principal components often exhibit spatial structure that beg to be interpreted. However, orthogonality of principal components severely constrains the structure of high-order components. It is sometimes argued that physical interpretation can be improved by transforming a subset of principal components. The most popular criterion for transforming the EOFs is the *varimax* criterion, which rotates the leading K EOFs into the new matrix $\tilde{\mathbf{E}}$ using the transformation

$$\underset{M \times K}{\tilde{\mathbf{E}}} = \underset{M \times K}{\mathbf{E}} \quad \underset{K \times K}{\mathbf{T}} \tag{12.63}$$

where $\mathbf{T}$ is an orthogonal matrix. The orthogonal matrix $\mathbf{T}$ is chosen to maximize a measure of "simplicity" termed *varimax*, which is

$$\sum_{k=1}^{K} \left[\sum_{m=1}^{M} \tilde{E}_{mk}^4 - \frac{1}{M} \left(\sum_{m=1}^{M} \tilde{E}_{mk}^2 \right)^2 \right]. \tag{12.64}$$

Loosely speaking, this quantity measures the spatial variance of the squared EOF coefficients. The resulting components are known as *rotated EOFs*. However, the interpretation of rotated EOFs can be subtle (see Dommenget and Latif, 2002; Jolliffe, 2003).

An individual EOF is sometimes claimed to be a "mode of oscillation" of the physical system. For instance, the leading EOF of SST, shown in Figure 12.1, is sometimes described as a mode of oscillation associated with El Niño and the Southern Oscillation. One problem with this interpretation is that it is not clear what is meant by a "mode." The climate system is nonlinear and no universally accepted definition of "mode" in a nonlinear system exists. Typically, one considers dynamical equations linearized about some background state, in which case the modes would be identified with the eigenmodes of the linearized system. Unfortunately, eigenmodes of linear models are either neutral, grow to infinity, or decay to zero. More realistic linear theories of climate variability tend to be based on stochastic models – that is, damped linear systems driven by random noise. However, eigenmodes of linear systems are not necessarily orthogonal, in contrast to EOFs. For instance, the eigenmodes of linearized fluid dynamical systems with background shear are not orthogonal with respect to an energy norm (Farrell and Ioannou, 1996). It turns out that there is only one class of stochastic models whose EOFs correspond to eigenmodes: a linear system with orthogonal eigenmodes driven by noise that is white in space and time. Unfortunately, such systems tend to be too simplistic to describe realistic climate variability.

12.7 Conceptual Questions

1 What is the purpose of principal component analysis (PCA)?
2 Ignoring complications like area weighting, explain in one sentence how to perform PCA.
3 How do you compute the variance explained by a principal component?
4 In what sense are principal components orthogonal?
5 In what sense is principal component analysis "the best"?
6 How many principal components should you choose?
7 When is principal component analysis most useful? When is it least useful?
8 What population quantity is being diagnosed by PCA?
9 When is it risky to give a physical explanation of an *individual* principal component?
10 How can you identify propagating waves from PCA?
11 PCA is related to a form of regression. How does this form differ from ordinary least squares?
12 Describe two ways to perform PCA given a data matrix $\mathbf{Y}$.
13 Explain why the variance explained by the leading principal component is biased.
14 When is the spacing between eigenvalues important?

13

Field Significance

... in testing 100 independent true null hypotheses one can almost be sure to get at least one false significant result.[1]

Bender and Lange (2001)

Consequences of the widespread and continued failure to address the issue of multiple hypothesis testing are overstatement and overinterpretation of the scientific results, to the detriment of the discipline.[2]

Wilks (2016)

Field significance is concerned with testing a large number of hypothesis *simultaneously*. Previous chapters have discussed methods for testing one hypothesis, such as whether *one* variable is correlated with *one* other variable. Field significance is concerned with whether one variable is related to a random *vector*. In statistics, this is an example of a *multiple testing* problem, since more than one statistical hypothesis is being tested. In climate applications, the variables in the random vector typically correspond to quantities at different *geographic locations*. As such, neighboring variables are correlated and therefore exhibit *spatial dependence*. This spatial dependence needs to be taken into account when testing hypotheses. This chapter introduces the concept of field significance and explains three hypothesis test procedures: a Monte Carlo method proposed by Livezey and Chen (1983) and an associated permutation test, a regression method proposed by DelSole and Yang (2011), and a procedure to control the false discovery rate, proposed in a general context by Benjamini and Hochberg (1995) and applied to field significance problems by Ventura et al. (2004) and Wilks (2006).

---

[1] Reprinted from *Journal of Clinical Epidemiology*, Bender and Lange (2001), with permission from Elsevier.
[2] Reprinted by permission from the American Meteorological Society.

13.1 The Problem

Is weather over North America related to ENSO? (ENSO stands for "El Niño-Southern Oscillation.") To investigate this question, let ENSO be identified with an index Y, and let North American weather be identified with winter-mean temperature over S grid points $X_1, \ldots, X_S$. Then, a common approach to investigating this question is to examine a *correlation map* or a *regression map*. A correlation map shows the correlation between Y and X_s at each spatial location $s = 1, \ldots, S$. A regression map shows the regression coefficients $a_1, \ldots, a_S$ for the model $X_s = a_s Y + b + \epsilon$. A correlation map is nondimensional and useful for seeing how well the linear model fits the data, while a regression map has units and is useful for seeing the change in X_s resulting from unit change in Y. The question of interest is: How do we use the S correlation coefficients or S regression coefficients to decide if ENSO and North American temperature are related?

As a concrete example, consider the correlation map shown in Figure 13.1. This map shows the correlation between the December to February (DJF) Niño 3.4 index and the DJF temperature over each location in North America. As can be seen in the figure, the Niño 3.4 index is positively correlated with temperatures in the Northern part of the United States, and negatively correlated with temperatures in

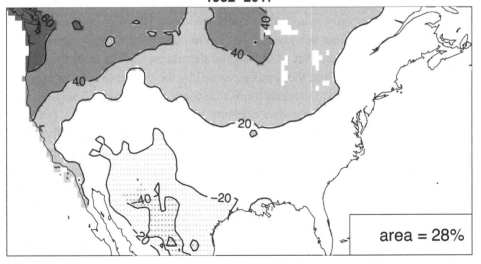

Correlation (in %) between DJF Niño 3.4 and DJF Temperature 1982−2017

Figure 13.1 Correlation (in percentages) between DJF Niño 3.4 and North American temperature during 1982–2017. Temperature data comes from the GHCN-CAMS data set ($0.5° \times 0.5°$, Fan and van den Dool, 2008), while Niño 3.4 comes from NOAA's standard climate indices. The percentage area of correlations exceeding the univariate 5% significance threshold is in the bottom right corner.

the Southern United States and Mexico. Does this figure demonstrate that North American temperatures are related to ENSO?

Not necessarily. Recall from Chapter 2 that two independent variables have vanishing population correlation, but often have nonzero *sample* correlation. Thus, the mere existence of a nonzero sample correlation does not imply a dependence. As discussed in Chapter 2, an approximate 5% significance test is to reject the hypothesis of vanishing correlation if the absolute value of the sample correlation exceeds $2/\sqrt{N}$. The correlation map in Figure 13.1 was computed from 33 winters, so the 5% significance threshold is about $2/\sqrt{33} \approx 0.35$. The correlation map reveals absolute correlations exceeding 0.35, so can we conclude from this that ENSO and winter temperatures over North America are related?

No! To understand why, one must recall the precise meaning of a significance test (see Chapter 2 for details). Suppose a correlation is statistically significant. What precisely does this mean? It means this: If the test were applied repeatedly on samples of independent and normally distributed variables, 5% of the absolute correlations would exceed the threshold. For the region shown in Figure 13.1, there are 4,445 grid points over land. If temperatures across grid points were independent, then about 5% of the grid points would be expected to exceed the significance threshold. In fact, 28% of the area exceeds the significance threshold. Does this result necessarily imply that ENSO is related to N. American temperatures?

No! Our expectation that 5% of the grid points would randomly exceed the significance threshold assumed that temperatures at different locations were *independent*. This assumption generally is not true: variability in one geographic region is correlated with variability in a neighboring region. In other words, temperature fields exhibit *spatial dependence*. Part of this spatial dependence arises from the fundamental laws of physics: Fluid motion is governed by the Navier–Stokes equations, which contain advection and pressure terms that connect one region to neighboring regions. To illustrate spatial dependence, Figure 13.2 shows the correlation between the temperature at a selected grid point, known as the *base point*, and temperature at other grid points. The left panel shows the correlation map for a base point in the Southeastern United States: 68% of the domain is correlated with that base point (more precisely, 68% of the absolute correlations exceed the significance threshold). Similarly, the right panel of Figure 13.2 shows the correlation map for a base point in the Southwestern United States: 37% of the domain is correlated with that base point. These results demonstrate two important properties of the temperature field: It is spatially correlated, and the geographical distribution of these correlations depends on location.

Because of spatial dependence, a variable that happens to have a significant *sample* correlation with temperature at a given location will tend to exhibit statistically significant correlations in neighboring regions. This property is illustrated

Correlation Maps of DJF Temperature
for Two Random Base Points

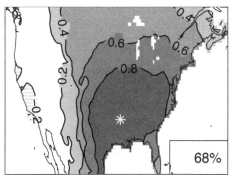

Figure 13.2 Correlation map between DJF temperature and temperature at a selected grid point indicated by a white asterisk. The percentage area of correlations exceeding the univariate 5% significance threshold is in the bottom right corner.

in Figure 13.3, which shows correlation maps with respect to *randomly generated numbers*. Since the numbers come from a computer's random number generator, the numbers cannot be related to temperature – the null hypothesis is assured. Nevertheless, in one case, more than 5% of the area has sample correlations exceeding the 5% threshold. Thus, for a spatially dependent process, significant correlations around a grid point provide no independent confirmation of a relation, but merely reflect the fact that neighboring grid points co-vary with the selected grid point.

Spatial dependence reduces the *effective* number of independent variables. To understand this concept, consider S variables on a line. If the variables are independent, then there are S independent variables. On the other hand, if the variables are perfectly correlated with each other, then they fluctuate in unison and, in effect, only *one* variable exists. In real data, the correlation is neither zero nor one, so the effective number of independent variables is between 1 and S. Estimates of the effective number are discussed in Jones et al. (1997) and Bretherton et al. (1999). One measure is based on the distance over which the average correlation between two points decays by $1/e$. This distance is known as a *length scale* and typical values are listed in Table 13.1. Among the variables listed, geopotential height has the longest length scale, followed by temperature and winds. Length scales tend to be longer in tropics than mid-latitudes, and longer at upper levels than near the surface.

Thus, while 5% of variables may randomly exceed a threshold *on average*, spatial dependence implies that the *variance* is larger than would be the case for independent variables. Hence, the actual percentage can differ greatly from 5% (see Figure 13.3).

Table 13.1. *Typical length scale (in km) for selected monthly mean variable fields. From DelSole and Shukla (2009).*

Variable	Midlatitude	Tropics
500-hPa geopotential height	1,900	3,800
200-hPa geopotential height	2,200	4,200
Mean sea level pressure	2,100	2,900
850-hPa Temperature	1,600	2,700
200-hPa zonal wind	1,200	1,500
1000-hPa zonal wind	800	1,000
200-hPa meridional wind	1,300	1,500

Correlation Maps for Random Time Series

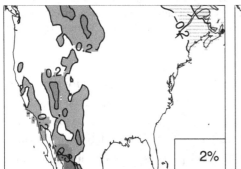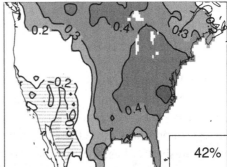

Figure 13.3 Correlation map between DJF North American temperature and a *randomly generated time series*. The two figures show results for two selected realizations. The percentage area of correlations exceeding the univariate 5% significance threshold is in the bottom right corner.

Of course, if the data are correlated in time, then this too should be taken into account. The winter-to-winter correlation of Niño 3.4 and N. American temperatures is small. Hereafter, *we assume that all variables in question are independent from year to year.*

In the remainder of this chapter, we discuss three approaches to deciding whether one variable is related to a field of other variables. Such procedures are known as *field significance tests* in the climate literature. Before reviewing these tests, let us discuss their interpretation. Specifically, what should one conclude if a local region has a significant correlation with ENSO but the map of correlations in which it is embedded is not field significant? The interpretation depends very much on how the location was chosen. If the location was chosen *after* looking at the correlation map, then the local significance test is not informative. After all, the local region was chosen *because* of its high correlation, so the null hypothesis that the samples

were drawn blindly from two independent populations is known to be wrong before
the test is performed. In contrast, the field significance test adds information about
this situation – it says that the local region of significant correlations, surrounded by
a larger region of insignificant correlations, is of the kind that can be explained by
random chance *when multiple testing associated with neighboring variables is taken
into account*. However, if the location was chosen without using any information
from the data, for example, the location was chosen because of its geographical
proximity to a mountain or a lake, then the local significance test retains its usual
(valid) interpretation.

13.2 The Livezey–Chen Field Significance Test

Livezey and Chen (1983) proposed a procedure for testing field significance based
on *Monte Carlo methods*. A Monte Carlo method is any computational method in
which random numbers are generated repeatedly to solve a statistics problem. The
basic idea is to analyze a sample from a population in which the null hypothesis is
true, and then repeat this analysis on many independently generated samples. The
procedure proposed by Livezey and Chen has been illustrated partly in Figure 13.3.
Specifically, in place of the Niño 3.4 index, random numbers were generated by
a computer and used to compute a correlation map. Because the random numbers
were generated independently of observations, the null hypothesis is necessarily
true. Importantly, the observed temperature field remains the same, hence *the spatial
dependence between grid points is preserved*. After generating a correlation map,
one can then compute any field statistic, such as the area exceeding a univariate
threshold. For instance, in the two examples shown in Figure 13.3, 2% and 42% of
the area had local correlations exceeding the significance threshold. However, this
is just two examples. To estimate a *distribution* of possible outcomes, this procedure
can be repeated many times, each time using a different random time series.

A limitation of Monte Carlo methods is that the population from which random
numbers are drawn must be specified *precisely*. For instance, one might assume the
Niño 3.4 index has a Gaussian distribution with known mean and variance (recall
serial correlation is small and therefore ignored here). But what if the Gaussian
assumption was questionable? An alternative approach that does not require specifying the population is to *randomly permute* the Niño 3.4 index. That is, if the
reference time series were merely *shuffled*, then all relevant statistical attributes
(e.g., the mean, variance, skewness) would be preserved. At the same time, shuffling
the reference time series scrambles any relation between it and the field, hence the
null hypothesis is assured. After the reference time series has been shuffled, one can
compute a field statistic, say θ. Next, the reference time series can be shuffled again

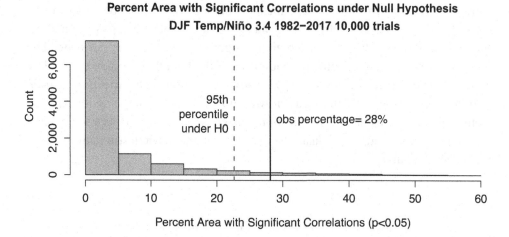

Figure 13.4 Histogram of the percent of area over North America with correlations exceeding the 5% univariate threshold, where the correlation is between DJF temperature and 10,000 random permutations of the Niño 3.4 index. The upper 5% tail of the histogram is indicated by a vertical dashed line. The observed percent of area of correlations between DJF temperature and Niño 3.4 during 1982–2017 exceeding the univariate threshold is indicated by the vertical solid line.

and a new field statistic θ computed for the new time series. Repeating this process M times yields $\theta_1, \theta_2, \ldots, \theta_M$ drawn from a population in which the null hypothesis is true. These values can then be used to estimate the significance threshold for rejecting the null hypothesis. The value of M should be chosen large enough to ensure that the results do not depend on the particular realization of shuffles. This test is known as a *permutation test*, and it is exact whether the population is Gaussian or not.

A histogram of the results from the permutation test for 10,000 repetitions is shown in Figure 13.4. The 95th percentile is indicated by the vertical dashed line and corresponds to percentage area of 22%. This result means that if the null hypothesis is true, 95% of the time fewer than 22% of the correlations exceed the univariate threshold. In contrast, the observed area with correlations exceeding the univariate threshold is 28%. Therefore, the observed statistic exceeds the 5% significance threshold under the null hypothesis. In this case, the correlation map is said to have "field significance."

How large does the domain have to be to apply these methods? The domain can be *any* size, including just one grid point, in which case the field significance test reduces to a standard univariate test. Importantly, the domain should be chosen independently of the sample. A very bad idea would be to compute a correlation map and then, *after looking at the map*, choose the domain to include only the regions with significant correlations. This procedure is a form of *data fishing* and renders the test uninformative. For this reason, it is routine to explain the basis for choosing the particular domain being analyzed.

The permutation test and Livezey–Chen test have important limitations. For one, the tests are based on *counts* and therefore do not account for the *location* of the significant correlations. In reality, spatial dependence varies with location. For instance, Figure 13.2 shows that points in the Southeast United States have longer-range spatial dependence than points in the Southwest. In contrast, these tests count the eastern and western points equally. Also, the tests do not account for the *magnitude* by which a local threshold is exceeded. For instance, an area with an average correlation of 0.4 counts the same as the same area with average correlation of 0.9. Yet intuitively, stronger correlations ought to imply greater significance. Finally, there exist other methods for finding dependencies between two data sets, that is, Canonical Correlation Analysis (CCA; see Chapter 15). Ideally, we would like a field significance test that accounts for both geographic dependence of spatial correlations and their magnitudes, and has a well-defined relation to other methods for finding dependencies. Such a test is reviewed in the next section.

13.3 Field Significance Test Based on Linear Regression

In this section we discuss a field significance test proposed by DelSole and Yang (2011). Recall that a field significance test considers the hypothesis that Y (e.g., ENSO) is independent of $X_1, X_2, \ldots, X_M$ (e.g., temperature at different locations). To define this hypothesis precisely, it is worthwhile to review the definition of independence. For any two random variables A and B, all aspects are described by the joint distribution $p(a, b)$. Furthermore, the conditional distribution $p(a|b)$ describes the distribution of A *given* that B has the value b. If A and B are independent, then the following relations hold

$$p(a|b) = p(a) \tag{13.1}$$
$$p(b|a) = p(b) \tag{13.2}$$
$$p(a, b) = p(a)p(b). \tag{13.3}$$

These relations hold even if A represents more than one variable. Thus, we may identify $A = \{X_1, X_2, \ldots, X_M\}$ and $B = Y$, and express the null hypothesis as

$$p(x_1, \ldots, x_M|y) = p(x_1, \ldots, x_M) \tag{13.4}$$
$$p(y|x_1, \ldots, x_M) = p(y) \tag{13.5}$$
$$p(x_1, \ldots, x_M, y) = p(x_1, \ldots, x_M)p(y). \tag{13.6}$$

An important concept is that specifying a regression model is tantamount to specifying a conditional distribution. For instance, suppose

$$Y = \beta_1 X + \beta_0 + \epsilon, \tag{13.7}$$

where β_0 and β_1 are regression coefficients and ϵ is a random variable distributed as $\mathcal{N}(0, \sigma^2)$. In effect, this regression model specifies the conditional distribution $p(y|x)$. To see this, recall that $p(y|x)$ gives the distribution of Y when $X = x$. Because x is *constant*, it follows that $\beta_1 x + \beta_0$ is also constant and therefore

$$p(y|x) \sim \mathcal{N}(\beta_1 x + \beta_0, \sigma^2). \tag{13.8}$$

Accordingly, we make the following associations between regression models and distributions:

$$Y = \epsilon + \beta_0 + \beta_1 X_1 + \cdots + \beta_M X_M \quad \Rightarrow \quad p(y|x_1, \ldots, x_M) \tag{13.9}$$

$$Y = \epsilon + \beta_0 \quad \Rightarrow \quad p(y). \tag{13.10}$$

If (13.5) is true, then

$$H_0 : \beta_1 = \beta_2 = \cdots = \beta_M = 0. \tag{13.11}$$

For normal distributions, the arrows in (13.9) and (13.10) can be reversed. That is, the hypothesis that Y and $\{X_1, X_2, \ldots, X_M\}$ are independent and normally distributed is equivalent to the hypothesis H_0 (13.11) in the univariate regression model (13.9).

The other expressions for independence (13.4) and (13.6) also have an interpretation in regression models. Specifically, $p(x_1, \ldots, x_M|y)$ is equivalent to specifying the models

$$X_1 = \gamma_1 Y + \mu_1 + \epsilon_1$$
$$X_2 = \gamma_2 Y + \mu_2 + \epsilon_2$$
$$\vdots$$
$$X_M = \gamma_M Y + \mu_M + \epsilon_M.$$

Hypothesis (13.4) corresponds to hypothesis $\gamma_1 = \cdots = \gamma_M = 0$. This collection of models constitutes a *multivariate regression model* and the procedure discussed in Chapter 14 can be used to test the hypothesis $\gamma_1 = \cdots = \gamma_M = 0$. Physically, these models might be derived from the idea that ENSO (Y) *causes* variability in temperature (X), e.g., through changes in the atmospheric circulation. Alternatively, hypothesis (13.6) constitutes a hypothesis on the *structure of the covariance matrix*, and the test for this hypothesis is equivalent to a procedure known as *Canonical Correlation Analysis* (see Chapter 15). All three hypothesis tests lead to precisely the same conclusion, hence it does not matter which formulation is used. The simplest approach is to consider the regression model (13.9).

The procedure for testing hypothesis (13.11) has been discussed in Chapter 9; therefore only a summary will be given here. First one fits the regression model (13.9) to data and computes the sum square error, which we denote as $SSE_{Y|X}$. Similarly, one fits the regression model (13.10) and computes the corresponding sum square error, which we denote as SSE_Y. Then the statistic for testing hypothesis H_0 is

$$F = \frac{SSE_Y - SSE_{Y|X}}{SSE_{Y|X}} \frac{N - M - 1}{M}, \tag{13.12}$$

where N is the total sample size. If the null hypothesis is true and the errors are Gaussian, then F has an F distribution with M and $N - M - 1$ degrees of freedom. To see the reasonableness of this test, suppose Y is independent of the Xs. Then including the Xs in a regression model will not improve the fit very much, hence $SSE_Y \approx SSE_{Y|X}$, and F will be "small." Conversely, if Y depends on the Xs, then including the Xs could give a much better fit, implying a "large" difference $SSE_Y - SSE_{Y|X}$, and therefore a large F. The denominator $SSE_{Y|X}$ in (13.12) provides a reference for defining "large" and "small."

A problem arises when one attempts to fit regression model (13.9) to the data used to generate Figure 13.1. Specfically, there are 4,445 grid points but only 35 winters of data. The resulting least squares problem is under-specified and has no unique solution. This situation reflects the fact that we are attempting to test 4,445 distinct hypotheses using only 35 samples, which is difficult. DelSole and Yang (2011) suggest that instead of using the original grid-point data for X, one can use the first few principal components of X (see Chapter 12 for details on principal component analysis). To explain this approach, let the temperature field be represented by the $N \times S$ data matrix $\mathbf{X}$, where N is the number of years and S is the number of grid points. The data is centered, so each column of $\mathbf{X}$ has zero mean. Then, principal component analysis decomposes $\mathbf{X}$ as

$$\begin{matrix} \mathbf{X} & = & \mathbf{F} & \mathbf{E}^T \\ N \times S & & N \times N & N \times S, \end{matrix} \tag{13.13}$$

where the columns of $\mathbf{F}$ and $\mathbf{E}$ are the PCs and EOFs, respectively. Now we simply use the first T principal components of $\mathbf{X}$ to test field significance, hence we consider the model

$$\begin{matrix} \mathbf{y} & = & \dot{\mathbf{F}} & \boldsymbol{\beta} & + & \beta_0 \mathbf{j} & + & \boldsymbol{\epsilon} \\ N \times 1 & & N \times T & T \times 1 & & N \times 1 & & N \times 1, \end{matrix} \tag{13.14}$$

where $\dot{\mathbf{F}}$ is a matrix containing only the first T PC time series and $\mathbf{j}$ is a vector of all ones. If T is small, then standard univariate regression techniques can be applied.

The question arises as to how many PC time series T should be included in the model. Here, we use a criterion known as Mutual Information Criterion (MIC), which is discussed in Chapter 14. For the regression model (13.14), MIC is

$$\text{MIC} = N \log \left(1 - R^2\right) + \left(\frac{N(N+1)(T+1)}{N - (T+1) - 2} - \frac{N(N+1)}{N - 1 - 2} - \frac{N(N+1)T}{N - T - 2}\right),$$

(13.15)

where R^2 is the fraction of variance of Y explained by $X_1, \ldots, X_M$ and is defined as

$$R^2 = \frac{SSE_Y - SSE_{Y|X}}{SSE_Y}.$$

(13.16)

Figure 13.5 shows MIC as a function of T. The minimum occurs for $T = 3$ PCs. For comparison, 5% critical values of MIC for testing H_0 are also shown in Figure 13.5. These critical values were derived by finding the critical F, then using (13.16) and (13.12) to derive

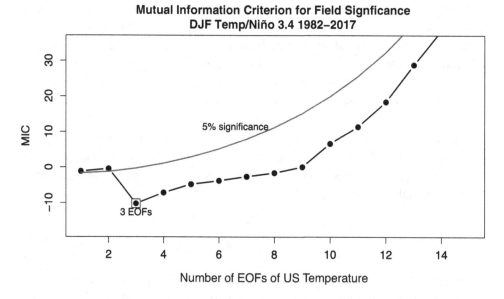

Mutual Information Criterion for Field Signficance
DJF Temp/Niño 3.4 1982–2017

Figure 13.5 Mutual Information Criterion (curve with dots) for the regression model (13.14), where **y** is DJF Niño 3.4 and $\dot{\mathbf{F}}$ contains the leading principal components of DJF North American temperature over the period 1982–2017 (curve with dots), and the 5% significance threshold for hypothesis H_0 (thin curve). The minimum MIC is indicated by a box.

$$R^2 = \frac{F}{F + \frac{N-M-1}{M}}, \tag{13.17}$$

which gives the critical R-square, then substituting the result into (13.15). MIC is significant when it is *less* than the critical value. The figure shows that 3 PCs is indeed significant.

This significance test does not account for the fact that T is chosen to minimize MIC. One approach to accounting for the selection procedure is cross-validation: leave one year out, use MIC to select T, then use the resulting regression model to predict the withheld year. Then repeat this procedure for each year in turn until all years have been used for withholding. A scatter plot of the observed versus predicted values is shown in Figure 13.6, with the symbol showing each T selected by MIC. As can be seen, MIC selects 3 PCs in each case. The correlation of the results is significant, implying field significance.

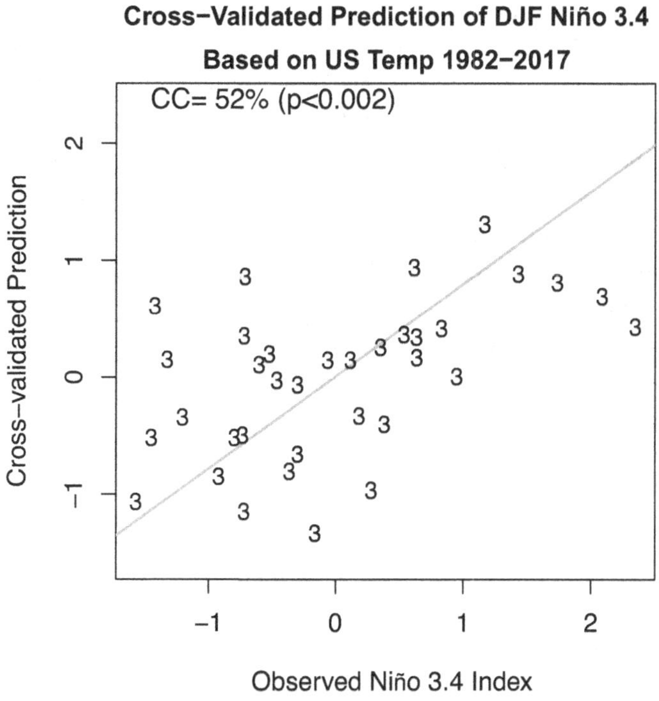

Figure 13.6 Scatter plot of the observed versus predicted DJF Niño 3.4 over the period 1982–2017, where the prediction for a given year is based on withholding that year from data, using MIC to select the number of predictors (PCs of US temperature), and then using the resulting regression model to predict the withheld year. Each number in the scatter plot shows the number of predictors selected by MIC for that prediction. The correlation coefficient (CC) and associated p-value is shown in the top left. The best-fit line is shown as the solid line.

13.4 False Discovery Rate

A shortcoming of these tests is that they give little information about *where* significant relations exist. Typically, the location of significant correlations is of crucial interest. One approach to identifying such locations is to control the family-wise error rate, that is, control the probability of *at least one* false rejection (see Chapter 11). Unfortunately, when the number of tests is in the thousands, controlling the family-wise error rate leads to such stringent criteria that only a small number of departures from the null hypothesis can be detected. In fact, controlling the family-wise error rate may not even be a reasonable goal when thousands of hypotheses are tested. After all, a small fraction of erroneous rejections out of thousands may not seriously affect an investigator's interpretation of the results.

Benjamini and Hochberg (1995) proposed a different approach that controls the expected fraction of rejected hypotheses that are incorrect, that is, the False Discovery Rate (FDR). For instance, suppose N tests are performed and N_R of them lead to rejections. Then, FDR = 10% would imply that $N_R/10$ of the rejections are incorrect. Hopefully, 10% of the rejections being wrong will not mislead us to the wrong conclusion. The most straightforward procedure for controlling the FDR assumes that the tests are independent. However, Ventura et al. (2004) showed that, in simulated data with correlation structure similar to observations, the procedure still works well. Unfortunately, there is no simple diagnostic for deciding when spatial dependence is too strong to apply FDR procedures.

The procedure for controlling the FDR is very simple, See Benjamini and Hochberg (1995) for proof. To explain the procedure, let $\hat{\rho}_s$ be the sample correlation at the s'th grid point and let H_s denote the hypothesis $\rho_s = 0$. Recall from Chapter 2 that if H_s is true, then

$$t_s = \frac{\hat{\rho}_s \sqrt{N-2}}{\sqrt{1 - \hat{\rho}_s^2}}, \tag{13.18}$$

has a t-distribution with $N - 2$ degrees of freedom. From this, one can compute the p-value of the test H_s, which we denote as p_s. Then, the Benjamini–Hochberg procedure for controlling the FDR at the rate γ is the following:

1 Order the p-values in increasing order. Denote them by $p_{(1)}, p_{(2)}, \ldots, p_{(S)}$.
2 Let the corresponding ordered hypotheses be denoted as $H_{(1)}, H_{(2)}, \ldots, H_{(S)}$.
3 Compare the ordered p-values to the sequence $\gamma/S, 2\gamma/S, \ldots, \gamma$.
4 Identify the largest K such that $p_{(K)} \leq K\gamma/S$.
5 Reject $H_{(1)}, H_{(2)}, \ldots, H_{(K)}$.

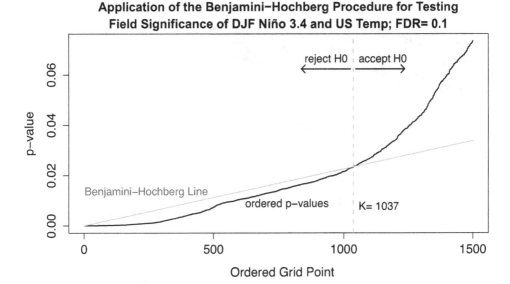

Figure 13.7 Result of the Benjamini–Hochberg procedure applied to correlation map in Figure 13.1. First, S different hypothesis tests are performed, resulting in S p-values. These p-values are ordered from smallest to largest and plotted as the black curve (only the first 1,500 are plotted). Next, the ordered p-values are compared to the "the Benjamini–Hochberg line," $s\gamma/S$, where γ is the False Discovery Rate (here 10%) and $s = 1, \ldots, S$. This line is indicated by the grey line. The two curves intersect at K. The procedure is to reject hypotheses $1, \ldots, K$ associated with the smallest K p-values.

This procedure is illustrated and discussed further in Figure 13.7. A popular choice for the FDR is $\gamma = 10\%$. For our numerical example, regions for which the Benjamini–Hochberg procedure rejects $\rho = 0$ are shown in Figure 13.8. This test suggests winter temperatures over Washington/British Columbia and Mexico are influenced by ENSO.

13.5 Why Different Tests for Field Significance?

This chapter presented three tests for field significance: the Livezey–Chen test and associated permutation test, a linear regression test, and a test based on False Discovery Rate. Why three tests? This is typical of multivariate tests – the null hypothesis can be violated in a variety of ways, and no single test has the most power across all possible violations. Each test has its strengths and limitations. The Livezey–Chen test/permutation test is extremely simple but ignores the magnitude and location of significant correlations. The regression test has a well-defined connection to other hypothesis tests in fundamentally equivalent problems, but it replaces the field by

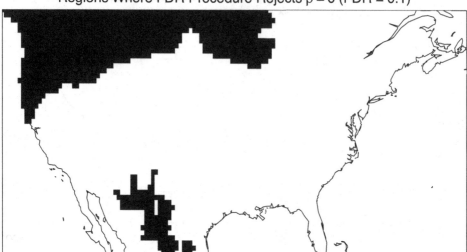

Figure 13.8 Regions where the False Discovery Rate criterion rejects the null hypothesis of vanishing correlation between DJF Niño 3.4 and North American temperature, with FDR = 10%. The corresponding correlation map for this data is shown in Figure 13.1.

its first few principal components. The False Discovery Rate is extremely simple and identifies specific locations of significant correlations, but it is not valid when the field is *highly* spatially correlated. In particular, the results of the False Discovery Rate depend on resolution – a high-resolution field will have many correlated neighboring points. Also, the False Discovery Rate is unlike the other tests because it controls for the expected proportion of rejected null hypotheses that were true. A prudent approach is to apply different tests and see if they agree. If they do not agree, then field significance is sensitive to the manner in which the null hypothesis is violated.

13.6 Further Topics

The study of time-varying spatial fields is known as spatio-temporal data analysis and has an extensive literature. A good entry into this literature is Cressie and Wikle (2011).

13.7 Conceptual Questions

- What does the term "field significance" mean?
- How does field significance differ from univariate significance?

- What is the difference between a correlation map and regression map? When do you use one vs. other?
- What *exactly* is the problem with isolating significant local correlations and using those correlations to conclude that a relation exists in those regions?
- Could you use field significance to investigate whether ENSO is related to <u>two</u> fields (e.g., temperature and pressure)?
- If a 5% significance level is used to mask out local correlations, why can't we assume that only 5% of the area will exceed the significance threshold when the null hypothesis is true?
- Why do we expect *more* than 5% of the area of a correlation map to exceed the univariate 5% significance threshold?
- It is sometimes argued that the field significance of a correlation map can be judged by the extent to which the spatial patterns are coherent. Is this correct? Why or why not?
- What if a person looks at just one grid point and concludes something different from a field significance test?
- How does serial correlation affect a field significance test?
- Describe the Livezey–Chen field significance test in your own words.
- How would you perform the Livezey–Chen test if your time series is non-Gaussian? Serially correlated?
- What are some limitations of the Livezey–Chen Test?
- Using your own words, describe the regression approach to field significance.
- The regression approach to field significance uses a model of the form $X = YB + E$, where the direction of causality may be in the wrong direction (that is, we believe X causes Y, but the model has Y causing X). Why is this acceptable?
- What are some limitations of the regression approach to field significance?
- Using your own words, describe the False Discovery Rate procedure.
- What is the interpretation of the False Discovery Rate?
- If the gridded data set is very high resolution (1km versus 100 km), how does this impact the aforementioned tests?
- Why is there no *single* way to test field significance?
- How big does the domain have to be to apply field significance?

14

Multivariate Linear Regression

The linear model involves the simplest and seemingly most restrictive statistical properties: independence, normality, constancy of variance, and linearity. However, the model and the statistical methods associated with it are surprisingly versatile and robust.[1]

Rencher and Schaalje (2008)

Multivariate linear regression is a method for modeling linear relations between two random vectors, say X and Y. Common reasons for using multivariate regression include (1) *predicting* Y given X, (2) *testing* hypotheses about the relation between X and Y, and (3) *projecting* Y onto prescribed time series or spatial patterns X. Special cases of multivariate regression include Linear Inverse Models (LIMs) and Vector Autoregressive Models. Multivariate regression is also fundamental to other statistical techniques, including canonical correlation analysis, discriminant analysis, and predictable component analysis. This chapter introduces multivariate linear regression and discusses estimation, measures of association, hypothesis testing, and model selection. In climate studies, model selection often involves selecting Y as well as X. For instance, Y may be a set of principal components that need to be chosen. This is a nonstandard selection problem. A criterion known as Mutual cross-entropy is introduced here to select X and Y simultaneously.

---

[1] Reprinted by permission from John Wiley and Sons, *Linear Models in Statistics, 2nd Edition*, Rencher and Schaalje (2008).

14.1 The Problem

Univariate regression considers a scalar y. Multivariate regression considers a vector $\mathbf{y}$. To illustrate, we use multivariate regression to predict Sea Surface Temperature (SST). Specifically, we use the SST at one time to predict SST at a later time based on the model

$$\mathbf{y} = \mathbf{B}^T\mathbf{x} + \boldsymbol{\mu} + \boldsymbol{\epsilon}, \tag{14.1}$$

where $\mathbf{x}$ and $\mathbf{y}$ are SST fields at the initial and target times, respectively, $\boldsymbol{\mu}$ is an intercept term, $\boldsymbol{\epsilon}$ is a mean-zero error term, and $\mathbf{B}$ is a matrix. (The transpose of $\mathbf{B}$ is used so that it agrees with equations that appear later.)

Our data set is the ERSSTv5 of Huang et al. (2017).[2] This data set provides monthly mean estimates of observed temperature on a $2° \times 2°$ longitude-latitude grid. We focus on the Pacific Ocean between $30°S$ and $60°N$ and the period 1950–2018.[3] For this domain and time span, there are 2,976 grid points and 69 years, hence the number of variables in $\mathbf{x}$ and $\mathbf{y}$ exceeds the number of time steps. As a result, estimation of $\mathbf{B}$ in (14.1) has no unique solution. In climate studies, a standard approach is to represent the SST anomalies by its first few PCs (see Chapter 12). For reasons discussed in Section 14.6, we select the first five PCs of SST for both $\mathbf{x}$ and $\mathbf{y}$. Therefore, $\mathbf{x}$ and $\mathbf{y}$ are five-dimensional vectors containing the PC amplitudes at the initial and target times, and $\mathbf{B}$ is a 5×5 matrix.

The matrix $\mathbf{B}$ is estimated using the least squares method. As will be shown in Section 14.3, the result is equivalent to estimating a univariate regression model separately for each PC in $\mathbf{y}$. That is, y_1 is predicted from the five PCs for $\mathbf{x}$, then y_2 is predicted from the same five PCs of $\mathbf{x}$, and so on. The regression coefficients are collected into the rows of $\hat{\mathbf{B}}^T$ and $\hat{\boldsymbol{\mu}}$. The associated prediction model for $\mathbf{y}$ is then $\hat{\mathbf{y}} = \hat{\mathbf{B}}^T\mathbf{x} + \hat{\boldsymbol{\mu}}$.

To judge prediction performance, the data set is partitioned into two parts: a *training* set for computing Empirical Orthogonal Function (EOFs) and estimating $\mathbf{B}$ and $\boldsymbol{\mu}$, and a *validation* set for assessing the prediction model. The predicted SST field is obtained by multiplying each predicted PC time series by its corresponding EOF and then summing over all five components. Because SST exhibits a significant seasonal cycle, X and Y correspond to fixed calendar months. The details are discussed in Section 14.7.

To illustrate, we predict December-mean SST based on prior June SSTs (that is, 6-month lead). Figure 14.1 shows the observed SST and corresponding prediction for three years characterized by neutral, warm, and cold ENSO conditions. The prediction captures the large-scale structure of the anomaly during strong ENSO years

---

[2] www.ncdc.noaa.gov/data-access/marineocean-data/extended-reconstructed-sea-surface-temperature-ersst-v5

[3] The domain boundaries are chosen to avoid the observation-sparse regions in the deep Southern Hemisphere and the sea-ice regions in the North Pacific. The data prior to 1950 is excluded because of its lower quality. Temperature anomalies are defined by regressing out the mean and linear trend from each grid point.

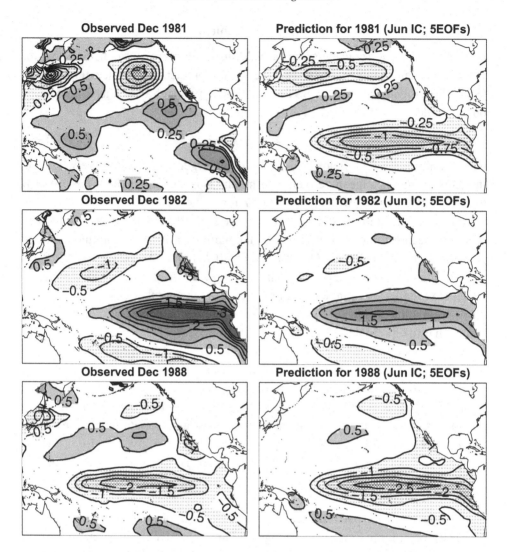

Figure 14.1 Comparison between observed (left) and predicted (right) December SST anomalies for a neutral (top; 1981), warm (middle; 1982), and cold (bottom; 1988) ENSO year. Prediction is based on five PCs using June initial condition. Contour label is degrees Celcius.

(1982 and 1988), but not during the neutral year (for instance, the prediction for 1981 has the wrong sign along the equatorial Pacific). Nevertheless, the root mean square errors are similar: 0.52, 0.65, and $0.47°C$ for 1981, 1982, 1988, respectively. The similarity of MSEs is consistent with the fact that the regression model (14.1) assumes errors are *independent of the initial condition*.

Our prediction is said to have skill if its errors are significantly less than those based on (14.1) with $\mathbf{B} = \mathbf{0}$. Equivalently, we want to know whether to reject the hypothesis $\mathbf{B} = \mathbf{0}$. This test is discussed in Section 14.3 and $\mathbf{B} = \mathbf{0}$ is rejected with p-value less than 10^{-10}.

14.2 Review of Univariate Regression

Multivariate regression can be developed in terms of univariate regression, hence it is instructive to briefly review univariate regression. Consider the model

$$Y = X_1\beta_1 + X_2\beta_2 + \cdots + X_M\beta_M + \epsilon, \tag{14.2}$$

where $X_1, \ldots, X_2$ are known as *predictors* and Y is known as the *predictand*. The intercept is included as one of the predictors. The model is univariate in the sense that only one Y variable is considered. The case of one Y is referred to as *univariate multiple regression*, or just *univariate regression*, while the case of many Y's is known as *multivariate* regression.

The Model

We assume N samples of $(Y, X_1, \ldots, X_M)$ are generated from (14.2). A model for each sample is specified by adding an n-index to the model (14.2):

$$Y_n = X_{n,1}\beta_1 + X_{n,2}\beta_2 + \cdots + X_{n,M}\beta_M + \epsilon_n. \tag{14.3}$$

The set of equations for $n = 1, \ldots, N$ can be written in matrix form as

$$\begin{array}{ccccc} \mathbf{y} & = & \mathbf{X} & \boldsymbol{\beta} & + & \boldsymbol{\epsilon}. \\ N \times 1 & & N \times M & M \times 1 & & N \times 1 \end{array} \tag{14.4}$$

In the example of Section 14.1, the number of predictors (that is, grid points) exceeded the number of samples (that is, years), hence $M > N$ and the problem was ill-posed. After replacing X and Y by five PC time series, then $M < N$, where N is the number of years, M is the number of PC time series plus one (for the intercept term), hence $M = 6$, and the matrix equation (14.4) has the following pattern.

$$\begin{array}{ccccc} \mathbf{y} & = & \mathbf{X} & \boldsymbol{\beta} & + & \boldsymbol{\epsilon} \\ N \times 1 & & N \times M & M \times 1 & & N \times 1 \end{array}$$

Least-Squares Estimate

The least-squares estimate of $\boldsymbol{\beta}$ is obtained by minimizing the sum square residual $\|\mathbf{y} - \mathbf{X}\boldsymbol{\beta}\|^2$. The resulting estimate is

$$\hat{\boldsymbol{\beta}} = \left(\mathbf{X}^T\mathbf{X}\right)^{-1}\mathbf{X}^T\mathbf{y}. \tag{14.5}$$

Orthogonality Constraint

The residuals $\mathbf{y} - \mathbf{X}\hat{\boldsymbol{\beta}}$ are orthogonal to the predictors:

$$\mathbf{X}^T\left(\mathbf{y} - \mathbf{X}\hat{\boldsymbol{\beta}}\right) = \mathbf{0}. \tag{14.6}$$

Variance of the Noise

The noise term ϵ_n is assumed to be *iid*,

$$\text{cov}[\epsilon_n, \epsilon_{n'}] = \delta_{nn'}\, \sigma^2_{Y|X}, \tag{14.7}$$

where $\sigma^2_{Y|X}$ is the variance of ϵ and $\delta_{nn'}$ is the Kronecker-delta function (it equals 1 when $n = n'$ and vanishes otherwise). The fact that the covariance vanishes when $n \neq n'$ reflects the independence assumption. The fact that the variance of ϵ_n is independent of n reflects the "identically distributed" assumption. The variance of ϵ is the *conditional* variance of Y given $X_1, \ldots, X_M$, and can be estimated from the mean square residuals

$$\overline{\sigma}^2_{Y|X} = \frac{(\mathbf{y} - \mathbf{X}\hat{\boldsymbol{\beta}})^T(\mathbf{y} - \mathbf{X}\hat{\boldsymbol{\beta}})}{N}. \tag{14.8}$$

In contrast to the unbiased estimate (9.13), the above estimate uses a divisor N rather than $N - M$ and therefore is a *biased* estimate of noise variance. There is no loss here: The unbiased estimate can be recovered by rescaling the biased estimate. The biased estimate (14.8) is used in this chapter because certain standard distributions and selection criteria are tailored to the biased version and therefore yield simpler equations in what follows.

R-square

A measure of linear association between Y and $X_1, \ldots, X_M$ is

$$R^2 = 1 - \frac{\overline{\sigma}^2_{Y|X}}{\overline{\sigma}^2_Y}. \tag{14.9}$$

This quantity lies between 0 and 1, with 1 indicating a perfect linear fit and 0 indicating that no linear combination of predictors is correlated with Y.

Prediction

After estimating the parameters in the model (14.2), the model may be used to make predictions. Specifically, given the M predictors $X_1, \ldots, X_M$, the corresponding prediction for Y would be

$$\hat{Y} = \hat{\beta}_1 X_1 + \cdots + \hat{\beta}_M X_M. \tag{14.10}$$

No error term ϵ is included in the prediction equation – ϵ is purely random and adding it to a prediction would only make the prediction worse. A quick way to see this is to suppose that all parameters in the regression model (14.2) are known. Then, the best prediction is $\beta_1 X_1 + \cdots + \beta_M X_M$, and the associated mean square is

$$\text{MSE} = \mathbb{E}\left[(Y - (\beta_1 X_1 + \cdots + \beta_M X_M))^2\right] = \mathbb{E}\left[\epsilon^2\right] = \sigma_{Y|X}^2. \tag{14.11}$$

In contrast, if an independent error term ϵ_0 with the same distribution as ϵ were added to the prediction equation, then

$$\begin{aligned}
\text{MSE} &= \mathbb{E}\left[(Y - (\beta_1 X_1 + \cdots + \beta_M X_M + \epsilon_0))^2\right] \\
&= \mathbb{E}\left[(\epsilon - \epsilon_0)^2\right] = \mathbb{E}\left[\epsilon^2\right] + \mathbb{E}\left[\epsilon_0^2\right] = 2\sigma_{Y|X}^2. \tag{14.12}
\end{aligned}$$

Thus, adding an error term doubles the MSE.

Distribution of the Sample Estimates
The distributions of $\hat{\boldsymbol{\beta}}$ and $\overline{\sigma}_{Y|X}^2$ depend on the distribution of ϵ. Assuming

$$\epsilon \sim \mathcal{N}_N\left(\mathbf{0}, \sigma_{Y|X}^2 \mathbf{I}\right), \tag{14.13}$$

then Chapter 9 showed that $\hat{\boldsymbol{\beta}}$ and $\overline{\sigma}_{Y|X}^2$ are independent and

$$\hat{\boldsymbol{\beta}} \sim \mathcal{N}_M\left(\boldsymbol{\beta}, \sigma_{Y|X}^2 \left(\mathbf{X}^T \mathbf{X}\right)^{-1}\right) \quad \text{and} \quad N\left(\frac{\overline{\sigma}_{Y|X}^2}{\sigma_{Y|X}^2}\right) \sim \chi_{N-M}^2. \tag{14.14}$$

Hypothesis Testing
We often want to know if certain predictors can be dropped. This question can be addressed by testing the hypothesis that the associated elements in $\boldsymbol{\beta}$ vanish. To do this, partition the predictors as $\mathbf{X} = \begin{bmatrix} \mathbf{K} & \mathbf{R} \end{bmatrix}$, where K denotes variables to keep and R denotes variables either to remove or retain. Then, the regression model is

$$\underset{N \times 1}{\mathbf{y}} = \underset{N \times M_K}{\mathbf{K}} \underset{M_K \times 1}{\boldsymbol{\beta}_K} + \underset{N \times M_R}{\mathbf{R}} \underset{M_R \times 1}{\boldsymbol{\beta}_R} + \underset{N \times 1}{\boldsymbol{\epsilon}}. \tag{14.15}$$

If $\boldsymbol{\beta}_R = \mathbf{0}$, then the predictors $\mathbf{R}$ can be removed. Therefore, we want to test the hypothesis $\boldsymbol{\beta}_R = \mathbf{0}$. The procedure is to first fit (14.15) and compute the sample variance $\overline{\sigma}_{Y|K,R}^2$. Next, the model without $\mathbf{R}$ (that is, $\mathbf{y} = \mathbf{K}\boldsymbol{\beta}_K + \boldsymbol{\epsilon}$) is fitted to the same data and the associated sample variance $\overline{\sigma}_{Y|K}^2$ is computed. Then, the ratio of noise variances

$$\lambda = \frac{\overline{\sigma}_{Y|K,R}^2}{\overline{\sigma}_{Y|K}^2}, \tag{14.16}$$

measures the improvement in fit resulting from adding $\mathbf{R}$. If $\boldsymbol{\beta}_R = \mathbf{0}$ and (14.13) holds, then

$$F = \frac{1 - \lambda}{\lambda} \frac{N - M_R - M_K}{M_R},$$

(14.17)

has an F distribution with M_R and $N - M_K - M_R$ degrees of freedom. The hypothesis $\beta_R = 0$ is rejected for sufficiently large values of F. To see the reasonableness of this test, suppose the hypothesis is true. Then, adding $\mathbf{R}$ to a regression model that already contains $\mathbf{K}$ does not improve the fit much. In this case, the two noise variances are similar, $\lambda \approx 1$, and F is small. Conversely, if the hypothesis is false, then adding $\mathbf{R}$ will improve the fit. The improved fit means that $\overline{\sigma}^2_{Y|K,R} \ll \overline{\sigma}^2_{Y|K}$ and $\lambda \approx 0$, hence F is large.

Model Selection Typically, one knows the Y-variable to be predicted, but the predictors are not specified and therefore must be selected from a pool of candidate variables. In such applications, *model selection* refers to the process of selecting a subset of predictors to use in a regression model. Various approaches to model selection were discussed in Chapter 10, including cross-validation and Akaike's Information Criteria (AIC).

14.3 Estimating Multivariate Regression Models

Now consider modeling more than one Y. In climate applications, the Y variables are often data values on a spatial grid, hence we use the label "s" for "space," although later s will identify the PC index of Y (in the same way that M may start out indicating the number of grid points in X, but later it indicates the number of PCs in X). The model for each sample n and variable s is

$$Y_{n,s} = X_{n,1}B_{1,s} + X_{n,2}B_{2,s} + \cdots + X_{n,M}B_{M,s} + \Xi_{n,s},$$

(14.18)

where Ξ is the Greek letter Xi chosen for its similarity to ϵ. The system of equations for $n = 1, \ldots, N$ and $s = 1, \ldots, S$ can be written concisely as

$$\begin{array}{ccccc} \mathbf{Y} & = & \mathbf{X} & \mathbf{B} & + & \Xi. \\ N \times S & & N \times M & M \times S & & N \times S \end{array}$$

(14.19)

The regression coefficients form a *matrix* $\mathbf{B}$ because each X-Y pair requires its own coefficient. For a single time step n, the predictand in $\mathbf{Y}$ is a *row vector*. (In the population model (14.1), the state vector was a column vector, hence the use of *transpose* $\mathbf{B}$). Figure 14.2 shows a schematic of the pattern associated with the matrix regression equations (14.19) when used to model four time series A, B, C, D. If $\mathbf{X}$ and $\mathbf{Y}$ contain the same variables but shifted in time, such that $\mathbf{Y} = \mathbf{Z}_t$ and $\mathbf{X} = \mathbf{Z}_{t-1}$, then (14.19) can be written as

$$\mathbf{Z}_t = \mathbf{Z}_{t-1}\mathbf{B} + \Xi,$$

then in climate studies, this model is known as a Linear Inverse Model (LIM), while in the statistical literature it is known as a first-order Vector Autoregressive Model, or VAR(1).

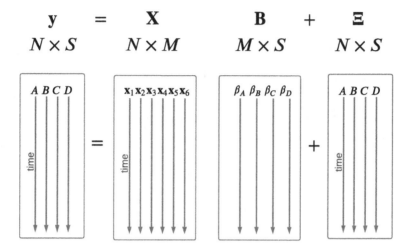

$$
\begin{array}{ccccccc}
\mathbf{y} & = & \mathbf{X} & & \mathbf{B} & + & \Xi \\
N \times S & & N \times M & & M \times S & & N \times S
\end{array}
$$

Figure 14.2 Schematic illustrating the pattern of multivariate regression model (14.19) for modeling time series at grid points A, B, C, D as a linear combination of time series $\mathbf{x}_1, \mathbf{x}_2, \mathbf{x}_3, \mathbf{x}_4, \mathbf{x}_5, \mathbf{x}_6$.

The natural approach to estimating $\mathbf{B}$ is to find values that minimize the sum square residual (SSE) *over all variables*. This measure is known as the Frobenious norm and is defined

$$
\text{SSE} = \|\mathbf{Y} - \mathbf{XB}\|_F^2 = \sum_{s=1}^{S} \sum_{n=1}^{N} \left(Y_{ns} - \sum_{m=1}^{M} X_{nm} B_{ms} \right)^2 . \tag{14.20}
$$

An efficient way to solve this minimization problem is to partition $\mathbf{Y}$ as

$$
\begin{array}{cc}
\mathbf{Y} & = \begin{bmatrix} \mathbf{y}_1 & \mathbf{y}_2 & \cdots & \mathbf{y}_S \\ N \times 1 & N \times 1 & \cdots & N \times 1 \end{bmatrix}, \\
N \times S &
\end{array} \tag{14.21}
$$

and partition $\mathbf{B}$ as

$$
\begin{array}{cc}
\mathbf{B} & = \begin{bmatrix} \boldsymbol{\beta}_1 & \boldsymbol{\beta}_2 & \cdots & \boldsymbol{\beta}_S \\ M \times 1 & M \times 1 & \cdots & M \times 1 \end{bmatrix}. \\
M \times S &
\end{array} \tag{14.22}
$$

Then, the sum square error (14.20) can be written equivalently as

$$
\text{SSE} = \|\mathbf{y}_1 - \mathbf{X}\boldsymbol{\beta}_1\|^2 + \|\mathbf{y}_2 - \mathbf{X}\boldsymbol{\beta}_2\|^2 + \cdots + \|\mathbf{y}_S - \mathbf{X}\boldsymbol{\beta}_S\|^2 . \tag{14.23}
$$

Note that $\boldsymbol{\beta}_1$ appears only in the first term on the right-hand side. Therefore, we may estimate $\boldsymbol{\beta}_1$ simply by minimizing just the first term, which is exactly equivalent to univariate regression. Applying this reasoning to the s'th term, the least squares estimate for $\boldsymbol{\beta}_s$ is

$$\hat{\boldsymbol{\beta}}_s = \left(\mathbf{X}^T\mathbf{X}\right)^{-1}\mathbf{X}^T\mathbf{y}_s, \quad \text{for } s = 1, \dots, S, \tag{14.24}$$

which can be written in matrix form as

$$\hat{\mathbf{B}} = \begin{bmatrix} \hat{\boldsymbol{\beta}}_1 & \cdots & \hat{\boldsymbol{\beta}}_S \end{bmatrix} = \left(\mathbf{X}^T\mathbf{X}\right)^{-1}\mathbf{X}^T\mathbf{Y}. \tag{14.25}$$

Note that this is identical to (14.5) after replacing the vector $\mathbf{y}$ by the matrix $\mathbf{Y}$. Similarly, each solution $\boldsymbol{\beta}_1, \dots, \boldsymbol{\beta}_S$ satisfies the orthogonality constraint

$$\mathbf{X}^T\left(\mathbf{y}_s - \mathbf{X}\hat{\boldsymbol{\beta}}_s\right) = \mathbf{0} \quad \text{for } s = 1, \dots, S, \tag{14.26}$$

which can be written in matrix form as

$$\mathbf{X}^T\left(\mathbf{Y} - \mathbf{X}\mathbf{B}\right) = \mathbf{0}, \tag{14.27}$$

which is the multivariate generalization of the orthogonality condition (14.6). In general, much of our intuition from univariate regression carries over to multivariate regression because each column of $\hat{\mathbf{B}}$ is a univariate regression estimate.

Incidentally, the same estimate $\hat{\mathbf{B}}$ in (14.25) is obtained if the Frobenius norm weighs different s variables differently, and even for more general norms.

The generalization of the noise variance $\sigma_{Y|X}^2$ is the *noise covariance matrix* $\boldsymbol{\Sigma}_{Y|X}$:

$$\text{cov}[\Xi_{ns}, \Xi_{n's'}] = \delta_{nn'}\left(\boldsymbol{\Sigma}_{Y|X}\right)_{ss'}. \tag{14.28}$$

This equation states that the rows of Ξ are independent realizations from a population with covariance matrix $\boldsymbol{\Sigma}_{Y|X}$. The latter matrix can be estimated from the residuals as

$$\overline{\boldsymbol{\Sigma}}_{Y|X} = \frac{\left(\mathbf{Y} - \mathbf{X}\hat{\mathbf{B}}\right)^T\left(\mathbf{Y} - \mathbf{X}\hat{\mathbf{B}}\right)}{N}. \tag{14.29}$$

This expression looks exactly like (14.8), except matrices replace vector quantities. The reason for using the divisor N here is the same as in the case of univariate regression. For future reference, this equation is related to other sample covariance matrices as

$$\overline{\boldsymbol{\Sigma}}_{Y|X} = \overline{\boldsymbol{\Sigma}}_{YY} - \overline{\boldsymbol{\Sigma}}_{YX}\overline{\boldsymbol{\Sigma}}_{XX}^{-1}\overline{\boldsymbol{\Sigma}}_{XY}. \tag{14.30}$$

If the X and Y variables are centered, then these covariance matrices are defined as

$$\overline{\Sigma}_{YY} = \frac{1}{N}\mathbf{Y}^T\mathbf{Y}, \quad \overline{\Sigma}_{XX} = \frac{1}{N}\mathbf{X}^T\mathbf{X}, \quad \text{and} \quad \overline{\Sigma}_{XY} = \overline{\Sigma}_{YX}^T = \frac{1}{N}\mathbf{X}^T\mathbf{Y}. \quad (14.31)$$

For future reference, we define

$$\overline{\Sigma}_{(XY)} = \frac{1}{N}(\mathbf{X} \quad \mathbf{Y})^T \begin{pmatrix} \mathbf{X} \\ \mathbf{Y} \end{pmatrix} = \begin{pmatrix} \overline{\Sigma}_{XX} & \overline{\Sigma}_{XY} \\ \overline{\Sigma}_{YX} & \overline{\Sigma}_{YY} \end{pmatrix}. \quad (14.32)$$

In univariate regression, a measure of association is R-square (14.9). A natural generalization of R-square that emerges naturally in hypothesis testing and model selection is

$$\mathcal{R}^2 = 1 - \frac{|\overline{\Sigma}_{Y|X}|}{|\overline{\Sigma}_{YY}|}. \quad (14.33)$$

$\mathcal{R}^2$ vanishes if $\overline{\Sigma}_{YX} = \mathbf{0}$; that is, when the sample correlation between each Y and X vanishes. This fact can be seen from (14.30), which shows that in this case $\overline{\Sigma}_{Y|X} = \overline{\Sigma}_{YY}$, which in turns implies $\mathcal{R}^2 = 0$. At the opposite extreme, $\mathcal{R}^2$ equals one if each Y variable is a linear combination of X's (that is, if $\mathbf{Y} = \mathbf{XB}$), because in this case $\overline{\Sigma}_{Y|X} = \mathbf{0}$.

14.4 Hypothesis Testing in Multivariate Regression

We often want to know if certain predictors can be dropped from a model. This question can be addressed by testing the hypothesis that certain regression parameters vanish. In multivariate regression, dropping one X-variable affects the prediction of *all* the Y-variables. This fact needs to be taken into account in the hypothesis test. The most widely used approach is the *Likelihood Ratio Test* (LRT), which is covered in most upper-level statistics texts. The derivation of LRTs for multivariate regression requires high-level math, but the final procedure can be explained relatively simply. Accordingly, we simply summarize the LRT for hypotheses in multivariate regression, and refer readers to other texts for proofs (e.g., Mardia et al., 1979; Anderson, 1984; Johnson and Wichern, 2002).

Consider the following multivariate generalization of (14.15):

$$\underset{N \times S}{\mathbf{Y}} = \underset{N \times M_K}{\mathbf{K}} \underset{M_K \times S}{\mathbf{B}_K} + \underset{N \times M_R}{\mathbf{R}} \underset{M_R \times S}{\mathbf{B}_R} + \underset{N \times S}{\mathbf{\Xi}},$$

where $M = M_R + M_K$. This model can be written in the standard form (14.19), and therefore the noise covariance matrix can be estimated from (14.29), which we denote as $\overline{\Sigma}_{Y|K,R}$. Next, we consider a model that excludes $\mathbf{R}$, namely,

$$\mathbf{Y} = \mathbf{KB}_K + \mathbf{\Xi}, \quad (14.34)$$

and estimate the noise covariance matrix of this model, which we denote as $\overline{\Sigma}_{Y|K}$. If the noise term has distribution

$$\text{rows of } \Xi \overset{iid}{\sim} \mathcal{N}_S\left(\mathbf{0}, \Sigma_{Y|X}\right), \tag{14.35}$$

then the Likelihood Ratio Test is based on the statistic

$$\Lambda = 1 - \mathcal{R}^2 = \frac{|\overline{\Sigma}_{Y|K,R}|}{|\overline{\Sigma}_{Y|K}|}. \tag{14.36}$$

This statistic is the multivariate generalization of (14.16) and always lies between 0 and 1, for reasons that are essentially similar to those in univariate regression.

If the hypothesis $\mathbf{B}_R = \mathbf{0}$ is true and (14.35) holds, then Λ has a distribution known as *Wilks' Lambda Distribution with degrees of freedom* $(S, M_R, N - M_R - M_K)$. The hypothesis is rejected for sufficiently *small* values of Λ. Wilks' Lambda Distribution is not widely available in standard software packages. In the special case of $S = 1$ or $M_R = 1$, then Λ can be transformed into another statistic with an exact F distribution (Mardia et al., 1979). Alternatively, if N is sufficiently large, then

$$\chi^2 = -\left(\nu_E - \frac{S - \nu_H + 1}{2}\right) \log \Lambda, \tag{14.37}$$

where $\nu_E = N - M$ and $\nu_H = M_R$, has an approximate chi-squared distribution with $S\nu_H$ degrees of freedom. Hypothesis $\mathbf{B}_R = \mathbf{0}$ is rejected when $\chi^2 > \chi^2_{0.05,\,S\nu_H}$. Lastly, if more accuracy is required, quantiles of Wilks' Lambda can be estimated by Monte Carlo techniques.

14.5 Selecting X

The Likelihood Ratio Test assumes that two (and only two) candidate models are under consideration and are *nested*. By nested, we mean that the X-variables of one model are a subset of the X-variables in the other. In practice, however, we often need to explore many models, including nonnested ones. Generalizing hypothesis testing to the comparison of multiple nonnested models is not straightforward.

To find a reasonable selection criterion, recall that the goal of model selection is to choose the model that is "closest" to the model that generated the data. For simplicity, suppose the data is $\mathbf{z}$. Regardless of how $\mathbf{z}$ was generated, the result can be interpreted as samples from a probability density function (PDF) $p(\mathbf{z})$. $p(\mathbf{z})$ is known as the *true* PDF and is not known. Instead, we have candidate PDFs $q_1(\mathbf{z}; \boldsymbol{\phi}_1), q_2(\mathbf{z}; \boldsymbol{\phi}_2), \ldots$, where $\boldsymbol{\phi}_1, \boldsymbol{\phi}_2$ denote the corresponding model parameters. We want to choose the candidate PDF that is "closest" to the true PDF. This goal requires defining a measure of the difference between PDFs. For reasons that will become clear shortly (also see Burnham and Anderson, 2002), an attractive measure of the difference in PDFs is the Kullback–Leibler (KL) Divergence

$$\mathbb{D}(Z) = \int p(\mathbf{z}) \log \left(\frac{p(\mathbf{z})}{q(\mathbf{z}; \boldsymbol{\phi})} \right) d\mathbf{z}. \tag{14.38}$$

This expression may be written equivalently as

$$\mathbb{D}(Z) = \mathbb{E}_Z \left[\log \left(\frac{p(\mathbf{z})}{q(\mathbf{z}; \boldsymbol{\phi})} \right) \right], \tag{14.39}$$

where $\mathbb{E}_Z[\cdot]$ denotes the expectation over $p(\mathbf{z})$. It is obvious that KL Divergence vanishes if $p(\mathbf{z}) = q(\mathbf{z}; \boldsymbol{\phi})$, since $\log 1 = 0$. A less obvious fact is that KL Divergence vanishes if *and only if* $p(\mathbf{z}) = q(\mathbf{z}; \boldsymbol{\phi})$, otherwise it is *positive* (Cover and Thomas, 1991). Thus, KL Divergence behaves like a "distance" measure – it vanishes when the two PDFs are equal, otherwise $\mathbb{D}(Z) \geq 0$. KL Divergence has numerous other properties (e.g., it is invariant to invertible nonlinear transformations of $\mathbf{z}$), but these properties are not critical here. By expanding the logarithm, the inequality $\mathbb{D}(Z) \geq 0$ can be expressed equivalently as

$$\mathbb{IC}(Z) \geq 2\,\mathbb{H}(Z), \tag{14.40}$$

where we define

$$\mathbb{IC}(Z) = -2\,\mathbb{E}_Z \left[\log q(\mathbf{z}; \boldsymbol{\phi}) \right] \quad \text{and} \quad \mathbb{H}(Z) = -\mathbb{E}_Z \left[\log p(\mathbf{z}) \right]. \tag{14.41}$$

$\mathbb{IC}(Y|X)$ is known as the *cross entropy*. $\mathbb{H}(Z)$ is known as the *entropy* of $\mathbf{z}$ and is unobservable since it requires knowing the true PDF. Fortunately, entropy does not depend on $q(\cdot)$. Therefore, the model that minimizes $\mathbb{IC}(Z)$ also minimizes KL divergence. In this way, we can obtain a selection criterion that does not require knowing the true PDF ($\mathbb{IC}(Z)$ requires averaging over $p(\mathbf{z})$, but this average can be estimated from data). The minimum possible value of $\mathbb{IC}(Z)$ is obtained if and only if the candidate PDF equals the true PDF. This property means that the log-score is a *proper score* (Gneiting and Raftery, 2007). In fact, $\log q(\mathbf{z}; \boldsymbol{\phi})$ is the *only* local function of $q(\mathbf{z}; \boldsymbol{\phi})$ with this property (Gneiting and Raftery, 2007). For these reasons, $\mathbb{IC}$ is a very compelling criterion for model selection.

In practice, $\boldsymbol{\phi}$ is unknown and must be estimated from samples. This estimation can lead to overfitting, which must be taken into account. Let $\hat{\boldsymbol{\phi}}(\mathbf{z}_*)$ denote an estimate of $\boldsymbol{\phi}$ based on sample $\mathbf{z}_*$ from $p(\mathbf{z})$. Substituting $\hat{\boldsymbol{\phi}}(\mathbf{z}_*)$ in (14.41) yields a cross entropy that is random through its dependence on $\mathbf{z}_*$. A fundamental principle in model selection is to judge model performance based on how well the model predicts an *independent* sample $\mathbf{z}_0$ from the same distribution. Following Akaike (1973), we average the log-score over $\mathbf{z}_*$ and $\mathbf{z}_0$, which have identical distributions but are independent of each other, and multiply by 2 (for historical reasons, but this multiplication does not affect the final selection), which leads to the criterion

$$\mathbb{AIC}(Z) = -2\,\mathbb{E}_{Z_*}\left[\mathbb{E}_{Z_0}\left[\log q(\mathbf{z}_0; \hat{\boldsymbol{\phi}}(\mathbf{z}_*))\right]\right].\tag{14.42}$$

Following Akaike (1973), we term $\mathbb{AIC}(Z)$ *Akaike's Information Criterion* for $q(\mathbf{z}; \phi)$.

Specifying the regression model (14.4) with (14.13) is equivalent to specifying the PDF

$$q\left(\mathbf{y}|\mathbf{X}; \boldsymbol{\beta}, \sigma_{Y|X}^2\right) = \mathcal{N}_N\left(\mathbf{X}\boldsymbol{\beta}, \sigma_{Y|X}^2\mathbf{I}\right),\tag{14.43}$$

or more explicitly,

$$q\left(\mathbf{y}|\mathbf{X}; \boldsymbol{\beta}, \sigma_{Y|X}^2\right) = \left(\frac{1}{2\pi\,\sigma_{Y|X}^2}\right)^{N/2}\exp\left[\|\mathbf{y} - \mathbf{X}\boldsymbol{\beta}\|^2/(2\sigma_{Y|X}^2)\right].\tag{14.44}$$

Substituting $q(\mathbf{x}, \mathbf{y}) = q(\mathbf{x})q(\mathbf{y}|\mathbf{x})$ into (14.42) shows that $\mathbb{AIC}$ satisfies the *chain rule*

$$\mathbb{AIC}(X, Y) = \mathbb{AIC}(X) + \mathbb{AIC}(Y|X).\tag{14.45}$$

One of the achievements of Hirotugu Akaike was to show that $\mathbb{AIC}$ could be estimated for arbitrary PDFs, provided the candidate PDF includes the true model and N is large (Akaike, 1973). An estimate of $\mathbb{AIC}(Y|X)$ for the univariate regression model (14.4) is AIC in (10.27). However, it has been shown that AIC is *not asymptotically consistent*; that is, AIC does not select the true model with probability one in the limit $N \to \infty$, assuming a finite number of parameters in the true PDF. Other criteria that are asymptotically consistent have been derived, including BIC in (10.29). However, in climate applications, asymptotic consistency is a dubious requirement since any candidate PDF invented by humans likely requires an infinite number of parameters to capture the true PDF. Rather, the goal is to select a candidate PDF that makes good predictions, even if it is not perfect.

In practice, AIC tends to select overfitted models for finite N. Hurvich and Tsai (1989) argued that the cause of this overfitting tendency lies in the asymptotic approximations used to derive AIC. To derive a bias-corrected version of AIC, Hurvich and Tsai (1989) evaluated the Kullback–Leibler (KL) divergence *exactly* for normal distributions. The resulting criterion, AICc, often outperforms its competitors (McQuarrie and Tsai, 1998) and has become a standard criterion recommended by many investigators (e.g., Burnham and Anderson, 2002, 66). However, a subtle assumption in the derivation of AICc is that predictor values are the *same* in the training and validation samples (DelSole and Tippett, 2021b). This contrasts with our application, in which the predictors are random and thus differ between training and validation samples. To derive a criterion for general predictors, suppose the predictors are a combination of random and deterministic predictors, so that

$$\mathbf{X}_* = \begin{bmatrix} \mathbf{F} & \mathbf{R}_* \end{bmatrix}, \qquad (14.46)$$

where $\mathbf{F}$ is a fixed $N \times M_F$ matrix, and $\mathbf{R}_*$ is a random $N \times M_R$ matrix, where $M = M_F + M_R$. The rows of $\mathbf{R}_*$ are independent realizations from a multivariate normal distribution. One column of $\mathbf{F}$ is a vector of ones corresponding to the intercept, hence $1 \leq M_F \leq M$. Let $(\mathbf{y}_*, \mathbf{X}_*)$ denote the *training* sample for estimating model parameters, and let $(\mathbf{y}_0, \mathbf{X}_0)$ denote the *validation* sample for measuring out-of-sample prediction error, where

$$\mathbf{X}_0 = \begin{bmatrix} \mathbf{F} & \mathbf{R}_0 \end{bmatrix}. \qquad (14.47)$$

$\mathbf{R}_0$ is from the same distribution as $\mathbf{R}_*$ but is independent of $\mathbf{R}_*$. $\mathbf{F}$ is the same in (14.46) and (14.47). Then, DelSole and Tippett (2021b) prove the following.

> **Theorem 14.1** *An unbiased estimate of* AIC *in* (14.42) *for the regression model* (14.19), *under the predictor models* (14.46) *and* (14.47), *is*
>
> AICm$(Y|X)$
>
> $$= N \log |\overline{\boldsymbol{\Sigma}}_{Y|X}| + NS \log 2\pi + \frac{SN(N + M_F)(N - M_F - 1)}{(N - M - S - 1)(N - M - 1)}. \qquad (14.48)$$
>
> *The "m" in AICm emphasizes that the regression model contains a mix of random and deterministic explanatory variables.*

The case $M_R = 0$ is known as *Fixed-X* (Rosset and Tibshirani, 2020). Evaluating Theorem 14.1 for $M_R = 0$ and $M = M_F$ gives the Fixed-X AICc,

$$\text{AICc}(Y|X) = N \log |\overline{\boldsymbol{\Sigma}}_{Y|X}| + SN \log 2\pi + \frac{SN(N + M)}{N - M - S - 1}. \qquad (14.49)$$

Note that this reduces to (10.28) for $S = 1$. In climate studies, Fixed-X occurs in such situations as estimating the annual or diurnal cycle, because $\mathbf{X}$ specifies a periodic function of time that repeats itself in training and independent samples. This case was considered in the examples in Chapter 10, hence AICc was used. However, in many other applications, including the SST-prediction problem at the beginning of this chapter, the predictor values differ from training to validation samples, in which case Random-X is more appropriate. We use the term *Random-X* to mean $M_F = 1$, corresponding to the intercept, and the remaining $M_R = M - 1$ predictors are random. (This definition of Random-X differs from that of Rosset and Tibshirani (2020) by including the intercept in the regression model.) The criterion for Random-X can be derived from Theorem 14.1 using $M_F = 1$ (for the intercept), which gives a new criterion that we refer to as AICr:

$$\text{AICr}(Y|X) = N \log |\overline{\boldsymbol{\Sigma}}_{Y|X}| + SN \log 2\pi + \frac{SN(N+1)(N-2)}{(N-M-S-1)(N-M-1)}.$$
(14.50)

Importantly, AICr satisfies the *sample* chain rule $\text{AICr}(XY) = \text{AICr}(X) + \text{AICr}(Y|X)$. To see this, note that $\text{AICr}(X)$ and $\text{AICr}(XY)$ are criteria for regression models

$$\mathbf{X} = \mathbf{j}\boldsymbol{\mu}_X^T + \mathbf{E}_X \qquad\qquad \text{rows of } \mathbf{E}_X \sim \mathcal{N}(\mathbf{0}, \boldsymbol{\Sigma}_{XX})$$
$$[\mathbf{X}\ \mathbf{Y}] = \mathbf{j}[\boldsymbol{\mu}_X^T\ \boldsymbol{\mu}_Y^T] + \mathbf{E}_{XY} \qquad \text{rows of } \mathbf{E}_{XY} \sim \mathcal{N}(\mathbf{0}, \boldsymbol{\Sigma}_{(XY)}),$$

where $\boldsymbol{\mu}_X$ and $\boldsymbol{\mu}_Y$ are M_R- and S-dimensional vectors giving regression coefficients for the intercept, $\mathbf{j}$ is a vector of ones, and $\mathbf{E}_X$ and $\mathbf{E}_{XY}$ are random matrices with distributions as indicated earlier. The AICr for these models can be derived from (14.50) after identifying the proper number of predictors and predictands. The result is

$$\text{AICr}(X) = N \log \left|\overline{\boldsymbol{\Sigma}}_{XX}\right| + NM_R \log 2\pi + \frac{M_R(N+1)N}{N - M_R - 2} \qquad (14.51)$$

$$\text{AICr}(X, Y) = N \log \left|\overline{\boldsymbol{\Sigma}}_{(XY)}\right| + (M_R + S)N \log 2\pi + \frac{(M_R + S)(N+1)N}{N - M_R - S - 2},$$
(14.52)

where covariance matrices are defined in (14.31) and (14.32). It is straightforward to verify $\text{AICr}(Y|X) = \text{AICr}(X, Y) - \text{AICr}(X)$, using $M_R = M - 1$ and the standard identity

$$|\overline{\boldsymbol{\Sigma}}_{(XY)}| = |\overline{\boldsymbol{\Sigma}}_{Y|X}| \cdot |\overline{\boldsymbol{\Sigma}}_{XX}|.$$
(14.53)

14.6 Selecting Both X and Y

The previous discussion assumed that only the X-variables were being selected; that is, the Y-variables have been chosen and are not open to selection. However, in the SST-prediction problem discussed at the beginning of this chapter, Y needs to be selected too. AICr(Y|X) does not provide a meaningful criterion for selecting Y because AICr is a proxy for prediction error, and comparing prediction errors for different quantities is not meaningful. In fact, our problem is to select *both* X and Y. We call this the *simultaneous selection problem*. This is a nonstandard problem in the statistics literature. This section derives a criterion for the simultaneous selection problem. The resulting criterion can also be used to select X and Y separately, since these are mere special cases of simultaneous selection.

Whatever the criterion for simultaneous selection, (1) it cannot depend on which variable sets are labeled X and Y, since these labels are arbitrary, and (2) it should be consistent with $\mathbb{AIC}$ for selecting X- or Y-variables separately. Only one criterion satisfies these requirements, as described in the following proposition (DelSole and Tippett, 2021a).

Proposition 14.1 *To within an additive constant, the only criterion for selecting X and Y variables that is a symmetric functional of the distributions of X and Y, and whose differences equal the corresponding differences in $\mathbb{AIC}$, is $\mathbb{MICa}$, defined as*

$$\mathbb{MICa}(X;Y) = \mathbb{AIC}(X,Y) - \mathbb{AIC}(Y) - \mathbb{AIC}(X) \tag{14.54}$$

$$= \mathbb{AIC}(Y|X) - \mathbb{AIC}(Y) \tag{14.55}$$

$$= \mathbb{AIC}(X|Y) - \mathbb{AIC}(X). \tag{14.56}$$

All three expressions are equivalent because of the chain rule (14.45). Accordingly, an unbiased estimate of $\mathbb{MICa}$ for Random-X is

$$MIC(Y;X) = AICr(Y|X) - AICr(Y) \tag{14.57}$$

$$= AICr(X|Y) - AICr(X) \tag{14.58}$$

$$= AICr(Y,X) - AICr(X) - AICr(Y). \tag{14.59}$$

In terms of regression model (14.19),

$$MIC(Y;X) = N \log |\overline{\Sigma}_{Y|X}| - N \log |\overline{\Sigma}_{YY}| + \mathcal{P}(N, M_X, S), \tag{14.60}$$

where

$$\mathcal{P}(N, M_X, S) = (N+1)N \left(\frac{M_X + S}{N - M_X - S - 2} - \frac{M_X}{N - M_X - 2} - \frac{S}{N - S - 2} \right).$$

$MIC(Y;X)$ is a proper score for simultaneous selection (DelSole and Tippett, 2021a).

To gain insight into how MIC works, note that (14.60) can be written equivalently as

$$MIC(Y;X) = N \log \left(1 - \mathcal{R}^2\right) + \mathcal{P}(N, M_X, S). \tag{14.61}$$

Recall that $\mathcal{R}$ is a measure of goodness-of-fit and tends to increase as more variables are included in the model. This increase occurs whether the added variables are X, Y, or both. Therefore, the first term, $\log \left(1 - \mathcal{R}^2\right)$ *decreases* with the number of variables while the second term, $\mathcal{P}$, increases. In essence, minimizing MIC represents a balance between maximizing fit and minimizing number of parameters.

By definition, *negative* MIC means that the model's AICr is less than that of the intercept-only model. To ensure that the selected model is *significantly* better than the intercept-only model, the Likelihood-Ratio Test can be used. Recall from

Section 14.4 that the Likelihood-Ratio Test for the hypothesis that all predictor coefficients vanish leads to the statistic $\Lambda = 1 - \mathcal{R}^2$. Combining this relation with (14.61) and using the asymptotic distribution (14.37) leads to the criterion MIC $<$ MIC$_{\text{crit}}$ where, for 5% significance,

$$\text{MIC}_{\text{crit}} = -\frac{\chi^2_{0.05,\,SM_X}}{N - (S + M_X + 3)/2} + \frac{\mathcal{P}(N, M_X, S)}{N}. \tag{14.62}$$

Note that significance requires MIC to be *less* than MIC$_{\text{crit}}$.

Returning to the SST-prediction problem in Section 14.1, a plot of MIC versus the number of PCs is shown in Figure 14.3. In this example, S is the number of PCs for Y, and M_X is the number of PCs for X. We have constrained the number of PCs for X and Y to be the same (this is done for simplicity and is not necessary). The minimum MIC occurs at five PCs. Hence, we select five PCs for X and Y. Any useful prediction model should yield better forecasts than the model with $\mathbf{B} = 0$, which is considered a "no-skill" baseline (also known as "the climatological model"). The fact that MIC lies below the 5% significance curve implies that the hypothesis $\mathbf{B} = 0$ is rejected for all candidate models.

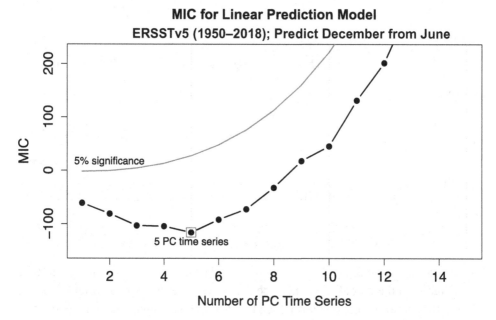

Figure 14.3 Mutual Information Criterion (MIC) versus number of PC time series of a linear regression model for predicting December-mean SSTs based on June SSTs over the Pacific Ocean ($30°\,S - 60°\,N$). The predictor and predicand have the same number of PC time series. The minimum MIC is indicated by a box. The 5% significance threshold for the hypothesis that all regression coefficients (except for the intercept) vanish is shown as the solid curve.

For future reference, we note that $\mathbb{MICa}$ is an extension of $\mathbb{MIC}$, defined as

$$\mathbb{MIC}(X;Y) = \mathbb{IC}(Y|X) - \mathbb{IC}(Y) = -2\mathbb{E}_{XY}\left[\log\frac{q(\mathbf{x},\mathbf{y})}{q(\mathbf{x})q(\mathbf{y})}\right]. \tag{14.63}$$

If $p(\mathbf{x},\mathbf{y}) = q(\mathbf{x},\mathbf{y})$, then $\mathbb{AIC}(Y|X) = 2\mathbb{H}(Y|X)$ and $\mathbb{MIC}(X;Y) = -2\mathbb{M}(X;Y)$, where $\mathbb{M}(X;Y)$ is the **mutual information** of $p(\mathbf{x},\mathbf{y})$:

$$\mathbb{M}(X;Y) = \mathbb{E}_{XY}\left[\log\frac{p(\mathbf{x},\mathbf{y})}{p(\mathbf{x})p(\mathbf{y})}\right].$$

Thus, $\mathbb{MIC}$ is related to (negative) mutual information in the same way that $\mathbb{IC}$ is related to entropy. We call $\mathbb{MIC}$ *Mutual Information Criterion*.

14.7 Some Details about Regression with Principal Components

In the prediction example of Section 14.1, several issues related to the EOFs arose. Specifically, EOFs were computed from the training sample, yet the associated time series in the verification data need to be derived through projection. Also, the regression model predicted only the PCs, hence the PCs need to be multiplied by the appropriate EOFs to construct spatial maps. Details of these and other calculations are discussed in this section.

Let $\mathbf{e}_1, \mathbf{e}_2, \ldots, \mathbf{e}_T$ be the first T EOFs of the (anomaly) training data set. In the SST example of Section 14.1, each EOF vector has 2,976 grid points over the ocean domain. These EOFs can be collected into the $S \times T$ matrix

$$\mathbf{E} = \begin{pmatrix} \mathbf{e}_1 & \mathbf{e}_2 & \cdots & \mathbf{e}_T \end{pmatrix}. \tag{14.64}$$

This matrix has a corresponding pseudoinverse $\mathbf{E}^i$, which has the property

$$\mathbf{E}^T\mathbf{E}^i = \mathbf{I}. \tag{14.65}$$

The derivations of EOFs and the pseudoinverse are described in Chapter 12.

Let $\mathbf{X}$ denote the initial anomaly and $\mathbf{Y}$ denote the subsequent target anomaly. Using dots to denote truncated EOF matrices, the PC time series can be obtained by projection as

$$\underset{N \times T}{\tilde{\dot{\mathbf{F}}}_X} = \underset{N \times S}{\mathbf{X}} \quad \underset{S \times T}{\dot{\mathbf{E}}^i} \tag{14.66}$$

$$\underset{N \times T}{\tilde{\dot{\mathbf{F}}}_Y} = \underset{N \times S}{\mathbf{Y}} \quad \underset{S \times T}{\dot{\mathbf{E}}^i}, \tag{14.67}$$

where T denotes the "truncation" (that is, the number of PCs in X and Y) and tildes indicate quantities in EOF space. The idea now is to predict the PC time series $\tilde{\dot{\mathbf{F}}}_Y$

based on $\dot{\tilde{\mathbf{F}}}_X$ (instead of predicting $\mathbf{Y}$ from $\mathbf{X}$). Accordingly, consider the regression model

$$\dot{\tilde{\mathbf{F}}}_Y = \dot{\tilde{\mathbf{F}}}_X \tilde{\mathbf{B}} + \mathbf{j}\tilde{\mu}^T + \tilde{\mathbf{\Xi}}, \tag{14.68}$$

where $\mathbf{j}$ is an N-dimensional vector of ones. This model can be written in the form

$$\mathbf{Z} = \mathbf{W}\mathbf{C} + \mathbf{\Xi}, \tag{14.69}$$

where

$$\mathbf{Z} = \dot{\tilde{\mathbf{F}}}_Y, \quad \mathbf{W} = \begin{bmatrix} \mathbf{j} & \dot{\tilde{\mathbf{F}}}_X \end{bmatrix} \quad \text{and} \quad \mathbf{C} = \begin{pmatrix} \tilde{\mu}^T \\ \tilde{\mathbf{B}} \end{pmatrix}. \tag{14.70}$$

This model is in the form (14.19), hence from (14.25) the least squares estimate of $\mathbf{C}$ is

$$\hat{\mathbf{C}} = \left(\mathbf{W}^T\mathbf{W}\right)^{-1}\mathbf{W}^T\mathbf{Z}. \tag{14.71}$$

Given validation time series $\tilde{\mathbf{F}}_{X,V}$, the corresponding prediction in the validation data is

$$\dot{\hat{\tilde{\mathbf{F}}}}_{Y,V} = \tilde{\mathbf{F}}_{X,V}\hat{\mathbf{C}}, \tag{14.72}$$

which is the multivariate generalization of (14.10). To express this prediction as a spatial pattern, each PC amplitude is multiplied by its corresponding EOF and summed. These operations can be summarized concisely as

$$\hat{\mathbf{Y}} = \dot{\hat{\tilde{\mathbf{F}}}}_{Y,V}\mathbf{E}^T = \tilde{\mathbf{F}}_{X,V}\hat{\mathbf{C}}\mathbf{E}^T. \tag{14.73}$$

14.8 Regression Maps and Projecting Data

Another application of multivariate regression is constructing regression maps. As discussed in Chapter 13, regression maps are used to investigate the relation between a time-varying climate index $r(t)$ and other climate variables in different geographical locations (possibly lagged in time). Mathematically, a regression map is associated with the model

$$y(t,s) = \mu + p(s)r(t) + \epsilon(t,s), \tag{14.74}$$

where $y(t,s)$ is the time-varying spatial field in question and s is a parameter indicating spatial location. The unknown vector $p(s)$ is identified with the regression map, since it is a set of regression coefficients that can displayed as a map. This model can be expressed in the form (14.19), with the identifications

$$y(t, s) \rightarrow \mathbf{Y}, \quad p(s) \rightarrow \mathbf{p}, \quad r(t) \rightarrow \mathbf{r}, \quad \mathbf{X} = \begin{pmatrix} \mathbf{j} & \mathbf{r} \end{pmatrix}, \quad \text{and} \quad \mathbf{B} = \begin{pmatrix} \mathbf{\mu}^T \\ \mathbf{p}^T \end{pmatrix}.$$

$$(14.75)$$

The least squares estimate of $\mathbf{B}$ is obtained from (14.25), the second row of which gives $\mathbf{p}$.

This procedure is often described as "constructing a regression map given $r(t)$." In other problems, the pattern $p(s)$ is given and the problem is to determine $r(t)$ (e.g., the pattern may come from some other statistical analysis of some other data set). This problem can be converted into the form (14.19) merely by swapping the meaning of time and space. This procedure is often described as "projecting the data onto $p(s)$."

14.9 Conceptual Questions

1 What is the problem with making predictions like those in Figure 14.1 using the original data for X? In other words, why were only the leading EOFs of the X-variables used?

2 Are you impressed by the predictions shown in Figure 14.1? If so, can you think of reasons why they might not be as impressive as they seem? If you are not impressed, why?

3 When do you use AICc? AICr? MIC?

4 EOFs depend on the data and so there are risks to using them as a basis set for statistical prediction. Are there reasons to believe that the leading EOFs may bias a linear regression model? How would you test this?

5 What are the standard assumptions on the noise term in univariate regression? How do these assumptions change in multivariate regression?

6 Explain the sense in which multivariate regression is essentially the same as univariate regression. Explain how it is different.

7 Define R-square for a univariate regression model. Describe how the multivariate version $\mathcal{R}^2$ (14.33) is a natural generalization of this?

8 What is the difference between a regression fit and a prediction? Why is this distinction important.

9 Can you show when $\mathcal{R}^2$ in (14.33) is zero, and when it is one?

10 Why would one test the hypothesis $\mathbf{B} = \mathbf{0}$? Does the hypothesis $\mathbf{B} = \mathbf{0}$ mean that X and Y are independent?

11 Explain how to test if a coefficient vanishes in a univariate regression model. How does this test differ in multivariate regression? The multivariate regression

test must reduce to the univariate version when only one Y variable exists. Given this, show how F and Λ are related in the case of one Y variable.

12 Explain how to use AIC to select predictors for a univariate model. Explain how to use MIC to select predictors *and predictands* for multivariate regression.

13 In what way is model selection more challenging in multivariate regression than in univariate regression?

14 Comparing univariate and multivariate regression, why is estimation the same, but hypothesis testing is different?

15 What is the relation between hypothesis testing and model selection based on MIC?

15

Canonical Correlation Analysis

> Marksmen side by side firing simultaneous shots at targets, so that the deviations are in part due to independent individual errors and in part to common causes such as wind, provide a familiar introduction to the theory of correlation; but only the correlation of horizontal components is ordinary discussed, whereas the complex consisting of horizontal and vertical deviations may be even more interesting. [1]
>
> *Hotelling (1936)*

The correlation coefficient measures the linear relation between scalar X and scalar Y. How do you measure the linear relation between *vector X* and *vector Y*? Canonical Correlation Analysis (CCA) provides a way. Specifically, CCA finds a linear combination of X, and a (separate) linear combination of Y, that maximizes the correlation. The resulting maximized correlation is known as a *canonical correlation* and generalizes the correlation coefficient to vector X and vector Y. The associated linear combinations define *canonical components* that describe the spatial and temporal structure of the correlated variables. CCA does more than find the two components with maximal correlation – it decomposes two sets of variables into an ordered sequence of component pairs ordered such that the first pair has maximum correlation, the second has maximum correlation subject to being uncorrelated with the first, and so on. The entire decomposition can be derived from a Singular Value Decomposition of a suitable matrix of data values. This decomposition provides an insightful interpretation of a multivariate regression model between X and Y and is very useful when only a few components explain most of the linear relation. If Y is a scalar, then CCA is equivalent to linear regression, and testing the significance of a canonical correlation is equivalent to a field significance test. When the dimension of the X and Y vectors is too large, overfitting becomes a problem. In this case, CCA is often computed using a few principal components of X and Y. The criterion for selecting the number of principal components is not standard. The Mutual Information Criterion (MIC) introduced in Chapter 14 is used in this chapter.

---

[1] Reprinted by permission from Oxford University Press, *Biometrika*, Hotelling (1936).

15.1 The Problem

Suppose you want to predict a set of Y-variables from a set of X-variables. To explore this possibility, you compute the correlation between every pair of Xs and Ys. Unfortunately, all pairwise correlations are small (e.g., less than 0.01). Does this mean that the Xs cannot give good predictions of the Ys? No!

As a simple numerical example, suppose there are only two variables for X and Y, so $\mathbf{x} = (x_1, x_2)^T$ and $\mathbf{y} = (y_1, y_2)^T$. Suppose the corresponding covariance matrices are

$$\mathrm{cov}[\mathbf{x}, \mathbf{x}] = \begin{pmatrix} 1 & 0 \\ 0 & 1 \end{pmatrix} \tag{15.1}$$

$$\mathrm{cov}[\mathbf{y}, \mathbf{y}] = \begin{pmatrix} 0.50005 & -0.49995 \\ -0.49995 & 0.50005 \end{pmatrix} \tag{15.2}$$

$$\mathrm{cov}[\mathbf{y}, \mathbf{x}] = \begin{pmatrix} 0.005 & 0.005 \\ 0.005 & 0.005 \end{pmatrix}. \tag{15.3}$$

Therefore, the pairwise correlations are

$$\mathrm{cor}[y_i, x_j] = \frac{\mathrm{cov}[y_i, x_j]}{\sqrt{\mathrm{var}[y_i]\,\mathrm{var}[x_j]}} = \frac{0.005}{\sqrt{0.5}} \approx 0.007 \quad \text{for all } i, j. \tag{15.4}$$

Clearly, all pairwise correlations are small.

Now, instead of considering variables *individually*, consider the sums $y_1 + y_2$ and $x_1 + x_2$. These sums can be denoted by $\mathbf{j}^T \mathbf{y}$ and $\mathbf{j}^T \mathbf{x}$, where $\mathbf{j}$ is a 2-dimensional vector of ones. From (7.73) we know $\mathrm{cov}[\mathbf{j}^T \mathbf{y}, \mathbf{j}^T \mathbf{x}] = \mathbf{j}^T \mathrm{cov}[\mathbf{y}, \mathbf{x}]\mathbf{j}$ and $\mathrm{var}[\mathbf{j}^T \mathbf{x}] = \mathbf{j}^T \mathrm{cov}[\mathbf{x}, \mathbf{x}]\mathbf{j}$, which amounts to summing all matrix elements. Therefore,

$$\mathrm{var}[\mathbf{j}^T \mathbf{x}] = 2, \quad \mathrm{var}[\mathbf{j}^T \mathbf{y}] = 0.0002, \quad \mathrm{cov}[\mathbf{j}^T \mathbf{y}, \mathbf{j}^T \mathbf{x}] = 0.02.$$

It follows that

$$\mathrm{cor}[\mathbf{j}^T \mathbf{x}, \mathbf{j}^T \mathbf{x}] = \frac{0.02}{\sqrt{2 * 0.0002}} = 1. \tag{15.5}$$

Thus, despite small pair-wise correlations, certain *linear combinations* of variables are *perfectly correlated*. What is going on is that there is a component of $\mathbf{y}$ that is linearly related with $\mathbf{x}$, but it has relatively small variance and is obscured by another component of $\mathbf{y}$ that has no dependence on $\mathbf{x}$. By taking a particular linear combination of the elements of $\mathbf{y}$, the component with high variance is filtered from $\mathbf{y}$, which exposes the linear relation.

In climate applications, it is unrealistic to expect perfect correlations, but certain linear combinations still may have high correlations with linear combinations of other variables. If there are thousands of variables in X and Y, how do you find the specific linear combinations that maximize the correlation? *Canonical Correlation Analysis (CCA)* is a procedure for finding such combinations.

15.2 Summary and Illustration of Canonical Correlation Analysis

In 1899, India experienced a severe drought and famine that killed over 4.5 million people. A contributing factor was that rainfall from the annual monsoon was below normal. Because "monsoon failures" were often associated with famines, great interest was attached to predicting Indian monsoon rainfall a few months in advance. Gilbert Walker became director of the Indian Meteorological Department in 1904 and devoted the next two decades to this problem. As G. I Taylor explains in his biography of Walker (Taylor, 1962):

When Walker started work he realized at once that there was little scientific basis for the production of seasonal forecasts. His whole previous scientific career had been devoted to problems in which results could be produced by mathematical analysis from an assured basis of simple principles. It is a sign of the high quality and flexibility of his mind that he could realize that all the skill he had acquired in the past would be of little use to him in his new situation. He decided that since he saw no prospect of treating the weather as a subject to which mathematical reasoning from well established premises could be applied, he would collect all the relevant information which had been recorded and treat it statistically without attempting to trace physical connexions between cause and effect.[2]

By the 1920s, Gilbert Walker and others had established three "big swayings or oscillations" in the climate system, one of which is the *Southern Oscillation* (Walker, 1928). Walker and Bliss (1932) presented evidence that the Southern Oscillation influenced temperature and precipitation around the world. This was one of the first examples of a *teleconnection*. A teleconnection refers to climate anomalies that are related to each other over long distances (e.g., thousands of kilometers). In the 1960s, it was realized that the Southern Oscillation is connected to El Niño, leading to the name El Niño-Southern Oscillation, or *ENSO*. Because ENSO persists from month to month, teleconnections with ENSO provide a scientific basis for predicting shifts in weather a month or two in advance.

How can teleconnections be identified from data? Walker's approach was basically a two-step procedure: first identify a "big swaying or oscillation," then compute the correlation between it and other variables. Walker's first step was closely related to PCA (see Chapter 12), although PCA had not been invented during this

---

[2] Reprinted by permission from The Royal Society, *Biographical memoirs of Fellows of the Royal Society*, Taylor (1962).

period of research. This strategy can work if the teleconnection happens to be a principal component. But, if the teleconnection is not a pure principal component, then this approach is not guaranteed to give the best teleconnection. After all, performing PCA on one variable in isolation from the other cannot be expected to give the best *predictive* relation. To best identify a teleconnection, both variables need to inform the procedure from the beginning.

Canonical Correlation Analysis is a procedure for finding highly correlated components between two data sets. The procedure was invented by Hotelling (1936) (who also invented PCA; Hotelling, 1933). The procedure has an enormous number of applications in such diverse fields as economics, psychology, sociology, medicine. Here, we illustrate its use for identifying ENSO teleconnections. Accordingly, one group of variables will be sea surface temperatures (SSTs) in the Pacific, which is the center of action for ENSO. The second group of variables will be land temperatures over North America. Brief details of the data sets are listed next.

- SST data: Extended Reconstruction SST of Huang et al. (2017), called ERSSTv5.[3]
- U.S. Temperature data: the GHCN-CAMS data of Fan and van den Dool (2008).[4]
- Time Period: 1950–2017
- Time Means: December-January-February (DJF)
- Land domain: 130W–50W, 30N–55N (the contiguous United States plus part of Canada)
- SST domain: Pacific Ocean within 55E–60W, 30S–60N
- Pre-processing: the mean and linear trend from each grid point is regressed out.

CCA finds a linear combination of X-variables, and a (separate) linear combination of Y-variables, such that the two combinations have maximum correlation. For our data, there are 2,976 grid points for SST and 4,445 grid points for US temperature, but only 68 years of data. Because the sample size is much smaller than the number of variables, the maximization problem is under-determined, and linear combinations with exactly unit correlation can be found. Such large correlations are unrealistic – computing the same linear combinations in independent data will invariably produce poor correlations. The problem is overfitting (see Chapter 10). In climate studies, the standard approach to mitigate overfitting is to project the data onto the leading principal components, and then perform the analysis using just those PCs. This approach assumes that the maximally correlated component is a *linear combination* of leading principal components. For reasons discussed in Section 15.7, we choose four PCs for SST and three PCs of US temperature.

[3] www.ncdc.noaa.gov/data-access/marineocean-data/extended-reconstructed-sea-surface-temperature-ersst-v5
[4] www.esrl.noaa.gov/psd/data/gridded/data.ghcncams.html

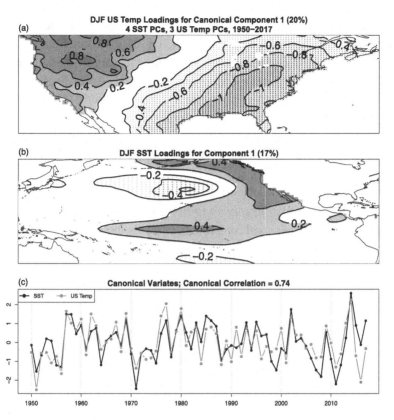

Figure 15.1 Leading Loading Vectors for DJF US temperature (a) and Pacific SST (b), and their variates (c), during 1950–2017, based on four PC time series of SST and three PC time series of temperature. The percent of variance explained by each pattern is indicated in the respective titles.

The result of applying CCA to this data set is illustrated in Figure 15.1. The top two panels show the *canonical loading vectors*, and the bottom panel shows corresponding time series, known as *canonical variates*. Multiplying one of the loading vectors by its variate produces a sequence of patterns, one for each year, known as a *canonical component*. This component best approximates the data in some objective sense (see Section 15.3). The correlation between the two variates is 0.74 and known as the *canonical correlation*. By construction, this correlation is the largest of any linear combination of these PCs. The SST canonical loading vector has loadings concentrated in the central equatorial Pacific. The corresponding loading vector of US temperature is a dipole structure (Figure 15.1a). From this result, we learn that when the SST variate is in the positive phase (e.g., central Pacific SSTs are warm), the northwest United States experiences anomalously warm temperature and the southeast United States experiences anomalously cool temperatures.

This result can be used to make "predictions" of one field given the other. To illustrate, suppose the SST variate equals one. The regression coefficient for predicting Y from X, or X from Y, is the correlation coefficient, namely 0.74. Therefore, the predicted amplitude for US temperature variate is 0.74. Accordingly, we multiply 0.74 by the US temperature loading vector. Over the Alabama-Mississipi area, the loadings are around -1, hence we predict about a -0.74 degree C cooling in this area *due to this canonical component*. In contrast, over the Wyoming-Idaho area, the loadings are around 0.8, hence we predict about $0.74 * 0.8 \approx 0.6$ degree C warming in this area due to this component.

After finding this component, we may seek the linear combination of variables that maximizes the correlation *subject to being uncorrelated in time with the first*. This yields the *second canonical component*, which is shown in Figure 15.2. Repeating this procedure yields subsequent canonical components, ordered by correlation

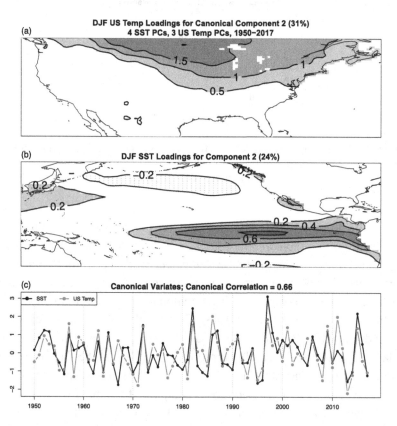

Figure 15.2 Same as Figure 15.1, but for the *second* canonical component

from largest to smallest. In practice, it is not necessary to derive components itera-
tively as just described. Instead, these components can be derived all at once through
a Singular Vector Decomposition of a suitable matrix. Summing the resulting canonical components recovers the data that went into the analysis, demonstrating that the
canonical components *decompose* the data sets. This decomposition is analogous to
that in Principal Component Analysis, except that instead of decomposing one data
set into components ordered by explained variance, CCA decomposes *two* data sets
into pairs of components ordered by their correlation. Canonical components are
orthogonal in time (that is, uncorrelated), but not orthogonal in space.

The percent of variance explained by each canonical component is indicated in
the titles of Figures 15.1 and 15.2. In this example, the second component explains
more variance than the first. This is not an error: CCA ranks components by correlation, not variance. In this example, the only guarantee on the variance explained by
the first component is that it must be at least as large as that of the last PC included in
the CCA. Finally, it should be recognized that the explained variance quantifies the
total variance of the components, not the *predictable variance*, which is a fraction of
the total. The predictable variance is the total variance multiplied by the square of the
canonical correlation. Thus, the leading US temperature pattern explains 20% of the
total variance, but only $0.74^2 \approx 0.55$ of this variance is predictable. The total variance that is predictable by the two components is $0.2 * 0.74^2 + 0.31 * 0.66^2 \approx 24\%$.

The ENSO correlation map in Figure 13.1 differs from the loading vectors in
Figures 15.1 and 15.2. One reason for this is that Figure 13.1 shows a *correlation
map* whereas the loading vectors are *regression maps*. Correlation and regression
coefficients differ by a factor that depends on the ratio of standard deviations, hence
the spatial structures differ because variance depends on location. Also, in terms of
spatial structure, the second loading vector more closely resembles the leading EOF
(compare Figures 15.1b and 12.1). A more direct way to relate these results to ENSO
is to correlate the canonical variates with the Niño 3.4 index. The strongest correlation occurs for the second SST variate ($\rho \approx 0.70$) while the weakest correlation
occurs for the first US temperature variate ($\rho \approx 0.32$). The fact that the second rather
than the first variate is more strongly correlated with an ENSO index represents no
contradiction: the goal of CCA is to find teleconnections, which depends on both
SST and US temperature, whereas the goal of the Niño 3.4 index or the leading EOF
of SST is to diagnose ENSO variability and depends only on SST.

Can we reject the null hypothesis that SST and US temperature are independent?
This question cannot be answered using the correlation test discussed in Chap-
ter 2 because that test does not account for the maximization process used in CCA.
Instead, we need to know the sampling distribution of the canonical correlations.

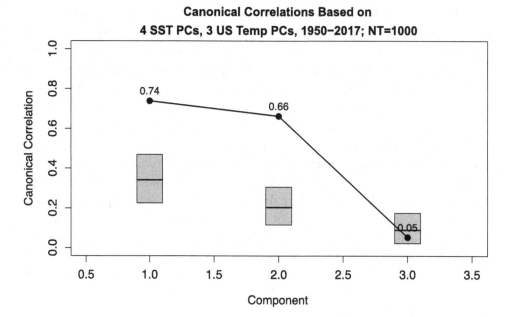

Figure 15.3 Canonical correlations for DJF Pacific SST and DJF US temperature during 1950–2017, based on four PCs of SST and three PCs of US temperature (filled circles connected by lines). The boxes show results from 10,000 applications of CCA on random data. Specifically, the lower and upper edges of each box show the 5th and 95th percentiles from the Monte Carlo simulations. The horizontal line in the middle of each box shows the median.

This distribution can be estimated from Monte Carlo techniques. Specifically, we use a random number generator to create synthetic PC time series, perform CCA, and then repeat many times. By construction, the random numbers come from a population in which the null hypothesis is true. Further details are discussed in Section 15.8. The canonical correlations computed from Monte Carlo simulations are compared to the canonical correlations derived from the data set in Figure 15.3. The figure shows that the first *two* canonical correlations computed from observational data exceed the 95th percentiles from the Monte Carlo simulations. For reasons discussed in Section 15.8, the interpretation of the second correlation is not straightforward, but nevertheless further analysis (and Monte Carlo simulation) reveals that the second canonical correlation is significant as well. Accordingly, we reject the null hypothesis of independence and conclude that *two* significant linear relations exist. The fact that CCA identifies *two* canonical components suggests that ENSO teleconnections between SST and US temperature are more complicated than a single pattern and therefore cannot be captured by a single index.

15.3 Population Canonical Correlation Analysis

As mentioned in the previous section, climate applications of CCA involve project-ing data onto the leading principal components. There are some complexities with this approach, particularly with transforming results from one space to another, but these have nothing to do with CCA itself. In this section, we derive the popula-tion version of CCA, which avoids these complexities. The sample version, which involves dimension reduction and SVD, is discussed in Section 15.6.

In most climate applications, CCA is applied to time-varying fields. Accordingly, CCA will be described in the context of analyzing two time-dependent fields $\mathbf{x}(t)$ and $\mathbf{y}(t)$, where t is a time index. Note that the fields need not be simultaneous – for instance, one field may be separated from the other by a fixed lag time τ, say $\mathbf{y}(t)$ could be $\mathbf{x}(t + \tau)$. Each field is described by a set of scalar time series,

$$\mathbf{x}(t) = \begin{pmatrix} x_1(t) \\ x_2(t) \\ \vdots \\ x_{S_X}(t) \end{pmatrix} \quad \text{and} \quad \mathbf{y}(t) = \begin{pmatrix} y_1(t) \\ y_2(t) \\ \vdots \\ y_{S_Y}(t) \end{pmatrix}, \quad (15.6)$$

where S_X and S_Y are the dimensions of $\mathbf{x}(t)$ and $\mathbf{y}(t)$, respectively. S_X and S_Y correspond to the number of grid points in each field, or to the number of prin-cipal components, as discussed in Section 15.6. The goal is to determine a linear combination of the elements of $\mathbf{x}(t)$, a linear combination of the elements of $\mathbf{y}(t)$, such that the resulting combinations have maximum correlation. This goal assumes both vectors are defined at the same number of time steps $t = 1, \ldots, N$. A linear combination of the elements of $\mathbf{x}(t)$ is

$$r_X(t) = q_1 x_1(t) + q_2 x_2(t) + \cdots + q_{S_X} x_{S_X}(t), \quad (15.7)$$

where $q_1, \ldots, q_{S_X}$ are weighting coefficients. Note that the linear combination is a *scalar* $r_X(t)$ that varies in time. This linear combination can also be written as

$$r_X(t) = \mathbf{q}_X^T \mathbf{x}(t), \quad (15.8)$$

where $\mathbf{q}_X$ is a vector containing the weighting coefficients. Similarly,

$$r_Y(t) = \mathbf{q}_Y^T \mathbf{y}(t). \quad (15.9)$$

The vectors $\mathbf{q}_X$ and $\mathbf{q}_Y$ are known as *projection vectors*, and the associated linear combinations r_X and r_Y are known as *variates*. We want the projection vectors $\mathbf{q}_X$ and $\mathbf{q}_Y$ that maximize the correlation between r_X and r_Y.

Define the Correlation

By definition, the correlation between r_X and r_Y is

$$\rho = \text{cor}[r_X, r_Y] = \frac{\text{cov}[r_X, r_Y]}{\sqrt{\text{var}[r_X]\,\text{var}[r_Y]}}. \tag{15.10}$$

Using (15.8) and (15.9), the covariance between r_X and r_Y can be written as

$$\text{cov}[r_X, r_Y] = \text{cov}[\mathbf{q}_X^T\mathbf{x}, \mathbf{q}_Y^T\mathbf{y}] = \mathbf{q}_X^T\,\text{cov}[\mathbf{x}, \mathbf{y}]\mathbf{q}_Y = \mathbf{q}_X^T\mathbf{\Sigma}_{XY}\mathbf{q}_Y, \tag{15.11}$$

where the fact that the **q**-vectors are not random has been used, and we define

$$\mathbf{\Sigma}_{XY} = \mathbb{E}[(\mathbf{x} - \mathbb{E}[\mathbf{x}])\,(\mathbf{y} - \mathbb{E}[\mathbf{y}])^T]. \tag{15.12}$$

Similarly, we have

$$\text{var}[r_X] = \mathbf{q}_X^T\mathbf{\Sigma}_{XX}\mathbf{q}_X \quad\text{and}\quad \text{var}[r_Y] = \mathbf{q}_Y^T\mathbf{\Sigma}_{YY}\mathbf{q}_Y, \tag{15.13}$$

where $\mathbf{\Sigma}_{XX}$ is defined in (7.56), $\mathbf{\Sigma}_{YY}$ is defined by obvious analogy. Substituting these expressions into the correlation (15.10) gives

$$\rho = \frac{\mathbf{q}_X^T\mathbf{\Sigma}_{XY}\mathbf{q}_Y}{\sqrt{\left(\mathbf{q}_X^T\mathbf{\Sigma}_{XX}\mathbf{q}_X\right)\left(\mathbf{q}_Y^T\mathbf{\Sigma}_{YY}\mathbf{q}_Y\right)}}. \tag{15.14}$$

Maximize the Correlation

An extremum of ρ is found by solving $\partial\rho/\partial\mathbf{q}_X = \mathbf{0}$ and $\partial\rho/\partial\mathbf{q}_Y = \mathbf{0}$. Using results from (A.3), these derivatives yield

$$\frac{\partial\rho}{\partial\mathbf{q}_X} = \frac{\mathbf{\Sigma}_{XY}\mathbf{q}_Y}{\sqrt{\left(\mathbf{q}_X^T\mathbf{\Sigma}_{XX}\mathbf{q}_X\right)\left(\mathbf{q}_Y^T\mathbf{\Sigma}_{YY}\mathbf{q}_Y\right)}} - \rho\frac{\mathbf{\Sigma}_{XX}\mathbf{q}_X}{\mathbf{q}_X^T\mathbf{\Sigma}_{XX}\mathbf{q}_X} = \mathbf{0}, \tag{15.15}$$

$$\frac{\partial\rho}{\partial\mathbf{q}_Y} = \frac{\mathbf{\Sigma}_{YX}\mathbf{q}_X}{\sqrt{\left(\mathbf{q}_Y^T\mathbf{\Sigma}_{YY}\mathbf{q}_Y\right)\left(\mathbf{q}_X^T\mathbf{\Sigma}_{XX}\mathbf{q}_X\right)}} - \rho\frac{\mathbf{\Sigma}_{YY}\mathbf{q}_Y}{\mathbf{q}_Y^T\mathbf{\Sigma}_{YY}\mathbf{q}_Y} = \mathbf{0}, \tag{15.16}$$

Solving $\mathbf{q}_Y$ from (15.16) gives

$$\mathbf{q}_Y = \frac{1}{\rho}\sqrt{\frac{\mathbf{q}_Y^T\mathbf{\Sigma}_{YY}\mathbf{q}_Y}{\mathbf{q}_X^T\mathbf{\Sigma}_{XX}\mathbf{q}_X}}\,\mathbf{\Sigma}_{YY}^{-1}\mathbf{\Sigma}_{YX}\mathbf{q}_X. \tag{15.17}$$

Substituting this into (15.15) and re-arranging gives

$$\mathbf{\Sigma}_{XY}\mathbf{\Sigma}_{YY}^{-1}\mathbf{\Sigma}_{YX}\mathbf{q}_X = \rho^2\mathbf{\Sigma}_{XX}\mathbf{q}_X. \tag{15.18}$$

Alternatively, $\mathbf{q}_X$ could have been solved from (15.15) and substituted into (15.16), giving

$$\mathbf{\Sigma}_{YX}\mathbf{\Sigma}_{XX}^{-1}\mathbf{\Sigma}_{XY}\mathbf{q}_Y = \rho^2\mathbf{\Sigma}_{YY}\mathbf{q}_Y. \tag{15.19}$$

Solution

Equations (15.18) and (15.19) are *generalized eigenvalue problems*. Solution of such problems are discussed in Section A.2 and will be merely summarized here. First for each nontrivial $\mathbf{q}_X$ satisfying (15.18), there exists a corresponding eigenvalue ρ^2, which is real and nonnegative. The (positive) square root of the eigenvalue is known as the *canonical correlation* ρ. The maximum number of nonzero canonical correlations is $\min[S_X, S_Y]$. It is convention to order canonical correlations from largest to smallest, hence $\rho_1 \geq \rho_2 \geq \cdots$. For each canonical correlation $\rho_1, \rho_2, \ldots$, there exists a corresponding eigenvector $\mathbf{q}_{X,1}, \mathbf{q}_{X,2}, \ldots$ It can be shown that the nonzero eigenvalues from (15.18) and (15.19) are identical. Therefore, it is not necessary to solve both eigenvalue problems; rather, one set of eigenvectors may be obtained from (15.18), and the other may be obtained from (15.17), yielding $\mathbf{q}_{Y,1}, \mathbf{q}_{Y,2}, \ldots$ The eigenvectors $\mathbf{q}_{X,i}$ and $\mathbf{q}_{Y,i}$ are known as *canonical projection vectors*. Substituting these vectors into (15.8) and (15.9) yields the *canonical variates* $r_{X,i}$ and $r_{Y,i}$. The correlation between canonical variates,

$$\rho_i = \mathrm{cor}\left[r_{X,i}, r_{Y,i}\right], \tag{15.20}$$

is precisely equal to the i'th canonical correlation ρ_i obtained by solving the eigenvalue problem (15.18). The maximum correlation of any linear combination of $\mathbf{x}$ and of $\mathbf{y}$ is the first canonical correlation ρ_1. A proof that the leading component actually maximizes the correlation, and that subsequent components maximize correlation after regressing out preceding components, is given in Section 15.9.

Canonical Loading Vectors

These results can be used to *decompose* $\mathbf{x}$ and $\mathbf{y}$ into a complete set of component pairs ordered such that the first pair has maximum correlation, the second has maximum correlation subject to being uncorrelated with the first, and so on. To show this, it is convenient to collect the canonical projection vectors into matrices

$$\mathbf{Q}_X = \begin{bmatrix} \mathbf{q}_{X,1} & \mathbf{q}_{X,2} & \cdots & \mathbf{q}_{X,S_X} \end{bmatrix} \tag{15.21}$$

$$\mathbf{Q}_Y = \begin{bmatrix} \mathbf{q}_{Y,1} & \mathbf{q}_{Y,2} & \cdots & \mathbf{q}_{Y,S_Y} \end{bmatrix}, \tag{15.22}$$

and to collect the variates into vectors

$$\mathbf{r}_X = \begin{bmatrix} r_{X,1} & r_{X,2} & \cdots & r_{X,S_X} \end{bmatrix}^T \tag{15.23}$$

$$\mathbf{r}_Y = \begin{bmatrix} r_{Y,1} & r_{Y,2} & \cdots & r_{Y,S_Y} \end{bmatrix}^T. \tag{15.24}$$

When analyzing finite samples, a particular variate $r_{X,i}$ will be identified with a sample vector of length N, but here it is identified as a random variable represented by a scalar. The relations (15.8) and (15.9) can then be expressed equivalently as

$$\mathbf{Q}_X^T \mathbf{x} = \mathbf{r}_X \quad \text{and} \quad \mathbf{Q}_Y^T \mathbf{y} = \mathbf{r}_Y. \tag{15.25}$$

Inverting the **Q**-matrices yields

$$\mathbf{x} = \mathbf{P}_X \mathbf{r}_X \quad \text{and} \quad \mathbf{y} = \mathbf{P}_Y \mathbf{r}_Y, \tag{15.26}$$

where the columns of $\mathbf{P}_X$ and $\mathbf{P}_Y$ are the *canonical loading vectors* derived from

$$\mathbf{P}_X = \mathbf{Q}_X^{-1T} \quad \text{and} \quad \mathbf{P}_Y = \mathbf{Q}_T^{-1T}. \tag{15.27}$$

Equation (15.26) can be written equivalently as

$$\mathbf{x} = \mathbf{p}_{X,1} r_{X,1} + \mathbf{p}_{X,2} r_{X,2} + \cdots + \mathbf{p}_{X,S_X} r_{X,S_X} \tag{15.28}$$

$$\mathbf{y} = \mathbf{p}_{Y,1} r_{Y,1} + \mathbf{p}_{Y,2} r_{Y,2} + \cdots + \mathbf{p}_{Y,S_X} r_{Y,S_Y}. \tag{15.29}$$

The product $\mathbf{p}_{X,i} r_{X,i}$ is known as the i'th *canonical component* of $\mathbf{x}$. Similarly, $\mathbf{p}_{Y,i} r_{Y,i}$ is known as the i'th *canonical component* of $\mathbf{y}$. These equations show that $\mathbf{x}$ and $\mathbf{y}$ can be decomposed in terms of canonical components.

Normalization of Projection Vectors

Recall that eigenvectors are unique up to a multiplicative factor. This nonuniqueness is a consequence of the fact that the correlation (15.14) is invariant to re-scaling of the projection vectors (that is, the correlation for $\mathbf{q}_X$ is the same as that for $\alpha \mathbf{q}_X$ for any $\alpha > 0$, because the parameter α cancels in (15.14)). Without loss of generality, we choose this factor to normalize the variates to unit variance:

$$\text{var}[r_{X,i}] = 1 \quad \text{and} \quad \text{var}[r_{Y,i}] = 1. \tag{15.30}$$

As indicated in (15.13), this normalization implies that the projection vectors satisfy

$$\mathbf{q}_{X,i}^T \Sigma_{XX} \mathbf{q}_{X,i} = 1 \quad \text{and} \quad \mathbf{q}_{Y,i}^T \Sigma_{YY} \mathbf{q}_{Y,i} = 1. \tag{15.31}$$

With these normalizations, the square root term in (15.17) becomes one. Furthermore, multiplying on the left of (15.17) by $\mathbf{q}_Y^T \Sigma_{YY}$ and taking the transpose of the result yields

$$\mathbf{q}_{X,i}^T \Sigma_{XY} \mathbf{q}_{Y,i} = \rho_i. \tag{15.32}$$

Properties of the Projection Vectors

The eigenvectors from symmetric matrices possess orthogonality properties that have been discussed in Section A.2. These properties can be expressed in the present context as

$$\mathbf{Q}_X^T \Sigma_{XX} \mathbf{Q}_X = \mathbf{I} \quad \text{and} \quad \mathbf{Q}_Y^T \Sigma_{YY} \mathbf{Q}_Y = \mathbf{I}, \tag{15.33}$$

where the normalizations (15.31) have also been used. Also, the same steps used to derive (15.32) can be used to derive

$$\mathbf{Q}_Y^T \Sigma_{YX} \mathbf{Q}_X = \boldsymbol{\Lambda}, \tag{15.34}$$

where Λ is a diagonal matrix whose diagonal elements are the canonical correlations.

Variates are Uncorrelated in Three Different Ways
These identities for the projection vectors have implications for the canonical variates. Specifically, (15.33) implies that the covariance matrix for the variates r_X is

$$\text{cov}[r_X, r_X] = \text{cov}\left[Q_X^T x, Q_X^T x\right] = Q_X^T \text{cov}[x, x] Q_X = Q_X^T \Sigma_{XX} Q_X = I. \quad (15.35)$$

This identity summarizes two distinct properties. First, because the off-diagonal elements of I vanish, different canonical variates of x are uncorrelated. Second, because the diagonal elements of I equal one, each canonical variate is normalized to one. Similarly,

$$\text{cov}[r_Y, r_Y] = I. \quad (15.36)$$

Finally, the cross covariance matrix between the variates of x and y is

$$\text{cov}[r_Y, r_X] = \text{cov}\left[Q_Y^T y, Q_X^T x\right] = Q_Y^T \text{cov}[y, x] Q_X = \Lambda, \quad (15.37)$$

where (15.34) has been used. The off-diagonal elements of (15.35)–(15.37) can be expressed equivalently as

$$\text{cor}\left[r_{X,i}, r_{X,j}\right] = \text{cor}\left[r_{Y,i}, r_{Y,j}\right] = \text{cor}\left[r_{X,i}, r_{Y,j}\right] = 0 \quad \text{for } i \neq j. \quad (15.38)$$

That is, canonical variates are uncorrelated in three senses: each variate of x is uncorrelated with the other variates of x, each variate of y is uncorrelated with the other variates of y, and the i'th variate of x is uncorrelated with the j'th variate of y, where $i \neq j$.

15.4 Relation between CCA and Linear Regression

Surely CCA and linear regression are connected, but how? To answer this question, consider a linear prediction of y based on x, which has the form

$$\hat{y} = Ax, \quad (15.39)$$

where A is a constant matrix. We determine A by finding the matrix A that minimizes mean square error,

$$\text{MSE} = \mathbb{E}\left[\|y - Ax\|^2\right]. \quad (15.40)$$

For simplicity, assume x and y have zero mean. Using results from Section A.3, we obtain

$$\frac{\partial \text{MSE}}{\partial A} = 2\mathbb{E}\left[(y - Ax)x^T\right] = 2\left(\Sigma_{YX} - A\Sigma_{XX}\right). \quad (15.41)$$

Setting this to zero and solving for **A** yields

$$\mathbf{A} = \Sigma_{YX}\Sigma_{XX}^{-1}. \tag{15.42}$$

The covariance matrix Σ_{YX} is related to the canonical projection vectors through (15.34), which can be manipulated into the form

$$\Sigma_{YX} = \mathbf{Q}_Y^{T-1}\mathbf{\Lambda}\mathbf{Q}_X^{-1}. \tag{15.43}$$

Substituting this into (15.42) gives

$$\mathbf{A} = \mathbf{Q}_Y^{T-1}\mathbf{\Lambda}\mathbf{Q}_X^{-1}\Sigma_{XX}^{-1}.$$

It follows from (15.33) that $\mathbf{Q}_X^{-1}\Sigma_{XX}^{-1} = \mathbf{Q}_X^T$. Therefore, using (15.27),

$$\mathbf{A} = \mathbf{P}_Y\mathbf{\Lambda}\mathbf{Q}_X^T. \tag{15.44}$$

This identity shows that CCA *diagonalizes* the prediction operator from multivariate regression. To see this more explicitly, substitute (15.44) into (15.39) to obtain

$$\hat{\mathbf{y}} = \mathbf{P}_Y\mathbf{\Lambda}\mathbf{Q}_X^T\mathbf{x}. \tag{15.45}$$

This is the same prediction model as in (15.39), except expressed in terms of canonical components. Multiplying both sides of (15.45) by $\mathbf{Q}_Y^T$ and using definitions (15.25) yields

$$\hat{\mathbf{r}}_Y = \mathbf{\Lambda}\mathbf{r}_X. \tag{15.46}$$

Because $\mathbf{\Lambda}$ is diagonal, the canonical variates are *decoupled* and can be written as

$$\hat{r}_{Y,i} = \rho_i\, r_{X,i} \quad \text{for } i = 1, 2, \ldots, \min[S_X, S_Y]. \tag{15.47}$$

This result shows that the least squares prediction of the i'th canonical variate for **y** depends only on the i'th canonical variate for **x**, *with no cross terms*. Thus, predictions for canonical components are *decoupled* from each other. There is no difference between regression prediction of **y** based on **x**, and regression prediction of canonical components, provided all components are included. In essence, then, CCA decomposes the predictive relations between **x** and **y** in terms of patterns that can be studied pair-wise. Such pattern-based study is often more insightful than studying the prediction matrix **A** directly.

Because $\mathbf{\Lambda}$ is a diagonal matrix, the decomposition (15.45) leads to a simple interpretation of multivariate regression. Specifically, reading (15.45) from right to left corresponds to the following sequence of operations:

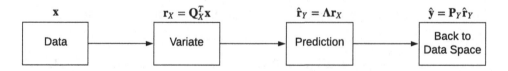

The diagonality of $\mathbf{\Lambda}$ implies that the prediction step is performed separately variate-by-variate, as in (15.47). This decomposition means that we can think of *multivariate* prediction as a set of *univariate* predictions. If only a few components of $\mathbf{y}$ are predictable, CCA will reveal this fact by producing only a few nonzero canonical correlations. In this case, the other components can be ignored, yielding a simple picture of the prediction process. This picture is often simpler than attempting to interpret the matrix $\mathbf{A}$ directly, which may be large-dimensional and not sparse. Although other matrix decompositions exist, CCA is special because it decomposes multivariate prediction into a set of independent univariate regressions ordered by correlation. This section has demonstrated this connection for population quantities. The analogous proof for sample quantities is presented in Section 15.6.

Given these connections, several consequences are immediate. First, if only one Y-variable exists, then $\mathbf{q}_X$ is merely the least squares regression coefficients for predicting Y given X, and the square of the canonical correlation is merely R^2 defined in (8.44). Second, Chapter 13 showed that univariate regression can be used to assess *field significance*. Thus the field significance test is a special case of testing the significance of a canonical correlation. Finally, if only one Y-variable and one X-variable exists, then their correlation is merely the canonical correlation ρ_1.

In Chapter 18, we discuss a procedure known as Predictable Component Analysis for finding the most predictable components of an ensemble forecast. If the ensemble forecast is generated by a linear regression model of the form

$$\mathbf{y} = \mathbf{A}\mathbf{x} + \boldsymbol{\epsilon}, \tag{15.48}$$

then the most predictable components are the canonical components.

15.5 Invariance to Affine Transformation

Canonical correlations satisfy a remarkable property. Recall that the correlation coefficient is invariant to re-scaling and additive constants of the variables. More precisely, the correlation satisfies the invariance property

$$\mathrm{cor}[X, Y] = \mathrm{cor}\left[(aX + b), (cY + d)\right], \tag{15.49}$$

where a, b, c, d are arbitrary constants with $ac > 0$. Canonical correlations generalize this invariance to a much wider class of transformations. Specifically, canonical correlations are invariant to *invertible affine transformations*. An affine transformation is of the form

$$\mathbf{x}' = \mathbf{L}_X\mathbf{x} + \mathbf{c}_X \quad \text{and} \quad \mathbf{y}' = \mathbf{L}_Y\mathbf{y} + \mathbf{c}_Y, \tag{15.50}$$

where $\mathbf{L}_X$ and $\mathbf{L}_Y$ are matrices and $\mathbf{c}_X$ and $\mathbf{c}_Y$ are arbitrary constant vectors. An *invertible* affine transformation is one in which $\mathbf{L}_X$ and $\mathbf{L}_Y$ are *nonsingular* (that is, their inverses exist). Such transformations are said to be invertible because the original vectors $(\mathbf{x}, \mathbf{y})$ can be recovered from transformed vectors $(\mathbf{x}', \mathbf{y}')$ by reversing the transformation.

To prove this invariance property, we invoke Theorem 7.6, which implies that covariance matrices for $(\mathbf{x}', \mathbf{y}')$ are related to those for $(\mathbf{x}, \mathbf{y})$ as

$$\boldsymbol{\Sigma}_{X'Y'} = \mathbf{L}_X \boldsymbol{\Sigma}_{XY} \mathbf{L}_Y^T, \quad \boldsymbol{\Sigma}_{X'} = \mathbf{L}_X \boldsymbol{\Sigma}_{XX} \mathbf{L}_X^T, \quad \text{and} \quad \boldsymbol{\Sigma}_{Y'} = \mathbf{L}_Y \boldsymbol{\Sigma}_{YY} \mathbf{L}_Y^T. \quad (15.51)$$

Therefore, the generalized eigenvalue problem (15.18) for the transformed variables is

$$\boldsymbol{\Sigma}_{X'Y} \boldsymbol{\Sigma}_{Y'Y'}^{-1} \boldsymbol{\Sigma}_{Y'X'} \mathbf{q}'_X = \rho^2 \boldsymbol{\Sigma}_{X'X'} \mathbf{q}'_X \quad (15.52)$$

$$\mathbf{L}_X \boldsymbol{\Sigma}_{XY} \mathbf{L}_Y^T \left(\mathbf{L}_Y \boldsymbol{\Sigma}_{YY} \mathbf{L}_Y^T \right)^{-1} \mathbf{L}_Y \boldsymbol{\Sigma}_{YX} \mathbf{L}_X^T \mathbf{q}'_X = \rho^2 \mathbf{L}_X \boldsymbol{\Sigma}_{XX} \mathbf{L}_X^T \mathbf{q}'_X$$

$$\mathbf{L}_X \boldsymbol{\Sigma}_{XY} \boldsymbol{\Sigma}_{YY}^{-1} \boldsymbol{\Sigma}_{YX} \mathbf{L}_X^T \mathbf{q}'_X = \rho^2 \mathbf{L}_X \boldsymbol{\Sigma}_{XX} \mathbf{L}_X^T \mathbf{q}'_X$$

$$\boldsymbol{\Sigma}_{XY} \boldsymbol{\Sigma}_{YY}^{-1} \boldsymbol{\Sigma}_{YX} \mathbf{L}_X^T \mathbf{q}'_X = \rho^2 \boldsymbol{\Sigma}_{XX} \mathbf{L}_X^T \mathbf{q}'_X$$

$$\boldsymbol{\Sigma}_{XY} \boldsymbol{\Sigma}_{YY}^{-1} \boldsymbol{\Sigma}_{YX} \mathbf{q}_X = \rho^2 \boldsymbol{\Sigma}_{XX} \mathbf{q}_X, \quad (15.53)$$

where $\mathbf{q}'_X$ is the projection vector for $\mathbf{x}'$, and we have defined the variable

$$\mathbf{q}_X = \mathbf{L}_X^T \mathbf{q}'_X. \quad (15.54)$$

Importantly, all the algebraic steps from (15.52) to (15.53) are reversible and *do not alter the canonical correlation ρ*. Therefore, the eigenvalues of (15.52) are identical to those of (15.53), which proves the invariance property.

Because of this invariance, variables with different units and different natural variances can be included in the same vector without normalizing the variables, since such normalizations are special cases of linear transformations.

In climate applications, data typically are projected onto a few principal components. Such transformations are *not invertible* because information is lost when other principal components are discarded. In general, noninvertible transformations tend to *reduce* the leading canonical correlation because the optimization is performed in a space of smaller dimensional. However, once the principal components are selected, then invertible transformations *within that space* preserve the canonical correlations.

15.6 Solving CCA Using the Singular Value Decomposition

To perform CCA, a natural idea is to replace population matrices with sample matrices and then solve eigenvalue problems (15.18) and (15.19). While this approach works, there exists an alternative but equivalent solution based on the Singular Value Decomposition (SVD) that is more efficient and more insightful because it clarifies the orthogonality properties of the solution in a way that makes

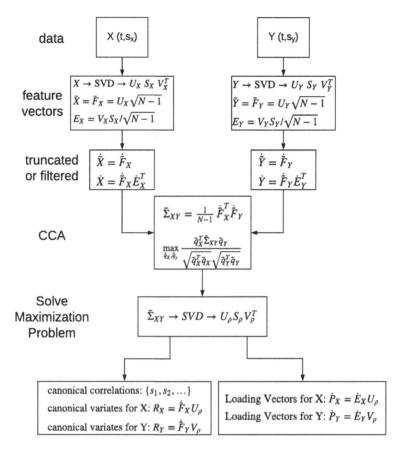

Figure 15.4 A flow chart of Canonical Correlation Analysis using Singular Value Decompositions.

them seem natural and less mysterious. This procedure also provides a natural way to reduce the dimensionality of the problem, which is important to mitigate overfitting. This procedure is discussed later in this section. The procedure may seem intricate because two data sets and *three* SVDs are involved, so to facilitate understanding we provide a flowchart in Figure 15.4 and an equation summary in Table 15.1.

Data Matrices

In practice, the available data are discretized in both space and time. Let N denote the number of time steps, and let S_X and S_Y denote the number of spatial elements in **x** and **y**, respectively. The data sets can be collected into *data matrices* as

$$\begin{array}{ccc} \mathbf{X} & \text{and} & \mathbf{Y} \\ N \times S_X & & N \times S_Y \end{array}, \tag{15.55}$$

Table 15.1. *Summary of Equations for Canonical Correlation Analysis.*

N	Number of independent samples
$\mathbf{X}$ and $\mathbf{Y}$	$[N \times S_X]$ and $[N \times S_Y]$ data matrices
$\mathbf{X} = \tilde{\mathbf{F}}_X \mathbf{E}_X^T$ and $\mathbf{Y} = \tilde{\mathbf{F}}_Y \mathbf{E}_Y^T$	data decomposition
$\dfrac{\tilde{\mathbf{F}}_X^T \tilde{\mathbf{F}}_X}{N-1} = \mathbf{I}$ and $\dfrac{\tilde{\mathbf{F}}_Y^T \tilde{\mathbf{F}}_Y}{N-1} = \mathbf{I}$	uncorrelated time series
$\dfrac{1}{N-1}\dot{\mathbf{F}}_X^T \dot{\mathbf{F}}_Y = \mathbf{U}_\rho \mathbf{S}_\rho \mathbf{V}_\rho^T$	SVD of filtered cross-cov. matrix
$\mathbf{S}_\rho$	canonical correlations
$\tilde{\mathbf{Q}}_X = \mathbf{U}_\rho$ and $\tilde{\mathbf{Q}}_Y = \mathbf{V}_\rho$	canonical projection vectors
$\tilde{\mathbf{R}}_X = \dot{\mathbf{F}}_X \tilde{\mathbf{Q}}_X$ and $\tilde{\mathbf{R}}_Y = \dot{\mathbf{F}}_Y \tilde{\mathbf{Q}}_Y$	canonical variates
$\dfrac{\tilde{\mathbf{R}}_X^T \tilde{\mathbf{R}}_X}{N-1} = \mathbf{I}$ and $\dfrac{\tilde{\mathbf{R}}_Y^T \tilde{\mathbf{R}}_Y}{N-1} = \mathbf{I}$	variates are uncorrelated, normalized
$\dfrac{\tilde{\mathbf{R}}_X^T \tilde{\mathbf{R}}_Y}{N-1} = \mathbf{S}_\rho$	canonical correlation matrix
$\mathbf{P}_X = \dfrac{\mathbf{X}^T \tilde{\mathbf{R}}_X}{N-1}$ and $\mathbf{P}_Y = \dfrac{\mathbf{Y}^T \tilde{\mathbf{R}}_Y}{N-1}$	canonical regression patterns
$\dfrac{\mathrm{tr}\left[\mathbf{X}\mathbf{W}_X\mathbf{X}^T\right]}{N-1}$ and $\dfrac{\mathrm{tr}\left[\mathbf{Y}\mathbf{W}_Y\mathbf{Y}^T\right]}{N-1}$	total variance of X and Y
$\mathbf{p}_{X,k}^T \mathbf{W}_X \mathbf{p}_{Y,k}$ and $\mathbf{p}_{Y,k}^T \mathbf{W}_Y \mathbf{p}_{Y,k}$	variance explained by k'th component
$\hat{\mathbf{Y}} = \tilde{\mathbf{R}}_X \Lambda \mathbf{P}_Y^T$	Least Squares prediction of $\dot{\mathbf{Y}}$
$\hat{\mathbf{X}} = \tilde{\mathbf{R}}_Y \Lambda \mathbf{P}_X^T$	Least Squares prediction of $\dot{\mathbf{X}}$

where the dimension of the matrices are indicated below each matrix. The data is assumed to be *centered* – that is, the time mean has been subtracted from each row separately, so each time series has zero mean. The time dimension N must be equal for the two data matrices, but the spatial dimensions usually differ. Incidentally, the random vectors $\mathbf{x}$ and $\mathbf{y}$ that appear in the population version are *rows* of $\mathbf{X}$ and $\mathbf{Y}$. As a result, some sample quantities will be the *transpose* of the population quantities of earlier sections.

Orthogonalizing the Time Series

First, the X-variables are transformed into an uncorrelated set of variables. To perform this transformation, we compute the SVD of $\mathbf{X}$,

$$\mathbf{X} = \mathbf{U}_X \mathbf{S}_X \mathbf{V}_X^T, \tag{15.56}$$

where $\mathbf{U}_X, \mathbf{S}_X, \mathbf{V}_X$ have their usual interpretations (see Theorem 12.1). $\mathbf{U}_X$ is a new data matrix whose columns are N-dimensional and mutually orthogonal. Multiplying it by $\mathbf{S}_X \mathbf{V}_X^T$ recovers the original data, so there is no loss of information. It will prove convenient to normalize the new variables to unit variance, hence we define

$$\tilde{\mathbf{F}}_X = \sqrt{N-1} \mathbf{U}_X \quad \text{and} \quad \mathbf{E}_X = \mathbf{V}_X \mathbf{S}_X^T / \sqrt{N-1}, \tag{15.57}$$

so that the time series are uncorrelated and normalized to unit variance,

$$\frac{1}{N-1} \tilde{\mathbf{F}}_X^T \tilde{\mathbf{F}}_X = \mathbf{I}, \tag{15.58}$$

and so that the original data matrix can be written as

$$\begin{array}{ccc} \mathbf{X} & = & \tilde{\mathbf{F}}_X & \mathbf{E}_X^T \\ N \times S_X & & N \times M_X & M_X \times S_X. \end{array} \tag{15.59}$$

In this decomposition, $\tilde{\mathbf{F}}_X$ contain uncorrelated time series and $\mathbf{E}_X$ contain spatial patterns. The purpose of this SVD is to obtain *uncorrelated time series* $\tilde{\mathbf{F}}_X$ that satisfies (15.58). Recall that principal component analysis also decomposes data into the form (15.59) with the time series satisfying (15.58), hence it could be used instead of the SVD (the two decompositions would in fact be the same if $\mathbf{W} = \mathbf{I}$ in 12.39). Similarly, we assume that $\mathbf{Y}$ is decomposed in the form

$$\begin{array}{ccc} \mathbf{Y} & = & \tilde{\mathbf{F}}_Y & \mathbf{E}_Y^T \\ N \times S_Y & & N \times M_Y & M_Y \times S_Y. \end{array} \tag{15.60}$$

where the columns of $\tilde{\mathbf{F}}_Y$ are uncorrelated and standardized,

$$\frac{1}{N-1} \tilde{\mathbf{F}}_Y^T \tilde{\mathbf{F}}_Y = \mathbf{I}. \tag{15.61}$$

Hereafter, we assume that data matrices $\mathbf{X}$ and $\mathbf{Y}$ have been decomposed into the forms (15.59) and (15.60), which, crucially, satisfy (15.58) and (15.61).

Dimension Reduction

Next, to mitigate overfitting, the number of components in the decomposition is truncated. Let T_X and T_Y denote the number of columns of $\tilde{\mathbf{F}}_X$ and $\tilde{\mathbf{F}}_Y$ that are retained, respectively. Then

$$\begin{array}{ccc} \dot{\mathbf{X}} & = & \dot{\mathbf{F}}_X & \dot{\mathbf{E}}_X^T \\ N \times S_X & & N \times T_X & T_X \times S_X, \end{array} \tag{15.62}$$

and

$$
\begin{array}{ccc}
\dot{\mathbf{Y}} & = & \dot{\tilde{\mathbf{F}}}_Y & \dot{\mathbf{E}}_Y^T \\
N \times S_Y & & N \times T_Y & T_Y \times S_Y,
\end{array}
\tag{15.63}
$$

where dots indicate quantities in which certain components of variability have been removed. To avoid confusion, the space in which the original data is represented will be termed the *data space*, and the reduced dimensional space to which the data are projected will be termed the *feature space* (here it could also be termed *EOF space* or *PC space*). Tilde $(\tilde{)}$ indicates a quantity in feature space. The matrices $\dot{\mathbf{E}}_X$ and $\dot{\mathbf{E}}_Y$ transform vectors from feature space to data space.

Solution by SVD

In feature space, the data matrices are $\dot{\tilde{\mathbf{F}}}_X$ and $\dot{\tilde{\mathbf{F}}}_Y$, and the covariance matrices are

$$
\tilde{\mathbf{\Sigma}}_{XY} = \frac{1}{N-1} \dot{\tilde{\mathbf{F}}}_X^T \dot{\tilde{\mathbf{F}}}_Y
\tag{15.64}
$$

$$
\tilde{\mathbf{\Sigma}}_{XX} = \frac{1}{N-1} \dot{\tilde{\mathbf{F}}}_X^T \dot{\tilde{\mathbf{F}}}_X = \mathbf{I}
\tag{15.65}
$$

$$
\tilde{\mathbf{\Sigma}}_{YY} = \frac{1}{N-1} \dot{\tilde{\mathbf{F}}}_Y^T \dot{\tilde{\mathbf{F}}}_Y = \mathbf{I}.
\tag{15.66}
$$

The fact that $\tilde{\mathbf{\Sigma}}_{XX}$ and $\tilde{\mathbf{\Sigma}}_{YY}$ are identity matrices is important and is the reason for choosing $\dot{\tilde{\mathbf{F}}}_X$ and $\dot{\tilde{\mathbf{F}}}_Y$ to satisfy (15.58) and (15.61). Substituting these matrices in the eigenvalue problems (15.18) and (15.19) gives

$$
\tilde{\mathbf{\Sigma}}_{XY} \tilde{\mathbf{\Sigma}}_{XY}^T \tilde{\mathbf{q}}_X = \rho^2 \tilde{\mathbf{q}}_X,
\tag{15.67}
$$

and

$$
\tilde{\mathbf{\Sigma}}_{XY}^T \tilde{\mathbf{\Sigma}}_{XY} \tilde{\mathbf{q}}_Y = \rho^2 \tilde{\mathbf{q}}_Y.
\tag{15.68}
$$

Although solutions could be obtained by standard eigenvalue solvers, it is much more efficient to obtain the solution by SVD. Specifically, if the SVD of $\tilde{\mathbf{\Sigma}}_{XY}$ is

$$
\begin{array}{ccccc}
\tilde{\mathbf{\Sigma}}_{XY} & = & \mathbf{U}_\rho & \mathbf{S}_\rho & \mathbf{V}_\rho \\
T_X \times T_Y & & T_X \times T_X & T_X \times T_Y & T_Y \times T_Y,
\end{array}
\tag{15.69}
$$

then $\tilde{\mathbf{\Sigma}}_{XY}^T \tilde{\mathbf{\Sigma}}_{XY}$ and $\tilde{\mathbf{\Sigma}}_{XY} \tilde{\mathbf{\Sigma}}_{XY}^T$ both yield an eigenvector decomposition (try it). Substituting (15.69) into (15.67) and (15.68) reveals that the canonical projection vectors are

$$
\tilde{\mathbf{Q}}_X = \mathbf{U}_\rho \quad \text{and} \quad \tilde{\mathbf{Q}}_Y = \mathbf{V}_\rho,
\tag{15.70}
$$

and that $\mathbf{S}_\rho = \mathbf{\Lambda}$, hence the i'th canonical correlation $\hat{\rho}_i$ is the i'th singular value

$$
\hat{\rho}_i = s_i.
\tag{15.71}
$$

Thus, SVD of $\tilde{\Sigma}_{XY}$ yields *both* sets of canonical projection vectors simultaneously. Furthermore, the variates are obtained from the sample analog of (15.25), namely

$$\tilde{\mathbf{R}}_X = \dot{\mathbf{F}}_X \tilde{\mathbf{Q}}_X \quad \text{and} \quad \tilde{\mathbf{R}}_Y = \dot{\mathbf{F}}_Y \tilde{\mathbf{Q}}_Y. \tag{15.72}$$

The i'th columns of $\tilde{\mathbf{R}}_X$ and $\tilde{\mathbf{R}}_Y$ are the i'th *sample* canonical variates of $r_{X,i}$ and $r_{Y,i}$, respectively. Since the original data matrices $\mathbf{X}$ and $\mathbf{Y}$ were centered, the canonical variates are also centered. It is readily verified that the variates satisfy the orthogonality relations

$$\frac{1}{N-1}\tilde{\mathbf{R}}_X^T\tilde{\mathbf{R}}_X = \mathbf{I} \tag{15.73}$$

$$\frac{1}{N-1}\tilde{\mathbf{R}}_Y^T\tilde{\mathbf{R}}_Y = \mathbf{I} \tag{15.74}$$

$$\frac{1}{N-1}\tilde{\mathbf{R}}_X^T\tilde{\mathbf{R}}_Y = \mathbf{S}_\rho, \tag{15.75}$$

which are the sample analogs of (15.35), (15.36), and (15.37), respectively.

Loading Vectors

The loading vectors are obtained by finding the matrix $\dot{\mathbf{P}}_X$ such that $\tilde{\mathbf{R}}_X\dot{\mathbf{P}}_X^T$ approximates the data matrix $\dot{\mathbf{X}}$ as closely as possible, in the sense of minimizing

$$\|\dot{\mathbf{X}} - \tilde{\mathbf{R}}_X\dot{\mathbf{P}}_X^T\|_F^2. \tag{15.76}$$

This is a standard multivariate regression problem (see Chapter 14) with solution

$$\dot{\mathbf{P}}_X = \dot{\mathbf{X}}^T\tilde{\mathbf{R}}_X\left(\tilde{\mathbf{R}}_X^T\tilde{\mathbf{R}}_X\right)^{-1} = \frac{1}{N-1}\dot{\mathbf{X}}^T\tilde{\mathbf{R}}_X. \tag{15.77}$$

Substituting (15.72) and (15.62) yields the equivalent expression

$$\dot{\mathbf{P}}_X = \dot{\mathbf{E}}_X\tilde{\mathbf{Q}}_X. \tag{15.78}$$

The i'th column of $\dot{\mathbf{P}}_X$ is loading vector $p_{X,i}$. Thus, the i'th loading vector is obtained by multiplying $\mathbf{E}_X$ by the i'th column of $\tilde{\mathbf{Q}}_X$. Similar expressions hold for loading vectors of $\mathbf{Y}$ (just replace X-subscripts with Y-subscripts). These solutions do not depend on the norm used in (15.76) (e.g., a norm based on an area weighting gives the same solution).

Regression Patterns

Once a variate time series is obtained, a regression pattern can be computed between it and *any* time series. The *regression patterns* that approximate the original *unfiltered* data $\mathbf{X}$ are

$$\mathbf{P}_X = \frac{1}{N-1}\mathbf{X}^T\tilde{\mathbf{R}}_X, \tag{15.79}$$

which differs from (15.77) by replacing $\dot{\mathbf{X}}$ by $\mathbf{X}$. Loading vectors and regression patterns are identical if $\dot{\mathbf{X}}$ is based on principal components of the entire data set, because the retained columns of $\tilde{\mathbf{F}}_X$ are uncorrelated with discarded ones. In general, the patterns will differ if the feature vectors are not the EOFs of the full data set. When they differ, regression patterns $\mathbf{P}_X$ tend to be more physically interpretable because they are not constrained to lie in feature space, in contrast to loading vectors $\dot{\mathbf{P}}_X$.

Linear Prediction

A linear prediction of $\tilde{\mathbf{R}}_Y$ given $\tilde{\mathbf{R}}_X$ is based on

$$\hat{\tilde{\mathbf{R}}}_Y = \tilde{\mathbf{R}}_X \tilde{\mathbf{B}}_{XY}, \tag{15.80}$$

where $\tilde{\mathbf{B}}_{XY}$ is a matrix of regression coefficients. As discussed in Chapter 8, the least squares estimate of $\tilde{\mathbf{B}}_{XY}$ is

$$\hat{\tilde{\mathbf{B}}}_{XY} = \left(\tilde{\mathbf{R}}_X^T \tilde{\mathbf{R}}_X \right)^{-1} \tilde{\mathbf{R}}_X^T \tilde{\mathbf{R}}_Y = \hat{\mathbf{\Lambda}}, \tag{15.81}$$

where (15.75) has been used and the caret ˆ indicates a sample estimate. This result shows that the regression operator is *diagonal*, with diagonal elements equal to the canonical correlations. Consequently, the least squares prediction of the kth canonical variate $\mathbf{r}_Y^k$ is

$$\hat{\mathbf{r}}_Y^k = \hat{\rho}_k \mathbf{r}_X^k. \tag{15.82}$$

This is the sample version of (15.46), which shows that the canonical components can be predicted independently of each other, component-by-component, in the sample version too. Substituting (15.81) into (15.80) yields the least squares regression fit of $\tilde{\mathbf{R}}_Y$

$$\hat{\tilde{\mathbf{R}}}_Y = \tilde{\mathbf{R}}_X \mathbf{\Lambda}. \tag{15.83}$$

In practice, we want the regression fit in data space, which is obtained by post-multiplying the regression fit $\hat{\tilde{\mathbf{R}}}_Y$ by $\mathbf{P}_Y$. Finally, using (15.72), we obtain

$$\hat{\mathbf{Y}} = \dot{\mathbf{F}}_X \tilde{\mathbf{Q}}_X \mathbf{\Lambda} \mathbf{P}_Y^T, \tag{15.84}$$

which is the (transposed) sample version of (15.45). See Section 15.4 for further discussion about the significance of this decomposition.

Explained Variance

We usually want to know how much variance is explained by the canonical components, and what fraction of variance can be predicted. This requires defining a measure of variance for a multivariate vector. It is natural to use a measure of variance consistent with the decomposition (15.59). For instance, principal component analysis discussed in Chapter 12 uses a matrix $\mathbf{W}_X$ to define an area-weighted norm. Accordingly, let us define the sum total variance as

$$\|\mathbf{X}\|_X^2 = \frac{1}{N-1} \operatorname{tr}\left[\mathbf{X}\mathbf{W}_X\mathbf{X}^T\right], \tag{15.85}$$

where the X-subscript identifies a norm appropriate for X-variables with a suitable choice of $\mathbf{W}_X$. The sum total variance of the canonical components $\tilde{\mathbf{R}}_X\mathbf{P}_X^T$ is

$$\|\tilde{\mathbf{R}}_X\mathbf{P}_X^T\|_X^2 = \frac{1}{N-1} \operatorname{tr}\left[\tilde{\mathbf{R}}_X\mathbf{P}_X^T\mathbf{W}\mathbf{P}_X\tilde{\mathbf{R}}_X^T\right] \tag{15.86}$$

$$= \operatorname{tr}\left[\mathbf{P}_X^T\mathbf{W}\mathbf{P}_X\left(\frac{\tilde{\mathbf{R}}_X^T\tilde{\mathbf{R}}_X}{N-1}\right)\right] \tag{15.87}$$

$$= \operatorname{tr}\left[\mathbf{P}_X^T\mathbf{W}\mathbf{P}_X\right] \tag{15.88}$$

$$= \sum_{k=1}^{T_X} \mathbf{p}_{X,k}^T\mathbf{W}\mathbf{p}_{X,k}. \tag{15.89}$$

This equation implies that the term $\mathbf{p}_{X,k}^T\mathbf{W}\mathbf{p}_{X,k}$ can be identified as the variance of the k'th canonical component. The *fraction* of variance explained by the k'th component is

$$\frac{\mathbf{p}_{X,k}^T\mathbf{W}\mathbf{p}_{X,k}}{\|\mathbf{X}\|_X^2}. \tag{15.90}$$

The sum over all T_X components is less than 100% because CCA was applied only to the T_X components of $\tilde{\mathbf{F}}_X$, and therefore the canonical components can explain only an amount of variance explained by these T_X components. Similar relations hold for $\mathbf{Y}$.

15.7 Model Selection

How many PCs should be chosen for X and Y? That is, what values of T_X and T_Y should be chosen? Despite the popularity of CCA, no standard criterion for selecting variables is currently used in the climate literature. Fujikoshi (1985) proposed a criterion for selecting T_X based on Akaike's Information Criterion (AIC). Later, Fujikoshi et al. (2010) generalized this criterion to selecting both T_X and T_Y. However, this generalization was based on a slightly different measure known as Distance Information Criterion (DIC). Also, DIC effectively compares prediction models relative to some maximal set of variables, but the choice of maximal variable set might seem to introduce arbitrariness into the criterion. DelSole and Tippett (2021a) resolved these issues and showed that the correct AIC criterion could be computed much more directly using Mutual Information Criterion (MIC), discussed in Chapter 14. In terms of the present notation, this criterion is

$$\operatorname{MIC}(T_X, T_Y) = N \log \Lambda + \mathcal{P}(N, T_X, T_Y), \tag{15.91}$$

where

$$P(N, T_X, T_Y) = N \left(\frac{(T_X + T_Y)(N + 1)}{N - T_X - T_Y - 2} - \frac{T_X(N + 1)}{N - T_X - 2} - \frac{T_Y(N + 1)}{N - T_Y - 2} \right),$$

$$(15.92)$$

and

$$\Lambda = \frac{|\overline{\mathbf{\Sigma}}_{Y|X}|}{|\overline{\mathbf{\Sigma}}_{YY}|} \quad \text{and} \quad \overline{\mathbf{\Sigma}}_{Y|X} = \overline{\mathbf{\Sigma}}_{YY} - \overline{\mathbf{\Sigma}}_{YX} \overline{\mathbf{\Sigma}}_{XX}^{-1} \overline{\mathbf{\Sigma}}_{XY}. \tag{15.93}$$

The sample covariance matrices are defined in (14.31) in terms of the (centered) data matrices $\mathbf{X}$ and $\mathbf{Y}$, which in the present example are identified with the matrices $\tilde{\mathbf{F}}_X$ and $\tilde{\mathbf{F}}_Y$, since we are selecting feature vectors. Just to be clear, although the latter matrices satisfy the orthogonality properties (15.58) and (15.61), these properties are not required in MIC.

To clarify the connection between MIC and CCA, recall the standard fact $|\mathbf{A}|/|\mathbf{B}| = |\mathbf{B}^{-1}\mathbf{A}|$, hence (15.93) can be written as

$$\Lambda = \left| \overline{\mathbf{\Sigma}}_{YY}^{-1} \overline{\mathbf{\Sigma}}_{Y|X} \right| = \left| \overline{\mathbf{\Sigma}}_{YY}^{-1} \left(\overline{\mathbf{\Sigma}}_{YY} - \overline{\mathbf{\Sigma}}_{YX} \overline{\mathbf{\Sigma}}_{XX}^{-1} \overline{\mathbf{\Sigma}}_{XY} \right) \right| = |\mathbf{I} - \overline{\mathbf{\Sigma}}_S|, \tag{15.94}$$

where

$$\overline{\mathbf{\Sigma}}_S = \overline{\mathbf{\Sigma}}_{YY}^{-1} \overline{\mathbf{\Sigma}}_{YX} \overline{\mathbf{\Sigma}}_{XX}^{-1} \overline{\mathbf{\Sigma}}_{XY}. \tag{15.95}$$

Recall that the determinant of a matrix equals the product of eigenvalues. It follows from (15.19) that the eigenvalues of $\overline{\mathbf{\Sigma}}_S$ are the squared canonical correlations, hence the eigenvalues of $\mathbf{I} - \overline{\mathbf{\Sigma}}_S$ are one minus the squared canonical correlations, and therefore,

$$\Lambda = (1 - \hat{\rho}_1^2)(1 - \hat{\rho}_2^2) \ldots (1 - \hat{\rho}_T^2), \tag{15.96}$$

where $T = \min[T_X, T_Y]$. It follows that MIC in (15.91) can be written equivalently as

$$\text{MIC}(T_X, T_Y) = N \sum_{k=1}^{T} \log\left(1 - \hat{\rho}_k^2\right) + P(N, T_X, T_Y). \tag{15.97}$$

Thus, MIC depends on the sample values only through the canonical correlations.

In the case of CCA, MIC is a two-dimensional function that depends on T_X and T_Y (it also depends on N, but this parameter is always the same). The value of MIC for a range of choices for T_X and T_Y for the example discussed in Section 15.2 is shown in Figure 15.5. The procedure is to choose T_X and T_Y that minimizes MIC. According to Figure 15.5, the minimum MIC occurs at four SST EOFs and three US temperature EOFs. This was the basis for choosing these truncations in the example presented in Section 15.2.

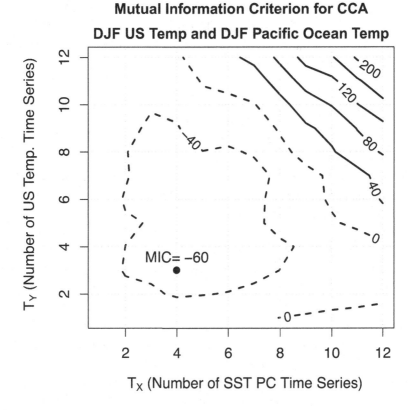

Figure 15.5 Values of MIC as a function T_X, the number of PC time series of DJF Pacific sea surface temperature, and as a function of T_Y, the number of PC time series of DJF US Temperature. The minimum MIC is indicated by a dot at $T_X = 4$ and $T_Y = 3$. See text for further details.

15.8 Hypothesis Testing

The natural null hypothesis for CCA is that **x** and **y** are independent. If **x** and **y** are independent, then $\Sigma_{YX} = \mathbf{0}$, and (15.14) implies that all population canonical correlations vanish. Because CCA yields so much information in the form of canonical correlations and canonical projection vectors, it is unclear what statistic should be used for testing independence. A basic principle in hypothesis testing is that the decision cannot depend on the physical units of the variables. For instance, a test performed on temperature data cannot depend on whether temperature is in units of Celsius or Fahrenheit, because there is no justification to chose one unit over another. Similarly, a test cannot depend on basis set since there is no reason to prefer one basis over another in the same space. This principle implies that the statistic $f(\cdot)$ for testing hypotheses must be invariant to invertible affine transformations of

$\mathbf{x}$ and $\mathbf{y}$. Section 15.5 showed that canonical correlations satisfy this property. It can be shown that *any* such statistic can depend only on the canonical correlations $\hat{\rho}_1, \ldots, \hat{\rho}_T$, where $T = \min[T_X, T_Y]$ (Muirhead, 2009, Theorem 11.2.2). Therefore, the test statistic must be of the form

$$f(\hat{\rho}_1, \ldots, \hat{\rho}_T). \tag{15.98}$$

The question arises as to what function $f(\cdot)$ should be chosen. There is no unique choice. The null hypothesis can be violated in many ways, and different functions have different abilities to detect departures from the null hypothesis (that is, they have different statistical power). In Section 15.2, we effectively used the leading canonical correlation as the statistic for testing the null hypothesis. This is known as the *Union-Intersection Test*. To derive the sampling distribution of $\hat{\rho}_1$, note that the null hypothesis of independence implies $\Sigma_{XY} = \mathbf{0}$. Since a multivariate normal distribution is completely specified by Σ_{XY}, Σ_{XX}, and Σ_{YY} (we ignore population means because they are irrelevant in CCA), the only relevant population parameters for testing independence are Σ_{XX} and Σ_{YY}. Because canonical correlations are invariant to linear invertible transformations (see Section 15.5), we may transform $\mathbf{x}$ so that its covariance matrix equals the identity matrix (see Section 15.5). Similarly for $\mathbf{y}$. The resulting transformed variables still have vanishing cross-covariance matrix. This shows that the sampling distribution for arbitrary normal distributions is the same as the sampling distributions of *iid* normal distributions. Accordingly, one can generate independent random numbers from a standardized *univariate* normal distribution, use those numbers to fill data matrices, compute the canonical correlations, and repeat many times to create a sample of canonical correlations under the null hypothesis. This procedure was used to generate the boxes in Figure 15.3. As shown in Figure 15.3, the maximum canonical correlation fell well outside the 5–95% interval obtained from 10,000 Monte Carlo trials, hence we reject the hypothesis that the variables are independent.

The question arises as to whether the second canonical correlation is significant. The difficulty with testing this hypothesis is that the null hypothesis of independence has been rejected, hence it is clearly an incorrect assumption for the population. A more appropriate hypothesis is that one component in the population has a linear relation. The existence of a *single* component is tantamount to assuming that Σ_{XY} is rank-1. In this case, CCA provides a transformation such that $\Sigma_{XX} = \mathbf{I}$, $\Sigma_{YY} = \mathbf{I}$, and Σ_{XY} is diagonal with only one nonzero diagonal element, namely the population canonical correlation ρ_1. To explore the sensitivity of the sampling distribution with respect to ρ_1, we perform the same Monte Carlo simulations as previously, except the first elements of $\mathbf{x}$ and $\mathbf{y}$ are related as

$$y_1 = \rho_1 x_1 + \epsilon_1 \sqrt{1 - \rho_1^2}, \tag{15.99}$$

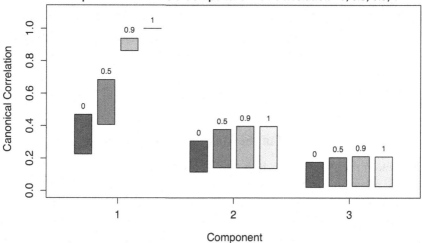

Figure 15.6 Canonical correlations from Monte Carlo simulations in which **x** and **y** are independent except for one component satisfying (15.99). The upper and lower edges of each box show the 5th and 95th percentiles for the four values of $\rho_1 = 0, 0.5, 0.9, 1$. The other parameters are chosen to be consistent with the data example in Section 15.2, namely $T_X = 4, T_Y = 3, N = 68$.

where ϵ_1 is a random variable with a standardized normal distribution. This model simulates pairs (x_1, y_1) such that x_1 and y_1 have unit variance and their correlation is ρ_1 (see Example 1.7). The 5th and 95th percentile range of the canonical correlations for different values of ρ_1 are shown in Figure 15.6. Not surprisingly, the leading correlation marches upward with increasing ρ_1 until it reaches 1. In contrast, the second and third canonical correlations increase only slightly even for $\rho_1 = 1$. In effect, when the population has one nonzero canonical correlation, the second and higher sample canonical correlations tend to be "dragged up," which creates a bias relative to the distribution when the null hypothesis is true. However, this upward bias is small. Comparing Figure 15.6 to Figure 15.3 shows that the second canonical correlation in the observational example is significant.

Instead of tests based on individual canonical correlations, one can conceive of tests based on *all* the canonical correlations. The most common hypothesis test for independence of **x** and **y** is based on the likelihood ratio test, which yields the statistic Λ defined in (15.96). In Chapter 14, we discussed the fact that $\log \Lambda$ as the asymptotic distribution (14.37), which in the present notation is

$$\chi^2 = -\left(N - \frac{T_X + T_Y + 3}{2}\right) \log \Lambda. \tag{15.100}$$

When the null hypothesis of independence is true, then (15.100) has an approximate chi-squared distribution with $T_X T_Y$ degrees of freedom. For our example in sec. 15.2, the sample value is 86 while the 5% threshold is 21.2, hence we reject the hypothesis of independence, consistent with our conclusions in that section.

Incidentally, these significance tests do not account for the fact that the number of PC time series were chosen based on a selection criterion that required comparing many choices. Model selection is difficult to take into account in hypothesis testing.

15.9 Proof of the Maximization Properties

The derivation presented in Section 15.3 showed merely that canonical projection vectors are associated with zero derivatives of the correlation; it did not show that they actually *maximize* the correlation. Also, the second and higher components are claimed to maximize the correlation subject to being uncorrelated with the previous components, but this fact has not been proven explicitly. This section fills these gaps.

Because $\mathbf{Q}_X$ and $\mathbf{Q}_Y$ are invertible, we may define the following vectors:

$$\mathbf{z}_X = \mathbf{Q}_X^{-1}\mathbf{q}_X \quad \text{and} \quad \mathbf{z}_Y = \mathbf{Q}_Y^{-1}\mathbf{q}_Y, \tag{15.101}$$

which are simply a different way of representing $\mathbf{q}_X$ and $\mathbf{q}_Y$. Inverting these formulas,

$$\mathbf{q}_X = \mathbf{Q}_X \mathbf{z}_X \quad \text{and} \quad \mathbf{q}_Y = \mathbf{Q}_Y \mathbf{z}_Y. \tag{15.102}$$

Substituting these into (15.14), and making use of the identities (15.33) and (15.34), gives

$$\rho^2 = \frac{\left(\mathbf{z}_X^T \mathbf{\Lambda} \mathbf{z}_Y\right)^2}{\left(\mathbf{z}_X^T \mathbf{z}_X\right)\left(\mathbf{z}_Y^T \mathbf{z}_Y\right)}. \tag{15.103}$$

Now consider the *generalized* Cauchy–Schwartz inequality (A.7), which implies

$$\left(\mathbf{z}_X^T \mathbf{\Lambda} \mathbf{z}_Y\right)^2 \leq \left(\mathbf{z}_X^T \mathbf{\Lambda} \mathbf{z}_X\right)\left(\mathbf{z}_Y^T \mathbf{\Lambda} \mathbf{z}_Y\right). \tag{15.104}$$

Because $\mathbf{\Lambda}$ is diagonal, we have

$$\mathbf{z}_X^T \mathbf{\Lambda} \mathbf{z}_X = \sum_{i=1}^{M_{\min}} \rho_i \, (\mathbf{z}_X)_i^2 = \rho_1 \mathbf{z}_{X,1}^2 + \rho_2 \mathbf{z}_{X,2}^2 + \cdots + \rho_{M_{\min}} \mathbf{z}_{X,\min}. \tag{15.105}$$

Each term in this sum is positive. The canonical correlations are ordered from largest to smallest, hence $\rho_1 \geq \rho_i$ for any i. Therefore, if we replace $\rho_2, \rho_3, \ldots$ by ρ_1, we obtain a new sum that is greater than or equal to the original sum. Thus

$$\mathbf{z}_X^T \Lambda \mathbf{z}_X = \sum_{i=1}^{M_{\min}} \rho_i \left(\mathbf{z}_X\right)_i^2 \leq \rho_1 \sum_{i=1}^{M_{\min}} \left(\mathbf{z}_X\right)_i^2 = \rho_1 \left(\mathbf{z}_X^T \mathbf{z}_X\right). \tag{15.106}$$

Similarly, we have the bound

$$\mathbf{z}_Y^T \Lambda \mathbf{z}_Y = \sum_{i=1}^{M_{\min}} \rho_i \left(\mathbf{z}_Y\right)_i^2 \leq \rho_1 \sum_{i=1}^{M_{\min}} \left(\mathbf{z}_Y\right)_i^2 = \rho_1 \left(\mathbf{z}_Y^T \mathbf{z}_Y\right). \tag{15.107}$$

Substituting these bounds in (15.104) and using (15.103) yields the bound

$$\rho^2 \leq \rho_1^2. \tag{15.108}$$

This proves that the maximum possible correlation is the first canonical correlation ρ_1. It is straightforward to see that if we substitute

$$\mathbf{z}_X = \begin{pmatrix} 1 \\ 0 \\ \vdots \\ 0 \end{pmatrix} \quad \text{and} \quad \mathbf{z}_Y = \begin{pmatrix} 1 \\ 0 \\ \vdots \\ 0 \end{pmatrix}, \tag{15.109}$$

into (15.103), we obtain $\rho^2 = \rho_1^2$. Finally, substituting (15.109) into (15.102) shows that the maximum is achieved by selecting the first canonical projection vectors $\mathbf{q}_{X,1}$ and $\mathbf{q}_{Y,1}$. These considerations prove that the maximum correlation is the first canonical correlation, and that this correlation is achieved by choosing the first canonical projection vectors. Next, consider the maximum correlation among all vectors $\mathbf{z}_X$ and $\mathbf{z}_Y$ orthogonal to (15.109). Such vectors have a zero in the first element but otherwise are unconstrained. Under these constraints, the summations in (15.106) and (15.107) start at $i = 2$ instead of $i = 1$, in which case the upper bound becomes ρ_2, thereby proving that the maximum correlation among all choices of $\mathbf{z}_X$ and $\mathbf{z}_Y$ that are orthogonal with the first pair is ρ_2. Furthermore, this maximum is achieved by choosing $\mathbf{q}_{X,2}$ and $\mathbf{q}_{Y,2}$. Finally, from (15.35), we have

$$\text{cov}[r_{X,i}, r_{X,j}] = \mathbf{q}_{X,i}^T \Sigma_{XX} \mathbf{q}_{X,j} = \mathbf{z}_{X,i}^T \mathbf{z}_{X,j}, \tag{15.110}$$

which shows that choosing $\mathbf{z}$-vectors to be orthogonal is equivalent to choosing variates to be *uncorrelated*. Thus, the second canonical correlation ρ_2 is the maximum correlation among all variates that are uncorrelated with the first canonical variates. Similarly, the third canonical correlation ρ_3 is the maximum correlation among all choices that are uncorrelated with the first two canonical variate pairs, and so on.

15.10 Further Topics

In climate studies, one of the most challenging aspects of CCA is dealing with the fact that the spatial dimension often exceeds the temporal dimension by orders of magnitude. In such cases, the covariance matrices are singular and overfitting is catastrophic – it will be possible to find canonical correlations equal to one, but these perfect correlations never verify in independent data sets. This problem was handled in the present chapter by projecting data onto a lower-dimensional space spanned by the leading EOFs of the data sets, which is by far the standard approach in climate studies. This procedure is known as a *regularization*. A completely different approach to regularization is *canonical ridge*, which replaces the covariance matrices for X and Y as follows (Vinod, 1976):

$$\Sigma_{XX} \to \Sigma_{XX} + \lambda_X \mathbf{I} \quad \text{and} \quad \Sigma_{YY} \to \Sigma_{YY} + \lambda_Y \mathbf{I}. \tag{15.111}$$

The new covariance matrices have inverses even when the original covariance matrices do not. However, this approach still involves a model selection problem in that the values of the regularization parameters λ_X and λ_Y need to be chosen. Other approaches to regularization assume that the projection vectors are sparse (Witten et al., 2009).

15.11 Conceptual Questions

1 Explain in one sentence what Canonical Correlation Analysis does.
2 What is the problem with CCA when the number of Xs and Ys exceeds the sample size?
3 Explain why the univariate test for zero correlation is not an appropriate test for assessing the significance of the leading canonical correlation.
4 What is the interpretation of the second canonical component? In what sense does the second maximize something? What about the third component?
5 Are the canonical components orthogonal in space or orthogonal in time? Explain all the ways in which canonical correlations are orthogonal.
6 Sometimes, the second canonical component explains more variance than the first. Why?
7 CCA can be computed from the SVD of what matrix?
8 If $\mathbf{x}$ and $\mathbf{y}$ are independent, why doesn't the sampling distribution of the canonical correlations depend on Σ_{XX} or Σ_{YY}?
9 CCA can be viewed as a way to diagonalize a multivariate regression operator. Why is this important? How does it help you interpret the linear relations?
10 Can you improve the canonical correlations by linearly transforming the data?

11 If you have only one Y variable, explain how CCA is equivalent to linear regression. How many canonical correlations are there? What is another name for ρ^2?

12 Why can't you just "throw all the data" at CCA and use it to find all the linear relations in a data set?

13 What exactly is a canonical projection vector? A canonical variate? A canonical loading vector? A canonical component?

16

Covariance Discriminant Analysis

… comparing two covariance matrices by analyzing certain linear combinations can be an interesting method itself, giving much more information than just the mere decision about equality or inequality. [1]

Flury (1985)

This chapter discusses a procedure for quantifying differences between two covariance matrices. Despite being applicable to a range of statistical problems, the general procedure has no standard name. In this chapter, we call it *Covariance Discriminant Analysis* (CDA). The term *discriminant analysis* often refers to procedures for comparing population *means*, but here the term will be used more generally to refer to comparing *covariance matrices*, which includes the difference-in-means test as a special case. The basic idea of CDA is to find a linear combination of variables that maximizes the ratio of variances. This *discriminant function* becomes a useful tool for distinguishing between populations and testing hypotheses about equality of populations. This technique is used in numerous other multivariate techniques, including canonical correlation analysis, predictable component analysis, and multivariate ANOVA. In fact, some calculations are so closely related that it will prove efficient to present these calculations once here, and then in other chapters refer to this chapter for the relevant calculations. In the statistics literature, the main purpose of CDA is to *test* hypotheses about covariance matrices. The associated sampling distribution is not standard, but can be estimated easily using Monte Carlo techniques. In climate studies, the main purpose is to *diagnose* differences between covariance matrices. CDA does this by decomposing two multivariate time series, separately, into components ordered such that the variance ratio of the first component is maximized, and each succeeding component maximizes the variance ratio under the constraint that it is uncorrelated with the preceding components. For example, CDA has been used to identify components

---

[1] Reprinted by permission from Taylor & Francis, *Journal of the American Statistical Association*, Flury (1985).

of low-frequency variability by maximizing the ratio of low-frequency to high-frequency variance (Schneider and Held, 2001; DelSole and Tippett, 2009; Wills et al., 2018). To mitigate overfitting, the standard approach is to apply CDA to a few principal components. No standard criterion exists for choosing the number of principal components. A new criterion is proposed in this chapter. Importantly, CDA is sensitive to departures from normality, especially to distributions with thick tails (Box and Watson, 1962; Olson, 1974), so the results of CDA need to be interpreted with caution for non-Gaussian distributions.

16.1 The Problem

In Chapter 2, we discussed a procedure for testing the hypothesis that two random variables have the same variance. If the two random variables are X and Y, then the hypothesis is

$$H_0' : \text{var}[X] = \text{var}[Y]. \tag{16.1}$$

The standard test for this hypothesis is to compute the respective sample variances, $\hat{\sigma}_X^2$ and $\hat{\sigma}_Y^2$, and then reject hypothesis H_0' if the ratio

$$F = \frac{\hat{\sigma}_X^2}{\hat{\sigma}_Y^2}, \tag{16.2}$$

is sufficiently far from one (using thresholds derived from the appropriate F distribution). It is assumed that X and Y are independent. Now consider two random *vectors* **x** and **y**, also assumed to be independent. The natural generalization of hypothesis H_0' is that the *covariance matrices* Σ_X and Σ_Y are equal. This hypothesis can be stated as

$$H_0 : \Sigma_X = \Sigma_Y. \tag{16.3}$$

Note that H_0 states that *each element* of Σ_X equals the corresponding element of Σ_Y. The technique described here can be generalized further as a comparison of two sample covariance matrices $\hat{\Sigma}_X$ and $\hat{\Sigma}_Y$ whose expectations are equal under a null hypothesis:

$$H_0'' : \mathbb{E}\left[\hat{\Sigma}_X\right] = \mathbb{E}\left[\hat{\Sigma}_Y\right]. \tag{16.4}$$

This generalization leads to a flexible framework for comparing data sets.

In this chapter, **x** and **y** refer to the same variables, but drawn from possibly different populations. Accordingly, **x** and **y** have the same dimensions, and their corresponding covariance matrices, Σ_X and Σ_Y, also have the same dimensions. However, the number of samples of **x** and **y** may differ. A few examples should clarify.

Difference in climate variability

It is well established that *seasonal mean* temperature over certain regions may depend on whether El Niño or La Niña is occurring, but does *variability* depend on El Niño or La Niña? To investigate this question, one could collect temperature data during El Niño and separately during La Niña and test if the respective covariance matrices differ. In this case, $\mathbf{x}$ could denote a vector of temperatures during El Niño, and $\mathbf{y}$ could denote the *same* vector but during La Niña.

Difference in Variances and Correlations

H_0 implies that the elements of $\mathbf{x}$ and $\mathbf{y}$ have the same variances *and* the same pairwise correlations. To see this, suppose that $\mathbf{x}$ is a two-dimensional vector, in which case its 2×2 covariance matrix $\boldsymbol{\Sigma}_X$ can be written as

$$\boldsymbol{\Sigma}_X = \begin{pmatrix} \sigma_1^2 & \rho_{12}\sigma_1\sigma_2 \\ \rho_{12}\sigma_1\sigma_2 & \sigma_2^2 \end{pmatrix}, \tag{16.5}$$

where σ_1^2 and σ_2^2 are the variances of the two elements of $\mathbf{x}$, and ρ_{12} is the correlation between those two elements. Knowledge of the variances σ_1^2 and σ_2^2, *and the correlation* ρ_{12}, is sufficient to specify this 2×2 matrix. More generally, for an S-dimensional vector $\mathbf{x}$, knowledge of each of the S variances, and each of the $S(S-1)/2$ pair-wise correlations, uniquely specifies $\boldsymbol{\Sigma}_X$. It follows that H_0 is equivalent to the hypothesis that the variance of each element of $\mathbf{x}$ equals the variance of the corresponding element in $\mathbf{y}$, and that the correlation between any pair of elements of $\mathbf{x}$ equals the correlation between the corresponding pair in $\mathbf{y}$.

Difference in Principal Components

In climate science, the leading principal components are often major targets of analysis. A common question is whether the EOFs change with time, or whether the EOFs of one climate model are the same as those of a different climate model. Unfortunately, direct comparison of EOFs can be misleading because *individual* EOFs can be sensitive to the sample used to calculate them. However, the hypothesis that two populations have the same EOF patterns and variances is equivalent to H_0. To see this, recall that an arbitrary covariance matrix $\boldsymbol{\Sigma}$ has the eigenvector decomposition

$$\boldsymbol{\Sigma} = \mathbf{EDE}^T, \tag{16.6}$$

where the column vectors of $\mathbf{E}$ are the EOF patterns, and $\mathbf{D}$ is a diagonal matrix whose diagonal elements specify the EOF variances. The EOF patterns and variances uniquely specify a covariance matrix, and vice versa. Thus, deciding H_0 is false is equivalent to deciding that the EOF patterns or variances differ between two data sets.

Difference between Models and Observations

A frequent question in climate studies is whether a climate model generates realistic simulations. To characterize variability, one could select M climate indices (e.g., indices of ENSO, PNA, NAO, AMO, etc. – the definitions are not important here) and then estimate the corresponding $M \times M$ covariance matrix from observations and from climate model simulations. The covariance matrix summarizes all variances and pair-wise correlations between indices. Then, the question of whether a climate model simulates these M indices realistically can be investigated by testing H_0.

Low-Frequency Component Analysis

Climate time series often vary over a range of time scales. Special interest is attached to components that vary on the longest time scales because of their predictability and/or relation to climate change. One approach to finding such components is the following. If the time series were *iid*, then a running mean over W steps would have 1/W times the variance of the unfiltered time series. Thus, if $\mathbf{z}(t)$ is a time series and $\tilde{\mathbf{z}}(t)$ the corresponding running mean, then the *iid* hypothesis implies

$$\mathrm{cov}[\tilde{\mathbf{z}}] = \mathrm{cov}[\mathbf{z}]/W. \tag{16.7}$$

In contrast, if some components are serially correlated, then their variances in $\tilde{\mathbf{z}}(t)$ should be larger than expected from (16.7) (see Section 5.3). This effect is illustrated in Figure 16.1. Components that deviate most strongly from this hypothesis are

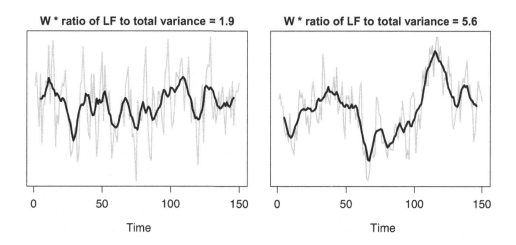

Figure 16.1 Random time series (thin grey) and corresponding nine-step running mean (thick black) for small (left) and large (right) serial correlation. The ratio of the variance of the nine-step running mean over the total variance, times 9, is shown in the titles. For *iid*, variance ratios should be close to one.

the most promising candidates for significant low-frequency variability. This is the essence of a technique known as Low-Frequency Component Analysis (Schneider and Held, 2001; Wills et al., 2018).

Maximize Signal-to-Noise Ratio

The previous example is a special case of a much more general idea. Specifically, the *average* of a quantity often estimates a "signal" of interest. Without averaging, the signal may be difficult to detect because it is obscured by random noise. With a sufficient amount of averaging, the noise is diminished and a signal is said to be detected when the variability of the averages exceeds a threshold that depends on the variability of the noise. The variance of the noise can be estimated from residuals about the average. CDA can be used to find the linear combination of variables that maximizes the signal-to-noise ratio, which is therefore the component that is most likely to be detectable.

Other Multivariate Techniques

Many multivariate procedures can be framed in terms of equality of covariance matrices, including Canonical Correlation Analysis, Multivariate Analysis of Variance, and even a test for equality of multivariate *means* (see Section 16.7).

In the previous examples, hypothesis testing is most straightforward when the samples are independent in time. If, instead, the samples are serially correlated, then hypothesis testing in the presence of serial correlation requires more complex techniques (see DelSole and Tippett 2020 and subsequent parts 2 and 3 for further details).

16.2 Illustration: Most Detectable Climate Change Signals

In this section, we summarize Covariance Discriminant Analysis (CDA) and illustrate it with an example. Proofs and further details will be given in subsequent sections.

The example we choose concerns the response of surface air temperature to a climate forcing (e.g., increasing greenhouse gas concentration). This example is interesting because the technique will estimate the impact of climate forcing while making fairly mild assumptions about the nature of those impacts. Even without climate change, weather variability would still exist and cause random climate variability. Variability that arises spontaneously as a result of dynamical instabilities in the atmosphere–ocean fluid system is known as *internal variability* and denoted by a time-dependent vector $\mathbf{u}_t$, where we use "u" to remind us that it refers to "unforced" variability. In general, internal variability is serially correlated, and accounting for this effect requires a fairly complex procedure. For illustration purposes, we neglect serial correlation and assume that $\mathbf{u}_t$ is *iid*:

$$\mathbf{u}_t \overset{iid}{\sim} \mathcal{N}_S \left(\boldsymbol{\mu}_U, \boldsymbol{\Sigma}_U \right). \tag{16.8}$$

To estimate $\boldsymbol{\mu}_U$ and $\boldsymbol{\Sigma}_U$, a climate model is run with climate forcing containing only annual and diurnal cycles. This is known as a *control run* and gives $\mathbf{y}_t$ for $t = 1, \ldots, N_Y$, which is modeled as

$$\mathbf{y}_t = \mathbf{u}_t. \tag{16.9}$$

Then, unbiased estimates of $\boldsymbol{\mu}_U$ and $\boldsymbol{\Sigma}_U$ are obtained in the standard way as

$$\hat{\boldsymbol{\mu}}_Y = \frac{1}{N_Y} \sum_{t=1}^{N_Y} \mathbf{y}_t \quad \text{and} \quad \hat{\boldsymbol{\Sigma}}_Y = \frac{1}{N_Y - 1} \sum_{t=1}^{N_Y} \left(\mathbf{y}_t - \hat{\boldsymbol{\mu}}_Y \right) \left(\mathbf{y}_t - \hat{\boldsymbol{\mu}}_Y \right)^T. \tag{16.10}$$

In particular, from Theorem 7.9,

$$\hat{\boldsymbol{\mu}}_Y \sim \mathcal{N}_S \left(\boldsymbol{\mu}_U, \frac{1}{N_Y} \boldsymbol{\Sigma}_U \right). \tag{16.11}$$

Next, the model is run with a climate forcing that differs from year to year. This is known as the *forced* run and it produces a new set of vectors $\mathbf{x}_t$ for $t = 1, \ldots, N_X$. Without going into detailed physics, a common paradigm is to assume that the response to climate forcing $\mathbf{f}_t$ is additive and does not alter the distribution of internal variability. Temperature approximately satisfies this assumption on long spatial and temporal scales. Accordingly, the model for this run is

$$\mathbf{x}_t = \mathbf{u}_t^* + \mathbf{f}_t, \tag{16.12}$$

where $\mathbf{u}_t^*$ has the same distribution as $\mathbf{u}_t$ but is independent of $\mathbf{u}_t$. Since our goal is to compare forced and control simulations, including *changes in the mean*, we measure variability relative to the *same reference*, namely the mean in the absence of forced variability. Accordingly, we compute covariance matrices based on deviations relative to the *mean of the control run*. For reasons that will become clear shortly, we use the estimate

$$\hat{\boldsymbol{\Sigma}}_X = \left(1 + \frac{1}{N_Y} \right)^{-1} \frac{1}{N_X} \sum_{t=1}^{N_X} \left(\mathbf{x}_t - \hat{\boldsymbol{\mu}}_Y \right) \left(\mathbf{x}_t - \hat{\boldsymbol{\mu}}_Y \right)^T. \tag{16.13}$$

The forced response $\mathbf{f}_t$ is deterministic, so it is treated as a constant when computing expectations over internal variability. Thus, substituting (16.12) into (16.13) and taking expectations gives

$$\mathbb{E}\left[\hat{\boldsymbol{\Sigma}}_X \right] = \boldsymbol{\Sigma}_U + \boldsymbol{\Sigma}_F, \tag{16.14}$$

where

$$\Sigma_F = \left(1 + \frac{1}{N_Y}\right)^{-1} \frac{1}{N_X} \sum_{t=1}^{N_X} \mathbf{f}_t \mathbf{f}_t^T. \qquad (16.15)$$

It is simple to show that Σ_F is positive semi-definite (that is, $\mathbf{z}^T \Sigma_F \mathbf{z} \geq 0$ for all $\mathbf{z}$).

The question of interest is whether forcing has caused a response. Accordingly, we consider the null hypothesis of no forced response, that is, $\mathbf{f} = \mathbf{0}$, hence $\Sigma_F = \mathbf{0}$. In this case,

$$\mathbf{x}_t \sim \mathcal{N}_S\left(\boldsymbol{\mu}_U, \Sigma_U\right), \qquad (16.16)$$

and since $\mathbf{x}_x$ and $\hat{\boldsymbol{\mu}}_Y$ are independent and the latter has distribution (16.11),

$$\mathbf{x}_t - \hat{\boldsymbol{\mu}}_Y \sim \mathcal{N}_S\left(\mathbf{0}, \left(1 + \frac{1}{N_Y}\right)\Sigma_U\right). \qquad (16.17)$$

It follows that $\hat{\Sigma}_X$ defined in (16.13) is an unbiased estimate of Σ_U, provided the null hypothesis is true. This is the reason for using (16.13) as an estimate. Unfortunately, $\mathbf{f}$ itself cannot be observed. Instead, we observe only $\mathbf{x}_t$, which is the sum $\mathbf{u}_t^* + \mathbf{f}_t$. Based on (16.14) and the fact that $\mathbb{E}\left[\hat{\Sigma}_Y\right] = \Sigma_U$, the hypothesis $\Sigma_F = \mathbf{0}$ implies

$$\mathbb{E}\left[\hat{\Sigma}_X\right] = \mathbb{E}\left[\hat{\Sigma}_Y\right]. \qquad (16.18)$$

Thus, our question has been reduced to testing a hypothesis of the form (16.4). We call this hypothesis "equality of covariance matrices," even though $\mathbb{E}[\hat{\Sigma}_X]$ is not strictly a covariance matrix of a stationary random variable (because $\mathbf{f}_t$ is deterministic).

To test equality of covariance matrices, one is confronted immediately with the problem of *measuring* differences between covariance matrices. After all, a covariance matrix may have hundreds of elements. One strategy is to examine a *linear combination of variables*. For instance, if $\mathbf{x}$ is an S-dimensional vector, then

$$r_X = q_1 x_1 + q_2 x_2 + \cdots + q_S x_S, \qquad (16.19)$$

is a linear combination of the elements of $\mathbf{x}$, where $q_1, \ldots, q_S$ are coefficients to be determined. Importantly, the combination converts the vector $\mathbf{x}$ into the *scalar* r_X. Similarly, a linear combination of elements of the S-dimensional vector $\mathbf{y}$ is

$$r_Y = q_1 y_1 + q_2 y_2 + \cdots + q_S y_S. \qquad (16.20)$$

The scalars r_X and r_Y are known as *variates*. In many climate applications, S corresponds to the number of spatial locations. In this case, the linear combination represents a weighted sum over space, and therefore r_X and r_Y correspond to time-varying scalars. Differences in variability can be measured by the ratio of variances,

$$\lambda = \frac{\text{var}[r_X]}{\text{var}[r_Y]}. \tag{16.21}$$

To appreciate the merit of using λ to measure differences in covariance matrices, let us collect the coefficients $q_1, \ldots, q_S$ into the vector $\mathbf{q}$, known as a *projection vector*. Then, the linear combinations (16.19) and (16.20) may be expressed equivalently as

$$r_X = \mathbf{q}^T \mathbf{x} \quad \text{and} \quad r_Y = \mathbf{q}^T \mathbf{y}. \tag{16.22}$$

According to (7.68), the variance of the random variables r_X and r_Y can be expressed as

$$\text{var}[r_X] = \mathbf{q}^T \boldsymbol{\Sigma}_X \mathbf{q} \quad \text{and} \quad \text{var}[r_Y] = \mathbf{q}^T \boldsymbol{\Sigma}_Y \mathbf{q}, \tag{16.23}$$

and therefore the *variance ratio* is

$$\lambda = \frac{\text{var}[r_X]}{\text{var}[r_Y]} = \frac{\mathbf{q}^T \boldsymbol{\Sigma}_X \mathbf{q}}{\mathbf{q}^T \boldsymbol{\Sigma}_Y \mathbf{q}}. \tag{16.24}$$

This expression shows that λ is a ratio of quadratic forms. Such a ratio is known as a *Rayleigh quotient*. Obviously, if H_0 is true, then $\lambda = 1$ for *any* choice of $\mathbf{q}$. On the other hand, if $\boldsymbol{\Sigma}_X \neq \boldsymbol{\Sigma}_Y$, then λ differs from unity for *some* choice of $\mathbf{q}$.

Substituting (16.14) and $\boldsymbol{\Sigma}_Y = \boldsymbol{\Sigma}_U$ into (16.24) gives the ratio

$$\lambda = \frac{\mathbf{q}^T (\boldsymbol{\Sigma}_U + \boldsymbol{\Sigma}_F) \mathbf{q}}{\mathbf{q}^T \boldsymbol{\Sigma}_U \mathbf{q}} = 1 + \frac{\mathbf{q}^T \boldsymbol{\Sigma}_F \mathbf{q}}{\mathbf{q}^T \boldsymbol{\Sigma}_U \mathbf{q}}. \tag{16.25}$$

The matrix $\boldsymbol{\Sigma}_U$ is positive definite while $\boldsymbol{\Sigma}_F$ is positive semi-definite, which means the last term on the right is nonnegative. Clearly, $\lambda = 1$ if the null hypothesis $\boldsymbol{\Sigma}_F = 0$ is true. However, if the null hypothesis is false, then $\lambda > 1$ *regardless of the details of how* $\mathbf{f}$ *differs from zero*. In particular, the forced response has not been assumed to have a trend or any particular spatial structure (although we will eventually assume that the forced response projects on the leading EOFs). Values greater than one indicate a response to climate forcing, hence we seek projection vectors that make λ *as large as possible*.

As will be shown later in this chapter, the projection vector that optimizes the variance ratio is found by solving the generalized eigenvalue problem

$$\boldsymbol{\Sigma}_X \mathbf{q} = \lambda \boldsymbol{\Sigma}_Y \mathbf{q}. \tag{16.26}$$

Solutions to such equations are discussed in Section A.2 and will be merely summarized here. First, for each nontrivial eigenvector $\mathbf{q}$ satisfying (16.26), there exists a corresponding eigenvalue λ, known as a *discriminant ratio*.[2] The eigenvalues are real and positive. We assume eigenvalues are distinct, in which case the eigenvectors are unique up to a multiplicative constant. By convention, discriminant ratios are ordered from largest to smallest, $\lambda_1 > \lambda_2 > \cdots > \lambda_S$. The corresponding eigenvectors are known as *discriminant projection vectors* and denoted $\mathbf{q}_1, \mathbf{q}_2, \ldots,$ $\mathbf{q}_S$. The linear combination based on projection vector $\mathbf{q}_i$ is known as the i'th *discriminant function*, and its evaluation yields the *discriminant variates*

$$r_{X,i} = \mathbf{q}_i^T \mathbf{x} \quad \text{and} \quad r_{Y,i} = \mathbf{q}_i^T \mathbf{y}. \tag{16.27}$$

The corresponding discriminant ratio is derived from (16.24) as

$$\lambda_i = \frac{\text{var}[r_{X,i}]}{\text{var}[r_{Y,i}]} = \frac{\mathbf{q}_i^T \Sigma_X \mathbf{q}_i}{\mathbf{q}_i^T \Sigma_Y \mathbf{q}_i}. \tag{16.28}$$

The solution to our optimization problem is the following: The largest possible variance ratio is the first eigenvalue λ_1, and this value is obtained when the projection vector equals the first eigenvector $\mathbf{q}_1$. The second eigenvalue λ_2 is the largest variance ratio *among all variates that are uncorrelated with the first*, the third eigenvalue λ_3 is the largest variance ratio among all variates uncorrelated with the first *two*, and so forth (these properties are demonstrated in Section 16.4). In addition, it follows from the symmetry of Σ_X and Σ_Y that the discriminant functions are *uncorrelated*, in the sense

$$\text{cov}[r_{X,i}, r_{X,j}] = 0 \quad \text{and} \quad \text{cov}[r_{Y,i}, r_{Y,j}] = 0 \quad \text{if } i \neq j. \tag{16.29}$$

The correlation between $r_{X,i}$ and $r_{Y,j}$ vanishes because $\mathbf{x}$ and $\mathbf{y}$ are independent.

To illustrate this procedure, we use covariance matrices estimated from model simulations. The data in this example is as follows:

- Simulations: phase 5 of the Coupled Model Intercomparison Project (Taylor et al., 2012).
- variable: annual-mean, 2m air temperature
- Model: NCAR CCSM4 model[3]
- control simulation: 300 years with pre-industrial ($\sim$ 1850) forcings
- forced simulation: 156-year period 1850–2005 with observation-based forcings.
- ensembles: two forced simulations start from random states from the control.
- grid: global 1.25° × 0.94° (288 × 192 grid points).

---

[2] There is no standard nomenclature for this test. We adopt a terminology that is a hybrid between statistical textbooks and terms used in related multivariate methods such as canonical correlation analysis and discriminant analysis.

[3] For further documentation, see www.cesm.ucar.edu/models/ccsm4.0/ccsm/

Given sample data, it is natural to substitute sample covariance matrices into the generalized eigenvalue problem (16.26). Then, the equation to solve is

$$\hat{\boldsymbol{\Sigma}}_X \hat{\mathbf{q}} = \hat{\lambda} \hat{\boldsymbol{\Sigma}}_Y \hat{\mathbf{q}}. \tag{16.30}$$

Numerically, this eigenvalue problem is solved using standard linear algebra packages.

In this example, the number of grid points is 55,296, which far exceeds the number of time steps. As a result, the covariance matrices in (16.30) are singular. This means it is possible to find a vector $\mathbf{q}$ such that $\mathbf{q}^T \hat{\boldsymbol{\Sigma}}_Y \mathbf{q} = 0$, which makes the variance ratio (16.24) infinity, or undefined if the numerator also vanishes. This situation always occurs when the spatial dimension exceeds the temporal dimension, which is common in climate applications. Even if the covariance matrices were not singular, optimization techniques such as this tend to overfit if there are too many parameters to estimate. By overfitting, we mean that the procedure unavoidably fits some random variability that is peculiar to the sample and not representative of the population. The standard approach to this problem is to *regularize* the problem by introducing constraints on the weights. In climate science applications, the most common regularization is to project the data onto a lower dimensional space, and then maximize the variance ratio in that subspace. This projection reduces the number of adjustable parameters in the projection vector $\mathbf{q}$, and thus reduces the number of parameters that can be varied for optimization, thereby mitigating overfitting.

The most common approach is to project the data onto the leading Empirical Orthogonal Functions (EOFs; see Chapter 12). To avoid biases, the EOFs are usually derived by pooling equal sample sizes of $\mathbf{x}$ and $\mathbf{y}$. Choosing the number of EOFs is known as *variable selection*. There is no general agreement as to the best criterion for variable selection. We recommend a criterion related to Mutual Information Criterion discussed in Section 14.6. Applying this criterion indicates choosing $T = 6$ EOFs (see Section 16.6 for details).

Figure 16.2 shows the discriminant ratios obtained by solving (16.30) in the space of the six leading EOFs. A log-scale is recommended for plotting discriminant ratios so that the same proportional difference above and below one yields the same distance on the y-axis. Note that discriminant ratios can be less than one because (16.25) applies only to population quantities, not sample quantities. The first discriminant ratio is well above the others, but is it significant? Because (16.28) is a variance ratio, one might be tempted to use the univariate F-test (see Chapter 2) to test equality of variances. Unfortunately, $\hat{\lambda}_1$ does not have an F distribution under H_0 because $\hat{\mathbf{q}}_1$ *was chosen to maximize the sample variance ratio.* Instead, we need the sampling distribution that takes into account the maximization. This distribution can be estimated using *Monte Carlo techniques*, as discussed in Section 16.3. Specifically, one draws samples from the *same* population, hence

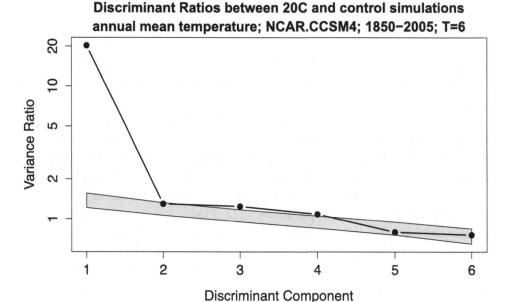

Figure 16.2 Discriminant ratios between twentieth century and control simulations of annual mean temperature from the NCAR CCSM4 model based on truncation parameter $T = 6$ (dots connected by lines). The shaded area shows the 5th to 95th percentile range of discriminant ratios computed from Monte Carlo simulations under the hypothesis of equal covariance matrices.

H_0 is true, performs CDA, and then repeats many times to estimate quantiles of the discriminant ratios. The resulting 5th to 95th percentile range is shown as the shaded region in Figure 16.2. The first discriminant ratio is well above the shaded, implying that the null hypothesis should be rejected at the 5% level. As discussed in Section 16.3, the second and higher ratios are consistent with the hypothesis that only one discriminant ratio differs significantly from one.

This analysis reveals more than a mere difference between covariance matrices. In particular, it shows that *a single component* dominates the difference in variability between the two simulations. The variates $r_{X,1}(t)$ and $r_{Y,1}(t)$ describe the time variation of the component, and the loading vector $\mathbf{p}_1$ describes the spatial structure. The variates are computed from (16.27) and shown in Figure 16.3b. The variate from the control simulation is plotted on the time axis before 1850, while variates for the forced simulation are plotted after 1850. While the variates from both the control simluation and forced simulation exhibit similar randomness on year-to-year time scales, a positive trend is present in the variate from the forced simulations. The fact that two realizations of variate 1 from the forced simulation are so close to each other indicates that discriminant analysis has removed substantial internal variability, leaving mostly the response to external forcing. The i'th loading

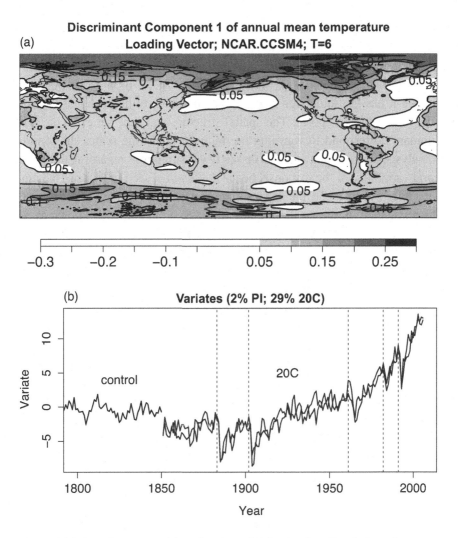

Figure 16.3 Loading vector (a) and variates (b) for the first discriminant between the forced and control simulations of annual-mean near-surface air temperature from the NCAR CCSM4 model. The component was derived from the first six EOFs of the total data. Time series for years before 1850 come from the control, while time series after 1850 come from the forced simulation. Variables are expressed as anomalies relative to the control mean. The dashed vertical lines indicate years of the five most significant volcanic eruptions (in terms of the change in visible optical depth): Krakatoa in 1883, Santa Maria in 1902, Mt. Agung in 1961, El Chichon in 1982, and Mt. Pinatubo in 1991. The units of the loading vector is degrees Kelvin per standard deviation of the variate. The percentage variance explained in the control and forced runs is indicated in the title of the bottom panel.

vector is a vector $\mathbf{p}_i$ such that $\mathbf{p}_i r_{X,i}(t)$ best approximates $\mathbf{x}(t)$ in a certain sense (see Sections 16.4 and 16.5 for this and other interpretations of loading vectors.) The product of a loading vector and its variate is called a *discriminant component*. The leading loading vector $\mathbf{p}_1$ is shown in Figure 16.3a. Its most distinctive feature is that it has the same sign everywhere and tends to have largest amplitudes over the continents and poles. The spatial structure and time series of the leading discriminant correspond to warming on multi-decadal time scales. In contrast to trend analysis, discriminant analysis makes few assumptions regarding the spatial or temporal variability of the response, hence the features seen in Figure 16.3 emerge from the data. For instance, the discriminant pattern implies that the warming trend is strongest locally in the northern polar latitudes, a phenomenon known as *Arctic amplification*. The variates decrease abruptly following major volcanic eruptions and then recover after a few years, a behavior that is expected following the injection of stratospheric aerosols. The discriminant component explains about 2% of the variability in the control simulation but almost 30% of the variability in the forced simulation. Such a change in variability is very unlikely to occur by random chance.

There are some caveats worth emphasizing about these results. First, the significance test did not account for the fact that the choice $T = 6$ was made by optimizing a selection criterion. Also, it did not account for serial correlation in the time series. Second, the space of variability was based on EOFs, which may create biases even though both data sets were pooled evenly to compute the EOFs. Third, it is known that Gaussian-based hypothesis tests on variances are sensitive to departures from normality (Box, 1953; Olson, 1974). If the population is suspected of being non-Gaussian, permutation techniques can be used instead to test significance (LaJoie and DelSole, 2016).

16.3 Hypothesis Testing

The differences identified by CDA are meaningful only if those differences reflect real differences between populations, rather than sampling variability. A significance test for differences in covariance matrices was described in Section 16.2 and illustrated in Figure 16.2. This section discusses further details and other related issues.

A desirable property of any hypothesis test is that it should not depend on the physical units used to measure data. For example, suppose CDA is applied to temperature data in which all temperatures are measured in Celsius units. Next, the same analysis is repeated in its entirety, except all temperatures are measured in Fahrenheit units. If the two tests yield different outcomes, then the test would be meaningless because there is no reason to prefer one physical unit over another. By similar reasoning, a test should not depend on the *basis set* used to represent the data. After all, a vector space can be represented by different basis sets, but there is no reason to prefer one over another. Since a basis set can be transformed into another by linear transformation, it follows that the test should be invariant to *invertible affine transformations*. An affine transformation is of the form

$$\mathbf{x}^* = \mathbf{L}\mathbf{x} + \mathbf{c} \quad \text{and } \mathbf{y}^* = \mathbf{L}\mathbf{y} + \mathbf{c}, \tag{16.31}$$

where $\mathbf{L}$ is an invertible matrix and $\mathbf{c}$ is an arbitrary vector. Note that unit changes, or any change in scale, correspond to the special case in which $\mathbf{L}$ is a diagonal matrix.

Importantly, the *discriminant ratios* are invariant to invertible affine transformations. To show this, let the eigenvalue problem associated for CDA of $\mathbf{x}^*$ and $\mathbf{y}^*$ in (16.31) be

$$\boldsymbol{\Sigma}_{X^*}\mathbf{q}^* = \lambda\boldsymbol{\Sigma}_{Y^*}\mathbf{q}^*. \tag{16.32}$$

Recall that adding a constant $\mathbf{c}$ to a random vector has no impact on covariances. Furthermore, transformations of the form (16.31) will transform $\boldsymbol{\Sigma}_X$ and $\boldsymbol{\Sigma}_Y$ into

$$\boldsymbol{\Sigma}_{X^*} = \mathbf{L}\boldsymbol{\Sigma}_X\mathbf{L}^T \quad \text{and} \quad \boldsymbol{\Sigma}_{Y^*} = \mathbf{L}\boldsymbol{\Sigma}_Y\mathbf{L}^T, \tag{16.33}$$

(see Chapter 7). Substituting (16.33) into (16.32) yields

$$\mathbf{L}\boldsymbol{\Sigma}_X\mathbf{L}^T\mathbf{q}^* = \lambda\mathbf{L}\boldsymbol{\Sigma}_Y\mathbf{L}^T\mathbf{q}^*. \tag{16.34}$$

Since $\mathbf{L}$ is nonsingular, we may multiply by $\mathbf{L}^{-1}$ to eliminate $\mathbf{L}$ on the left. Moreover, define a new vector $\mathbf{L}^T\mathbf{q}^* = \mathbf{q}$, so that eigenvalue problem (16.34) can be written as

$$\boldsymbol{\Sigma}_X\mathbf{q} = \lambda\boldsymbol{\Sigma}_Y\mathbf{q}. \tag{16.35}$$

Since the matrices in this equation are identical to those in (16.26), it follows that the eigenvalues must be identical. Therefore, the eigenvalues of the generalized eigenvalue problem are invariant to invertible affine transformations of the variables.

It can be shown that *any* measure of the difference between covariance matrices which is invariant to affine transformations must depend *only* on discriminant ratios. Thus, it suffices to determine the sampling distribution of the discriminant ratios. The sampling distribution of any function of the discriminant ratios can be estimated by Monte Carlo techniques. In this technique, one evaluates the test statistic repeatedly using random numbers drawn from a population in which the null hypothesis is true. The 5% significance threshold is then estimated from the 95th percentile of the numerically generated test statistics. In our problem, the null hypothesis is that the population covariance matrices are equal. But what covariance matrix should be chosen? Recall that the test statistic should be invariant to invertible linear transformations. Any positive definite covariance matrix can be transformed into the identity matrix by suitable linear transformation.

To see this, note that if the eigenvector decomposition of the covariance matrix is $\boldsymbol{\Sigma} = \mathbf{U}\mathbf{S}\mathbf{U}^T$, then

$$\text{choosing} \quad \mathbf{L} = \mathbf{S}^{-1/2}\mathbf{U}^T \quad \text{gives} \quad \mathbf{L}\boldsymbol{\Sigma}\mathbf{L}^T = \mathbf{I}. \tag{16.36}$$

Therefore, realizations of $\mathbf{x}$ and $\mathbf{y}$ may be generated by drawing independent random numbers from $\mathcal{N}(\mathbf{0}, \mathbf{I})$, or equivalently, from a standardized *univariate* normal distribution. In particular, the sampling distribution *cannot depend on the population covariance matrix* $\boldsymbol{\Sigma}$, which is important since $\boldsymbol{\Sigma}$ is unknown. For multivariate normal distributions, the sampling distribution of the discriminant ratios depends only on the sample sizes and dimension.

What function of the discriminant ratios should be chosen? In the case of two variances, the only function of two variances that satisfies this invariance property is the ratio of variances, which leads to the F-statistic for hypothesis testing. In the multivariate case, the most common measure is based on the likelihood ratio test, which leads to

$$\Lambda = \left(\frac{|\overline{\boldsymbol{\Sigma}}_X|^{N_X} |\overline{\boldsymbol{\Sigma}}_Y|^{N_Y}}{|\overline{\boldsymbol{\Sigma}}_P|^{N_P}} \right)^{1/N_P}, \tag{16.37}$$

where the matrices are defined in (14.31) and (16.108). This measure is invariant to invertible affine transformations and can also be written in terms of discriminant ratios. Also, the sampling distribution can be estimated by exactly the same Monte Carlo procedure described previously, since that procedure merely generates realizations of $\overline{\boldsymbol{\Sigma}}_X$ and $\overline{\boldsymbol{\Sigma}}_Y$, which can be substituted repeatedly in (16.37) to estimate quantiles of the sampling distribution.

Other test statistics that satisfy the invariance property could be chosen. No single measure is best suited for all possible situations. For each measure, the threshold value for a Type I error rate can be determined from Monte Carlo techniques. However, when the null hypothesis is false, the tests have different *powers* depending on how the null hypothesis is violated; that is, depending on the difference in covariance matrices

$$\Delta = \boldsymbol{\Sigma}_X - \boldsymbol{\Sigma}_Y. \tag{16.38}$$

In seasonal and climate predictability studies, the difference in populations is often dominated by a single pattern. For example, in seasonal forecasting, a single ENSO teleconnection pattern often dominates. In climate change, a single pattern often dominates the response to human influences. In such situations, the difference in covariance matrices is dominated by a rank-1 matrix, and in such cases the most powerful test is based on the leading discriminant ratio $\hat{\lambda}_1$. This statistic is known as *Roy's largest root* and testing its significance is known as the *union-intersection test* (Flury, 1985). Applying this test to Figure 16.2, we see that the leading discriminant ratio lies well above the 95th percentile of the leading discriminant ratios under the null hypothesis, hence we reject the null hypothesis.

What about the second discriminant ratio? In Figure 16.2, the second discriminant ratio lies below the 95th percentile, but the third and fourth ratios lie above the threshold. What should we conclude? It would be inconsistent to reject H_0 based on the first ratio and then to assume it when deciding the significance of the remaining ratios. Therefore, the shaded area in Figure 16.2 is not relevant to deciding significance of the second and higher ratios. Instead, because H_0 was rejected based on a test for rank-1 differences, it is natural to hypothesize that the populations differ by a rank-1 matrix. A rank-1 covariance matrix can be represented as $\Delta = \mathbf{w}\mathbf{w}^T$ for some vector $\mathbf{w}$. Accordingly, we consider the hypothesis

$$H_1 : \quad \mathbf{w}\mathbf{w}^T = \mathbf{\Sigma}_X - \mathbf{\Sigma}_Y. \tag{16.39}$$

We want the sampling distribution of the discriminant ratios under the new hypothesis H_1. This distribution depends on $\mathbf{w}$, which is unknown. Under hypothesis H_1, the generalized eigenvalue problem (16.26) becomes

$$\mathbf{\Sigma}_X \mathbf{q} = \lambda \mathbf{\Sigma}_Y \mathbf{q} \tag{16.40}$$

$$\left(\mathbf{w}\mathbf{w}^T + \mathbf{\Sigma}_Y\right) \mathbf{q} = \lambda \mathbf{\Sigma}_Y \mathbf{q} \tag{16.41}$$

$$\mathbf{w}\mathbf{w}^T \mathbf{q} = (\lambda - 1) \mathbf{\Sigma}_Y \mathbf{q}. \tag{16.42}$$

Because the matrix on the left is rank-1, only one nontrivial eigenvalue exists. It is straightforward to verify that $\mathbf{q} = \mathbf{\Sigma}_Y^{-1} \mathbf{w}$ is an eigenvector and the corresponding eigenvalue is

$$\lambda = 1 + \mathbf{w}^T \mathbf{\Sigma}_Y^{-1} \mathbf{w}. \tag{16.43}$$

This result shows two things about the *population* discriminant ratios under H_1: only one ratio differs from unity, and this ratio depends only on $\mathbf{w}^T \mathbf{\Sigma}_Y^{-1} \mathbf{w}$, which is a measure of the "length" of $\mathbf{w}$. The sampling distribution of the discriminant ratios under hypothesis H_1 can be quantified by Monte Carlo techniques. Specifically, we draw samples from

$$\mathbf{x} = \sqrt{\lambda_0 - 1} z \, \mathbf{w} + \boldsymbol{\epsilon} \quad \text{and} \quad \mathbf{y} = \boldsymbol{\epsilon}^*, \tag{16.44}$$

where $z \sim \mathcal{N}(0, 1)$, $\mathbf{w}$ is chosen to yield $\mathbf{w}^T \mathbf{\Sigma}_Y^{-1} \mathbf{w} = 1$, and $\boldsymbol{\epsilon}$ and $\boldsymbol{\epsilon}^*$ are independent random vectors from a population with covariance matrix $\mathbf{\Sigma}_Y$ (which can be arbitrary). Model (16.44) has one (population) discriminant ratio equal to λ and all other discriminant ratios equal to 1. The 95th percentiles of the eigenvalues for selected values of λ are shown in Figure 16.4. As expected, the 95th percentile of the first eigenvalue increases with λ. However, the 95th percentile of the second and higher eigenvalues increases slightly with λ. This slight inflation is consistent with the third and fourth discriminant ratios seen in Figure 16.2. Overall, then, the results

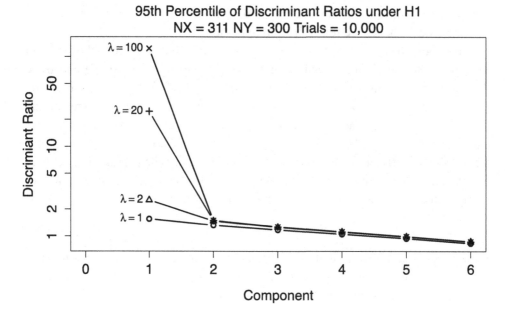

Figure 16.4 The 95th percentile of discriminant ratios from 10,000 random samples from model (16.44), for parameter values consistent with the example in Section 16.2. The four curves show results for four selected *population* values $\lambda = 1, 2, 20, 100$.

shown in Figure 16.2 are consistent with hypothesis H_1, that is, the two population covariance matrices differ by a rank-1 matrix.

16.4 The Solution

This section proves that the generalized eigenvalue problem (16.26) solves the discriminant analysis problem and proves various properties of the discriminant components.

An extremum of λ in (16.24) can be found by solving $\partial\lambda/\partial\mathbf{q}=0$ (see Section A.3):

$$\frac{\partial\lambda}{\partial\mathbf{q}} = \frac{2\boldsymbol{\Sigma}_X\mathbf{q}}{\mathbf{q}^T\boldsymbol{\Sigma}_Y\mathbf{q}} - 2\frac{\mathbf{q}^T\boldsymbol{\Sigma}_X\mathbf{q}}{(\mathbf{q}^T\boldsymbol{\Sigma}_Y\mathbf{q})^2}\boldsymbol{\Sigma}_Y\mathbf{q}$$

$$= \frac{2}{\mathbf{q}^T\boldsymbol{\Sigma}_Y\mathbf{q}}(\boldsymbol{\Sigma}_X\mathbf{q} - \lambda\boldsymbol{\Sigma}_Y\mathbf{q}) = 0. \tag{16.45}$$

We assume that $\boldsymbol{\Sigma}_X$ and $\boldsymbol{\Sigma}_Y$ are positive definite (Section 16.5 discusses the case when this assumption does not hold). In this case, the derivative vanishes if and only if

$$\boldsymbol{\Sigma}_X\mathbf{q} = \lambda\boldsymbol{\Sigma}_Y\mathbf{q}. \tag{16.46}$$

Normalization of the Projection Vectors

In general, eigenvectors are not unique. For instance, if $\mathbf{q}$ is an eigenvector solution of (16.26), then $\alpha\mathbf{q}$ (where α is nonzero) is also an eigenvector, because it satisfies (16.26) too. To select a particular vector, it is convention to normalize the projection vector so that $r_{Y,i}$ has unit variance; that is, select $\mathbf{q}$ so that $\mathrm{var}[r_{Y,i}] = 1$. This convention, plus the definition (16.28), imply

$$\mathrm{var}[r_{Y,i}] = 1 \quad \text{and} \quad \mathrm{var}[r_{X,i}] = \lambda_i. \tag{16.47}$$

Note that the *sign* of the projection vector is not constrained by this convention.

Properties of the Projection Vectors

The matrices in the generalized eigenvalue problem (16.26) are symmetric. This symmetry imparts special properties on the discriminant coefficients and variates. In particular, the eigenvectors from the generalized eigenvalue problem (16.26) satisfy the following orthogonality properties (proven in Section A.2)

$$\mathbf{q}_i^T \boldsymbol{\Sigma}_X \mathbf{q}_j = 0 \quad \text{and} \quad \mathbf{q}_i^T \boldsymbol{\Sigma}_Y \mathbf{q}_j = 0, \quad \text{for } i \neq j. \tag{16.48}$$

These properties can be expressed in more concise form by defining

$$\mathbf{Q} = [\mathbf{q}_1 \quad \mathbf{q}_2 \quad \cdots \quad \mathbf{q}_S]. \tag{16.49}$$

Using this matrix, the eigenvalue problem (16.26) can be written equivalently as

$$\boldsymbol{\Sigma}_X \mathbf{Q} = \boldsymbol{\Sigma}_Y \mathbf{Q} \boldsymbol{\Lambda}, \tag{16.50}$$

where $\boldsymbol{\Lambda}$ is a diagonal matrix whose diagonal elements are the eigenvalues of (16.26). Then, the orthogonality properties (16.48) and normalization (16.47) can be expressed as

$$\mathbf{Q}^T \boldsymbol{\Sigma}_Y \mathbf{Q} = \mathbf{I}. \tag{16.51}$$

Also, pre-multiplying both sides of (16.50) by $\mathbf{Q}^T$ and invoking (16.51) shows that

$$\mathbf{Q}^T \boldsymbol{\Sigma}_X \mathbf{Q} = \boldsymbol{\Lambda}. \tag{16.52}$$

Since both covariance matrices are positive definite, the eigenvectors can be chosen to be linearly independent, which ensures that $\mathbf{Q}$ is invertible.

Variates Are Uncorrelated

Identities (16.51) and (16.52) for the projection vectors imply that the variates are uncorrelated. To see this, collect the variates into a single vector as

$$\mathbf{r}_X = \begin{bmatrix} r_{X,1} & r_{X,2} & \cdots & r_{X,S} \end{bmatrix}^T, \tag{16.53}$$

and similarly for $\mathbf{r}_Y$. Then, the definitions (16.27) imply

$$\mathbf{r}_X = \mathbf{Q}^T \mathbf{x} \quad \text{and} \quad \mathbf{r}_Y = \mathbf{Q}^T \mathbf{y}. \tag{16.54}$$

The covariance matrix for $\mathbf{r}_Y$ is thus

$$\text{cov}[\mathbf{r}_Y, \mathbf{r}_Y] = \text{cov}[\mathbf{Q}^T \mathbf{y}, \mathbf{Q}^T \mathbf{y}] = \mathbf{Q}^T \, \text{cov}[\mathbf{y}, \mathbf{y}] \mathbf{Q} = \mathbf{Q}^T \boldsymbol{\Sigma}_Y \mathbf{Q} = \mathbf{I}. \tag{16.55}$$

This identity summarizes two distinct properties. First, the vanishing off-diagonal elements imply that the variates are uncorrelated, because

$$\text{cov}[r_{Y,i}, r_{Y,j}] = \mathbf{q}_i^T \boldsymbol{\Sigma}_Y \mathbf{q}_j = 0 \quad \text{for } i \neq j. \tag{16.56}$$

Second, because the diagonal elements of the covariance matrix equal one, the variates have unit variance, consistent with (16.47). Similarly, the covariance matrix for $\mathbf{r}_X$ is

$$\text{cov}[\mathbf{r}_X, \mathbf{r}_X] = \text{cov}[\mathbf{Q}^T \mathbf{x}, \mathbf{Q}^T \mathbf{x}] = \mathbf{Q}^T \, \text{cov}[\mathbf{x}, \mathbf{x}] \mathbf{Q} = \mathbf{Q}^T \boldsymbol{\Sigma}_X \mathbf{Q} = \boldsymbol{\Lambda}. \tag{16.57}$$

Again, the diagonal nature of $\boldsymbol{\Lambda}$ implies the variates for X are uncorrelated, and that the variance of $r_{X,i}$ is λ_i, consistent with (16.47).

Loading Vectors Since $\mathbf{Q}$ is invertible, (16.51) and (16.52) can be written equivalently as

$$\boldsymbol{\Sigma}_X = \mathbf{Q}^{T-1} \boldsymbol{\Lambda} \mathbf{Q}^{-1} \quad \text{and} \quad \boldsymbol{\Sigma}_Y = \mathbf{Q}^{T-1} \mathbf{Q}^{-1}. \tag{16.58}$$

If we define

$$\mathbf{P} = \left(\mathbf{Q}^T \right)^{-1}, \tag{16.59}$$

then the covariance matrices can be written as

$$\boldsymbol{\Sigma}_X = \mathbf{P} \boldsymbol{\Lambda} \mathbf{P}^T \quad \text{and} \quad \boldsymbol{\Sigma}_Y = \mathbf{P} \mathbf{P}^T. \tag{16.60}$$

Also, multiplying (16.54) by $\mathbf{P}$ yields

$$\mathbf{x} = \mathbf{P} \mathbf{r}_X \quad \text{and} \quad \mathbf{y} = \mathbf{P} \mathbf{r}_Y. \tag{16.61}$$

The columns of $\mathbf{P}$, denoted $\mathbf{p}_1, \ldots, \mathbf{p}_S$, are known as *loading vectors*. The expression $\mathbf{P} \mathbf{r}_X$ represents a *sum of discriminant components*:

$$\mathbf{P} \mathbf{r}_X = \mathbf{p}_1 r_{X,1} + \cdots + \mathbf{p}_S r_{X,S}. \tag{16.62}$$

Thus, the identities (16.61) show that $\mathbf{x}$ and $\mathbf{y}$ can be expressed separately as a linear combination of variates multiplied by their corresponding loading vector. In other words, each random vector can be *decomposed* into a sum of components ordered by their variance ratios. This decomposition is useful for interpreting differences in variability.

Loading Vectors Are Regression Patterns

The loading vectors may also be derived as *regression patterns*. To do this, we find the matrix $\mathbf{P}$ that minimizes the difference measure

$$\gamma_Y = \mathbb{E}\left[(\mathbf{y}' - \mathbf{Pr}_Y)^T \mathbf{W} (\mathbf{y}' - \mathbf{Pr}_Y) \right], \qquad (16.63)$$

where $\mathbf{y}' = \mathbf{y} - \mathbb{E}[\mathbf{y}]$ and $\mathbf{W}$ is a matrix that defines the distance measure (typically, it is a diagonal matrix whose diagonal elements are proportional to the local area). The solution is found by differentiating with respect to $\mathbf{P}$:

$$\frac{d\gamma_Y}{d\mathbf{P}} = (-2)\mathbb{E}\left[\mathbf{W}(\mathbf{y}' - \mathbf{Pr}_Y)\, \mathbf{r}_Y^T \right] = (-2)\mathbf{W} \left(\text{cov}[\mathbf{y}, \mathbf{r}_Y] - \mathbf{P}\,\text{cov}[\mathbf{r}_Y, \mathbf{r}_Y] \right).$$

Setting this to zero and solving yields

$$\mathbf{P} = \text{cov}[\mathbf{y}, \mathbf{r}_Y], \qquad (16.64)$$

where (16.55) has been used. The i'th column of $\mathbf{P}$ can be expressed as

$$\mathbf{p}_i = \text{cov}[\mathbf{y}, r_{Y,i}], \qquad (16.65)$$

which shows that $\mathbf{p}_i$ is the *regression pattern* between $\mathbf{y}$ and the standardized variate $r_{Y,i}$. The product $\mathbf{p}_i r_{Y,i}$, is known as the i'th *discriminant component* of Y. Substituting $r_{Y,i} = \mathbf{q}_i^T \mathbf{y}$ yields the relation

$$\mathbf{p}_i = \text{cov}[\mathbf{y}, \mathbf{q}_i^T \mathbf{y}] = \mathbf{\Sigma}_Y \mathbf{q}_i. \qquad (16.66)$$

The relation could have been derived from (16.58). Finally, note that the solution (16.64) is independent of $\mathbf{W}$. Thus, the loading vectors minimize the distance measure γ_Y for all choices of positive definite matrices $\mathbf{W}$.

Similarly, $\mathbf{x}$ can be decomposed by determining the matrix $\mathbf{P}$ that minimizes

$$\gamma_X = \mathbb{E}\left[(\mathbf{x}' - \mathbf{Pr}_X)^T \mathbf{W} (\mathbf{x}' - \mathbf{Pr}_X) \right]. \qquad (16.67)$$

Again, the solution is

$$\mathbf{P} = \text{cov}[\mathbf{x}, \mathbf{r}_X] \, (\text{var}[\mathbf{r}_X, \mathbf{r}_X])^{-1} = \text{cov}[\mathbf{x}, \mathbf{Q}^T \mathbf{x}] \mathbf{\Lambda}^{-1} = \mathbf{\Sigma}_X \mathbf{Q} \mathbf{\Lambda}^{-1} = \mathbf{\Sigma}_Y \mathbf{Q}, \quad (16.68)$$

where (16.50) has been used. Note that the loading vectors defined by (16.64) and (16.68) are identical despite being derived from two separate random variables $\mathbf{x}$ and $\mathbf{y}$. For this reason, $\mathbf{P}$ is not distinguished by subscript X or Y.

Explained Variance

This decomposition also decomposes total variance. Specifically, the sum total variance of the elements of **y** can be defined as

$$v_Y = \mathbb{E}\left[\mathbf{y}'^T \mathbf{W} \mathbf{y}'\right]. \tag{16.69}$$

Using the decomposition (16.60) and standard properties of the trace operator, the total variance can be written as

$$v_Y = \mathbb{E}\left[(\mathbf{y}')^T \mathbf{W} \mathbf{y}'\right] = \mathbb{E}\left[\text{tr}\left[\mathbf{W} \mathbf{y}'(\mathbf{y}')^T\right]\right] = \text{tr}\left[\mathbf{W} \mathbf{\Sigma}_Y\right]$$

$$= \text{tr}\left[\mathbf{W} \mathbf{P} \mathbf{P}^T\right] = \sum_{i=1}^{S} \mathbf{p}_i^T \mathbf{W} \mathbf{p}_i. \tag{16.70}$$

This expression demonstrates that the sum total variance can be expressed as a sum of terms involving loading vectors *with no cross terms*. Consequently, we may speak of the i'th discriminant component as explaining a certain amount of variance. More precisely,

$$\left(\begin{array}{c}\text{fraction of variance in } Y \text{ explained by} \\ i\text{'th discriminant component}\end{array}\right) = \frac{\mathbf{p}_i^T \mathbf{W} \mathbf{p}_i}{\text{tr}\left[\mathbf{\Sigma}_Y\right]}. \tag{16.71}$$

Similarly, we have

$$\left(\begin{array}{c}\text{fraction of variance in } X \text{ explained by} \\ i\text{'th discriminant component}\end{array}\right) = \lambda_i \frac{\mathbf{p}_i^T \mathbf{W} \mathbf{p}_i}{\text{tr}\left[\mathbf{\Sigma}_X\right]}. \tag{16.72}$$

Proof of the Optimality Properties

Technically, the discriminant ratios are stationary values of the variance ratio (that is, the derivative vanishes), but it has not yet been shown that they actually maximize or minimize the variance ratio. The fact that the projection vectors actually maximize or minimize the variance ratio can be proven using the aforementioned properties. Specifically, because **Q** is nonsingular, we may define the following vector

$$\mathbf{z} = \mathbf{Q}^{-1}\mathbf{q}. \tag{16.73}$$

The vector **z** is simply a different way of representing **q**. Inverting this formula gives

$$\mathbf{q} = \mathbf{Q}\mathbf{z}. \tag{16.74}$$

Substituting this projection vector into the variance ratio (16.24) and invoking orthogonality properties (16.51) and (16.52) yields

$$\lambda = \frac{\mathbf{z}^T \mathbf{\Lambda} \mathbf{z}}{\mathbf{z}^T \mathbf{z}}. \tag{16.75}$$

Because $\boldsymbol{\Lambda}$ is diagonal, we have

$$\mathbf{z}^T \boldsymbol{\Lambda} \mathbf{z} = \sum_{s=1}^{S} \lambda_s z_s^2 = \lambda_1 z_1^2 + \lambda_2 z_2^2 + \cdots + \lambda_S z_S^2. \tag{16.76}$$

Each term in Equation (16.76) is positive. Also, the λ's are positive and ordered from largest to smallest, hence $\lambda_1 \geq \lambda_s$ for any s. Therefore, if we replace $\lambda_2, \lambda_3, \ldots$ in the sum by λ_1, we obtain a new sum that is greater than or equal to the original sum. That is,

$$\mathbf{z}^T \boldsymbol{\Lambda} \mathbf{z} = \sum_{i=1}^{S} \lambda_i z_i^2 \leq \lambda_1 \sum_{i=1}^{S} z_i^2 = \lambda_1 \left(\mathbf{z}^T \mathbf{z} \right). \tag{16.77}$$

Dividing both sides of this equation by $\mathbf{z}^T \mathbf{z}$ yields the bound

$$\lambda \leq \lambda_1. \tag{16.78}$$

This inequality proves that the maximum variance ratio is the largest eigenvalue λ_1 of (16.26). It is straightforward to see that if we substitute

$$\mathbf{z} = \begin{pmatrix} 1 \\ 0 \\ \vdots \\ 0 \end{pmatrix}, \tag{16.79}$$

into (16.75), we obtain $\lambda = \lambda_1$. Thus, λ_1 is not only a maximum, but it can be achieved for suitable choice of $\mathbf{z}$. Substituting (16.79) into (16.74) yields the first column of $\mathbf{Q}$, namely $\mathbf{q}_1$. These considerations prove that the maximum variance ratio is λ_1 and is obtained when the projection vector is $\mathbf{q}_1$. This proof is rigorous and avoids calculus, but it requires knowing the representation (16.74) that simultaneously diagonalizes two covariance matrices.

Next, consider the maximum variance ratio among all vectors $\mathbf{z}$ that are orthogonal to (16.79). This constraint is equivalent to constraining the first element of $\mathbf{z}$ to be zero. Under this constraint, the summation in (16.77) starts at $s = 2$ instead of $s = 1$, in which case the upper bound becomes λ_2. Moreover, this variance ratio is obtained by setting the second element of $\mathbf{z}$ to one and the other elements to zero, which corresponds to selecting the second projection vector $\mathbf{q}_2$. Repeating this argument, we may prove that each succeeding component has the largest possible variance ratio under the constraint that it is orthogonal with the preceding components. Finally, from (16.56), we have

$$\operatorname{cov}[r_{Y,i}, r_{Y,j}] = \operatorname{cov}[\mathbf{q}_i^T \mathbf{y}, \mathbf{q}_j^T \mathbf{y}] = \mathbf{q}_i^T \boldsymbol{\Sigma}_Y \mathbf{q}_i = \mathbf{z}_i^T \mathbf{z}_j. \tag{16.80}$$

Thus, choosing **z**-vectors to be orthogonal is equivalent to choosing variates to be *uncorrelated*. Thus, the second discriminant ratio λ_2 is the maximum variance ratio among all variates that are uncorrelated with the first. Similarly, the third discriminant ratio λ_3 is the maximum variance ratio among all variates uncorrelated with the first two, and so on.

An argument similar to this can be used to prove that λ_S is the *minimum* possible variance ratio. In particular, we have the bound

$$\mathbf{z}^T \mathbf{\Lambda} \mathbf{z} = \sum_{i=1}^{S} \lambda_i z_i^2 \geq \lambda_S \sum_{i=1}^{S} z_i^2 = \lambda_S \left(\mathbf{z}^T \mathbf{z} \right). \tag{16.81}$$

Dividing both sides of this equation by $\mathbf{z}^T \mathbf{z}$ yields the bound

$$\lambda \geq \lambda_S. \tag{16.82}$$

Moreover, this bound is obtained when **z** has a one in the *last* element and a zero in the other elements. In a similar manner, we can prove that each successive component, *starting from the last*, has the *minimum* possible variance ratio among all choices that are uncorrelated to the preceding components.

16.5 Solution in a Reduced-Dimensional Subspace

This section discusses practical details of performing CDA, particularly the use of EOFs. Let **x** and **y** represent random variables on a spatial grid with S points, and let the number of samples be N_X and N_Y, respectively. The data sets can be collected in *data matrices* as

$$\begin{array}{ccc} \mathbf{X} & \text{and} & \mathbf{Y}, \\ N_X \times S & & N_Y \times S \end{array} \tag{16.83}$$

where the dimensions are indicated below each matrix. We assume that some reference has been subtracted from each row separately. Usually, this is the sample mean of the respective data sets, but it could be that the sample mean of one data set is subtracted from the other, as in Section 16.2. The spatial dimension S must be equal in the two data sets, but the time dimensions N_X and N_Y can differ.

Typically, S is much larger than either N_X or N_Y, so the sample covariance matrices are singular (hence not positive definite). As explained in Section 16.2, this yields singular matrices and overfitting problems. In climate applications, the standard procedure is to project the data onto a new space with lower dimension. To distinguish different spaces, let *data space* denote the original space of **X** and **Y**, and let *feature space* denote the space to which the data will be projected. To project data onto feature space, a basis set that spans the feature space is needed.

These basis vectors are known as *feature vectors* and denoted $\mathbf{e}_1, \mathbf{e}_2, \ldots, \mathbf{e}_T$, where T is the *feature dimension*. In the example discussed in Section 16.2, EOFs were used. In this section, we assume the feature vectors are linearly independent, but otherwise make no further assumptions about them. Collect the feature vectors into a single $S \times T$ matrix

$$\mathbf{E} = \begin{bmatrix} \mathbf{e}_1 & \mathbf{e}_2 & \cdots & \mathbf{e}_T \end{bmatrix}. \tag{16.84}$$

The amplitudes of the feature vectors are determined by projection methods, which amounts to multiplying data matrices by the *generalized inverse* $\mathbf{E}^i$

$$\mathbf{E}^i = \mathbf{WE} \left(\mathbf{E}^T \mathbf{WE} \right)^{-1}, \tag{16.85}$$

where $\mathbf{W}$ is a positive definite matrix that defines how grid points are weighted (e.g., often a diagonal matrix whose diagonal elements are proportional to the local area; see Chapter 14 for further details). Note that there is only one set of feature vectors because $\mathbf{x}$ and $\mathbf{y}$ must be in the same space. The amplitudes of the feature vectors are then obtained as

$$\begin{array}{ccc} \tilde{\mathbf{F}}_X & = & \mathbf{X} & \mathbf{E}^i, \\ N_X \times T & & N_X \times S & S \times T \end{array} \tag{16.86}$$

and

$$\begin{array}{ccc} \tilde{\mathbf{F}}_Y & = & \mathbf{Y} & \mathbf{E}^i, \\ N_Y \times T & & N_Y \times S & S \times T \end{array} \tag{16.87}$$

where tildes indicate quantities in feature space. $\tilde{\mathbf{F}}_X$ and $\tilde{\mathbf{F}}_Y$ are known as *feature time series*. Multiplying feature time series by feature vectors *does not recover the original data*, because feature vectors capture only part of the data as a result of truncation of the feature space dimension. This product will be denoted by

$$\begin{array}{ccc} \dot{\mathbf{X}} & = & \tilde{\mathbf{F}}_X & \mathbf{E}^T, \\ N_X \times S & & N_X \times T & T \times S \end{array} \tag{16.88}$$

and

$$\begin{array}{ccc} \dot{\mathbf{Y}} & = & \tilde{\mathbf{F}}_Y & \mathbf{E}^T, \\ N_Y \times S & & N_Y \times T & T \times S \end{array} \tag{16.89}$$

where the dot indicates that only the part that projects on the feature vectors is included. If the feature vectors are EOFs, then $\dot{\mathbf{X}}$ and $\dot{\mathbf{Y}}$ often explain a large fraction of the variance.

Feature time series $\tilde{\mathbf{F}}_X$ and $\tilde{\mathbf{F}}_Y$ are centered the same way as $\mathbf{X}$ and $\mathbf{Y}$ and have the same format as $\mathbf{X}$ and $\mathbf{Y}$, but have reduced dimension T instead of S. The idea now is to apply CDA to the feature time series. Sample covariances in feature space are

$$\tilde{\boldsymbol{\Sigma}}_X = \frac{1}{N_X - 1}\tilde{\mathbf{F}}_X^T\tilde{\mathbf{F}}_X \quad \text{and} \quad \tilde{\boldsymbol{\Sigma}}_Y = \frac{1}{N_Y - 1}\tilde{\mathbf{F}}_Y^T\tilde{\mathbf{F}}_Y. \tag{16.90}$$

The projection vectors that optimize variance ratios are then determined by solving the generalized eigenvalue problem

$$\tilde{\boldsymbol{\Sigma}}_X\tilde{\mathbf{q}} = \lambda\tilde{\boldsymbol{\Sigma}}_Y\tilde{\mathbf{q}}. \tag{16.91}$$

These covariance matrices are $T \times T$, hence there are T eigenvalues, which are ordered from largest to smallest, $\lambda_1 \geq \lambda_2 \geq \cdots \geq \lambda_T$. The corresponding eigenvectors $\tilde{\mathbf{q}}_1, \tilde{\mathbf{q}}_2, \ldots, \tilde{\mathbf{q}}_T$ are the projection vectors for the T feature vectors.

Covariance Discriminant Analysis then proceeds as discussed in Section 16.4. The steps are essentially the same, so the results are merely summarized in Table 16.1. Note that in the sample version the variates are matrices instead of random vectors:

$$\begin{array}{ccc} \tilde{\mathbf{R}}_X & = & \tilde{\mathbf{F}}_X & \tilde{\mathbf{Q}} \\ N \times T & & N_X \times T & T \times T \end{array}, \tag{16.92}$$

and

$$\begin{array}{ccc} \tilde{\mathbf{R}}_Y & = & \tilde{\mathbf{F}}_Y & \tilde{\mathbf{Q}} \\ N \times T & & N_Y \times T & T \times T \end{array}. \tag{16.93}$$

The only extra step is to transform the loading vector from feature space to data space, so that the loading vector can be interpreted. The loading vectors follow from (16.64), which in feature space is $\tilde{\mathbf{P}} = \tilde{\boldsymbol{\Sigma}}_Y\tilde{\mathbf{Q}}$. However, an equivalent approach to deriving loading vectors is as a regression pattern, as indicated by (16.64), which has the sample expression

$$\dot{\mathbf{P}} = \dot{\mathbf{Y}}^T\tilde{\mathbf{R}}_Y\left(\tilde{\mathbf{R}}_Y^T\tilde{\mathbf{R}}_Y\right)^{-1} = \frac{1}{N_Y - 1}\dot{\mathbf{Y}}^T\tilde{\mathbf{R}}_Y. \tag{16.94}$$

Substituting (16.89) for $\dot{\mathbf{Y}}$, and (16.93) for $\tilde{\mathbf{R}}_Y$, yields

$$\dot{\mathbf{P}} = \frac{1}{N_Y - 1}\mathbf{E}\tilde{\mathbf{F}}_Y^T\tilde{\mathbf{F}}_Y\tilde{\mathbf{Q}} = \mathbf{E}\tilde{\boldsymbol{\Sigma}}_Y\tilde{\mathbf{Q}} = \mathbf{E}\tilde{\mathbf{P}}. \tag{16.95}$$

This expression implies that the j'th loading vector is $\dot{\mathbf{p}}_j = \mathbf{E}\tilde{\mathbf{p}}_j$. In other words, the loading vector can be derived simply by summing feature vectors, each weighted by the amplitude given by the appropriate element of $\tilde{\mathbf{p}}$.

Once variates are obtained, they may be regressed with respect to *any* field. Loading vectors are merely a special case in which variates are regressed with

Table 16.1. *Summary of Equations in Covariance Discriminant Analysis.*

population	sample	
$\boldsymbol{\Sigma}_X = \text{cov}[\mathbf{x}, \mathbf{x}]$	$\tilde{\boldsymbol{\Sigma}}_X = \frac{1}{N_X-1}\tilde{\mathbf{F}}_X^T\tilde{\mathbf{F}}_X$	covariance matrix
$\boldsymbol{\Sigma}_Y = \text{cov}[\mathbf{y}, \mathbf{y}]$	$\tilde{\boldsymbol{\Sigma}}_Y = \frac{1}{N_Y-1}\tilde{\mathbf{F}}_Y^T\tilde{\mathbf{F}}_Y$	covariance matrix
$\boldsymbol{\Sigma}_X\mathbf{q} = \lambda\boldsymbol{\Sigma}_Y\mathbf{q}$	$\tilde{\boldsymbol{\Sigma}}_X\tilde{\mathbf{q}} = \lambda\tilde{\boldsymbol{\Sigma}}_Y\tilde{\mathbf{q}}$	generalized eigenvalue problem
$\mathbf{Q} = \begin{bmatrix}\mathbf{q}_1 & \cdots & \mathbf{q}_S\end{bmatrix}$	$\tilde{\mathbf{Q}} = \begin{bmatrix}\tilde{\mathbf{q}}_1 & \cdots & \tilde{\mathbf{q}}_T\end{bmatrix}$	eigenvectors are projection vectors
$\mathbf{Q}^T\boldsymbol{\Sigma}_X\mathbf{Q} = \boldsymbol{\Lambda}$	$\tilde{\mathbf{Q}}^T\tilde{\boldsymbol{\Sigma}}_X\tilde{\mathbf{Q}} = \boldsymbol{\Lambda}$	normalization and orthogonality
$\mathbf{Q}^T\boldsymbol{\Sigma}_Y\mathbf{Q} = \mathbf{I}$	$\tilde{\mathbf{Q}}^T\tilde{\boldsymbol{\Sigma}}_Y\tilde{\mathbf{Q}} = \mathbf{I}$	normalization and orthogonality
$\mathbf{r}_X = \mathbf{Q}^T\mathbf{x}$	$\tilde{\mathbf{R}}_X = \tilde{\mathbf{F}}_X\tilde{\mathbf{Q}}$	variates
$\mathbf{r}_Y = \mathbf{Q}^T\mathbf{y}$	$\tilde{\mathbf{R}}_Y = \tilde{\mathbf{F}}_Y\tilde{\mathbf{Q}}$	variates
$\text{cov}[\mathbf{r}_X, \mathbf{r}_X] = \boldsymbol{\Lambda}$	$\frac{1}{N_X-1}\tilde{\mathbf{R}}_X^T\tilde{\mathbf{R}}_X = \boldsymbol{\Lambda}$	variates are uncorrelated, normalized
$\text{cov}[\mathbf{r}_Y, \mathbf{r}_Y] = \mathbf{I}$	$\frac{1}{N_Y-1}\tilde{\mathbf{R}}_Y^T\tilde{\mathbf{R}}_Y = \mathbf{I}$	variates are uncorrelated, normalized
$\mathbf{P} = \boldsymbol{\Sigma}_Y\mathbf{Q}$	$\tilde{\mathbf{P}} = \tilde{\boldsymbol{\Sigma}}_Y\tilde{\mathbf{Q}}$	loading vectors
$\mathbf{P} = \boldsymbol{\Sigma}_Y\mathbf{Q}$	$\dot{\mathbf{P}} = \mathbf{E}\tilde{\mathbf{P}}$	loading vectors
$\mathbf{P} = \text{cov}[\mathbf{x}, \mathbf{r}_X]$	$\dot{\mathbf{P}} = \frac{1}{N_X-1}\dot{\mathbf{X}}^T\tilde{\mathbf{R}}_X$	loading vectors are regression patterns
$\mathbf{P} = \text{cov}[\mathbf{y}, \mathbf{r}_Y]$	$\dot{\mathbf{P}} = \frac{1}{N_Y-1}\dot{\mathbf{Y}}^T\tilde{\mathbf{R}}_Y$	loading vectors are regression patterns
$\boldsymbol{\Sigma}_X = \mathbf{P}\boldsymbol{\Lambda}\mathbf{P}^T$	$\tilde{\boldsymbol{\Sigma}}_X = \tilde{\mathbf{P}}\boldsymbol{\Lambda}\tilde{\mathbf{P}}^T$	loading vectors are matrix decompositions
$\boldsymbol{\Sigma}_Y = \mathbf{P}\mathbf{P}^T$	$\tilde{\boldsymbol{\Sigma}}_Y = \tilde{\mathbf{P}}\tilde{\mathbf{P}}^T$	loading vectors are matrix decompositions
$\mathbf{x} = \mathbf{P}\mathbf{r}_X$	$\dot{\mathbf{X}} = \tilde{\mathbf{R}}_X\dot{\mathbf{P}}^T$	random variable/data decomposition
$\mathbf{y} = \mathbf{P}\mathbf{r}_Y$	$\dot{\mathbf{Y}} = \tilde{\mathbf{R}}_Y\dot{\mathbf{P}}^T$	random variable/data decomposition
$\lambda_k\mathbf{p}_k^T\mathbf{W}\mathbf{p}_k$	$\lambda_k\dot{\mathbf{p}}_k^T\mathbf{W}\dot{\mathbf{p}}_k$	variance of $\mathbf{x}/\dot{\mathbf{X}}$ explained by k'th component
$\mathbf{p}_k^T\mathbf{W}\mathbf{p}_k$	$\dot{\mathbf{p}}_k^T\mathbf{W}\dot{\mathbf{p}}_k$	variance of $\mathbf{y}/\dot{\mathbf{Y}}$ explained by k'th component

respect to $\dot{\mathbf{X}}$ or $\dot{\mathbf{Y}}$. Alternative regression patterns may be obtained by regressing with the *original* data sets:

$$\mathbf{P}_Y = \frac{1}{N_Y - 1} \mathbf{Y}^T \tilde{\mathbf{R}}_Y \quad \text{and} \quad \mathbf{P}_X = \frac{1}{N_X - 1} \mathbf{X}^T \tilde{\mathbf{R}}_X. \tag{16.96}$$

In contrast to loading vectors, regression patterns $\mathbf{P}_X$ and $\mathbf{P}_Y$ may depend on whether they were derived from just $\mathbf{X}$ or $\mathbf{Y}$. This sensitivity arises because discriminant functions were derived from feature space and as a result have no guaranteed properties outside of that space. On the other hand, the latter regression patterns tend to be *less sensitive to the truncation parameter T*. The reason for this is that the leading variate time series tend not to be sensitive to T, hence regressing them with the same fields tends to produce the same results. Moreover, the regression patterns $\mathbf{P}_X$ and $\mathbf{P}_Y$ tend to be easier to interpret because they are not distorted by projecting through feature space. Accordingly, the regression patterns $\mathbf{P}_X$ and $\mathbf{P}_Y$ are the recommended quantities to study for physical insight.

16.6 Variable Selection

What criterion should be used to select the number of EOFs? In Chapter 14, we discussed the cross entropy $\mathbb{IC}$ for selecting PDF models (see Section 14.5). CDA can be framed in terms of selecting PDF models too. To see this, let us define the random variable

$$\mathbf{z} = \begin{pmatrix} \mathbf{x} \\ \mathbf{y} \end{pmatrix}.$$

In CDA, X and Y are independent, hence the covariance matrix of $\mathbf{z}$ has the general form

$$\boldsymbol{\Sigma}_Z = \begin{pmatrix} \boldsymbol{\Sigma}_{XX} & \mathbf{0} \\ \mathbf{0} & \boldsymbol{\Sigma}_{YY} \end{pmatrix}. \tag{16.97}$$

If H_0 is true, then $\mathbf{x}$ and $\mathbf{y}$ have common covariance matrix $\boldsymbol{\Sigma}_P$ and

$$\boldsymbol{\Sigma}_Z^\omega = \begin{pmatrix} \boldsymbol{\Sigma}_P & \mathbf{0} \\ \mathbf{0} & \boldsymbol{\Sigma}_P \end{pmatrix}. \tag{16.98}$$

We think of $\boldsymbol{\Sigma}_Z^\omega$ as a *constrained* version of $\boldsymbol{\Sigma}_Z$. A constrained matrix is said to have *structure*, and selecting between a constrained and unconstrained covariance matrix is said to be a problem of selecting *covariance structure*. Let $q(\mathbf{z})$ and $q_\omega(\mathbf{z})$ denote normal distributions with covariance matrix (16.97) and (16.98), respectively, and let

$$\mathbb{IC}(Z) = -2\,\mathbb{E}_Z\left[\log q(\mathbf{z})\right] \quad \text{and} \quad \mathbb{IC}_\omega(Z) = -2\,\mathbb{E}_Z\left[\log q_\omega(\mathbf{z})\right]. \tag{16.99}$$

Partition $\mathbf{z}$ as

$$\mathbf{z} = \begin{pmatrix} \mathbf{z}_K \\ \mathbf{z}_R \end{pmatrix}, \tag{16.100}$$

where $\mathbf{z}_K$ denotes variables to keep and $\mathbf{z}_R$ denotes variables either to remove or retain. Our criterion is this: We retain $\mathbf{z}_R$ in CDA if predictions of $\mathbf{z}_R$ are better from $q(\mathbf{z})$ than from $q_\omega(\mathbf{z})$. Therefore, the criterion to retain $\mathbf{z}_R$ is

$$\mathbb{IC}(Z_R|Z_K) - \mathbb{IC}_\omega(Z_R|Z_K) = \mathbb{E}_Z \left[\log \frac{q(\mathbf{z}_R|\mathbf{z}_K)}{q_\omega(\mathbf{z}_R|\mathbf{z}_K)} \right] < 0. \tag{16.101}$$

Let the difference in $\mathbb{IC}$ be denoted as

$$\mathbb{DIS}(q, q_\omega)(Z) = \mathbb{IC}(Z) - \mathbb{IC}_\omega(Z). \tag{16.102}$$

We use the name $\mathbb{DIS}$ because it measures the degree to which q and q_ω may be *discriminated*. Recall the probability identity

$$q(\mathbf{z}_K, \mathbf{z}_R) = q(\mathbf{z}_K)q(\mathbf{z}_R|\mathbf{z}_K). \tag{16.103}$$

It follows from this that criterion (16.101) is equivalent to

$$\mathbb{DIS}(q, q_\omega)(Z_R|Z_K) = \mathbb{DIS}(q, q_\omega)(Z_R, Z_K) - \mathbb{DIS}(q, q_\omega)(Z_K) < 0. \tag{16.104}$$

This result shows that the criterion for retaining $\mathbf{z}_R$ is equivalent to choosing the variable set that minimizes $\mathbb{DIS}$. In effect, variables are selected only if they *improve* our ability to discriminate between two populations relative to the variables already included.

$\mathbb{DIS}$ can be interpreted as a generalization of $\mathbb{MIC}$ discussed in Chapter 14. Specifically, if information criteria are used to select between the covariance matrices in the first row of Table 16.2, then $\mathbb{DIS}(q, q_\omega)(X, Y) = \mathbb{MIC}(X; Y)$. Thus, $\mathbb{MIC}$ is a special case of $\mathbb{DIS}$.

Table 16.2. *Schematic Showing the Covariance Structure That Is under Comparison in Canonical Correlation Analysis (CCA) and Covariance Discrimination Analysis (CDA)*

Procedure	Σ_Z		Σ_Z^ω	
CCA	$\begin{pmatrix} \Sigma_{XX} & \Sigma_{XY} \\ \Sigma_{YX} & \Sigma_{YY} \end{pmatrix}$		$\begin{pmatrix} \Sigma_{XX} & 0 \\ 0 & \Sigma_{YY} \end{pmatrix}$	
CDA	$\begin{pmatrix} \Sigma_{XX} & 0 \\ 0 & \Sigma_{YY} \end{pmatrix}$		$\begin{pmatrix} \Sigma_P & 0 \\ 0 & \Sigma_P \end{pmatrix}$	

Following Chapter 14, we estimate $\mathbb{IC}$ using AICr. Then, $\mathbb{DIS}$ is estimated from (16.102). Let the data matrices be (16.86) and (16.87). The criterion for $q(\mathbf{z})$ follows from the fact that $\tilde{\mathbf{F}}_X$ and $\tilde{\mathbf{F}}_Y$ are independent, hence $\mathrm{AICr}(X, Y) = \mathrm{AICr}(X) + \mathrm{AICr}(Y)$. Thus, by (14.51)

$$\mathrm{AICr}(X) = N_X \log |\overline{\Sigma}_{XX}| + T N_X \log 2\pi + \frac{T N_X(N_X + 1)}{N_X - T - 2} \tag{16.105}$$

$$\mathrm{AICr}(Y) = N_Y \log |\overline{\Sigma}_{YY}| + T N_Y \log 2\pi + \frac{T N_Y(N_Y + 1)}{N_Y - T - 2}, \tag{16.106}$$

where

$$\overline{\Sigma}_{XX} = \tilde{\mathbf{F}}_X^T \tilde{\mathbf{F}}_X / N_X \quad \text{and} \quad \overline{\Sigma}_{YY} = \tilde{\mathbf{F}}_Y^T \tilde{\mathbf{F}}_Y / N_Y.$$

On the other hand, the criterion for $q_\omega(\mathbf{z})$ follows from the fact that $\mathbf{x}$ and $\mathbf{y}$ are drawn from the *same* distribution, hence all the data can be pooled together as

$$\mathbf{Z} = \begin{pmatrix} \tilde{\mathbf{F}}_X \\ \tilde{\mathbf{F}}_Y \end{pmatrix}. \tag{16.107}$$

The corresponding criterion is

$$\mathrm{AICr}_\omega(X, Y) = N_P \log |\overline{\Sigma}_P| + T N_P \log 2\pi + \frac{T N_P(N_P + 1)}{N_P - T - 2},$$

where

$$\overline{\Sigma}_P = \frac{N_X \overline{\Sigma}_{XX} + N_Y \overline{\Sigma}_{YY}}{N_P} \quad \text{and} \quad N_P = N_X + N_Y. \tag{16.108}$$

Therefore, an unbiased estimate of $\mathbb{DIS}$ is

$$\mathrm{DIS}(q, q_\omega)(X, Y) = \mathrm{AICr}(X) + \mathrm{AICr}(Y) - \mathrm{AICr}_\omega(X, Y)$$

$$= N_P \log \Lambda + T\left(\frac{N_X(N_X + 1)}{N_X - T - 2} + \frac{N_Y(N_Y + 1)}{N_Y - T - 2} - \frac{N_P(N_P + 1)}{N_P - T - 2}\right)$$

where Λ is defined in (16.37).

In the example in Section 16.2, the full control run corresponds to $N_Y = 300$, and two forced ensemble simulations are pooled to give $N_X = 2 * 156 = 312$. The resulting values of DIS are shown in Figure 16.5. The procedure is to choose the truncation parameter T that minimizes DIS, which in this example is $T = 6$. The fact that the selected DIS is negative tells us that $\mathrm{AICr}(X, Y) < \mathrm{AICr}_\omega(X, Y)$, implying that AICr would select covariance matrix (16.97) over (16.98).

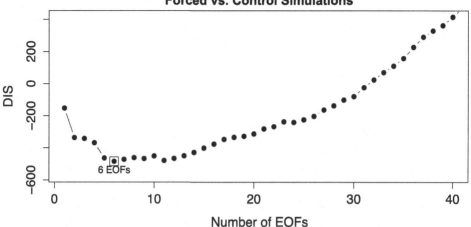

Figure 16.5 Values of DIS for comparing covariance matrices of annual mean 2m air temperature simulated by the NCAR CCSM4 model between the twentieth century and control simulations. The minimum value is indicated by a box.

16.7 Further Topics

CDA is related to other statistical procedures. For instance, Chapter 18 discusses Multivariate Analysis of Variance (MANOVA), which finds the most predictable component of an E-member ensemble forecast. The procedure is equivalent to testing the hypothesis

$$\Sigma_S = \frac{1}{E}\Sigma_N, \tag{16.109}$$

where Σ_S and Σ_N are so-called signal and noise covariance matrices (see Chapter 18 for definition of these terms). Another example concerns Canonical Correlation Analysis (CCA; see Chapter 15). To see the connection, consider the linear model

$$\mathbf{y} = \mathbf{L}\mathbf{x} + \boldsymbol{\epsilon}, \tag{16.110}$$

where $\mathbf{L}$ is a linear regression operator and $\boldsymbol{\epsilon}$ is random noise that is independent of $\mathbf{x}$. This linear model implies that the covariance of $\mathbf{y}$ is

$$\Sigma_{YY} = \mathbf{L}\Sigma_{XX}\mathbf{L}^T + \Sigma_{\epsilon}. \tag{16.111}$$

Substituting the least squares estimate $\mathbf{L} = \Sigma_{YX}\Sigma_{XX}^{-1}$ and rearranging yields

$$\Sigma_{\epsilon} = \Sigma_{YY} - \Sigma_{YX}\Sigma_{XX}^{-1}\Sigma_{XY} \tag{16.112}$$

(see Chapter 15 for definitions of these terms). Applying Covariance Discriminant Analysis to test the hypothesis $\Sigma_{\epsilon} = \Sigma_Y$ leads to CCA.

CDA bears some similarity to a procedure called *Factor Analysis*, but the two procedures are different. Factor analysis attempts to decompose a covariance matrix into the form

$$\Sigma_X = \Lambda\Lambda^T + \Psi, \qquad (16.113)$$

where Λ is low-rank and Ψ is diagonal. Superficially, this looks similar to

$$\Sigma_X = \Sigma_F + \Sigma_U, \qquad (16.114)$$

where Σ_F is also low-rank. However, in factor analysis Ψ is *diagonal* and estimated from the same sample as Σ_X, whereas in discriminant analysis there exists an independent estimate of Σ_U, and the off-diagonal elements are not restricted to be zero.

Combining Variables in CDA

Section 16.3 showed that discriminant ratios are invariant to invertible linear transformations. One consequence of this invariance property is that CDA can be applied to combinations of physically different variables without rescaling. For instance, one could define a vector $\mathbf{x}$ whose elements specify temperature and wind, and the resulting discriminant ratios would be independent of the units of temperature and wind. In contrast, a decomposition such as like Principal Component Analysis requires some sort of re-scaling to combine physically different variables (as discussed in Chapter 12), which introduces a dependence on the arbitrary choice of scaling. To discuss this further, suppose the state vector is a combination of temperature $\mathbf{t}$ and wind $\mathbf{u}$, hence

$$\mathbf{z} = \begin{pmatrix} \mathbf{t} \\ \mathbf{u} \end{pmatrix} \quad \Rightarrow \quad \mathrm{cov}[\mathbf{z}] = \begin{pmatrix} \Sigma_{TT} & \Sigma_{TU} \\ \Sigma_{UT} & \Sigma_{UU} \end{pmatrix}. \qquad (16.115)$$

Then, CDA seeks linear combinations of temperature *and* wind, which are of the form

$$r_X = \mathbf{q}_T^T \mathbf{t}_X + \mathbf{q}_U^T \mathbf{u}_X \quad \text{and} \quad r_Y = \mathbf{q}_T^T \mathbf{t}_Y + \mathbf{q}_U^T \mathbf{u}_Y, \qquad (16.116)$$

which maximizes the variance ratio $\mathrm{var}[r_X]/\mathrm{var}[r_Y]$. The solution is characterized by the leading discriminant ratio λ_1, the *joint* projection vector

$$\mathbf{q} = \begin{pmatrix} \mathbf{q}_T \\ \mathbf{q}_U \end{pmatrix}, \qquad (16.117)$$

associated variates r_X and r_Y, and a *joint* loading vector

$$\mathbf{p} = \begin{pmatrix} \mathbf{p}_T \\ \mathbf{p}_U \end{pmatrix}. \qquad (16.118)$$

Importantly, the variate r_X is multiplied by the *two* loading vectors, $\mathbf{p}_T$ and $\mathbf{p}_U$. Thus, in this decomposition, $\mathbf{p}_T$ and $\mathbf{p}_U$ fluctuate *in synchrony* with each other.

Equality of Means Test

At the beginning of this chapter, it was asserted that CDA contains traditional discriminant analysis as a special case. Traditional discriminant analysis finds linear combinations of variables that discriminate between the *means* of two populations. To show this equivalence, let the sample means of the two data sets be $\hat{\mu}_X$ and $\hat{\mu}_Y$, and the sample covariances be $\hat{\Sigma}_X$ and $\hat{\Sigma}_Y$. Consider a linear combination of random variables based on the projection vector $\mathbf{q}$. This gives the scalar quantities

$$\hat{\mu}_X = \mathbf{q}^T \hat{\mu}_X, \quad \hat{\mu}_Y = \mathbf{q}^T \hat{\mu}_Y, \quad \text{and} \quad \hat{\sigma}_{\text{pool}}^2 = \mathbf{q}^T \Sigma_{\text{pool}} \mathbf{q}, \tag{16.119}$$

where the pooled covariance matrix is

$$\Sigma_{\text{pool}} = \frac{1}{N_X + N_Y - 2} \left((N_X - 1)\hat{\Sigma}_X + (N_Y - 1)\hat{\Sigma}_Y \right). \tag{16.120}$$

The t-statistic for testing equality of means based on these quantities is

$$t = \frac{\hat{\mu}_X - \hat{\mu}_Y}{\hat{\sigma}_{\text{pool}}\sqrt{\frac{1}{N_X} + \frac{1}{N_Y}}} = \frac{\mathbf{q}^T \left(\hat{\mu}_X - \hat{\mu}_Y \right)}{\sqrt{\mathbf{q}^T \Sigma_{\text{pool}} \mathbf{q}}\sqrt{\frac{1}{N_X} + \frac{1}{N_Y}}}. \tag{16.121}$$

Squaring the t-statistic yields

$$t^2 = \frac{\mathbf{q}^T \left(\hat{\mu}_X - \hat{\mu}_Y \right) \left(\hat{\mu}_X - \hat{\mu}_Y \right)^T \mathbf{q}}{\mathbf{q}^T \Sigma_{\text{pool}} \mathbf{q} \left(\frac{1}{N_X} + \frac{1}{N_Y} \right)} = \frac{\mathbf{q}^T \Sigma_\Delta \mathbf{q}}{\mathbf{q}^T \Sigma_{\text{pool}} \mathbf{q}}, \tag{16.122}$$

where we have used the fact that $\mathbf{a}^T \mathbf{b} = \mathbf{b}^T \mathbf{a}$ for any two vectors $\mathbf{a}$ and $\mathbf{b}$, and

$$\Sigma_\Delta = \frac{1}{\frac{1}{N_X} + \frac{1}{N_Y}} \left(\hat{\mu}_X - \hat{\mu}_Y \right) \left(\hat{\mu}_X - \hat{\mu}_Y \right)^T. \tag{16.123}$$

Because t^2 is a Rayleigh quotient of the form (16.24), the projection vector that maximizes t^2 is obtained by solving the generalized eigenvalue problem

$$\Sigma_\Delta \mathbf{q} = \lambda \Sigma_{\text{pool}} \mathbf{q}. \tag{16.124}$$

It can be seen by inspection that Σ_Δ is a rank-1 matrix. This problem is equivalent to (16.42) and therefore has an analogous solution. Specifically, there is only one nontrivial eigenvalue, and it can be verified by substitution that the corresponding eigenvector is

$$\mathbf{q} = \Sigma_{\text{pool}}^{-1} \left(\hat{\mu}_X - \hat{\mu}_Y \right), \tag{16.125}$$

and that the associated eigenvalue is

$$T^2 = \frac{1}{\frac{1}{N_X} + \frac{1}{N_Y}} \left(\hat{\mu}_X - \hat{\mu}_Y \right)^T \Sigma_{\text{pool}}^{-1} \left(\hat{\mu}_X - \hat{\mu}_Y \right). \tag{16.126}$$

This eigenvalue is termed *Hotelling's T^2 statistic*. Under the null hypothesis of equal means, Hotelling's statistic is distributed as

$$\left(\frac{N_X + N_Y - 1 - S}{(N_X + N_Y - 2)S}\right) T^2 \sim F_{S, N_X + N_Y - 1 - S}. \tag{16.127}$$

The associated discriminant loading vector is simply the difference in means:

$$\mathbf{p} = \boldsymbol{\Sigma}_{\text{pool}} \mathbf{q} = \left(\hat{\boldsymbol{\mu}}_X - \hat{\boldsymbol{\mu}}_Y\right). \tag{16.128}$$

Incidentally, CDA can be used to generalize the test to more than two means, which leads to a procedure known as Multivariate Analysis of Variance (MANOVA; see Chapter 18).

16.8 Conceptual Questions

1 What is a discriminant ratio? What is a discriminant variate? What is a discriminant loading vector? What is a discriminant component?

2 Why can't you use the univariate F-test to test the significance of the discriminant ratio?

3 Explain how Canonical Correlation Analysis (see Chapter 15) is a special case of Covariance Discriminant Analysis.

4 Why is the sampling distribution of the discriminant ratios under H_0 independent of the population covariance matrix?

5 When comparing discriminant ratios to their respective sampling distribution under H_0, what potential biases are being neglected.

6 Why can't you just "throw all the data" at CDA and let it diagnose differences in covariance matrices?

7 Are discriminant components orthogonal in space? Orthogonal in time? Both?

8 The discriminant ratios were proven in (16.25) to be greater than one, yet Figure 16.2 shows discriminant ratios less than one. Explain why this is not a contradiction.

9 Show that *any* function of the difference in covariance matrix that is invariant to invertible affine transformations can depend only on the discriminant ratios.

17

Analysis of Variance and Predictability

Every treatment is applied to a different individual, so some of the variability in the data is due to the treatments and some of the variability is due to random variation among the experimental subjects. The purpose of an analysis of variance is to discern whether the differences associated with the treatments are large enough, compared to the underlying random variability, to warrant concluding that the treatments had an effect.[1]

Glantz and Slinker (2016)

… if separate solutions of the governing equations, nearly identical at some initial time, remained nearly identical as time progressed, the atmosphere would proceed to vary periodically. The observed absence of complete periodicity therefore implies that separate solutions of the atmospheric equations diverge from one another, ultimately becoming as different as two solutions chosen at the same time of day and year but otherwise randomly. It follows that good predictions at sufficiently long range are unattainable, if the initial state is imperfectly known.[2]

Lorenz (1969)

The correlation test is a standard procedure for deciding if two variables are *linearly* related. This chapter discusses a test for independence that avoids the linearity assumption. The basic idea is the following. If two variables are dependent, then changing the value of one them, say c, changes the distribution of the other. Therefore, if samples are collected for fixed value of c, and additional samples are collected for a different value of c, and so on for different values of c, then a dependence implies that the distributions for different c's should differ. It follows that deciding that some aspect of the distributions depend on c is equivalent to deciding that the variables are dependent. A special case of this approach is

---

[1] Republished with permission of McGraw Hill LLC, from *Primer Of Applied Regression & Analysis Of Variance*, Glantz and Slinker (2016), 468; permission conveyed through the Copyright Clearance Center, Inc.
[2] Reprinted by permission from the American Meterological Society: *Bulletin of the American Meteorological Society*, "Three approaches to predictability," Edward Lorenz, 1969.

the t-test, which tests if two populations have identical means (see Chapter 2). Generalizing this test to more than two populations leads to *Analysis of Variance (ANOVA)*, which is the topic of this chapter. ANOVA is a method for testing if two or more populations have the same means. Importantly, the test for equality of means can be formulated without specifying *how* the means depend on c, hence a nonlinear dependence on c can be considered. In weather and climate studies, ANOVA is commonly used to quantify the predictability of an *ensemble forecast*, hence this framing is discussed extensively in this chapter. In particular, this chapter defines weather and seasonal predictability, discusses measures of predictability (e.g., signal-to-noise ratio and mutual information), and explains how ANOVA provides a natural framework for diagnosing predictability. Numerous extensions of ANOVA exist, but only *one-factor analysis* is reviewed in this chapter. The multivariate extension of ANOVA, MANOVA, is discussed in Chapter 18.

17.1 The Problem

The scientific method involves proposing a hypothesis and then checking that hypothesis against experiment. However, sometimes comparing experiment with hypothesis is not as easy as one might think. Suppose a scientist performs the following experiment: a variable Y is measured five times under the same conditions. Then, the condition is changed, and five more measurements of Y are taken. Suppose the experimental outcomes for conditions $c = 1, 2, 3, 4$ are shown in Figure 17.1. Has the scientist found strong evidence that c influences Y? Or, can

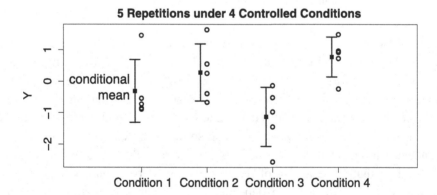

Figure 17.1 Results of a hypothetical experiment that is repeated five times for each of four controlled conditions. Open circles show the actual experimental results, filled squares show the sample mean for each individual condition, and error bars show one standard deviation about the mean computed from the five repetitions under each fixed condition.

the experimental results be explained by random variability alone? No more data can be collected – a decision must be made based on the data in hand.

This chapter discusses a procedure known as *Analysis of Variance* (ANOVA) for answering these questions. Notice that Y or c were never defined. ANOVA is general and has an enormous range of applications. For instance, ANOVA is applied in medical studies to decide if a drug is effective, material studies to decide if a chemical improves tensile strength, and in sociological studies to decide if an educational strategy improves test scores. In climate studies, ANOVA is used most often to diagnose predictability, hence this framing will be discussed extensively in this chapter.

Most texts on Analysis of Variance use a terminology derived historically from scientific experimentation in which a "factor" is applied at various "levels" for a certain number of "repetitions." In the weather and climate literature, the term "ensemble member" is used so frequently for repetitions that using a different word would invariably cause confusion. Accordingly, we adopt a terminology in which E repetitions at the same level are known as *ensemble members*, while the C different levels are known as *conditions*.

17.2 Framing the Problem

Experimental outcomes such as those illustrated in Figure 17.1 can be summarized in a matrix Y_{ec}, where $c = 1, 2, \ldots, C$ denotes the condition, and $e = 1, 2, \ldots, E$ denotes the ensemble member. The ensemble size E is assumed to be the same number for each c, which leads to a *balanced design*. Methods for dealing with unequal ensemble sizes are available but lead to more complex interpretations (see Glantz and Slinker, 2016, for further discussion).

Our goal is to decide if c influences Y. If Y is *normally* distributed, then c influencing Y implies that the mean or variance of Y depends on c. We denote this hypothesis as

$$\text{general alternative hypothesis:} \quad Y_{ec} \sim \mathcal{N}\left(\mu_{Y|c}, \sigma^2_{Y|c}\right), \quad (17.1)$$

where $\mu_{Y|c}$ and $\sigma^2_{Y|c}$ are, respectively, the *conditional mean and conditional variance* of Y, either of which may depend on c. If c does not influence Y, then the mean and variance do not depend on c. We denote this hypothesis as

$$\text{general null hypothesis:} \quad Y_{ec} \sim \mathcal{N}\left(\mu_Y, \sigma^2_Y\right), \quad (17.2)$$

where μ_Y and σ^2_Y denote the *unconditional* mean and variance, which do not depend on c.

Sample estimates of these parameters follow naturally from those defined in Chapter 1. For fixed c, the conditional mean is estimated as

$$\hat{\mu}_{Y|c} = \frac{1}{E} \sum_{e=1}^{E} Y_{ec}, \tag{17.3}$$

and the conditional variance is estimated using the unbiased estimate

$$\hat{\sigma}_{Y|c}^2 = \frac{1}{E-1} \sum_{e=1}^{E} \left(Y_{ec} - \hat{\mu}_{Y|c}\right)^2. \tag{17.4}$$

Under the general null hypothesis, the conditional means are equal and the conditional variances are equal. Therefore, unconditional quantities can be estimated by averaging conditional quantities. Accordingly, an estimate of the unconditional mean is

$$\hat{\mu}_Y = \frac{1}{EC} \sum_{e=1}^{E} \sum_{c=1}^{C} Y_{ec} = \frac{1}{C} \sum_{c=1}^{C} \hat{\mu}_{Y|c}, \tag{17.5}$$

and an estimate of the unconditional variance is

$$\hat{\sigma}_N^2 = \frac{\hat{\sigma}_{Y|1}^2 + \hat{\sigma}_{Y|2}^2 + \cdots + \hat{\sigma}_{Y|C}^2}{C} = \frac{1}{C(E-1)} \sum_{e=1}^{E} \sum_{c=1}^{C} \left(Y_{ec} - \hat{\mu}_{Y|c}\right)^2. \tag{17.6}$$

The subscript "N" stands for "noise." Table 17.1 shows the values of these estimates for the data displayed in Figure 17.1. By Theorem 1.5, $\hat{\mu}_{Y|c}$ and $\hat{\sigma}_{Y|c}^2$ are independent.

To decide between the two hypotheses, a two-step procedure is followed:

1 test if the variances depend on c,
2 test if the means depend on c, *assuming the variances are constant.*

If the first step indicates that variances depend on c, then we decide that the variance of Y depends on c and stop testing. However, if equality of variances is not rejected, then we proceed to the next step and test equality of means *assuming the variance does not depend on c.*

Table 17.1. *Sample Estimates of the Conditional Means, Conditional Variances, Unconditional Mean, and Unconditional Variance of the Samples Shown in Figure 17.1.*

	Condition 1	Condition 2	Condition 3	Condition 4	Unconditional
$\hat{\mu}_{Y|c}$	−0.31	0.28	−1.13	0.77	−0.10
$\hat{\sigma}_{Y|c}^2$	1.00	0.82	0.88	0.40	0.77

17.3 Test Equality of Variance

In Chapter 2, we discussed the F-test for testing equality of *two* variances. Here we want to test equality of C variances, where $C > 2$. That is, we want to test

$$\sigma^2_{Y|1} = \sigma^2_{Y|2} = \cdots = \sigma^2_{Y|C}. \tag{17.7}$$

To test this hypothesis, we use a test by Bartlett (1937), which is based on the statistic

$$\theta = \frac{C(E-1)}{\gamma} \log \left(\frac{\hat{\sigma}^2_N}{\left(\hat{\sigma}^2_{Y|1} \cdots \hat{\sigma}^2_{Y|C} \right)^{1/C}} \right), \tag{17.8}$$

where

$$\gamma = 1 + \frac{C+1}{3E(C-1)}. \tag{17.9}$$

The reasonableness of this statistic can be seen from the *arithmetic-geometric mean inequality* (AM-GM inequality), which states: for any C positive real numbers $X_1, \ldots, X_C$,

$$\frac{X_1 + \cdots + X_C}{C} \geq \sqrt[C]{X_1 \ldots X_C}. \tag{17.10}$$

Importantly, equality holds if and only if $X_1 = \cdots = X_C$. Since θ depends on the logarithm of the ratio of the arithmetic to geometric means, it follows that θ vanishes when the sample variances are equal, and is positive otherwise. In effect, θ measures *homogeneity of variances*. The decision rule is to reject hypothesis (17.7) if

$$\theta \geq \chi^2_{\alpha, C-1}. \tag{17.11}$$

The value of these parameters derived from Table 17.1 are

$$\theta = \frac{4 * (5-1)}{1 + \frac{4+1}{3*5*(4-1)}} * \log \left(\frac{0.77}{(1.00 * 0.82 * 0.88 * 0.40)^{1/4}} \right) \approx 0.8 \tag{17.12}$$

Since $\chi^2_{0.05, 4-1} = 7.81$, the decision rule (17.11) is not satisfied, hence no difference in variance has been detected. Accordingly, we proceed to test equality of means.

17.4 Test Equality of Means: ANOVA

Since no dependence between c and variance was detected in the last section, we now *assume the variances do not depend on c*. The null hypothesis is now

$$H_0: \quad \mu_{Y|1} = \mu_{Y|2} = \cdots = \mu_{Y|C} \quad \text{and} \quad \sigma_N^2 = \sigma_{Y|1}^2 = \sigma_{Y|2}^2 = \cdots = \sigma_{Y|C}^2.$$
(17.13)

The alternative hypothesis is that at least one pair of conditional means differ, under the constraint that the conditional variances are equal:

$$H_A: \quad \mu_{Y|i} \neq \mu_{Y|j} \text{ for some } i \neq j \quad \text{and} \quad \sigma_N^2 = \sigma_{Y|1}^2 = \sigma_{Y|2}^2 = \cdots = \sigma_{Y|C}^2.$$
(17.14)

These hypotheses are illustrated in Figure 17.2. This hypothesis is equivalent to the hypothesis

$$Y_{ec} \sim \mathcal{N}\left(\mu_{Y|c}, \sigma_N^2\right),$$
(17.15)

where σ_N does not depend on c. This is tantamount to assuming that the samples were generated by the model

$$Y_{ec} = \mu_{Y|c} + \epsilon_{ec},$$
(17.16)

where ϵ_{ec} is a random variable drawn from the normal distribution $\mathcal{N}(0, \sigma_N^2)$.

In Chapter 2, we learned about the t-test for testing equality of *two* means. To generalize beyond two means, a key idea in ANOVA is to consider the *variance of sample means*,

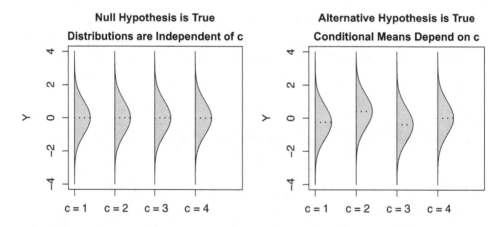

Figure 17.2 Schematic of the null and alternative hypothesis for ANOVA corresponding to Figure 17.1.

$$\hat{\sigma}_S^2 = \frac{1}{C-1} \sum_{c=1}^{C} (\hat{\mu}_{Y|c} - \hat{\mu}_Y)^2. \tag{17.17}$$

The subscript "S" stands for "signal." The logic for using $\hat{\sigma}_S^2$ for testing equality of means is simple: If the sample means are equal, then $\hat{\sigma}_S^2$ vanishes, otherwise $\hat{\sigma}_S^2$ is positive. Furthermore, $\hat{\sigma}_S^2$ increases as the sample means differ from their central value. Thus, $\hat{\sigma}_S^2$ measures the *homogeneity of sample means*. This partly explains where ANOVA gets its name: although we are testing equality of means, the statistic is based on variance.

To derive a decision rule, the distribution of $\hat{\sigma}_S^2$ under H_0 needs to be determined. It follows from (17.15) and Example 1.13 that the sample means are distributed as

$$\hat{\mu}_{Y|c} \sim \mathcal{N}(\mu_{Y|c}, \sigma_N^2/E). \tag{17.18}$$

Theorem 1.4 and Equation (17.18) imply

$$\frac{(C-1)\hat{\sigma}_S^2}{\left(\frac{\sigma_N^2}{E}\right)} \sim \chi_{C-1}^2 \quad \text{if } H_0 \text{ is true.} \tag{17.19}$$

Unfortunately, this quantity is not a statistic because σ_N^2 is unknown. An obvious step is to approximate σ_N^2 using (17.6). To derive the associated distribution, note that by Theorem 1.4, each (scaled) variance (17.4) has a chi-squared distribution with $E - 1$ degrees of freedom. Because independent chi-squared distributions are additive,

$$\frac{C(E-1)\hat{\sigma}_N^2}{\sigma_N^2} \sim \chi_{C(E-1)}^2. \tag{17.20}$$

By Theorem 1.5, sample variances $\hat{\sigma}_{Y|c}^2$ are independent of conditional means $\hat{\mu}_{Y|c}$; therefore, the statistics $\hat{\sigma}_S^2$ and $\hat{\sigma}_N^2$ are independent. Since these two variables also follow a chi-squared distribution when the null hypothesis is true, Theorem 2.2 implies that the ratio of (17.19) and (17.20), divided by their respective degrees of freedom,

$$F = E\frac{\hat{\sigma}_S^2}{\hat{\sigma}_N^2}, \tag{17.21}$$

has an F distribution with $C - 1$ and $C(E - 1)$ degrees of freedom, when H_0 is true. Recall that $\hat{\sigma}_S^2$ measures the homogeneity of sample means and therefore grows with differences between means. Thus, if H_A is true, then the distribution of F shifts to larger values relative that under H_0. Accordingly, the decision rule for $\alpha 100\%$ significance is the *one-sided* form

$$
\begin{aligned}
\text{Reject } H_0 & \quad \text{if} \quad F > F_{\alpha, C-1, C(E-1)} \\
\text{Accept } H_0 & \quad \text{if} \quad F \leq F_{\alpha, C-1, C(E-1)}.
\end{aligned}
\tag{17.22}
$$

Substituting the numerical values from Table 17.1 gives

$$
\hat{\sigma}_S^2 = \frac{(-0.31 + 0.1)^2 + (0.28 + 0.1)^2 + (-1.13 + 0.1)^2 + (0.77 + 0.1)^2}{4 - 1} \approx 0.67.
$$

Similarly, the estimated noise variance is $\hat{\sigma}_N^2 \approx 0.77$. Therefore, the F-statistic is

$$
F = 5 * \frac{0.67}{0.77} \approx 4.3.
\tag{17.23}
$$

By comparison, the 5% significance threshold for an F distribution with $C - 1 = 3$ and $C(E - 1) = 4 * 4 = 16$ is 3.24. Therefore, according to (17.22), we reject H_0, and decide that the mean of Y depends on c.

17.5 Comments about ANOVA

A few comments about ANOVA, as presented in the previous section, are in order:

- Typically, the term ANOVA refers solely to the test for equality of means based on the F statistic (17.21). In particular, it does not include a test for equality of variances, such as Bartlett's test based on θ (17.8). Nevertheless, some check on homogeneity of variances is implicit in ANOVA. For clarity, we have included this check as part of the procedure.
- If $C = 2$, then ANOVA is precisely equivalent to the t-test discussed in Chapter 2. On the other hand, if $C = 2$, Bartlett's test of equality of variances is not equivalent to the F-test, although the two tests are asymptotically equivalent for large sample sizes.
- ANOVA does not make any assumption about the functional dependence between $\mu_{Y|c}$ and c. In this sense, ANOVA is less restrictive than the correlation test. However, ANOVA is more restrictive because it requires *multiple repetitions* for the same condition.
- The ratio $\hat{\sigma}_S^2 / \hat{\sigma}_N^2$ is often termed an estimate of the *signal-to-noise ratio*. The signal-to-noise ratio is a nondimensional measure of the degree to which a distribution depends on condition. This measure as well as other common measures are discussed in Section 17.7.
- ANOVA can be generalized in a variety of ways. This section presented *one-way* ANOVA, in which there is only one *type* of condition, known as a *factor*. As a result, any variation that cannot be explained by the factor is assumed to be

random. However, in some cases it is clear that the variations can be explained by a second factor. For instance, suppose that in collecting the data shown in Figure 17.1, part of the data were collected in the morning and the rest were collected in the evening. The researcher may be curious about whether the time at which the experiments were conducted affects the results. This kind of question can be addressed by considering the *two-way* model

$$Y_{ecd} = \alpha_{Y|c} + \beta_{Y|d} + \gamma_{Y|cd} + \epsilon_{ecd}, \qquad (17.24)$$

where $\alpha_{Y|c}$ models the effect of c, $\beta_{Y|d}$ models the effect of time-of-day, and $\gamma_{Y|cd}$ models their interaction. Two-way ANOVA has been used to assess boundary conditions versus initial conditions in the predictability of a model (Zwiers, 1996). These and other generalizations can be found in texts devoted specifically to Analysis of Variance.

- A subtle issue concerns whether c is a random variable or not. Technically, independence was defined in Chapter 1 only for random variables. However, the concept of independence can be generalized to *parameters* of a distribution. Specifically, the distribution of Y may be derived from a statistical model involving a parameter c. In this case, the distribution of Y may depend on c, but the joint distribution of Y and c is not well defined because c is not a random variable. Nevertheless, the statement "Y is independent of c" still has meaning – it expresses the fact that the distribution of Y is the same for all values of c. It turns out that the ANOVA procedure is the same regardless of whether c is random or not. The model is known as a *random-effects* model if c is random and otherwise known as a *fixed-effects model*. Since the distinction does not affect the procedure, we will largely ignore this difference and consider c to be a random variable.

17.6 Weather Predictability

ANOVA provides an elegant framework for predictability. To explain the concept of predictability, consider a coin toss. Everyone knows that a coin toss is unpredictable, but what precisely does this mean? A common answer is "you cannot predict the outcome," but this is wrong! Here is a valid prediction: there is a 50% chance that the next coin toss will be heads. You might want to disqualify this prediction because it was *probabilistic*. However, no forecast of nature is perfect, and the clearest description of uncertainty is a probability distribution. Therefore, forecasts of nature *should* be probabilistic. Perhaps predicting heads with 50% probability should be disqualified because it is *not useful*. However, the definition of "useful" depends on the user – a prediction that is useless to one user may be useful to another.

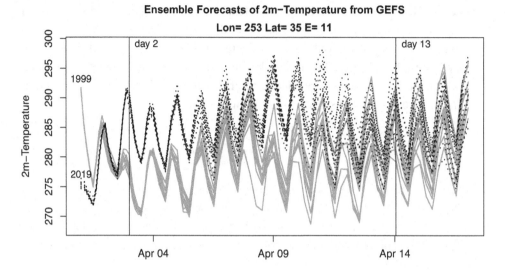

Figure 17.3 Forecasts of 2m-temperature by the Global Ensemble Forecast System (GEFS) at a particular point on the globe every six hours (see Hamill et al., 2013, for further details of the forecast system). The forecasts are initialized on April 1 of 1999 and 2019. For each year, 11 initial conditions are used to characterize the distribution of the initial state. Forecasts in 1999 are indicated by solid grey curves, while forecasts in 2019 are indicated by dots. The periodic oscillations every 24 hours reflects the diurnal cycle (e.g., afternoons tend to be warmer than evenings). For reference, days 2 and 13 are indicated by thin vertical lines.

Predictability attempts to quantify the degree to which a variable can be predicted without regard to whether that prediction is useful for a particular purpose.

The reason a coin toss is unpredictable is because *the distribution of outcomes has no discernible dependence on conditions that we can observe before the toss*. More generally, *a variable is defined to be unpredictable from a set of observations if it is independent of those observations* (DelSole and Tippett, 2017). It follows that a necessary condition for a variable to be predictable is that its distribution depends on the observations on which the prediction is based. This definition leads naturally to ANOVA, since ANOVA is a procedure for deciding whether a distribution depends on condition.

To illustrate these concepts, consider the weather forecasts shown in Figure 17.3. Two groups of forecasts are shown: one group initialized on April 1, 1999, and another on April 1, 2019. There are multiple forecasts from each start date because the initial state is uncertain. In practice, the globe is only partially observed at any given instant, hence observations do not fully specify the state. After combining everything we know from observations, their uncertainties, and prior information from the laws of physics (as represented in the forecast model), uncertainty still remains. Deriving a distribution of the state based on all this information is the goal of *data assimilation*, which is discussed in Chapter 20, but the details are not important for the present discussion. What is important is that forecasters have the ability to draw *random samples from the initial condition distribution*. In the specific example shown in Figure 17.3, the forecasts were generated from 11 initial states that are plausibly consistent with all available information at the start of the forecast. The 11 initial states are relatively close to each other at the initial time.

The 11 forecasts generated from 11 initial conditions form a *forecast ensemble*. At fixed lead time, ensemble members are interpreted as samples from the *conditional distribution*. Here, "condition" refers to all observations up to and including the start date that are used to initialize the forecast. Each condition is labeled uniquely by the start date. In predictability studies, it is the convention to term the conditional distribution the *forecast distribution*. Some characteristic features of the forecast distribution can be discerned from Figure 17.3. Initially, members of a forecast ensemble are relatively close to each other and separated from those initialized in a different year. As the forecast evolves, ensemble members spread apart. After about two weeks, the two ensembles overlap and become hard to distinguish. At this time, the forecast distribution exhibits no significant dependence on initial conditions and the *model* variable is said to be unpredictable (in a weather predictability sense). In essence, the forecast distributions merge into the same distribution, namely the unconditional distribution. In climate studies, the unconditional distribution is often known as the *climatological* distribution. A necessary condition for predictability is for the forecast and climatological distributions to differ. At sufficiently long lead times, the forecast and climatological distributions become indistinguishable and the variable is said to be unpredictable. The forecast and climatological distributions contain diurnal cycles, annual cycles, and multi-year variations associated with climate change, but these variations become the same in the two distributions at sufficiently long lead times and do not contribute to predictability (see DelSole and Tippett, 2017, for further discussion).

It is instructive to focus on forecasts at two lead times, say days 2 and 13, indicated by vertical lines in Figure 17.3. Forecasts at these two lead times are shown in Figure 17.4 using the same format as Figure 17.1, to emphasize the relation to ANOVA. For a given lead time, there Hence, $E = 11$ and $C = 2$. Hopefully, the correspondence between ANOVA and weather predictability is clear: different repetitions of an experiment under fixed condition correspond to different members of a forecast ensemble from the same start date; the conditional mean corresponds to the ensemble mean; the variance of the forecast ensemble corresponds to the conditional variance.

Deciding whether 2m-temperature is predictable is equivalent to deciding whether the forecast distribution depends on the initial observations. At two-day

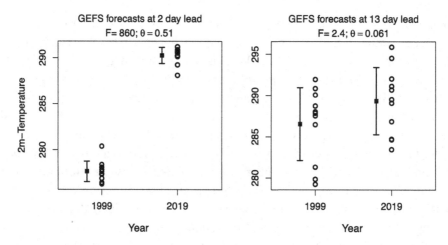

Figure 17.4 Forecasts of 2m-temperature at lead 2 days (left) and 13 days (right) from the GEFS forecasts shown in Figure 17.3. Circles show individual values of the 11-member forecasts, squares show the ensemble mean, and error bars show one standard deviation above and below the ensemble mean. The format of this figure is the same as Figure 17.1 to highlight the relation to ANOVA.

lead, the forecast values shown in Figure 17.4 are well separated from each other, and any reasonable analysis would conclude that the forecast distribution depends on year. At 13-day lead, the forecast values overlap considerably with each other, obscuring any dependence of the forecast distribution on initial observations. To decide whether a dependence exists, we perform ANOVA at each lead time separately. The resulting F and θ statistics are indicated at the top of each panel in Figure 17.4. For comparison, the 5% significance cutoff value for F is 4.35, and the cutoff value for θ is 3.8. As can be seen, the variance of each forecast ensemble shows no significant dependence on initial condition. On the other hand, the means show a significance dependence on initial conditions at two-day lead, but not at 13-day lead.

It should be understood that 2m-temperature may be predictable even if ANOVA does not detect it. After all, the predictability may just be *too small* to detect. Indeed, even after weather forecasts lose the ability to predict individual storms, they still may retain predictive information about *average* weather. For instance, ENSO can shift weather patterns for months. To isolate such predictability, a standard approach is to consider *monthly or seasonal averages*. If the predictability is persistent in time, then time averaging can filter out unpredictable weather variability and expose predictability. The resulting predictability is often known as *seasonal predictability*, to distinguish it from *weather predictability*, which was discussed earlier. Similarly, *spatial averages* can filter out unpredictable variability and yield predictable quantities, provided the predictability is spatially uniform over the region. Seasonal predictability and spatial filtering will be discussed in Chapter 18 on MANOVA.

17.7 Measures of Predictability

After detecting predictability, we often want to quantify it. For instance, we may be interested in how fast the predictability decays with lead time, or how the predictability compares between variables. There exist a variety of measures of predictability, including F, R-square, signal-to-noise ratio, predictive information, and correlation. It turns out that these measure are monotonic functions of each other. Thus, there is no information in one of them that cannot be obtained from the others. The purpose of this section is to present these different measures and show their equivalence. These equivalences also extend to multivariate measures (see Chapter 18).

The key to relating predictability measures is the *law of total variance*, which states

$$\mathbb{T} = \mathbb{S} + \mathbb{N}, \tag{17.25}$$

where

$$\mathbb{T} = \text{var}[Y], \quad \mathbb{S} = \text{var}\left[\mu_{Y|c}\right], \quad \text{and} \quad \mathbb{N} = \mathbb{E}[\sigma^2_{Y|c}]. \tag{17.26}$$

The law of total variance holds regardless of distribution. The sample version of the law of total variance is most simply stated in terms of "sum-square" quantities, defined here:

$$S = E \sum_{c=1}^{C} \left(\hat{\mu}_{Y|c} - \hat{\mu}_Y\right)^2 = E(C-1)\hat{\sigma}^2_S \tag{17.27}$$

$$N = \sum_{e=1}^{E} \sum_{c=1}^{C} \left(y_{ec} - \hat{\mu}_{Y|c}\right)^2 = C(E-1)\hat{\sigma}^2_N \tag{17.28}$$

$$T = \sum_{e=1}^{E} \sum_{c=1}^{C} \left(y_{ec} - \hat{\mu}_Y\right)^2 = (CE-1)\hat{\sigma}^2_T, \tag{17.29}$$

where S, N, T denote *signal, noise, and total*. Then, the (sample) law of total variance is

$$T = S + N. \tag{17.30}$$

If Y is unpredictable, then its distribution does not depend on c, hence $\mathbb{S}$ vanishes (because the conditional mean is constant) and the law of total variance gives $\mathbb{T} = \mathbb{N}$. Conversely, a sufficient condition for predictability is $\mathbb{T} \neq \mathbb{N}$. This suggests that predictability can be measured by the *difference* in variances (or better yet, a ratio).

Two of the most common measures of predictability are the signal-to-noise ratio and R-square, defined as

$$\text{SNR} = \frac{S}{N} \quad \text{and} \quad R^2 = \frac{S}{T} = \frac{\text{SNR}}{\text{SNR} + 1}. \tag{17.31}$$

Other ratios can be used, but these are related to SNR or $\mathbb{R}$. A natural estimate of the SNR is the ratio of the respective ANOVA estimates,

$$\text{SNR} = \frac{\hat{\sigma}_S^2}{\hat{\sigma}_N^2} = \frac{F}{E} = \frac{C(E-1)}{E(C-1)}\left(\frac{S}{N}\right). \tag{17.32}$$

However, this estimate is biased. It can be shown that

$$\mathbb{E}\left[\hat{\sigma}_S^2\right] = S + \frac{N}{E}, \tag{17.33}$$

which shows that $\hat{\sigma}_S^2$ *overestimates* the signal variance. Solving (17.33) for the signal variance S suggests that an unbiased estimate is $\hat{\sigma}_S^2 - \hat{\sigma}_N^2/E$, which in turn suggests defining an *adjusted* signal-to-noise ratio

$$\text{SNR}_{\text{ADJ}} = \frac{\hat{\sigma}_S^2 - \hat{\sigma}_N^2/E}{\hat{\sigma}_N^2} = \frac{\hat{\sigma}_S^2}{\hat{\sigma}_N^2} - \frac{1}{E}. \tag{17.34}$$

The adjusted signal-to-noise ratio is not guaranteed to be nonnegative. Similarly, a common measure of predictability is the proportion of variance explained by the signal, R-square. An estimate of R-square is S/T, which is guaranteed to lie between 0 and 1. This estimate can be expressed in terms of other quantities using the law of total variance; e.g.,

$$R^2 = \frac{S}{T} = \frac{S}{S+N} = \frac{S/N}{S/N+1} = 1 - \frac{N}{T} = \frac{F}{F + \frac{C(E-1)}{C-1}}. \tag{17.35}$$

These identities show that S/N, S/T, T/N, F are each *monotonically increasing functions* of the others. Consequently, if one of these quantities is known, then all of them are known. In analogy to the adjusted SNR, an adjusted R^2 may be defined as

$$R_{\text{ADJ}}^2 = \frac{F - 1}{F + \frac{C(E-1)}{C-1}}. \tag{17.36}$$

The adjusted R-square is not guaranteed to be positive.

Leung and North (1990) proposed measuring predictability by a quantity known as *mutual information*. Schneider and Griffies (1999) proposed measuring predictability by another quantity known as *predictive information*. DelSole (2004) showed that the two measures are identical when averaged over all conditions.

Denoting this averaged measure by $\mathbb{M}$, for Gaussian distributions, a natural estimate of this measure is

$$M = -\frac{k}{2} \log \frac{N}{T} = -\frac{k}{2} \log \left(1 - R^2\right), \tag{17.37}$$

where k is a scalar. This expression shows that predictive information is a monotonic function of R-square, hence it too has the same information content as the aforementioned measures.

Another approach is to imagine one of the ensemble members is "truth," but you do not know which one. Then, predictability can be measured by the correlation between ensemble member and ensemble mean. A natural estimate of this quantity is

$$\hat{\rho} = \frac{\sum_e \sum_c \left(y_{ec} - \hat{\mu}_Y\right) \left(\hat{\mu}_{Y|c} - \hat{\mu}_Y\right)}{\sqrt{\left(\sum_e \sum_c \left(y_{ec} - \hat{\mu}_Y\right)^2\right) \left(\sum_e \sum_c \left(\hat{\mu}_{Y|c} - \hat{\mu}_Y\right)^2\right)}}. \tag{17.38}$$

The numerator has a term that is independent of e, hence we may simplify it as follows:

$$\sum_c \left(\hat{\mu}_{Y|c} - \hat{\mu}_Y\right) \left(\sum_e \left(y_{ec} - \hat{\mu}_Y\right)\right) = E \sum_c \left(\hat{\mu}_{Y|c} - \hat{\mu}_Y\right)^2 = S. \tag{17.39}$$

Therefore,

$$\hat{\rho} = \frac{S}{\sqrt{ST}} = \sqrt{\frac{S}{T}} = \sqrt{R^2}. \tag{17.40}$$

Thus, the squared correlation is precisely equal to R-square.

These equivalences have an important consequence: The significance test for F can be converted immediately into a significance test for other measures. One simply finds the critical value of F from ANOVA, then uses the relations to convert that critical value into a critical value for signal-to-noise ratio, R-square, etc.

In summary, the F-statistic, signal-to-noise ratio, R-square, total-to-noise ratio, predictive information, and correlation are in one-to-one correspondence. Consequently, there is no advantage to using one measure over another. The choice of measure is often dictated by statistical reasons or physical interpretation. For instance, the signal-to-noise ratio often arises as a natural statistic for hypothesis testing while R-square is physically appealing because it corresponds to the fraction of predictable variance and lies between 0 and 1.

17.8 What Is the Difference between Predictability and Skill?

Note that the measures discussed in Section 17.7 did not reference *observations*. Predictability tells how close individual predictions are to each other; it does not tell how close the predictions are to observations. Instead, *forecast skill* tells how close forecasts are to observations. There is no necessary correspondence between predictability and skill. A model with high predictability may have poor skill, or a model with low (but nonzero) predictability may have high skill. The difference between predictability and skill is similar to the difference between precision and accuracy: accuracy measures how close a value is to its true value while precision measures how repeatable the measurement is, as illustrated in Figure 17.5.

To relate predictability and skill, consider a forecast system that predicts Y based on condition c. This forecast can be interpreted as some function of c, say $f(c)$. The mean square error of this forecast satisfies the identity

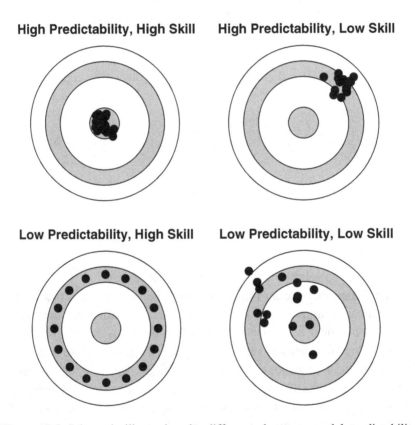

Figure 17.5 Schematic illustrating the difference between model predictability and skill. The bull's-eye center is the true value (that is, the "verification") and the dots are members of a forecast ensemble. The forecast is highly predictable when the dots are close *to each other* and has high skill when the *mean* of the dots is close to the true value. The case of low model predictability and high skill (bottom left) is illustrated with symmetrically distributed dots so that it is clear that their centroid is at the center.

$$\text{MSE} = \mathbb{E}\left[(Y - f(c))^2\right] = \mathbb{N} + \mathbb{B}, \tag{17.41}$$

where $\mathbb{N}$ is defined in (17.26) and

$$\mathbb{B} = \mathbb{E}\left[\left(f(c) - \mu_{Y|c}\right)^2\right]. \tag{17.42}$$

The term $\mathbb{B}$ measures the *forecast error due to conditional bias*. The conditional bias vanishes when $f(c) = \mu_{Y|c}$, otherwise it is *positive*. Therefore, we have the bound

$$\text{MSE} \geq \mathbb{N}. \tag{17.43}$$

The term $\mathbb{N}$ measures the *irreducible error* of a forecast and is a fundamental limit of the true dynamical system. This bound implies that the mean square error of a perfect prediction system is less than or equal to the mean square error of any forecast model.

If a prediction system has *no* predictability, then it cannot have skill. After all, an unpredictable variable is one whose forecast distribution is independent of initial condition. It follows that *predictability is a necessary condition for skill*.

MSE is oppositely related to skill (e.g., small MSE means large skill). A positively oriented measure of skill is *the mean squared error skill score*

$$\Gamma = 1 - \frac{\text{MSE}}{\mathbb{T}} = \left(1 - \frac{\mathbb{N}}{\mathbb{T}}\right) - \frac{\mathbb{B}}{\mathbb{T}} = \mathbb{R}^2 - \frac{\mathbb{B}}{\mathbb{T}} \leq \mathbb{R}^2. \tag{17.44}$$

The maximum value of Γ is 1 and occurs for a forecast in which $\text{MSE} = 0$. Negative values of Γ indicate a forecast that is "worse than climatology," in the sense that the MSE of the forecast is larger than the MSE of the climatological mean in which $\text{MSE} = \mathbb{T}$. Since real forecast systems are imperfect, $\mathbb{B} > 0$ and $\Gamma < \mathbb{R}^2$, which shows that model skill is a lower bound of the predictability of the true system.

These bounds can sometimes lead to confusion if one forgets that $\mathbb{N}$ and $\mathbb{R}^2$ are parameters of the *true* dynamical system. For instance, it is fairly standard to use *a forecast model* to estimate these parameters, giving N_M and R_M^2. One might inadvertently assume that the *model* predictability R_M^2 provides some bound on skill, but this is not the case. In general, there is no necessary correspondence between model predictability R_M^2 and true predictability $\mathbb{R}^2$– given positive R_M^2, any other value of R_M^2 can be obtained by increasing or decreasing the noise of a model (e.g., by defining $f(c)$ based on large ensemble sizes or by adding random numbers to $f(c)$). This kind of confusion explains certain apparent paradoxes. For instance, some dynamical models predict the observed NAO index better than they predict their own NAO index. This phenomenon is sometimes known as the "signal-to-noise paradox." This paradox effectively means $\Gamma > R_M^2$, but this does not contradict (17.44) because model predictability R_M^2 can be less than the true predictability R^2.

17.9 Chaos and Predictability

A fundamental property of the equations of atmospheric motion is that they exhibit *sensitive dependence on initial conditions*. This property is known as *chaos* and arises from the nonlinearities and instabilities inherent in the equations of motion. A common paradigm for investigating chaotic systems is to examine the difference of forecasts from slightly different initial states. It is left as an exercise to show that the mean square difference (MSD) between ensemble members is twice the noise variance estimated from ANOVA:

$$\text{MSD} = \frac{1}{CE(E-1)} \sum_{c=1}^{C} \sum_{e=1}^{E} \sum_{e'=1}^{E} (y_{ec} - y_{e'c})^2 = 2\hat{\sigma}_N^2. \tag{17.45}$$

Although small errors grow with lead time, the error growth cannot continue forever since the variables are bounded. When does error growth stop? As discussed earlier, variables become unpredictable at sufficiently long lead times. By the law of total variance (17.25), unpredictable variables satisfy $\mathbb{T} = \mathbb{N}$. Thus, the mean square difference between ensemble members is expected to grow until it "saturates" at twice the total variance. This reasoning motivates measuring error growth using the noise-to-total ratio

$$\text{NTR} = \frac{\mathbb{N}}{\mathbb{T}}. \tag{17.46}$$

NTR lies between 0 and 1, with zero indicating a zero-error forecast and one indicating unpredictability (see Section 17.7). One might estimate NTR using $\hat{\sigma}_N^2/\hat{\sigma}_T^2$, but this estimate is not guaranteed to be less than 1. An estimate of NTR that lies between 0 and 1 is

$$\text{NTR} = \frac{N}{T} = \left(1 + \frac{F(C-1)}{C(E-1)}\right)^{-1}. \tag{17.47}$$

NTR versus lead time is shown in Figure 17.6. As can be seen, NTR starts small and increases with lead time until it saturates after about two weeks. The critical value for NTR is found by substituting the critical F-value into Equation (17.47) and is indicated by the horizontal dashed line in Figure 17.6. Predictability is detected for all lead times in this example. It has been found empirically that NTR is often well fit by the logistic function

$$\text{NTR} = \frac{1}{1 + \exp[-k(t - t_0)]}, \tag{17.48}$$

where k and t_0 are parameters. Least squares estimates of these parameters are

$$k = \frac{1}{2.36 \text{ days}} \quad \text{and} \quad t_0 = 10 \text{ days}. \tag{17.49}$$

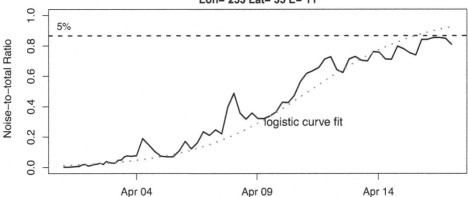

Figure 17.6 Estimate of noise-to-total ratio of the two ensemble forecasts shown in Figure 17.3 Figure 17.3 (solid curve). The NTR is estimated at each lead time separately using corresponding to the 21 years 1999–2019. The horizontal dashed line shows the 5% significance threshold for rejecting the hypothesis of equal means based on F, which was converted into NTR as described in the text. The hypothesis is rejected when NTR is *below* the significance curve. The dotted curve shows the best fit logistic curve.

This curve is illustrated in Figure 17.6. According to this equation, initial errors grow exponentially with an e-folding time of 2.36 days, and then enter a saturation phase after an inflection point at $t_0 = 10$ days.

17.10 Conceptual Questions

1 Why is it known as Analysis of Variance when the test is for a difference in means?

2 ANOVA and the correlation test are procedures for testing independence. Why are there two tests for the same hypothesis? When do you use which test?

3 In what sense is a coin toss unpredictable? After all, you can accurately predict that heads will appear 50% of the time.

4 What kind of uncertainty is an ensemble forecast used to quantify?

5 What is the forecast distribution? What is the climatological distribution?

6 After about two weeks, ANOVA detects no predictability of daily fields, yet we can make predictions of seasonal means. How do you reconcile these two facts?

7 What is the difference between predictability and skill?

8 Can a model have high predictability but make poor predictions of observations?

9 In what sense does it not matter whether you use signal-to-noise or correlation?

18

Predictable Component Analysis

> In this paper it is the purpose of the author to find the moments and distributions of some [hypothesis tests] generalized for samples from a multivariate normal population and to exhibit a method of attack which seems to be novel in its application.[1]
>
> *Wilks (1932)*

The previous chapter discussed Analysis of Variance (ANOVA), a procedure for deciding if populations have identical *scalar* means. This chapter discusses the generalization of this test to *vector* means, which is known as Multivariate Analysis of Variance, or MANOVA. MANOVA can detect predictability of random vectors and *decompose* a random vector into a sum of components ordered such that the first maximizes predictability, the second maximizes predictability subject to being uncorrelated with the first, and so on. This decomposition is known as *Predictable Component Analysis (PrCA)* or *signal-to-noise maximizing EOF analysis*. A slight modification of this procedure can decompose *forecast skill*. The connection between PrCA, Canonical Correlation Analysis, and Multivariate Regression is reviewed. In typical climate studies, the dimension of the random vector exceeds the number of samples, leading to an ill-posed problem. The standard approach to regularizing the problem is to project the data onto a small number of principal components and then perform PrCA on the selected PC time series. A key question in this approach is how many principal components should be used. This question can be framed as a model selection problem in multivariate linear regression. However, the selection problem is non-standard, and a new criterion is suggested in this chapter.

---

[1] Adapted by permission from Oxford University Press, Wilks (1932).

18.1 The Problem

One of the most exciting experiences in science is discovering that some seemingly random phenomena can actually be predicted. In the 1920s, Gilbert Walker established that atmospheric pressure tended to oscillate in at least three distinct ways, which he termed the North Atlantic Oscillation, the North Pacific Oscillation, and the Southern Oscillation (Walker, 1928). Walker was particularly interested in the Southern Oscillation because it seemed to be linked to Indian monsoon rainfall and therefore might be used to predict the latter, which would have enormous benefits to India. Unfortunately, the mechanism for this particular oscillation was obscure at the time. In the 1960s, Jacob Bjerknes studied El Niño, which at the time was thought to be a local phenomenon confined to the Peruvian coast. Bjerknes discovered that in fact El Niño is accompanied by changes over the entire tropical Pacific and was connected to the Southern Oscillation through a circulation he termed the Walker Circulation, in honor of Walker. Bjerknes proposed that this circulation maintained the temperature contrast along the equatorial Pacific, with cold waters on the eastern end near Peru and warm water at the western end near Australia, and that fluctuations in this wind explained El Niño and La Niña. Nevertheless, this mechanism seemed to be a strictly positive feedback, which could not explain its oscillatory nature. Wyrtki (1975) proposed that the oscillatory nature was due to the response of the ocean to changes in atmospheric surface winds, which excited equatorial Kelvin waves that propagated eastward, giving rise to a delayed response. Cane et al. (1986) put these elements together in a simple model and showed that El Niño is generally predictable one or two years ahead in this particular model. El Niño forecasts have developed considerably since that time and today are routine at operational forecasting centers around the world.

Today, El Niño is believed to be a manifestation of an unstable mode of the coupled atmosphere–ocean system, or perhaps a stable mode with a long decay time that is randomly excited (Battisti and Sarachik, 1995; Penland and Sardeshmukh, 1995). In either case, it is often suggested that El Niño is associated with a coherent *predictable component* in an otherwise noisy system. Can this predictable component be identified directly from a data set? That is, if you were given a large data set, could you identify a predictable component just by statistical analysis? One approach is to find highly correlated components between the initial and target times. This approach leads to Canonical Correlation Analysis, which is discussed in Chapter 15. This chapter approaches the problem differently. Given a *forecast* data set, can you identify a predictable component simply by analyzing the forecasts? This chapter discusses a method for doing this.

To solve this problem, a measure of predictability must be defined. The concept of predictability is summarized in Chapter 17, to which we refer readers for details. A basic paradigm is to consider a *forecast ensemble*, which is a set of forecasts initialized at plausible states of the system. Typically, the members of a forecast ensemble are clustered together at the beginning of the forecast, then they diverge from each other with time until, eventually, the individual forecasts are indistinguishable from a random draw from the past observational record (see Figure 17.3, for an example). This paradigm naturally suggests measuring predictability by the spread of the forecast ensemble. One such measure is the *noise-to-total* ratio – the variance of the ensemble members about the ensemble mean, divided by the total variance of the variable. Other measures include the signal-to-noise ratio, R-square between ensemble member and ensemble mean, and predictive information. Chapter 17 shows that each of these measures is a monotonic function of F, the statistic in ANOVA to test equality of means defined in (17.21). Accordingly, we use F to measure predictability in this chapter. In Section 18.6, we show that the same predictable components are obtained for each of the measures discussed in Chapter 17, so choosing F results in no loss of generality. An advantage of using F is that it relates immediately to ANOVA.

As an example, we consider forecasts from version 2 of the Climate Forecast System (CFSv2), a prediction model at the National Centers for Environmental Prediction (NCEP). The CFSv2 is a coupled ocean–land–atmosphere dynamical prediction system. Further details of the model are described in Saha et al. (2014).

We consider forecasts[2] of *April mean* 2m-temperature initialized in early November (that is, 6 month lead) for each year in the 28-year period 1982–2009. The initial start dates are Oct. 23, Oct. 28, Nov. 2, Nov. 7, and on each date the forecasts are initialized at 0, 6, 12, 18Z. For quantifying seasonal predictability, these start dates are sufficiently close to each other that they can be treated as $E = 16$ ensemble members under the same condition. This approach is known as a *lagged ensemble* because start dates are lagged over short intervals. Pooling forecasts with different start times makes sense in seasonal forecasting, provided the start dates are sufficiently close (e.g., a few days) to each other. Different conditions $c = 1, 2, \ldots, 28$ are distinguished uniquely by start date year. Thus, $E = 16$ and $C = 28$.

ANOVA can be applied to the data at each grid point. The F-value at each spatial location over North America is shown in Figure 18.1. Insignificant values at the 5% level are masked out. About 93% of the area contains significant F values. Forecast statistics at three selected locations are shown in the bottom three panels. Predictability is clearly evident in the time series: The dependence on initial condition is so strong that the 1-sigma interval of the ensemble members shows no overlap between certain years. It should be emphasized that these features of predictability are a property of a particular model (CFSv2) and may not reflect the predictability of the true climate system. Since ANOVA does not compare forecasts to observations,

---

[2] Technically, these are "hindcasts" rather than "forecasts." "Hindcasts" refer to predictions of past events and are used to estimate the performance of a prediction model. In contrast, "forecasts" refer to predictions of future events that are unknown at the time of the forecast. This distinction is not important in this analysis and therefore will be ignored.

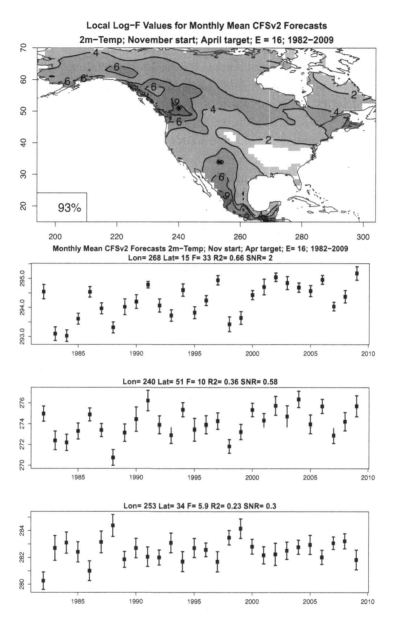

Figure 18.1 (top) Predictability of April mean 2m temperature in the CFSv2 for 16-member lagged ensemble forecasts initialized in late October and early November, as measured by log-F values computed from ANOVA at each spatial point over 1982–2009. F-values insignificant at the 5% level are masked out. The percentage land area with significant F values is indicated in the bottom left box. (bottom three panels) Mean plus/minus one standard deviation of the 16-member lagged ensemble at three grid points with locally large values of F. The location of the grid points, associated F-value, and *adjusted* R-square and SNR (defined in Section 17.7) are indicated in the title of each panel. The three grid points are indicated by asterisks in the top panel. The year in the x-axis corresponds to the year of the November initial conditions.

it does not say anything about *forecast skill* (this distinction is discussed in more detail in Section 17.8).

Since 93% of the area has significant F-values, 2m-temperature over North America is surely predictable in this model. Inspection of the time series shows that the first two time series are highly correlated with each other (correlation is 0.88), and are negatively correlated with the third (both correlations are about −0.58). Are these areas responding to the same source of predictability? For seasonal predictability, the natural candidate is ENSO, and indeed the correlations between the Niño 3.4 index and the ensemble mean forecasts in these areas are 0.76, 0.73, −0.74 (compared to the 5% significance threshold 0.38), strongly suggesting a connection to ENSO.

Although the above approach is valid, it has limitations. One limitation is that predictability is analyzed locally, point-by-point. Perhaps the predictability is more concisely expressed through *patterns*. Also, suppose there were *two* sources of predicability, each with different patterns. It would not be easy to identify these patterns from a local analysis. Instead, we need a multivariate approach that can extract *predictable patterns*. An approach for doing this is discussed in this chapter.

18.2 Illustration of Predictable Component Analysis

Before discussing the exact details of the procedure, we first illustrate it by applying it to the forecasts discussed in sec. 18.1[3]. For reasons discussed in Section 18.5, we select the first six PCs of April-mean 2m-temperature. PrCA finds a linear combination of the six PCs that maximizes predictability, as measured by the F-statistic. The resulting maximized F-values are shown in Figure 18.2. For comparison, the 98% confidence interval of each maximized F value under the null hypothesis of no predictability is indicated by the shaded region. The first three F values lie above the 99th percentile. Because the first value is significant, we reject H_0. The second and third components have similar F-ratios, suggesting they are "effectively degenerate" and therefore sensitive to the sample (this issue is discussed in more detail in Chapter 12). Accordingly, we focus on the first component.

For each maximized F-value, there corresponds a component with a particular spatial structure and time series. The time series for the first component is shown in the bottom panel of figure 18.3. This time series is strongly correlated with the Niño 3.4 index at the November start time (correlation coefficient is 0.88). Thus, the predictability of CFSv2 is strongly dominated by ENSO. We emphasize that we did not presume a relation to ENSO, instead PrCA *finds* a component with this

---

[3] see footnote 1.

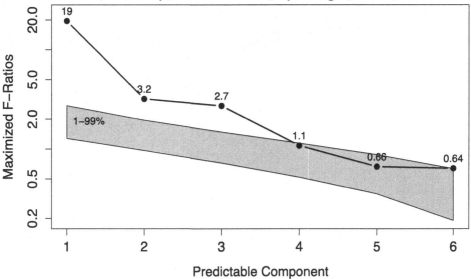

Figure 18.2 The spectrum of maximized F-ratios derived from CFSv2 hindcasts discussed in Section 18.2 using six principal components. Shaded area shows 1 to 99th percentile range of each maximized F-ratio from Monte Carlo simulations under H_0.

relation in the data. The spatial structure of the predictable component is described by the *loading vector* shown in the upper panel of Figure 18.3, and is interpreted the same way as in Principal Component Analysis – the loading vector times its time series, summed over all components, recovers the original data (that is, the six PCs in this case). The pattern is dominated by positive loadings in the north and negative loadings in the southwest United States. Thus, the predictability is characterized by an anti-correlation between the north and south. That is, when the variate is positive (e.g., in 1982), the north warms and the south cools, while if the variate is negative (e.g., in 1983), the north cools while the south warms. Note that the time series for the most predicable component is highly correlated with the three time series shown in Figure 18.1. Furthermore, the loading pattern correctly indicates the sign of the correlation. In this way, PrCA has elucidated the results shown in Figure 18.1.

The leading predictable component explains about 22% of the variance in the original data. Because each variate has unit total variance, the absolute value of the loading vector in a location gives the standard deviation of the component at that location. For instance, the standard deviation of this component is about 2 degrees in central Canada.

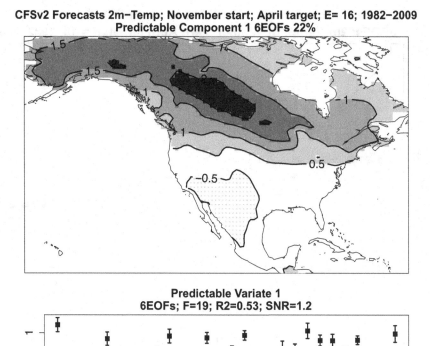

CFSv2 Forecasts 2m-Temp; November start; April target; E= 16; 1982–2009
Predictable Component 1 6EOFs 22%

Predictable Variate 1
6EOFs; F=19; R2=0.53; SNR=1.2

Figure 18.3 Regression pattern (top) and variate (bottom) of the most predictable component of the CFSv2 hindcasts discussed in Section 18.2. The component was computed using the leading six principal components of the data. The display format of the variate is the same as that of Figure 18.1. In the regression pattern, positive values are shaded while negative values are stippled.

18.3 Multivariate Analysis of Variance

The natural multivariate generalization of ANOVA is to consider an M-dimensional random *vector* $\mathbf{y}_{ec}$ with multivariate normal distribution

$$\mathbf{y}_{ec} \sim \mathcal{N}_M \left(\boldsymbol{\mu}_{Y|c}, \boldsymbol{\Sigma}_{Y|c} \right), \tag{18.1}$$

where $\boldsymbol{\mu}_{Y|c}$ and $\boldsymbol{\Sigma}_{Y|c}$ are the conditional mean and covariance matrix. We want to know if the mean or covariance matrix depend on c. Following the same logic as in ANOVA, we follow a two-step procedure. First, we decide if the covariance matrices depend on c. If not, then we decide if the mean vectors depend on c, conditioned on the assumption that the covariances are independent of c. The procedure discussed here depends on sample estimates of $\boldsymbol{\mu}_{Y|c}$ and $\boldsymbol{\Sigma}_{Y|c}$. These estimates follow from the

Table 18.1. *Multivariate Generalizations of the Conditional Mean (17.3), Unconditional Mean (17.5), Signal Variance (17.17), and Noise Variance (17.6).*

$\mathbf{y}_{ec}, \quad e = 1, \ldots E$ and $c = 1, \ldots, C$	M-dimensional vectors
$\hat{\boldsymbol{\mu}}_{Y\|c} = \dfrac{1}{E} \displaystyle\sum_{e=1}^{E} \mathbf{y}_{ec}$	conditional mean
$\hat{\boldsymbol{\mu}}_{Y} = \dfrac{1}{EC} \displaystyle\sum_{c=1}^{C}\sum_{e=1}^{E} \mathbf{y}_{ec}$	unconditional mean
$\hat{\boldsymbol{\Sigma}}_{Y\|c} = \dfrac{1}{E-1} \displaystyle\sum_{e=1}^{E} \left(\mathbf{y}_{ec} - \hat{\boldsymbol{\mu}}_{Y\|c}\right)\left(\mathbf{y}_{ec} - \hat{\boldsymbol{\mu}}_{Y\|c}\right)^{T}$	conditional covariance
$\hat{\boldsymbol{\Sigma}}_{N} = \dfrac{1}{C(E-1)} \displaystyle\sum_{c=1}^{C}\sum_{e=1}^{E} \left(\mathbf{y}_{ec} - \hat{\boldsymbol{\mu}}_{Y\|c}\right)\left(\mathbf{y}_{ec} - \hat{\boldsymbol{\mu}}_{Y\|c}\right)^{T}$	noise covariance
$\hat{\boldsymbol{\Sigma}}_{S} = \dfrac{1}{C-1} \displaystyle\sum_{c=1}^{C} \left(\hat{\boldsymbol{\mu}}_{Y\|c} - \hat{\boldsymbol{\mu}}_{Y}\right)\left(\hat{\boldsymbol{\mu}}_{Y\|c} - \hat{\boldsymbol{\mu}}_{Y}\right)^{T}$	signal covariance
$\hat{\boldsymbol{\Sigma}}_{T} = \dfrac{1}{CE-1} \displaystyle\sum_{c=1}^{C}\sum_{e=1}^{E} \left(\mathbf{y}_{ec} - \hat{\boldsymbol{\mu}}_{Y}\right)\left(\mathbf{y}_{ec} - \hat{\boldsymbol{\mu}}_{Y}\right)^{T}$	total covariance

corresponding ANOVA quantities simply by replacing scalar quantities by vectors, and by replacing squares by outer products. These estimates are summarized in Table 18.1.

18.3.1 Testing Equality of Covariance Matrices

We first consider the hypothesis that the covariance matrices are equal, namely

$$\boldsymbol{\Sigma}_{Y\|1} = \boldsymbol{\Sigma}_{Y\|2} = \cdots = \boldsymbol{\Sigma}_{Y\|C}. \tag{18.2}$$

The statistic for testing this hypothesis is a multivariate generalization of the Bartlett statistic (17.8), namely

$$\Theta = (1 - g)\, C(E - 1) \log\left(\frac{\left|\hat{\boldsymbol{\Sigma}}_{N}\right|}{\left(\left|\hat{\boldsymbol{\Sigma}}_{Y\|1}\right|\left|\hat{\boldsymbol{\Sigma}}_{Y\|2}\right|\cdots\left|\hat{\boldsymbol{\Sigma}}_{Y\|C}\right|\right)^{1/C}}\right). \tag{18.3}$$

where $|\cdot|$ denotes the determinant of a matrix and

$$g = \frac{(C+1)(2M^2 + 3M - 1)}{6C(M+1)(E-1)}. \tag{18.4}$$

Under the null hypothesis of equal conditional covariance matrices, Θ has an approximate chi-squared distribution with $(C-1)M(M+1)/2$ degrees of freedom.

For the example discussed in sec. 18.1, the statistic for testing equality of covariance matrices (18.3) is insignificant at the 1% level for all choices for the number of PCs (not shown). Therefore, we conclude that covariances do not depend on c and take them to be independent of c:

$$\Sigma_N = \Sigma_{Y|1} = \Sigma_{Y|2} = \cdots = \Sigma_{Y|C}. \tag{18.5}$$

18.3.2 Testing Equality of Multivariate Means: MANOVA

Next we test the hypothesis that the conditional means are independent of c:

$$H_0: \quad \mu_{Y|1} = \mu_{Y|2} = \cdots = \mu_{Y|C}. \tag{18.6}$$

This hypothesis is equivalent to assuming that samples are generated by the model

$$y_{ec} = \mu_{Y|c} + \epsilon_{ec}, \tag{18.7}$$

where ϵ_{ec} are random samples from $\mathcal{N}_M(0, \Sigma_N)$, and testing hypothesis (18.6).

The standard statistic for testing H_0 is based on

$$\overline{\Sigma}_S = E(C-1)\hat{\Sigma}_S/(CE) \quad \text{and} \quad \overline{\Sigma}_N = C(E-1)\hat{\Sigma}_N/(CE). \tag{18.8}$$

Then, H_0 is tested based on the statistic

$$\Lambda = \frac{|\overline{\Sigma}_N|}{|\overline{\Sigma}_N + \overline{\Sigma}_S|}. \tag{18.9}$$

This statistic lies between 0 and 1, with *small* values indicating rejection of H_0. As will be shown in Section 18.4, this statistic can be written in terms of the F ratios of the predictable components. In the scalar case $M = 1$, and this statistic reduces to

$$\Lambda = \left(1 + \frac{F(C-1)}{C(E-1)}\right)^{-1}, \tag{18.10}$$

where F is the F-statistic in ANOVA (17.21). Under the null hypothesis H_0, Λ has a distribution known as *Wilks' Lambda Distribution*. The critical values for Wilks' Lambda are not always included in standard statistics software, but for large sample sizes we may use the asymptotic approximation (14.37) with $\nu_E = C(E-1)$ and $\nu_H = C - 1$. In the statistics literature, this test is known as Multivariate Analysis of Variance (MANOVA).

In the example discussed in sec. 18.1, the statistic for testing equality of means (18.9) is significant at the 1% level for all choices for the number of PCs (this will be shown in more detail in Section 18.5. When H_0 is rejected, we say that *predictability has been detected.*

18.4 Predictable Component Analysis

The mere detection of predictability (by rejecting H_0) is unsatisfying because it tells us nothing about when or where this predictability occurs. To diagnose the space-time structure of the predictability, we explore projections of the form

$$r_{ec} = \mathbf{q}^T \mathbf{y}_{ec}, \tag{18.11}$$

where $\mathbf{q}$ is a projection vector. The projection transforms vector quantities into scalar quantities so that the random variable r_{ec} has the same format as Y_{ec}. Hence, ANOVA can be applied directly to r_{ec}, leading to the statistic

$$F = E \frac{\hat{\sigma}_S^2}{\hat{\sigma}_N^2}, \tag{18.12}$$

where $\hat{\sigma}_S^2$ and $\hat{\sigma}_N^2$ are defined as in (17.17) and (17.6), except y-variables are replaced by r-variables. For instance, the signal variance can be expressed as

$$\hat{\sigma}_S^2 = \frac{1}{C-1} \sum_{c=1}^{C} \left(\frac{1}{E} \sum_{e=1}^{E} r_{ec} - \frac{1}{EC} \sum_{e=1}^{E} \sum_{c=1}^{C} r_{ec} \right)^2 \tag{18.13}$$

$$= \frac{1}{C-1} \sum_{c=1}^{C} \left(\frac{1}{E} \sum_{e=1}^{E} \mathbf{q}^T \mathbf{y}_{ec} - \frac{1}{EC} \sum_{e=1}^{E} \sum_{c=1}^{C} \mathbf{q}^T \mathbf{y}_{ec} \right)^2 \tag{18.14}$$

$$= \frac{1}{C-1} \sum_{c=1}^{C} \left(\mathbf{q}^T \left(\hat{\mu}_{Y|c} - \hat{\mu}_Y \right) \right)^2 \tag{18.15}$$

$$= \mathbf{q}^T \hat{\mathbf{\Sigma}}_S \mathbf{q}. \tag{18.16}$$

A similar approach yields

$$\hat{\sigma}_N^2 = \mathbf{q}^T \hat{\mathbf{\Sigma}}_N \mathbf{q}. \tag{18.17}$$

Therefore, the F statistic can be written in terms of the projection vectors as

$$F = E \frac{\mathbf{q}^T \hat{\mathbf{\Sigma}}_S \mathbf{q}}{\mathbf{q}^T \hat{\mathbf{\Sigma}}_N \mathbf{q}}. \tag{18.18}$$

If $\hat{\boldsymbol{\Sigma}}_N$ is positive definite, then the denominator cannot vanish for $\mathbf{q} \neq \mathbf{0}$. For the moment, we assume $\hat{\boldsymbol{\Sigma}}_N$ is positive definite. We will return to this issue in Section 18.5.

If $\mathbf{q}^T \mathbf{y}$ is a pre-defined climate index, then we can test its predictability using the fact that the F-statistic defined in (18.18) has an F-distribution with $C - 1$ and $C(E - 1)$ degrees of freedom, when H_0 is true.

To find the most predictable component, we choose the projection vector $\mathbf{q}$ to *maximize F*. Happily, the solution to this problem has been presented in Chapter 16, namely, we perform Covariance Discriminant Analysis (CDA). Because the procedure is identical to CDA, the solution is merely summarized. Specifically, the projection vector that maximizes F is obtained by solving the generalized eigenvalue problem

$$E\hat{\boldsymbol{\Sigma}}_S\mathbf{q} = F\hat{\boldsymbol{\Sigma}}_N\mathbf{q}. \tag{18.19}$$

Since the matrices are $M \times M$, there will be M solutions to this equation. Each solution is characterized by an eigenvector $\mathbf{q}_i$ and eigenvalue F_i. The maximum eigenvalue is the largest possible F-value, and the corresponding eigenvector is the projection vector that gives this value. By convention, eigenvalues are ordered from largest to smallest, $F_1 \geq F_2 \geq \cdots \geq F_M$. The corresponding eigenvectors are denoted $\mathbf{q}_1, \ldots, \mathbf{q}_M$. Multiplying the i'th eigenvector $\mathbf{q}_i$ by the original data vector $\mathbf{y}_{ec}$ yields the i'th *predictable variate*

$$r_{ec,i} = \mathbf{q}_i^T \mathbf{y}_{ec}. \tag{18.20}$$

The variate for the most predictable component in our forecast example was shown in the bottom panel of Figure 18.3. The predictable variate $r_{ec,i}$ has the same format as Y_{ec} and hence can be interpreted similarly. Specifically, the signal and noise variances are derived from (18.13) and (18.17),

$$\hat{\sigma}_{S,i}^2 = \mathbf{q}_i^T \hat{\boldsymbol{\Sigma}}_S \mathbf{q}_i \quad \text{and} \quad \hat{\sigma}_{N,i}^2 = \mathbf{q}_i^T \hat{\boldsymbol{\Sigma}}_N \mathbf{q}_i, \tag{18.21}$$

and the corresponding F value is

$$F_i = E \frac{\mathbf{q}_i^T \hat{\boldsymbol{\Sigma}}_S \mathbf{q}_i}{\mathbf{q}_i^T \hat{\boldsymbol{\Sigma}}_N \mathbf{q}_i}. \tag{18.22}$$

The value of F_i derived from (18.22) is equal to the eigenvalue F_i.

Recall that the determinant of a matrix equals the product of its eigenvalues. A similar property holds for the ratio of determinants of matrices in a generalized eigenvalue problem. Based on this fact, it can be shown that the statistic (18.9) is

$$\Lambda = ((F_1/E + 1)(F_2/E + 1) \ldots (F_M/E + 1))^{-1}. \tag{18.23}$$

Thus, the test for equality of means can be performed conveniently from the eigenvalues computed in PrCA.

The complete set of eigenvectors can be collected into a single matrix as

$$\mathbf{Q} = \begin{bmatrix} \mathbf{q}_1 & \mathbf{q}_2 & \cdots & \mathbf{q}_M \end{bmatrix}. \tag{18.24}$$

Then, the variates from (18.20) are

$$\mathbf{r}_{ec} = \mathbf{Q}^T \mathbf{y}_{ec}. \tag{18.25}$$

Because the eigenvectors form a linearly independent set, $\mathbf{Q}$ is invertible and

$$\mathbf{P} \mathbf{r}_{ec} = \mathbf{y}_{ec}, \tag{18.26}$$

where

$$\mathbf{P} = (\mathbf{Q}^T)^{-1}. \tag{18.27}$$

The columns of $\mathbf{P}$ are the *loading vectors* $\mathbf{p}_1, \ldots, \mathbf{p}_M$. Equation (18.26) can be written as

$$\mathbf{p}_1 \mathbf{r}_{ec,1} + \cdots + \mathbf{p}_M \mathbf{r}_{ec,M} = \mathbf{y}_{ec}, \tag{18.28}$$

where the product $\mathbf{p}_i \mathbf{r}_{ec,i}$ is the i'th *predictable component*. Expression (18.28) shows that the original data can be *decomposed* into a sum of predictable components ordered by their predictability. In the climate literature, this procedure is known as *Predictable Component Analysis* (PrCA) or *Signal-to-Noise Maximizing EOF Analysis* (or similar phrasing), and the first component is known as *the most predictable component*.

It will be shown in Section 18.9 that the variates are uncorrelated in three different senses: the variates $r_{ec,i}$ themselves are uncorrelated, their conditional means are uncorrelated, and the residuals about the conditional mean are uncorrelated. Recall that eigenvectors are unique up to a multiplicative constant. It is helpful to choose this constant so that variates have unit *total variance*. That is, we re-scale the eigenvectors to satisfy

$$\mathbf{q}_i^T \hat{\boldsymbol{\Sigma}}_T \mathbf{q}_i = 1. \tag{18.29}$$

The advantages of this normalization are similar to those discussed in Chapter 12 in the context of Principal Component Analysis.

18.5 Variable Selection in PrCA

In the example discussed in Section 18.2, the data set comprised 2,664 grid points over North America, which far exceeded the total number of forecasts $EC = 448$. As a result, $\hat{\Sigma}_N$ in (18.18) is singular and projection vectors can be found to give $F = \infty$, which is clearly unrealistic. The problem is overfitting: PrCA finds the projection vector $\mathbf{q}$ that maximizes F *in the given sample*, but that same projection vector will not produce as large an F value when applied to independent data. The standard approach to dealing with this problem is to apply PrCA only to a few principal components of $\mathbf{y}_{ec}$. The question arises as to how many PCs should be chosen to maximize predictability without significant overfitting. Our strategy is to re-interpret this problem as a problem selecting regression models, and then apply model selection criteria discussed in Chapter 14 to decide the number of PCs.

First, re-express $\mathbf{y}_{ec}$ as the following matrix:

$$(\mathbf{y}_{ec})_m = (\mathbf{y})_{e+(c-1)E,m} . \tag{18.30}$$

On the right-hand side, ensemble members and conditions are "stacked" together in a single column for each m. Then, the MANOVA model (18.7) can be written as

$$
\begin{array}{ccccc}
\mathbf{Y} & = & \mathbf{X} & \mathbf{B} & + & \mathbf{\Xi} \\
CE \times M & & CE \times C & C \times M & & CE \times M'
\end{array} \tag{18.31}
$$

where

$$
\mathbf{X} =
\begin{pmatrix}
\mathbf{j}_E & \mathbf{0} & \cdots & \mathbf{0} \\
\mathbf{0} & \mathbf{j}_E & \cdots & \mathbf{0} \\
\vdots & \vdots & \ddots & \vdots \\
\mathbf{0} & \mathbf{0} & \cdots & \mathbf{j}_E
\end{pmatrix},
\quad \text{and} \quad
\mathbf{B} =
\begin{pmatrix}
\mu_{Y|1}^T \\
\mu_{Y|2}^T \\
\vdots \\
\mu_{Y|C}^T
\end{pmatrix},
\tag{18.32}
$$

and $\mathbf{j}_E$ is an E-dimensional vector of all ones. This model is of the same form as the multivariate regression model (14.19). Note, however, that the predictor matrix $\mathbf{X}$ consists of "dummy variables" that encode the condition labels. In particular, the regression model is Fixed-X; therefore we should use MIC based on AICc instead of AICr. To avoid confusion, we term the resulting criterion MICf, where "f" indicates Fixed-X, so that

$$\text{MICf}(X; Y) = \text{AICc}(Y|X) - \text{AICc}(Y). \tag{18.33}$$

It should be kept in mind that AICc does not satisfy the chain rule and therefore (18.33) cannot be written in alternative forms such as (14.58) and (14.59), in contrast to MIC.

An unbiased estimate of the noise covariance matrix for (18.31) is $\hat{\boldsymbol{\Sigma}}_N$. Thus, evaluating (14.49) for (18.31) yields

$$\text{AICc}(Y|X) = CE \log |\overline{\boldsymbol{\Sigma}}_N| + CEM \log 2\pi + \frac{CEM(CE+C)}{CE - C - M - 1}. \qquad (18.34)$$

On the other hand, under the null hypothesis, the MANOVA model can be written in the same form as (18.31), but using $\mathbf{X} = \mathbf{j}_{EC}$ and $\mathbf{B} = \boldsymbol{\mu}_Y^T$. An unbiased estimate of the noise covariance matrix for this model is $\hat{\boldsymbol{\Sigma}}_T$. Therefore,

$$\text{AICc}(Y) = CE \log |\overline{\boldsymbol{\Sigma}}_T| + CEM \log 2\pi + \frac{CEM(CE+1)}{CE - M - 2}, \qquad (18.35)$$

where

$$\overline{\boldsymbol{\Sigma}}_T = (CE - 1)\hat{\boldsymbol{\Sigma}}_T / (CE). \qquad (18.36)$$

Thus, the selection criterion is

$$\text{MICf}(X; Y) = CE \log \Lambda + \frac{CEM(CE+C)}{CE - C - M - 1} - \frac{CEM(CE+1)}{CE - M - 2}, \qquad (18.37)$$

where

$$\Lambda = \frac{|\overline{\boldsymbol{\Sigma}}_N|}{|\overline{\boldsymbol{\Sigma}}_T|}. \qquad (18.38)$$

In the next section, it will be shown that $\overline{\boldsymbol{\Sigma}}_T = \overline{\boldsymbol{\Sigma}}_S + \overline{\boldsymbol{\Sigma}}_N$, which implies that (18.38) is identical to (18.9). This equivalence shows that both MICf and the MANOVA test depend on the sample values only through Λ.

A plot of MICf as a function of the number of EOFs is shown in Figure 18.4. The minimum occurs at six EOFs, which justifies our using six PCs for PrCA in Section 18.2. Using the asymptotic distribution (14.37) under the null hypothesis, the criterion for detection of predictability (that is, rejecting H_0) at the 1% significance level can be written equivalently as

$$\text{MICf}(X; Y) < \frac{-CE\chi^2_{M\nu_H, 0.01}}{\nu_E - (S - \nu_H + 1)/2} + \frac{CEM(CE+C)}{CE - C - M - 1} - \frac{CEM(CE+1)}{CE - M - 2},$$

where $\nu_E = C(E-1)$ and $\nu_H = C-1$. The critical value (that is, the right-hand side of the equation) is shown in Figure 18.4 and lies entirely above MICf, indicating that predictability is detected at all EOF truncations.

Note that this test differs from the one discussed in Section 18.2. One distinctive difference is that this test depends on *all* the eigenvalues of (18.19), whereas the test discussed in Section 18.2 depends on only the *leading* eigenvalue. This illustrates a basic aspect of multivariate hypothesis testing – there are often many reasonable test statistics for the same null hypothesis. Although each test rejects the null hypothesis

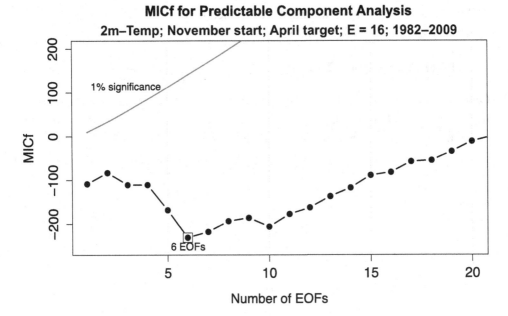

Figure 18.4 MICf versus the number of EOFs for Predictable Component Analysis of the CFSv2 ensemble forecasts discussed in Section 18.2 (curve with dots). The minimum MICf is indicated by a box. The 1% significance threshold is indicated by the grey curve. Significance occurs when MICf falls *below* the significance curve.

at the correct rate when the null hypothesis is true, they have different rejection rates when the null hypothesis is false (that is, they have different *powers*). Some tests are more powerful at detecting certain violations of the null hypothesis than other tests. Often, one examines different tests to see if the conclusions are sensitive to the test.

18.6 PrCA Based on Other Measures of Predictability

What happens if we maximize a *different* measure of predictability? That is, instead of maximizing F, we decided to maximize R-square. How would the resulting components compare? Answer: the components are exactly the same. In fact, exactly the same predictable components are obtained regardless of which measure discussed in Chapter 17 is maximized. This fact follows from the one-to-one correspondence between the measures. To prove this invariance, select any two measures in (17.35), say F and R^2. In Section 18.4, predictable components were derived by finding a projection that maximized F. Let us term the resulting value F_{max}. Using (17.35), we may convert this F-value into an R-square as

$$R_*^2 = \frac{F_{max}}{F_{max} + \frac{C(E-1)}{C-1}}. \tag{18.39}$$

Next, suppose we find the projection vector that maximizes R^2. Let the resulting value be denoted $R^2_{\max}$. Now suppose these two R-squares are not equal:

$$R^2_{\max} \neq R^2_*. \tag{18.40}$$

There are two possible cases. First, if

$$R^2_* > R^2_{\max}, \tag{18.41}$$

then we obtain a contradiction since no R^2 can exceed $R^2_{\max}$. On the other hand, suppose

$$R^2_* < R^2_{\max}. \tag{18.42}$$

Using (17.35), this may be re-stated as

$$F_{\max} < F_*, \tag{18.43}$$

where F_* is derived from $R^2_{\max}$ as

$$F_* = \frac{C(E-1)}{C-1} \left(\frac{R^2_{\max}}{1 - R^2_{\max}} \right). \tag{18.44}$$

Again, (18.43) contradicts the fact that $F_{\max}$ is the maximum value of F. Because both cases lead to a contradiction, then the original assumption that the $R^2_{\max}$ and R^2_* differ must be wrong, hence the measures are equal. Thus, we have shown that if F is maximized and (17.35) is used to convert it to R-square, the resulting R-square must also be the maximum R-square. The same argument can be applied to any two measures to show that if one measure is maximized and then converted into another measure using (17.35), then the latter measure must also be the maximum for that measure.

It is worthwhile to prove this equivalence *explicitly* in at least one case. The proof for other cases are similar. Accordingly, we choose to maximize F and show that the resulting projection vectors also maximize R-square. This equivalence is most easily proven by using the multivariate generalization of the law of total variance, which is

$$\overline{\Sigma}_T = \overline{\Sigma}_S + \overline{\Sigma}_N, \tag{18.45}$$

where $\overline{\Sigma}_S$, $\overline{\Sigma}_N$, and $\overline{\Sigma}_T$ are defined in (18.8) and (18.36). The projection vectors that maximize F are obtained by solving the generalized eigenvalue problem (18.19), which is

$$\overline{\Sigma}_S \mathbf{q} = \left(\frac{(C-1)}{C(E-1)} F \right) \overline{\Sigma}_N \mathbf{q}. \tag{18.46}$$

Next, we find the projection vectors that maximize R-square. R-square is the signal-to-total ratio and is related to the projection vector $\mathbf{q}$ as

$$R^2 = \frac{S}{T} = \frac{\mathbf{q}^T \overline{\boldsymbol{\Sigma}}_S \mathbf{q}}{\mathbf{q}^T \overline{\boldsymbol{\Sigma}}_T \mathbf{q}}. \tag{18.47}$$

Maximizing this Rayleigh quotient leads to the eigenvalue problem

$$\overline{\boldsymbol{\Sigma}}_S \mathbf{q} = R^2 \overline{\boldsymbol{\Sigma}}_T \mathbf{q}. \tag{18.48}$$

Eliminating $\overline{\boldsymbol{\Sigma}}_T$ using (18.45) and manipulating gives

$$\overline{\boldsymbol{\Sigma}}_S \mathbf{q} = R^2 \left(\overline{\boldsymbol{\Sigma}}_S + \overline{\boldsymbol{\Sigma}}_N \right) \mathbf{q} \tag{18.49}$$

$$\left(1 - R^2 \right) \overline{\boldsymbol{\Sigma}}_S \mathbf{q} = R^2 \overline{\boldsymbol{\Sigma}}_N \mathbf{q} \tag{18.50}$$

$$\overline{\boldsymbol{\Sigma}}_S \mathbf{q} = \left(\frac{R^2}{1 - R^2} \right) \overline{\boldsymbol{\Sigma}}_N \mathbf{q}. \tag{18.51}$$

Comparing this eigenvalue problem with (18.46) shows that the problems differ only by a multiplicative factor, which does not affect the eigenvectors. Therefore, the eigenvectors are identical. Moreover, the eigenvalues must be equal, hence

$$\frac{R^2}{1 - R^2} = \frac{(C - 1)}{C(E - 1)} F, \tag{18.52}$$

which is precisely the relation between R^2 and F derivable from (17.35). Because the projection vectors that maximize R^2 and F are identical, the corresponding variates and loading vectors also must be identical.

For completeness, we list the eigenvalue problems obtained by separately maximizing $F, \mathrm{SNR}, R^2, \rho^2, \Pi$:

$$
\begin{array}{lllll}
\text{maximize} & F & \Rightarrow & \hat{\boldsymbol{\Sigma}}_S \mathbf{q} = \lambda \hat{\boldsymbol{\Sigma}}_N \mathbf{q} & \Rightarrow & \lambda = F/E \\
\text{maximize} & \mathrm{SNR} & \Rightarrow & \hat{\boldsymbol{\Sigma}}_S \mathbf{q} = \lambda \hat{\boldsymbol{\Sigma}}_N \mathbf{q} & \Rightarrow & \lambda = \mathrm{SNR} \\
\text{maximize} & R^2 & \Rightarrow & \overline{\boldsymbol{\Sigma}}_S \mathbf{q} = \lambda \overline{\boldsymbol{\Sigma}}_T \mathbf{q} & \Rightarrow & \lambda = R^2 \\
\text{maximize} & \rho^2 & \Rightarrow & \overline{\boldsymbol{\Sigma}}_S \mathbf{q} = \lambda \overline{\boldsymbol{\Sigma}}_T \mathbf{q} & \Rightarrow & \lambda = \hat{\rho}^2 \\
\text{maximize} & \Pi & \Rightarrow & \overline{\boldsymbol{\Sigma}}_N \mathbf{q} = \lambda \overline{\boldsymbol{\Sigma}}_T \mathbf{q} & \Rightarrow & \lambda = \exp[-2\hat{\Pi}/k].
\end{array} \tag{18.53}
$$

Following the same arguments as in the previous paragraph, any one of these eigenvalue problems can be algebraically manipulated into any of the other forms. Thus, *each eigenvalue problem yields precisely the same eigenvectors as any of the other eigenvalue problems*. Furthermore, the eigenvalues are related to each other through (17.35). Framing PrCA as maximizing any one of these measures would lead to precisely the same predictable components as those derived by maximizing any of the other measures. Although the final results are equivalent, the specific procedure for deriving predictable components under one framing may *look* different from the derivation under a different framing, which can be confusing. For instance, PrCA could be described alternatively, and completely equivalently, as a procedure that finds the linear combination of variables that maximizes the signal-to-noise ratio. The resulting loading vectors are sometimes known as *signal-to-noise EOFs*, although they are identical to the predictable components derived from PrCA.

18.7 Skill Component Analysis

Given that predictability can be maximized, a natural question is whether *skill* can be maximized. The answer is yes. In fact, skill maximization appeared in the literature before predictability maximization. The essence of skill/predictability maximization was first described in the weather forecasting literature by Déqué (1988). Despite its pioneering nature, this paper had no impact on the forecast community. One reason for this is explained in Déqué et al. (1994), who wrote "Optimal statistical filtering as in Déqué (1988) cannot be applied yet, since it would require hundreds of independent forecasts; the presently available datasets are generally greater than several tens…" Today, this limitation no longer exists – forecasters have thousands of forecasts at their disposal! Déqué (1988) named the technique Predictable Component Analysis, but actually it decomposes forecast *skill*, not predictability. Later, Renwick and Wallace (1995) included this technique in their survey of techniques for identifying "well-predicted spatial patterns" and introduced the acronym PrCA. Schneider and Griffies (1999) proposed a similar technique to diagnose predictability, but were unaware of the previous papers. By remarkable coincidence, they gave their technique the same name. In that same year, Venzke et al. (1999) proposed the same technique, but they too were unaware of the previous papers. They named their technique "signal-to-noise maximizing analysis." Thus, PrCA has been used in the past to refer to maximizing skill and maximizing predictability. To avoid confusion, we will use PrCA to refer to maximizing predictability, and use *Skill Component Analysis* (SCA) to refer to maximizing skill. Regardless of whether skill or predictability is decomposed, the mathematics is exactly the same; only the quantity being decomposed differs.

To explain SCA, consider a vector forecast $\mathbf{f}(c)$ and corresponding verification $\mathbf{y}_c$. We desire the projection vector $\mathbf{q}$ that maximizes the mean squared error skill score Γ. The forecast error is $\boldsymbol{\epsilon} = \mathbf{f}(c) - \mathbf{y}_c$, hence the mean square error of the projected error $\mathbf{q}^T \boldsymbol{\epsilon}$ is

$$\text{MSE} = \frac{1}{C} \sum_{c=1}^{C} \left(\mathbf{q}^T \boldsymbol{\epsilon} \right)^2 = \mathbf{q}^T \hat{\boldsymbol{\Sigma}}_E \mathbf{q}, \tag{18.54}$$

where $\hat{\boldsymbol{\Sigma}}_E$ is the *error covariance matrix*

$$\hat{\Sigma}_E = \frac{1}{C} \sum_{c=1}^{C} \epsilon \epsilon^T. \tag{18.55}$$

(It is convention to center the verification and forecasts separately before computing errors, so ϵ also is centered.)

The corresponding total variance is

$$\hat{\sigma}_T^2 = \mathbf{q}^T \hat{\Sigma}_T \mathbf{q}. \tag{18.56}$$

Therefore, the estimated skill score is

$$\hat{\Gamma} = 1 - \frac{\text{MSE}}{\hat{\sigma}_T^2} = 1 - \frac{\mathbf{q}^T \hat{\Sigma}_E \mathbf{q}}{\mathbf{q}^T \hat{\Sigma}_T \mathbf{q}}. \tag{18.57}$$

Since the right-hand side is a Rayleigh quotient, we know from Chapter 16 that the projection vectors that optimize skill are obtained by solving the eigenvalue problem

$$\hat{\Sigma}_E \mathbf{q} = \lambda \hat{\Sigma}_T \mathbf{q}, \tag{18.58}$$

where the eigenvalue is $\lambda = 1 - \hat{\Gamma}$. Note that the largest skill is associated with the *smallest* eigenvalue. This procedure is formally equivalent to Covariance Discriminant Analysis (see Chapter 16), hence the variates, loading patterns, and regression patterns can be derived in precisely the same manner as in Predictable Component Analysis.

This derivation looks quite different from the derivation presented in Déqué (1988). As a result, the equivalence of the techniques may be difficult to see. Since a proof of this equivalence does not appear in the literature, we now prove that the two procedures yield the same components. In essence, Déqué argued that the most skillful components could be obtained from the eigenvectors of the error covariance matrix *in normalized principal component space*. To perform this calculation, one first computes the EOFs of the verification data and then projects all the forecasts onto these EOFs. This procedure yields the forecast and verification data matrices $\mathbf{F}_F$ and $\mathbf{F}_V$, respectively. To maximize skill, one solves (18.58), which is based on the covariance matrices

$$\hat{\Sigma}_E = (\mathbf{F}_F - \mathbf{F}_V)^T (\mathbf{F}_F - \mathbf{F}_V) / N \quad \text{and} \quad \hat{\Sigma}_T = \mathbf{F}_V^T \mathbf{F}_V / N. \tag{18.59}$$

However, because the EOFs were derived from the verification data, the PC time series $\mathbf{F}_V$ are uncorrelated, hence the covariance matrix $\hat{\Sigma}_T$ is diagonal. Furthermore, Déqué (1988) normalized the PCs to have unit variance, hence the diagonal elements of Σ_T are one. That is, $\hat{\Sigma}_T = \mathbf{I}$. Substituting these matrices into the skill measure (18.57) yields

$$\hat{\Gamma} = 1 - \frac{\mathbf{q}^T \hat{\Sigma}_E \mathbf{q}}{\mathbf{q}^T \mathbf{q}}. \tag{18.60}$$

Maximizing this measure leads to the *standard* eigenvector problem

$$\hat{\mathbf{\Sigma}}_E \mathbf{q} = \lambda \mathbf{q},\tag{18.61}$$

whose solutions are merely the eigenvectors of $\hat{\mathbf{\Sigma}}_E$. This shows that the eigenvectors of $\hat{\mathbf{\Sigma}}_E$ in normalized PC space give exactly the same results as Skill Component Analysis in normalized PC space.

18.8 Connection to Multivariate Linear Regression and CCA

A natural question is what happens when Predictable Component Analysis is applied to a multivariate regression model. After all, a regression model can make predictions and therefore can be analyzed as if it were a prediction system. The answer to this question reveals a fascinating connection between Predictable Component Analysis, Multivariate Linear Regression, and Canonical Correlation Analysis (CCA). This section assumes the reader is familiar with Chapter 14 on multivariate regression and Chapter 15 on CCA.

Consider the multivariate regression model

$$\underset{N \times M}{\mathbf{Y}} = \underset{N \times K}{\mathbf{X}} \quad \underset{K \times M}{\mathbf{B}} + \underset{N \times M}{\mathbf{\Xi}},\tag{18.62}$$

where the meaning of the symbols is the same as in Chapter 14. For this model, "fixed condition" is interpreted as "fixed X", and "multiple repetitions" is interpreted as multiple realizations of $\mathbf{\Xi}$. In fact, the conditional mean of this model is $\mathbf{XB}$, hence the signal and noise covariance matrices can be computed directly without generating ensemble forecasts. Since the intercept term is not important in predictability, it proves convenient to remove it by centering the columns of $\mathbf{X}$ and $\mathbf{Y}$. Then, the least squares estimate of $\mathbf{B}$ is

$$\hat{\mathbf{B}} = \left(\mathbf{X}^T \mathbf{X}\right)^{-1} \mathbf{X}^T \mathbf{Y},\tag{18.63}$$

a sample signal covariance matrix is

$$\hat{\mathbf{\Sigma}}_S = \left(\mathbf{X}\hat{\mathbf{B}}\right)^T \left(\mathbf{X}\hat{\mathbf{B}}\right) / N,\tag{18.64}$$

and a sample noise covariance matrix is

$$\hat{\mathbf{\Sigma}}_N = \left(\mathbf{Y} - \mathbf{X}\hat{\mathbf{B}}\right)^T \left(\mathbf{Y} - \mathbf{X}\hat{\mathbf{B}}\right) / N.\tag{18.65}$$

Chapter 15 described CCA in terms of the following sample covariance matrices:

$$\hat{\mathbf{\Sigma}}_{XX} = \frac{\mathbf{X}^T \mathbf{X}}{N}, \quad \hat{\mathbf{\Sigma}}_{YY} = \frac{\mathbf{Y}^T \mathbf{Y}}{N}, \quad \hat{\mathbf{\Sigma}}_{XY} = \hat{\mathbf{\Sigma}}_{YX}^T = \frac{\mathbf{X}^T \mathbf{Y}}{N}.\tag{18.66}$$

Substituting the least squares estimate (18.63) into the signal and noise covariance matrices and expressing all covariances in terms of (18.66) yields

$$\hat{\Sigma}_S = \hat{\Sigma}_{YX}\hat{\Sigma}_{XX}^{-1}\hat{\Sigma}_{XY} \tag{18.67}$$

$$\hat{\Sigma}_N = \hat{\Sigma}_{YY} - \hat{\Sigma}_{YX}\hat{\Sigma}_{XX}^{-1}\hat{\Sigma}_{XY}. \tag{18.68}$$

The most predictable components are obtained by solving the eigenvalue problem

$$\hat{\Sigma}_S\mathbf{q} = \lambda\hat{\Sigma}_N\mathbf{q}, \tag{18.69}$$

where λ is the signal-to-noise ratio. Substituting (18.67) and (18.68) into this eigen-value problem and algebraically manipulating it gives

$$\hat{\Sigma}_{YX}\hat{\Sigma}_{XX}^{-1}\hat{\Sigma}_{XY}\mathbf{q} = \left(\frac{\lambda}{1+\lambda}\right)\hat{\Sigma}_{YY}\mathbf{q}. \tag{18.70}$$

This eigenvalue problem is precisely the eigenvalue problem that arises in CCA (in particular, compare with (15.19)). It follows that the most predictable components of a linear regression model are identical to the canonical components between X and Y. This result should not be surprising. After all, correlation is the natural measure of dependence in a linear regression model; therefore the component with the largest correlation must be the most predictable component in the linear model. Furthermore, equating the eigenvalues in (18.70) and (15.19) gives

$$\frac{\lambda}{1+\lambda} = \rho^2, \tag{18.71}$$

which is the expected relation between R-square and signal-to-noise ratio λ (see 17.31).

In addition to these equivalences, the decision rule for the hypothesis $\mathbf{B} = \mathbf{0}$ based on (14.37) is identical to the decision rule for vanishing canonical correlations (15.100) (the hypothesis is slightly more complicated than $\mathbf{B} = \mathbf{0}$ because the regression model should include an intercept term). It is very satisfying that the decision rules based on predictability, linear regression, and CCA all agree with each other, since the underlying hypothesis about the population is the same (namely, that Y is independence of X). Further connections between PrCA, CCA, and regression are explored in DelSole and Chang (2003).

Given this equivalence, why not use CCA to find predictable components? The answer is that CCA requires the user to know the predictors X, whereas PrCA does not. In fact, PrCA can be used to *find* the predictors that go with Y.

18.9 Further Properties of PrCA

This section ties up a few loose ends about PrCA. In particular, this section will:

- clarify certain details about projecting data to a low-dimensional space and back.
- prove that predictable variates are uncorrelated in three different senses: they are uncorrelated among themselves, their conditional means are uncorrelated, and the residuals about the conditional means are uncorrelated.
- show that the variance explained by k'th predictable component is $\mathbf{p}_k^T \mathbf{W} \mathbf{p}_k$, where $\mathbf{W}$ is a matrix defining the variance norm.
- clarify the distinction between loading vectors and regression patterns.

Projecting to a Lower-Dimensional Space

In Section 18.2, predictability was maximized using a few principal components. The method described here is general and applies for arbitrary basis vectors (e.g., it does require the basis vectors to be EOFs), but for convenience we continue to call the basis vectors EOFs. Let the first M EOFs be denoted $\mathbf{e}_1, \mathbf{e}_2, \ldots, \mathbf{e}_M$. Collect these vectors as column vectors of the $S \times M$ matrix

$$\mathbf{E} = \begin{bmatrix} \mathbf{e}_1 & \mathbf{e}_2 & \cdots & \mathbf{e}_T \end{bmatrix}. \tag{18.72}$$

(Here, "S" stands for the number of spatial elements and not "signal.") Note that $\mathbf{E}$ is an EOF matrix while E denotes ensemble size. Time series for each basis vector is determined by finding the M-dimensional vector $\tilde{\mathbf{y}}_{ec}$ such that $\mathbf{E}\tilde{\mathbf{y}}_{ec}$ best approximates $\mathbf{y}_{ec}$. This is a standard regression problem with solution

$$\tilde{\mathbf{y}}_{ec} = \left(\mathbf{E}^T \mathbf{W} \mathbf{E} \right)^{-1} \mathbf{E}^T \mathbf{W} \mathbf{y}_{ec}, \tag{18.73}$$

where tilde denotes a quantity in EOF space, and $\mathbf{W}$ is a matrix that defines the norm for measuring differences (typically, it is a diagonal matrix with diagonal elements proportional to the local area). The transformation from PC time series to data space is

$$\dot{\mathbf{y}}_{ec} = \mathbf{E}\tilde{\mathbf{y}}_{ec}, \tag{18.74}$$

where the dot denotes a quantity in data space that is strictly a linear combination of EOF vectors. That is, $\mathbf{y}_{ec}$ and $\dot{\mathbf{y}}_{ec}$ have the same dimension S, but $\dot{\mathbf{y}}_{ec}$ is a *filtered* version of $\mathbf{y}_{ec}$ in the sense that variability orthogonal to the M EOFs has been removed from $\mathbf{y}_{ec}$.

The vector $\tilde{\mathbf{y}}_{ec}$ has lower dimension than $\mathbf{y}_{ec}$, but otherwise it has exactly the same format, hence Predictable Component Analysis can be applied to $\tilde{\mathbf{y}}_{ec}$. For completeness, the computations are summarized in Table 18.2. Additional properties of the predictable components will be derived next.

Table 18.2. *Predictable Component Analysis in EOF Space.*

$\mathbf{y}_{ec}, \quad e = 1, \ldots E \text{ and } c = 1, \ldots, C$	S-dimensional data vector
$\mathbf{E} = \begin{bmatrix} \mathbf{e}_1 & \cdots & \mathbf{e}_M \end{bmatrix}$	basis vectors for EOF space
$\tilde{\mathbf{y}}_{ec} = \left(\mathbf{E}^T \mathbf{W} \mathbf{E} \right)^{-1} \mathbf{E}^T \mathbf{W} \mathbf{y}_{ec}$	M-dimensional EOF time series
$\tilde{\boldsymbol{\mu}}_{Y|c} = \dfrac{1}{E} \sum_{e=1}^{E} \tilde{\mathbf{y}}_{ec}$	conditional mean
$\tilde{\boldsymbol{\mu}}_{Y} = \dfrac{1}{EC} \sum_{c=1}^{C} \sum_{e=1}^{E} \tilde{\mathbf{y}}_{ec}$	unconditional mean
$\tilde{\boldsymbol{\Sigma}}_S = \dfrac{1}{C-1} \sum_{c=1}^{C} \left(\tilde{\boldsymbol{\mu}}_{Y|c} - \tilde{\boldsymbol{\mu}}_Y \right) \left(\tilde{\boldsymbol{\mu}}_{Y|c} - \tilde{\boldsymbol{\mu}}_Y \right)^T$	signal covariance matrix
$\tilde{\boldsymbol{\Sigma}}_N = \dfrac{1}{C(E-1)} \sum_{c=1}^{C} \sum_{e=1}^{E} \left(\tilde{\mathbf{y}}_{ec} - \tilde{\boldsymbol{\mu}}_{Y|c} \right) \left(\tilde{\mathbf{y}}_{ec} - \tilde{\boldsymbol{\mu}}_{Y|c} \right)^T$	noise covariance matrix
$\tilde{\boldsymbol{\Sigma}}_T = \dfrac{1}{(CE-1)} \sum_{c=1}^{C} \sum_{e=1}^{E} \left(\tilde{\mathbf{y}}_{ec} - \tilde{\boldsymbol{\mu}}_Y \right) \left(\tilde{\mathbf{y}}_{ec} - \tilde{\boldsymbol{\mu}}_Y \right)^T$	total covariance matrix
$E \tilde{\boldsymbol{\Sigma}}_S \tilde{\mathbf{q}} = F \tilde{\boldsymbol{\Sigma}}_N \tilde{\mathbf{q}}$	eigenvalue problem
$r_{ec,i} = \tilde{\mathbf{q}}_i^T \tilde{\mathbf{y}}_{ec}$	predictable variate
$\widehat{\text{var}}[r_{ec,i}] = 1 \quad \Rightarrow \quad \tilde{\mathbf{q}}^T \tilde{\boldsymbol{\Sigma}}_T \tilde{\mathbf{q}} = 1$	normalization
$\tilde{\mathbf{p}}_i = \tilde{\boldsymbol{\Sigma}}_T \tilde{\mathbf{q}}_i$	loading vector in EOF space
$\dot{\mathbf{p}}_i = \mathbf{E} \tilde{\mathbf{p}}_i$	loading vector in data space
$\mathbf{p}_i^T \mathbf{W} \mathbf{p}_i$	variance explained by the i'th component

Matrix Notation

The properties discussed in the remainder of this section are most conveniently expressed in terms of matrices. Accordingly, let us define the eigenvector matrix

$$\tilde{\mathbf{Q}} = \begin{bmatrix} \tilde{\mathbf{q}}_1 & \tilde{\mathbf{q}}_2 & \cdots & \tilde{\mathbf{q}}_M \end{bmatrix}, \tag{18.75}$$

where $\tilde{\mathbf{q}}_1, \ldots, \tilde{\mathbf{q}}_M$ are the eigenvectors from (18.19) using covariance matrices defined in Table 18.2. With this notation, eigenvalue problem (18.19) can be written as

$$E \, \tilde{\boldsymbol{\Sigma}}_S \tilde{\mathbf{Q}} = \tilde{\boldsymbol{\Sigma}}_N \tilde{\mathbf{Q}} \boldsymbol{\Lambda}, \tag{18.76}$$

where $\boldsymbol{\Lambda}$ is a diagonal matrix whose diagonal elements equal the eigenvalues $F_1, \ldots, F_M$. Similarly, the variates are collected into a single vector as

$$\mathbf{r}_{ec} = \tilde{\mathbf{Q}}^T \tilde{\mathbf{y}}_{ec}. \tag{18.77}$$

For future reference, the noise covariance matrix of $\mathbf{r}_{ec}$ is

$$\boldsymbol{\Sigma}_N^{(r)} = \frac{1}{C(E-1)} \sum_{e=1}^{E} \sum_{c=1}^{C} \left(\mathbf{r}_{ec} - \frac{1}{E} \sum_{e=1}^{E} \mathbf{r}_{ec} \right) \left(\mathbf{r}_{ec} - \frac{1}{E} \sum_{e=1}^{E} \mathbf{r}_{ec} \right)^T \tag{18.78}$$

$$= \frac{1}{C(E-1)} \sum_{e=1}^{E} \sum_{c=1}^{C} \tilde{\mathbf{Q}}^T \left(\tilde{\mathbf{y}}_{ec} - \tilde{\boldsymbol{\mu}}_{Y|c} \right) \left(\tilde{\mathbf{y}}_{ec} - \tilde{\boldsymbol{\mu}}_{Y|c} \right)^T \tilde{\mathbf{Q}} \tag{18.79}$$

$$= \tilde{\mathbf{Q}}^T \tilde{\boldsymbol{\Sigma}}_N \tilde{\mathbf{Q}}. \tag{18.80}$$

Similarly, the signal covariance matrix of the variates is

$$\boldsymbol{\Sigma}_S^{(r)} = \tilde{\mathbf{Q}}^T \tilde{\boldsymbol{\Sigma}}_S \tilde{\mathbf{Q}}. \tag{18.81}$$

Variates Are Uncorrelated in a Signal and Noise Sense
The eigenvectors of symmetric matrices satisfy certain orthogonality properties. These properties are derived in Section A.2 and can be expressed as

$$\tilde{\mathbf{Q}}^T \tilde{\boldsymbol{\Sigma}}_N \tilde{\mathbf{Q}} = \mathbf{D}_N, \tag{18.82}$$

where $\mathbf{D}_N$ is a *diagonal* matrix. The diagonality of $\mathbf{D}_N$ and (18.80) implies that the noise variates are uncorrelated. That is, the residuals of the variates relative to the ensemble mean are uncorrelated. A second orthogonality relation can be derived by multiplying (18.76) on the left by $\tilde{\mathbf{Q}}^T$ and invoking (18.82), which yields

$$\mathbf{D}_S = \tilde{\mathbf{Q}}^T \tilde{\boldsymbol{\Sigma}}_S \tilde{\mathbf{Q}} = \mathbf{D}_N \boldsymbol{\Lambda}/E. \tag{18.83}$$

Because $\mathbf{D}_N$ is diagonal, $\mathbf{D}_S$ also is diagonal. Furthermore, diagonality of $\mathbf{D}_S$ and (18.81) implies that the ensemble mean variates are uncorrelated. Thus, we have shown that the variates are uncorrelated in a signal sense and in a noise sense.

Variates Are Uncorrelated in the Standard Sense
We now prove that variates are uncorrelated in a third sense. The proof is based on the law of total variance (18.45), which can be written in terms of the covariance matrices in Table 18.1 as

$$(CE - 1)\tilde{\boldsymbol{\Sigma}}_T = E(C - 1)\tilde{\boldsymbol{\Sigma}}_S + C(E - 1)\tilde{\boldsymbol{\Sigma}}_N. \tag{18.84}$$

This identity shows that the signal, noise, and total covariance matrices satisfy a linear constraint. As a result of this constraint, a linear transformation that diagonalizes any two of the matrices will automatically diagonalize the third. Specifically, multiplying (18.84) on the left by $\tilde{\mathbf{Q}}^T$ and on the right by $\tilde{\mathbf{Q}}$, and algebraically manipulating, yields

$$\mathbf{D}_T = \tilde{\mathbf{Q}}^T \tilde{\boldsymbol{\Sigma}}_T \tilde{\mathbf{Q}} = \left(\frac{E(C-1)}{CE-1} \right) \mathbf{D}_S + \left(\frac{C(E-1)}{CE-1} \right) \mathbf{D}_N. \qquad (18.85)$$

Since the right-hand side is a sum of diagonal matrices, so too is the left-hand side. Diagonality of $\mathbf{D}_T$ implies that the predictable variates are uncorrelated in the standard sense. In summary, predictable variates are uncorrelated in three difference senses: (1) the residuals about the ensemble mean are uncorrelated, (2) the conditional means are uncorrelated, and (3) the variates themselves are uncorrelated with other variates.

Recall that eigenvectors are unique up to a multiplicative constant. In this chapter, we recommend normalizing variates to give unit total variance, which implies.

$$\tilde{\mathbf{Q}}^T \tilde{\boldsymbol{\Sigma}}_T \tilde{\mathbf{Q}} = \mathbf{D}_T = \mathbf{I}. \qquad (18.86)$$

Loading Vectors
Loading vectors describe the spatial structure of the predictable component. To derive the loading vectors, multiply (18.77) on the left by the inverse of $\mathbf{Q}^T$:

$$\tilde{\mathbf{y}}_{ec} = \tilde{\mathbf{P}} \mathbf{r}_{ec}, \qquad (18.87)$$

where

$$\tilde{\mathbf{P}} = \left(\tilde{\mathbf{Q}}^T \right)^{-1} = \tilde{\boldsymbol{\Sigma}}_Y \tilde{\mathbf{Q}}. \qquad (18.88)$$

The columns of $\tilde{\mathbf{P}}$ are *loading vectors in EOF space*. Multiplying (18.87) by $\mathbf{E}$ gives

$$\dot{\mathbf{y}}_{ec} = \mathbf{P} \mathbf{r}_{ec} \qquad (18.89)$$

where (18.74) has been used, and we have defined

$$\dot{\mathbf{P}} = \mathbf{E} \tilde{\mathbf{P}}. \qquad (18.90)$$

Identity (18.89) shows that $\dot{\mathbf{y}}_{ec}$ can be written as a linear combination of predictable components. The columns of $\dot{\mathbf{P}}$ are the loading vectors in data space.

For future reference, we note that the covariance matrices (18.82), (18.83), and (18.86) can be expressed in terms of loading vectors (18.88) as

$$\tilde{\boldsymbol{\Sigma}}_T = \tilde{\mathbf{P}} \tilde{\mathbf{P}}^T, \quad \tilde{\boldsymbol{\Sigma}}_N = \tilde{\mathbf{P}} \mathbf{D}_N \tilde{\mathbf{P}}^T, \quad \text{and} \quad \tilde{\boldsymbol{\Sigma}}_S = \tilde{\mathbf{P}} \mathbf{D}_S \tilde{\mathbf{P}}^T. \qquad (18.91)$$

Variance Explained by a Predictable Component
It is often of interest to quantify the variance of a predicable component. The total variance is typically defined as

$$\|\mathbf{y}_{ec}\|_W^2 = \frac{1}{CE-1} \sum_{e=1}^{E} \sum_{c=1}^{C} (\mathbf{y}_{ec} - \hat{\boldsymbol{\mu}}_Y)^T \mathbf{W} (\mathbf{y}_{ec} - \hat{\boldsymbol{\mu}}_Y), \tag{18.92}$$

where $\mathbf{W}$ is a matrix that defines the variance norm and has an interpretation similar to that in (18.73). The total variance of $\dot{\mathbf{y}}_{ec}$ is therefore

$$\|\dot{\mathbf{y}}_{ec}\|_W^2 = \|\mathbf{E}\tilde{\mathbf{y}}_{ec}\|_W^2 = \frac{1}{CE-1} \sum_{e=1}^{E} \sum_{c=1}^{C} (\tilde{\mathbf{y}}_{ec} - \tilde{\boldsymbol{\mu}}_Y)^T \mathbf{E}^T \mathbf{W} \mathbf{E} (\tilde{\mathbf{y}}_{ec} - \tilde{\boldsymbol{\mu}}_Y)$$

$$= \operatorname{tr}\left[\mathbf{E}^T \mathbf{W} \mathbf{E} \tilde{\boldsymbol{\Sigma}}_T\right] = \operatorname{tr}\left[\mathbf{E}^T \mathbf{W} \mathbf{E} \tilde{\mathbf{P}} \tilde{\mathbf{P}}^T\right]$$

$$= \operatorname{tr}\left[\mathbf{P}^T \mathbf{W} \mathbf{P}\right] = \sum_{m=1}^{M} \mathbf{p}_m^T \mathbf{W} \mathbf{p}_m, \tag{18.93}$$

where we have used the definition of $\tilde{\boldsymbol{\Sigma}}_T$ in Table 18.2 and standard properties of the trace operator. This expression suggests that $\mathbf{p}_k^T \mathbf{W} \mathbf{p}_k$ specifies the amount of variance explained by the k'th predictable component. In particular, summing this variance over all predictable components gives the total variance in EOF space. Typically, one is interested in the *fraction* of variance explained by the k'th predictable component, which is

$$\begin{array}{c} \text{fraction of variance explained by} \\ k\text{'th predictable component} \end{array} = \frac{\mathbf{p}_k^T \mathbf{W} \mathbf{p}_k}{\|\mathbf{y}_{ec}\|_W^2}. \tag{18.94}$$

Regression Patterns

Loading vectors can also be derived as *regression patterns*. To see this, consider the problem of finding the matrix $\dot{\mathbf{P}}$ such that $\dot{\mathbf{P}} \mathbf{r}_{ec}$ best approximates $\mathbf{y}_{ec}$. We indicate this problem symbolically as

$$\dot{\mathbf{y}}_{ec} \approx \dot{\mathbf{P}} \mathbf{r}_{ec}. \tag{18.95}$$

This problem can be framed as a standard multivariate regression problem if the vectors $\dot{\mathbf{y}}_{ec}$ and $\mathbf{r}_{ec}$ are expressed as matrices. Accordingly, we define the $EC \times M$ matrix $\tilde{\mathbf{Y}}$,

$$\left(\tilde{\mathbf{Y}}\right)_{e+(c-1)E,m} = (\tilde{\mathbf{y}}_{ec})_m - (\tilde{\boldsymbol{\mu}}_Y)_m. \tag{18.96}$$

In essence, the m'th column of $\tilde{\mathbf{Y}}$ is a concatenation of all anomaly ensemble members e over all conditions c. Using this notation, $\dot{\mathbf{y}}_{ec}$ and $\mathbf{r}_{ec}$ are expressed as

$$\underset{EC \times S}{\dot{\mathbf{Y}}} = \underset{EC \times M}{\tilde{\mathbf{Y}}} \underset{M \times S}{\mathbf{E}^T} \tag{18.97}$$

and

$$
\underset{EC \times S}{\mathbf{R}} = \underset{EC \times M}{\tilde{\mathbf{Y}}} \ \underset{M \times S}{\tilde{\mathbf{Q}}}.
\tag{18.98}
$$

The associated covariance matrices are

$$
\frac{1}{EC-1}\mathbf{R}^T\mathbf{R} = \mathbf{I} \quad \text{and} \quad \tilde{\boldsymbol{\Sigma}}_Y = \frac{1}{EC-1}\tilde{\mathbf{Y}}^T\tilde{\mathbf{Y}}.
\tag{18.99}
$$

Under this notation, the approximation (18.95) can be expressed as

$$
\underset{EC \times S}{\dot{\mathbf{Y}}} \approx \underset{EC \times M}{\mathbf{R}} \ \underset{M \times S}{\dot{\mathbf{P}}^T}.
\tag{18.100}
$$

The least squares estimate of $\dot{\mathbf{P}}$ is

$$
\dot{\mathbf{P}} = \dot{\mathbf{Y}}^T\mathbf{R}\left(\mathbf{R}^T\mathbf{R}\right)^{-1} = \left(\tilde{\mathbf{Y}}\mathbf{E}^T\right)^T\left(\tilde{\mathbf{Y}}\tilde{\mathbf{Q}}\right)\frac{1}{EC-1} = \mathbf{E}\tilde{\boldsymbol{\Sigma}}_Y\tilde{\mathbf{Q}}.
\tag{18.101}
$$

This solution is identical to (18.90) because of (18.88).

Result (18.101) shows that the loading vectors in data space are a linear combination of EOF vectors. This constraint can sometimes complicate physical interpretation. Specifically, a predictable pattern represented by a few EOFs may be a visually distorted version of the true pattern. In practice, variates tend to be insensitive to M over a range of M. In such cases, physical interpretation is often facilitated by defining predictable components that are not constrained to be a linear combination of EOF vectors. Such patterns can be obtained by finding the matrix $\mathbf{P}$ such that

$$
\mathbf{y}_{ec} \approx \mathbf{P}\mathbf{r}_{ec}.
\tag{18.102}
$$

Note that here we are approximating the *original* data set $\mathbf{y}_{ec}$, in contrast to (18.95) where we are approximating the filtered data set $\dot{\mathbf{y}}_{ec}$. The least squares estimate of $\mathbf{P}$ is

$$
\mathbf{P} = \mathbf{Y}^T\mathbf{R}\left(\mathbf{R}^T\mathbf{R}\right)^{-1} = \frac{1}{EC-1}\mathbf{Y}^T\mathbf{R}.
\tag{18.103}
$$

This equation shows that the k'th column of $\mathbf{P}$ is a *regression map* between the k'th predictable variate and the original time series. We term the columns of $\mathbf{P}$ *regression patterns*. Regression patterns can exhibit spatial structures that are impossible to express as a linear combination of M EOF vectors. However, for the examples discussed in this chapter, no difference exists between regression patterns $\mathbf{P}$ and loading vectors $\dot{\mathbf{P}}$ because the EOF time series are PC time series. In particular, PC time series are uncorrelated with each other, hence $\mathbf{Y}^T\mathbf{R} = \dot{\mathbf{Y}}^T\mathbf{R}$. The distinction between regression patterns and loading vectors becomes relevant when the EOF vectors are not EOFs.

18.10 Conceptual Questions

1 Explain how MANOVA is effectively a "field significance test" for ANOVA?

2 What is a "predictable component?"

3 What is a predictable loading vector?

4 Explain how one can tell from visual inspection of Figure 18.1 that the grid point shown in the second panel is more predictable than the grid point shown in the fourth panel.

5 If MANOVA detects predictability, what precisely does this mean about $\mathbf{y}$?

6 In what way is the MANOVA hypothesis test unsatisfying, and how does PrCA help?

7 Why do PrCA and CCA use different criteria for selecting the number of PC time series?

8 PrCA finds components that maximize the signal-to-noise ratio. How are the resulting components related to those derived by maximizing signal-to-total, R^2, or predictive information Π. How are the eigenvalues related?

9 What's the difference between maximizing predictability and maximizing skill?

10 What are the predictable components of a linear regression model?

19

Extreme Value Theory

According to an old saying, a person standing with one foot on a hot stove and the other on a block of ice is comfortable on average. The corollary in climate science is that monthly averages smooth over a lot of important information...[1]

Zhang et al. (2011)

This chapter gives an introduction to extreme value theory. Unlike most statistical analyses, which are concerned with the *typical* properties of a random variable, extreme value theory is concerned with rare events that occur in the tail of the distribution. Also in contrast to other statistical analyses, the Gaussian distribution is not a fundamental distribution in extreme value theory. The cornerstone of extreme value theory is the *Extremal Types Theorem*. This theorem states that the *maximum* of N independent and identically distributed random variables can converge, after suitable normalization, only to a single distribution in the limit of large N. This limiting distribution is known as the *Generalized Extreme Value* (GEV) distribution. This theorem is analogous to the central limit theorem, except that the focus is on the *maximum* rather than the *sum* of random variables. The GEV provides the basis for estimating the probability of extremes that are more extreme than those that occurred in a sample. The GEV is characterized by three parameters, known as the location, scale, and shape. A procedure known as the maximum likelihood method can be used to estimate these parameters, quantify their uncertainty, and account for dependencies on time or external environmental conditions. This procedure leads to nonlinear sets of equations that can be solved by standard mathematical packages.

---

[1] Reprinted by permission from John Wiley & Sons, *Wiley Interdisciplinary Reviews: Climate Change*, Zhang et al. (2011).

19.1 The Problem and a Summary of the Solution

Any engineering structure will fail if it experiences a sufficiently extreme weather event. For instance, a tall building will collapse if sustained winds exceed a threshold. A levee will fail when water overtops the crest of the levee. For these reasons, many people are interested in the probability that some quantity of interest will exceed a threshold during the (intended) lifetime of a structure. Formally, given a random variable X, we are interested in the probability that X exceeds some threshold value x_p:

$$p = \Pr[X > x_p]. \tag{19.1}$$

The probability of exceedence p is closely related to the *cumulative probability distribution*, which is the probability that a random variable X is less or equal to x:

$$F(x) = \Pr[X \leq x]. \tag{19.2}$$

The probability of exceedance is merely one minus the cumulative distribution:

$$1 - F(x_p) = p. \tag{19.3}$$

Thus, while we are interested in probability of exceedence, we will often focus on its complement $F(x)$ because standard probability theory is framed in terms of $F(x)$.

As a concrete example, consider the average daily temperature during January over the mid-Atlantic United States. The daily values in each January during 1970–2015 are illustrated in Figure 19.1. Extreme cold temperatures are of most concern during winter, hence we focus on the monthly minima (dots in the figure). That is, given daily average temperatures for each January, we identify the minimum value and ignore all other values in that month. It is convention to frame the analysis based on maxima. Under this convention, minima can be analyzed simply by reversing the sign of the data. By reversing the sign, the minima become the maxima. Accordingly, we reverse the sign of the data and examine

$$M_N = \max\{X_1, \ldots, X_N\}, \tag{19.4}$$

where N is known as the *block size* and $X_1, \ldots, X_N$ are the 31 *negative* daily temperatures during January. There is one value of M_N per January. Let K be the number of blocks. In our example, $K = 46$ because the period 1970–2015 contains 46 Januaries. The cumulative distribution of M_N can be estimated by counting the number of data values below a specified threshold and dividing by $K + 1$ (see Makkonen, 2008, for why counts are divided by $K+1$ instead of K). More precisely, let sample values of M_N be $z_1, z_2, \ldots, z_K$, and let the ordered sample be denoted $z_{(1)} \leq z_{(2)} \leq \cdots \leq z_{(K)}$. There are exactly k observations less than or equal to $z_{(k)}$, so the empirical cumulative distribution is defined to be

$$\tilde{G}(z) = \frac{k}{K+1} \quad \text{for } z_{(k)} \leq z < z_{(k+1)}. \tag{19.5}$$

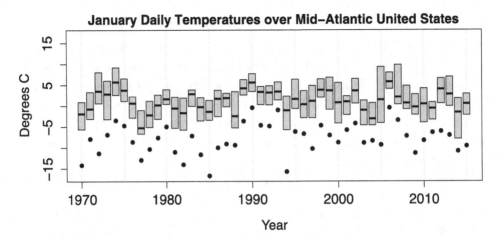

Figure 19.1 Box plot of daily January temperatures averaged over the mid-Atlantic states (North Carolina, Virginia, Maryland, West Virginia, Delaware) during 1970–2015 (46 years). The upper and lower edges of each box show the upper and lower quartiles, and the center line indicates the median. The most extreme daily temperature in each January is indicated by a dot. The data is the average temperature from the Global Historical Climatology Network.

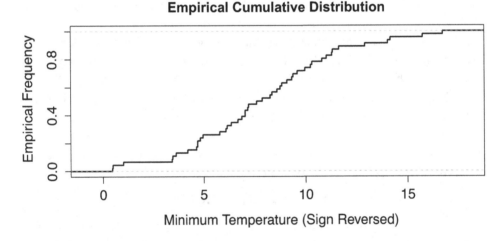

Figure 19.2 The empirical cumulative distribution function for the most extreme daily cold temperatures over the mid-Atlantic United States during January 1970–2015. For reasons discussed in the text, the x-axis shows the *negative* of minimum temperature (that is, the negative of the dots in Figure 19.1).

The resulting empirical cumulative distribution function is shown in Figure 19.2.

The empirical cumulative distribution shown in Figure 19.2 is not the best visual aid for studying extremes because the extreme values are concentrated in a small section at the top right corner. In extreme value analysis, p is a small number close to zero. It is often more helpful to work with its reciprocal, $1/p$, which is known as the *return period*. The return period is the *average* wait time between *iid* exceedences. The associated threshold x_p is known as the *return level*. For instance, if the probability of annual maxima to exceed x_p is $p = 0.01$, then the return period is $1/p = 100$ years and x_p is the 100-year return level. Note that return period is an *average* quantity: during a length of time equal to a return period, the event $X > x_p$ might not occur at all or it might occur multiple times.

A plot of the return period versus the return level is known as a *return level plot*. A return level plot based on the empirical return period $1/\tilde{G}(z_{(k)})$ and return level $z_{(k)}$ is shown in Figure 19.3. The plot suggests that a smooth curve could provide a reasonable fit to the points. The resulting curve could then be used to *extrapolate* from observed values to values more extreme than those that actually occurred in a sample. The question arises as to what function should be used for such extrapolations. Given the popularity of Gaussian distributions, one might try to fit the observed sample to a Gaussian distribution and then use that distribution to estimate the probability of an extreme value. However, the Gaussian distribution is often justified through the Central Limit Theorem, which states that the *sum* of independent and identically distributed random variables tends to a Gaussian

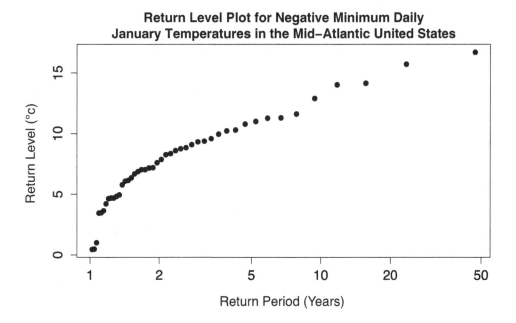

Figure 19.3 Return level plot of the most extreme daily cold temperatures over the mid-Atlantic United States during January 1970–2015. The y-axis shows the *negative* of minimum temperature (that is, the negative of the dots shown in Figure 19.1).

distribution for asymptotically large sample size. In contrast, the data points in Figure 19.3 are the *maxima* of a set of random variables, *not sums*. Even if samples are drawn from a Gaussian distribution, their maxima does not have a Gaussian distribution.

Fisher and Tippett (1928) proved that the maximum of N samples of *iid* random variables, properly standardized, can converge only to one of three distributions in the limit of large N. These three distributions can be modeled by a single distribution of the form

$$G(z) = \exp\left\{-\left[1 + \xi\left(\frac{z - \mu}{\sigma}\right)\right]^{-1/\xi}\right\}, \tag{19.6}$$

where μ, σ, ξ are known as the location, scale, and shape parameters, respectively. This distribution, known as the *Generalized Extreme Value (GEV) Distribution*, provides the basis for fitting a curve to the points in Figure 19.3. Specifically, the exceedance probability p is

$$p = 1 - G(z_p). \tag{19.7}$$

Solving this equation for z_p is a straightfoward algebra problem with solution

$$z_p = \mu - \frac{\sigma}{\xi}\left(1 - \{-\log(1 - p)\}^{-\xi}\right). \tag{19.8}$$

This equation relates the return level z_p to the return period $1/p$. For small p,

$$r_p = -\frac{1}{\log(1 - p)} \approx \frac{1}{p} = \text{return period.} \tag{19.9}$$

With this notation,

$$z_p = \mu - \frac{\sigma}{\xi}\left(1 - r_p^{\xi}\right). \tag{19.10}$$

Thus, plotting z_p as a function of r_p on a log-scale should follow a simple power law, as suggested in Figure 19.3. We term r_p the *approximate return period*.

To interpret the meaning of the location, scale, and shape parameters, note that the GEV distribution depends on μ and σ only in the form

$$G(z) = f\left(\frac{z - \mu}{\sigma}\right). \tag{19.11}$$

For *any* function $f(z)$, a graph of $f(z - \mu)$ is exactly the same as for $f(z)$, except shifted to the right or left, depending on whether the location parameter is positive or negative. This fact is illustrated in the left panel of Figure 19.4 and explains is why μ is known as the location parameter. Similarly, a graph of $f(z/\sigma)$ is the same

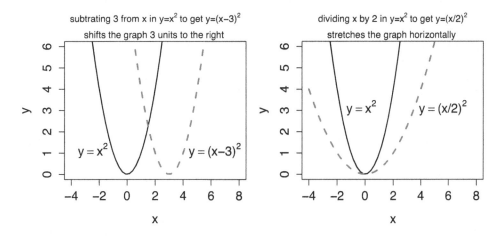

Figure 19.4 Illustration of how the graph of $y = f(x)$ shifts to the right when a positive constant is subtracted from x (left panel), or is stretched in the horizontal direction when x is divided by a positive constant (right panel). This particular example uses the parabola $y = x^2$.

as $f(z)$, except stretched or compressed in the horizontal direction, depending on whether the scale parameter is greater or less than one, respectively. This fact is illustrated in the right panel of Figure 19.4 and explains why σ is known as a scale parameter. Finally, ξ is known as a shape parameter because it produces changes in the GEV cumulative distribution function that cannot be produced by simple shifts or uniform stretching or compression.

The next natural question is how to estimate the parameters of the GEV, μ, σ, ξ. Here, we use the *Maximum Likelihood Method*, which is discussed in Section 19.3. The software package R has an easy-to-use package known as extRemes for estimating parameters of the GEV. The result of this procedure is summarized in Table 19.1 and gives the estimates

$$\mu \approx 6.5, \quad \sigma \approx 3.6, \quad \xi \approx -0.23. \tag{19.12}$$

The corresponding return level plot along with the GEV fit based on these estimates is shown in Figure 19.5. The figure suggests a reasonable fit.

We also want to describe the uncertainty of the estimates. The Maximum Likelihood Method also produces an estimate of the covariance matrix of the errors, which is given in Table 19.1. The diagonal elements of this matrix gives the variances of the estimates, hence the square root gives the standard errors. Thus, approximate 95% confidence intervals are

$$\mu \approx 6.5 \pm 1.14, \quad \sigma \approx 3.6 \pm 0.79, \quad \xi \approx -0.23 \pm 0.18. \tag{19.13}$$

Table 19.1. *An example of a GEV fit to cold extremes in mid-Atlantic states during*
January between 1970–2015 using extRemes *in R. Lines 3–9 show the data; line 19*
shows the maximum likelihood estimates of μ, σ, ξ, respectively; line 23 shows the
associated standard errors; lines 27–29 show the associated error covariance matrix.

```
1    library(extRemes) # LOAD THE LIBRARY FOR FITTING THE GEV
2    > t.cold  # SHOW THE NUMERICAL VALUES OF - MINIMUM TEMPERATURE
3     [1] 14.14900  7.86600 11.30500  6.86820  3.46800  4.69840  8.59410
4     [8] 12.90300 10.31000  7.59470  4.95460 11.00900 14.02500  7.16280
5    [15] 11.62400 16.71100  9.96940  9.08180  9.37610  3.63970  0.45510
6    [22]  4.64770  4.82900  1.00390 15.73100  6.14850  6.68200 10.23800
7    [29]  4.69930  7.02840  8.74390  5.78030  4.21900  8.83210  8.34690
8    [36]  9.31580  0.47943  3.44110  7.19450 11.27900  8.25640  6.35540
9    [43]  6.08500  7.02990 10.79600  9.58840
10   > cold.extremes = fevd(t.cold) # FIT TO GEV
11   > cold.extremes # PRINT OUT RESULTS
12
13   fevd(x = t.cold)
14
15   [1] "Estimation Method used: MLE"
16   ...
17    Estimated parameters:
18       location        scale        shape
19    6.5115237   3.5662767 -0.2347719
20
21    Standard Error Estimates:
22       location        scale        shape
23    0.58044433 0.40333521 0.09324019
24
25    Estimated parameter covariance matrix.
26                   location        scale        shape
27    location   0.33691562   0.01740978 -0.020276953
28    scale      0.01740978   0.16267929 -0.019456804
29    shape     -0.02027695  -0.01945680  0.008693734
```

Uncertainty in the parameters can be translated into uncertainties in the return level z_p using the *delta method*, which is discussed in Section 19.3. The result is the 95% confidence interval shown as the dashed curves in Figure 19.5. According to this figure, the 100-year level is about $16.6 \pm 2.8°$C. Note that the confidence interval widens considerably after the right-most data point. This widening is typical of extrapolation procedures – the uncertainty in return level grows quickly beyond the most extreme return period of the sample.

   One concern with this analysis is that it derived from an asymptotic theory that is valid only for large block size N. If the problem is restricted to January minimum temperatures, then one way to increase N is to make more frequent observations, say hourly, but this is not helpful because hourly observations are highly dependent and therefore do not satisfy the *iid* assumption. In fact, daily data tend to be serially correlated, so 31 days in January actually corresponds to *fewer* that 31 independent

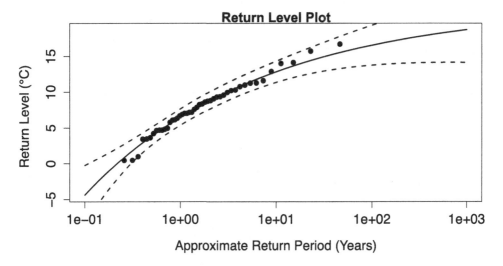

Figure 19.5 Return level plot for *negative* January temperature over mid-Atlantic United States during 1970–2015. The plot shows return levels derived from the empirical distribution (dots) and from the fitted GEV (solid curve). The dashed curves show the 95% confidence interval for the return level.

samples. If the problem can be extended to include months beyond January, then N could be increased by including the whole fall-winter-spring season. However, this is not helpful because the coldest temperatures tend to occur within several weeks of the peak of winter, so data in the fall or spring would never contribute to M_N. Despite these concerns, this procedure based on the GEV distribution often yields sensible results from climate data.

19.2 Distribution of the Maximal Value

This section derives the distribution of M_N. This derivation is not essential for practical applications, but is included here because it is interesting in itself and provides the foundation for more in-depth understanding of extreme value analysis.

We suppose $X_1, \ldots, X_N$ are *iid* random variables and consider their maximum M_N, defined in (19.4). The distribution of M_N can be written as

$$\Pr\left(M_N \leq z\right) = \Pr\left(\max\left\{X_1, \ldots, X_N\right\} \leq z\right) \tag{19.14}$$

$$= \Pr\left(X_1 \leq z \quad \text{and} \quad \cdots \quad \text{and} \quad X_N \leq z\right), \tag{19.15}$$

where the last equality follows from the fact that the largest of the X_n's is less than or equal to z if and only if each and every X_n is less than or equal to z. Because the random variables are independent, the joint distribution equals the product of the marginals:

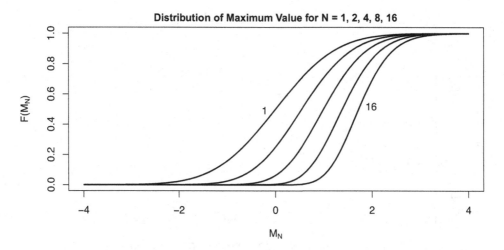

Distribution of Maximum Value for N = 1, 2, 4, 8, 16

Figure 19.6 The distribution of $M_N = \max\{X_1, \ldots, X_N\}$ for independent and identically distributed random variables $X_1, \ldots, X_N$ drawn from a normal distribution for $N = 1, 2, 4, 8, 16$.

$$\Pr\left(M_N \leq z\right) = \Pr\left(X_1 \leq z\right) \cdots \Pr\left(X_N \leq z\right). \tag{19.16}$$

Furthermore, because the random variables are identically distributed, then all the terms on the right-hand side are identical, in which case

$$\Pr\left(M_N \leq z\right) = \Pr\left(X_n \leq z\right)^N. \tag{19.17}$$

The behavior of this function as a function of N for random variables from a Gaussian distribution is illustrated in Figure 19.6. Since probability is between zero and one, the quantity $\Pr\left(X_n \leq z\right)^N$ tends to zero as N tends to infinity, for any fixed z. In essence, the distribution of the maximum shifts toward the right as N increases. This makes sense: as the sample size N increases, progressively larger maximum values are likely to occur.

If the random variable X_n has a finite upper limit, then the distribution of M_N approaches a delta function at that upper limit (that is, it "degenerates into a point mass"). If the variable has no upper limit, then the distribution *does not converge*, but rather moves to the right with increasing N. Also, the distribution changes scale as N increases. Fisher and Tippett (1928) had the idea of compensating for these changes by considering the standardized variable $(M_N - b_N)/a_N$, where a_N and b_N are scalars that depend on N in some way that counteracts the change in location and scale. Accordingly, we investigate whether

$$\Pr\left(\frac{M_N - b_N}{a_N} \leq z\right) \to G(z) \quad \text{as} \quad N \to \infty. \tag{19.18}$$

If, as in our example, we had K samples of M_N, we can consider the maximum of all KN samples, whose standardized values should have the same distribution, hence

$$\Pr\left(\frac{M_{NK} - b_{NK}}{a_{NK}} \le z\right) \to G(z) \quad \text{as} \quad NK \to \infty. \tag{19.19}$$

However, M_{NK} also has the equivalent definition

$$M_{NK} = \max\{(M_N)_1, \ldots, (M_N)_K\}. \tag{19.20}$$

Since each M_N has the same distribution, it follows that

$$
\begin{aligned}
(G(z))^K &= \left(\Pr\left(\frac{M_N - b_N}{a_N} \le z\right)\right)^K \\
&= \left(\Pr\left(\frac{\max\{(M_N)_1, \ldots, (M_N)_K\} - b_N}{a_N} \le z\right)\right) \\
&= \Pr\left(\frac{M_{NK} - b_N}{a_N} \le z\right) \\
&= G\left(\frac{(z - b_{NK} + b_N)a_N}{a_{NK}}\right),
\end{aligned}
$$

where the first line follows from (19.18), the second line follows from (19.17), the third line follows from (19.20), and the last line follows from (19.19). Because $G(z)$ is defined for asymptotically large N, $G(z)$ itself cannot depend on N; therefore the expression on the right-hand side cannot depend on N. Accordingly, we define

$$a'_K = \frac{a_{NK}}{a_N} \quad \text{and} \quad b'_K = (b_{NK} - b_N)/a'_K. \tag{19.21}$$

Using this notation, we have the identity

$$(G(z))^K = G(a'_K z + b'_K). \tag{19.22}$$

This identity is known as the *stability postulate*. A distribution satisfying this condition for $K = 1, 2, \ldots$ is said to be *max-stable*. Three different solutions can be derived according to whether a'_K is less than one, equal to one, or greater than one. We give an outline of the proof for the case $a'_K = 1$, and then summarize the solution for the other two cases. In the case $a'_K = 1$, we may evaluate (19.22) for KN samples to obtain

$$(G(z))^{KN} = G\left(z + b'_{KN}\right). \tag{19.23}$$

However, according to (19.22), this distribution can be written alternatively as

$$(G(z))^{KN} = \left((G(z))^K\right)^N = \left(G(z + b'_K)\right)^K = G(z + b'_K + b'_N). \tag{19.24}$$

These two equations will be equal if

$$b'_{NK} = b'_N + b'_K.$$ (19.25)

This equation has the same form as in (19.21) (for $a'_k = 1$), hence consistent solutions are obtained if $b'_K = b_K$. Accordingly, we drop the primes. Furthermore, a function $f(\cdot)$ with the property $f(NK) = f(N) + f(K)$ is the logarithm. Therefore,

$$b_K = c \log K,$$ (19.26)

where c is an arbitrary constant. This is the most general solution, but a proof relies on functional equations, which is beyond our scope (see Efthimiou, 2011, for proof).

To determine the function $G(z)$, take the logarithm of both sides to give

$$K \log G(z) = \log G(z + b_K).$$ (19.27)

This equation states that whenever z increases by b_K, $\log G(z)$ increases by a factor of K. It is easily verified that a function with this property is

$$\log G(z) = \alpha K^{z/b_K},$$ (19.28)

where α is an arbitrary constant. Substituting (19.26) into (19.28) gives

$$\log G(z) = \alpha \exp\left(\frac{z}{c}\right).$$ (19.29)

Solving this equation for $G(z)$ gives a solution in the form

$$G(z) = \exp\left\{-\exp\left[-\left(\frac{z-\mu}{\sigma}\right)\right]\right\} \quad \text{for } -\infty < z < \infty,$$ (19.30)

where μ and σ are merely re-parameterizations of the arbitrary constants α and c. This family of distributions is known as the *Gumbel* distribution.

Two other distribution families can be derived by considering $a'_K > 1$ or $a'_K < 1$. One of these distributions is the *Fréchet* distribution,

$$G(z) = \begin{cases} 0, & z \leq b \\ \exp\left\{-\left(\frac{z-\mu}{\sigma}\right)^{-\alpha}\right\} & z > b \end{cases}.$$ (19.31)

The other is the *Weibull* distribution

$$G(z) = \begin{cases} \exp\left\{-\left[-\left(\frac{z-\mu}{\sigma}\right)\right]^{\alpha}\right\} & z < b \\ 0, & z \geq b \end{cases}.$$ (19.32)

Here, $\sigma > 0$ and $\alpha > 0$. The three distributions, illustrated in the left panel of Figure 19.7, differ by the domain of Z: the Gumbel distribution has unbounded Z, the Weibull distribution has a finite upper bound on Z, and the Fréchet has

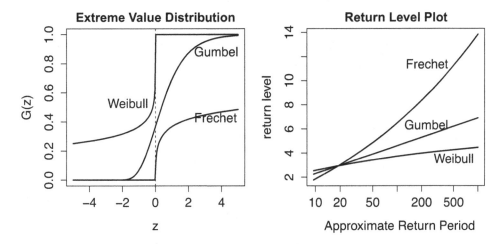

Figure 19.7 Illustration of the Gumbel, Fréchet, and Weibull distributions on linear scale (left) and on a return level plot (right).

finite lower bound on Z. Representative return level plots for the Gumbel, Fréchet, and Weibull distributions are shown in the right panel of Figure 19.7. The three distributions differ by the rate at which the return level increases with return period, with the Fréchet distribution leading to the strongest (and hence the most concerning) rate of increase in return level. The fit shown in Figure 19.5 corresponds to a Weibull distribution, which is concave downward on a return level plot.

All three families can be combined into a single family of distributions of the form

$$G(z) = \exp\left\{-\left[1 + \xi\left(\frac{z-\mu}{\sigma}\right)\right]^{-1/\xi}\right\}. \tag{19.33}$$

Specifically, the Gumbel distribution can be obtained by taking the limit $\xi \to 0$, while the Fréchet and Weibull distributions can be obtained for $\xi > 0$ and $\xi < 0$, respectively. A distribution of this form is known as the *Generalized Extreme Value* distribution.

These three distributions exhaust the possible limiting forms of the extreme value distribution. If X is distributed as a normal, exponential, or logistic, then the distribution of the maximal value converges to the Gumbel distribution (Kotz and Nadarajah, 2000, 6). A counter-intuitive fact is that even if the Gumbel is known to be the correct limit, in practice better approximations can be obtained using the GEV (Cohen, 1982).

These results are summarized in the following theorem:

Theorem 19.1 (Extremal Types Theorem) *Let $X_1, \ldots, X_N$ be iid random variables. Define the maximum of $X_1, \ldots, X_N$ as M_N:*

$$M_N = \max\{X_1, \ldots, X_N\}. \tag{19.34}$$

If there exists sequences of constants $\{a_N > 0\}$ and $\{b_N\}$ such that

$$Pr\left[\frac{M_N - b_N}{a_N} \le z\right] \to G(z) \quad as\ N \to \infty, \tag{19.35}$$

where $G(\cdot)$ is a nondegenerate cumulative distribution function, then $G(\cdot)$ is a member of the Generalized Extreme Value (GEV) distribution

$$G(z) = \exp\left\{-\left[1 + \xi\left(\frac{z - \mu}{\sigma}\right)\right]^{-1/\xi}\right\}, \tag{19.36}$$

for z defined on

$$1 + \xi(z - \mu)/\sigma > 0, \tag{19.37}$$

where the parameters of the distribution are

location parameter: μ $\quad -\infty < \mu < \infty$
scale parameter: σ $\quad \sigma > 0$
shape parameter: ξ $\quad -\infty < \xi < \infty.$

The limiting distribution as $\xi \to 0$ is Gumbel distribution (19.30).

A remarkable part of this theorem is that it states that the GEV is the *only* possible limit for the distribution of $(M_N - b_N)/a_N$, *regardless of the distribution of* X. Three points about this theorem should be noted. First, the theorem does not state that the distribution of $(M_N - b_N)/a_N$ *will* converge to the GEV, but only that *if* it converges, then the convergent distribution must be a member of the GEV family. Some distributions may not converge. Second, the location and scale parameters are not the mean and standard deviation of the GEV, although the symbols μ and σ often denote these quantities in other distributions. Third, the theorem does not specify how the parameters a_N and b_N are to be determined. Remarkably, this missing detail presents no practical barrier to estimation, because (19.35) is equivalent to

$$Pr[M_N \le z] = G(a_N z + b_N) = G^*(z), \tag{19.38}$$

where G^* is merely another member of the GEV family. Thus, the unknown parameters a_N and b_N can be subsumed into the location and scale, which will be estimated from data.

19.3 Maximum Likelihood Estimation

Having found the cumulative distribution function $G(z)$, the next step is to estimate its parameters. The familiar approach for Gaussian distributions is to estimate the mean and variance. This is an example of the *method of moments*. Note, however, that the mean and variance of the GEV distribution are not μ and σ^2, but rather some nonlinear functions of these parameters. There is a considerable literature on using the method of moments to estimate parameters of the GEV distribution (see Hosking et al., 1985).

However, in this chapter, we use a different approach to estimating parameters known as the *maximum likelihood method*. This method is preferred because it provides a natural framework for accounting for dependencies in time or environmental conditions. The method works for many problems, but to illustrate it we consider a Gaussian distribution. Although Gaussian distributions play a minor role in extreme value analysis, they provide a nice illustration of the maximum likelihood method because the computations are straightforward and lead to familiar results. Accordingly, consider the Gaussian distribution

$$p(x) = \frac{1}{\sqrt{2\pi}} \frac{1}{\sigma} \exp\left[-(x-\mu)^2/(2\sigma^2)\right], \tag{19.39}$$

which has mean μ and variance σ^2. Suppose $X_1, \ldots, X_N$ are independent random draws from distribution $p(x)$. Because the variables are independent, the joint distribution is

$$p(x_1, \ldots, x_N) = p(x_1) \ldots p(x_N) = \left(\frac{1}{\sqrt{2\pi}} \frac{1}{\sigma} e^{-\frac{(x_1-\mu)^2}{2\sigma^2}}\right) \ldots \left(\frac{1}{\sqrt{2\pi}} \frac{1}{\sigma} e^{-\frac{(x_N-\mu)^2}{2\sigma^2}}\right),$$

which may be simplified to

$$p(x_1, \ldots, x_N; \mu, \sigma^2) = \frac{1}{(2\pi)^{N/2}} \frac{1}{\sigma^N} \exp\left[-\frac{\sum_{n=1}^{N}(x_n-\mu)^2}{2\sigma^2}\right]. \tag{19.40}$$

The basic idea is to choose values of μ and σ that will make the probability density of the sample $p(x_1, \ldots, x_N; \mu, \sigma^2)$ a maximum. The resulting values of μ and σ are known as the *maximum likelihood estimates*. The merit of this method is discussed in numerous texts (e.g., Mardia et al., 1979), but we will focus only on those aspects that relate to extreme value analysis. To find the maximum likelihood estimates, we substitute the sample values $x_1, \ldots, x_N$ into Equation (19.40). After doing this, the sample values $x_1, \ldots, x_N$ are known, whereas μ and σ are the unknowns. In this situation, the function $p(x_1, \ldots, x_N; \mu, \sigma^2)$ is not a probability density function, so to avoid confusion it is given a new name, the *likelihood function*. The likelihood is a function of μ and σ. Our goal is to find the values of μ and σ that maximize the likelihood function.

To maximize the likelihood function, it proves convenient to maximize the log of the likelihood rather than the likelihood itself, which is acceptable because the derivative of $f(x)$ and $\log f(x)$ vanish at the same value x, since

$$\frac{d \log f}{dx} = \frac{1}{f}\frac{df}{dx} = 0, \tag{19.41}$$

The log of the likelihood function is

$$L = \log p(x_1, \ldots, x_N; \mu, \sigma^2) = -\frac{N}{2}\log 2\pi - \frac{N}{2}\log \sigma^2 - \frac{1}{2\sigma^2}\sum_{n=1}^{N}(x_n - \mu)^2. \tag{19.42}$$

Computing derivatives and setting the result to zero yields the equations

$$\frac{\partial L}{\partial \mu} = \frac{1}{\sigma^2}\sum_{n=1}^{N}(x_n - \mu) = \frac{1}{\sigma^2}\left(\left(\sum_{n=1}^{N}x_n\right) - N\mu\right) = 0 \tag{19.43}$$

$$\frac{\partial L}{\partial(\sigma^2)} = -\frac{N}{2}\frac{1}{\sigma^2} + \frac{1}{2(\sigma^2)^2}\sum_{n=1}^{N}(x_n - \mu)^2 = 0. \tag{19.44}$$

Solving the first equation for μ yields

$$\hat{\mu} = \frac{1}{N}\sum_{n=1}^{N}x_n,$$

which is the familiar sample mean. Solving the second equation, and substituting $\hat{\mu}$, yields

$$\overline{\sigma}^2 = \frac{1}{N}\sum_{n=1}^{N}(x_n - \hat{\mu})^2, \tag{19.45}$$

which is closely related to the unbiased estimate of the variance $\hat{\sigma}^2$.

This example illustrates an important fact: estimates of μ and σ^2 were *derived* from the maximum likelihood method. This is in contrast to previous chapters, where estimates were merely *asserted*. These estimates were obtained in closed form with the estimate on one side of the equation and sample quantities on the other. For the GEV distribution, however, estimates cannot be written in closed form. Despite this, it is still possible to estimate the parameters of the GEV distribution by maximizing the likelihood function *numerically* using standard optimization packages (e.g., the extRemes package in R; Gilleland and Katz, 2016). The procedure is the following. Recall that the GEV distribution defined in (19.36) depends on three parameters, μ, σ, ξ. For notational convenience, let the parameter values be consolidated into a single vector

$$\boldsymbol{\theta} = \begin{pmatrix} \mu & \sigma & \xi \end{pmatrix}^T. \tag{19.46}$$

The likelihood is derived from the probability density of the GEV as

$$g(z;\boldsymbol{\theta}) = \frac{dG}{dz} = \frac{1}{\sigma} \exp\left\{-\left[1 + \xi\left(\frac{z-\mu}{\sigma}\right)\right]^{-1/\xi}\right\} \left(1 + \xi\left(\frac{z-\mu}{\sigma}\right)^{-1/\xi}\right)^{-1/\xi-1}. \tag{19.47}$$

If the random variables $Z_1, \ldots, Z_N$ are independent, then the joint density is

$$g(z_1, \ldots, z_N; \boldsymbol{\theta}) = g(z_1; \boldsymbol{\theta})g(z_2; \boldsymbol{\theta}) \cdots g(z_N; \boldsymbol{\theta}). \tag{19.48}$$

We substitute sample values for $z_1, \ldots, z_N$, after which $g(z_1, \ldots, z_N; \boldsymbol{\theta})$ becomes a function of the parameters $\boldsymbol{\theta}$. The value of $\boldsymbol{\theta}$ that maximizes this likelihood can then be found by numerical optimization (typically, $\log g$ is maximized to avoid underflow problems).

Under certain conditions, the maximum likelihood estimate $\hat{\boldsymbol{\theta}}$ converges to the true parameter value in the limit of large sample size. In fact, if K is the number of samples and $\boldsymbol{\theta}_0$ is the true parameter vector, then as $K \to \infty$, it can be shown that

$$\left(\hat{\boldsymbol{\theta}} - \boldsymbol{\theta}_0\right) \to \mathcal{N}\left(\mathbf{0}, \mathcal{I}^{-1}\right), \tag{19.49}$$

where $\mathcal{I}$ is the scaled *Fisher Information Matrix*

$$\mathcal{I}_{ij} = K\mathbb{E}_Z\left[-\frac{\partial^2 \log g(z;\boldsymbol{\theta})}{\partial \theta_i \partial \theta_j}\bigg|_{\boldsymbol{\theta}_0}\right]. \tag{19.50}$$

Note the caveat "under certain conditions." One of the conditions for the asymptotic distribution to hold is that the domain of the distribution should not depend on the parameter $\boldsymbol{\theta}$. Unfortunately, this condition does not hold for the GEV: when $\xi \neq 0$, one end point of the GEV depends on $\mu - \sigma/\xi$. Nevertheless, Smith (1985) showed that the asymptotic distribution still holds provided $\xi > -1/2$. Fortunately, distributions that do not satisfy this condition (that is, $\xi \leq -1/2$) occur rarely in environmental applications.

This asymptotic distribution can be used to construct confidence intervals for the parameters. Ultimately, we want to construct confidence intervals on the return level. To illustrate how this is done, first consider some arbitrary linear function of the estimated parameters, say $h(\boldsymbol{\theta}) = \mathbf{w}^T \boldsymbol{\theta}$, where $\mathbf{w}$ are the coefficients of the linear function. Then (19.49) and Theorem 7.10 imply that the asymptotic distribution of this function is

$$\hat{h} = \mathbf{w}^T \hat{\boldsymbol{\theta}} \sim \mathcal{N}\left(\mathbf{w}^T \boldsymbol{\theta}_0, \mathbf{w}^T \mathcal{I}^{-1} \mathbf{w}\right). \tag{19.51}$$

Unfortunately, the return level depends nonlinearly on the parameters $\boldsymbol{\theta}$. For non-linear functions of the parameters, we can still derive an approximate distribution based on the fact that, for large K, the estimate will be close to the true value. In this case, $\hat{\boldsymbol{\theta}} - \boldsymbol{\theta}_0$ will be small and therefore $\hat{\boldsymbol{\theta}}$ can be approximated by the leading term in its Taylor series:

$$h(\hat{\boldsymbol{\theta}}) \approx h(\boldsymbol{\theta}_0) + \nabla_\theta h \cdot \left(\hat{\boldsymbol{\theta}} - \boldsymbol{\theta}_0\right). \tag{19.52}$$

Since the right-hand side is a linear function of the estimates and maximum likelihood estimates are asymptotically unbiased, that is, $E[\hat{\boldsymbol{\theta}}] = \boldsymbol{\theta}_0$, it follows that for large K

$$h(\hat{\boldsymbol{\theta}}) \sim \mathcal{N}\left(h(\boldsymbol{\theta}_0), (\nabla_\theta h)^T \mathcal{I}^{-1} (\nabla_\theta h)\right). \tag{19.53}$$

Because K is large, $\hat{\boldsymbol{\theta}}$ is close to $\boldsymbol{\theta}_0$ and Fisher's information matrix $\mathcal{I}$ can be evaluated at the sample estimates $\hat{\boldsymbol{\theta}}$. This approach to deriving a distribution for $h(\hat{\boldsymbol{\theta}})$ is known as the *delta method*. To apply the delta method to the return level z_p in (19.10), we use

$$\left(\nabla_\theta z_p\right)^T = \left(1, \quad -\xi^{-1}(1 - r_p^\xi), \quad \sigma\xi^{-2}(1 - r_p^\xi) - \sigma\xi^{-1}r_p^\xi \log r_p\right). \tag{19.54}$$

Thus, after the estimates $\hat{\boldsymbol{\theta}}$ are obtained and Fisher's information matrix is evaluated at $\hat{\boldsymbol{\theta}}$, (19.53) translates uncertainties in $\hat{\boldsymbol{\theta}}$ into uncertainties in return periods. Specifically, the $(1 - \alpha)100\%$ confidence interval for return periods is

$$\hat{z}_p \pm z_{\alpha/2}\sqrt{\left(\nabla_\theta \hat{z}_p\right)^T \mathcal{I}^{-1} \left(\nabla_\theta \hat{z}_p\right)} \tag{19.55}$$

where the estimated covariance matrix $\mathcal{I}^{-1}$ is given in Table. 19.1 and z_α is the upper α100-percentile of a standard normal distribution. This equation was used to construct the confidence intervals shown in Figure 19.5.

   In modern practice, confidence intervals are typically computed using a more accurate approach based on the *profile likelihood*, in which the likelihood is maximized while holding the return level fixed (Coles, 2001).

19.4 Nonstationarity: Changing Characteristics of Extremes

There is great interest in quantifying how the probability of extremes may be changing in time, or how they depend on environmental factors. This section discusses a framework to account for such dependencies. This framework assumes that the distributional changes between blocks are sufficiently slow that the *iid* assumption is satisfied sufficiently well within a block that the GEV distribution is applicable.

Just as changes in the average behavior of a random variable can be quantified through changes in mean and variance, changes in the extreme behavior of a variable can be quantified through changes in the location, scale, and shape parameter of the GEV distribution. As an example, consider the hypothesis that cold extremes are becoming less likely over long periods of time. One possible way to investigate this possibility is to consider the hypothesis that the location parameter μ depends on time:

$$Z_t \sim G(z; \mu(t), \sigma, \xi). \tag{19.56}$$

If the time variations in $\mu(t)$ are sufficiently slow, then they can be approximated as,

$$\mu(t) = \mu_0 + \mu_1 t, \tag{19.57}$$

where μ_1 quantifies the time rate of change of quantiles of the maximum value. In this model, each quantile changes at the same rate. Whether this is a good assumption needs to be checked against observations. For instance, over the past century, precipitation at the 99.9% quantiles appears to be increasing faster than precipitation at the 99% quantiles.

There is nothing special about the time parameter t in (19.57). We could just as easily substitute any function of time $f(t)$, giving

$$\mu(t) = \mu_0 + \mu_1 f(t). \tag{19.58}$$

The function $f(t)$ is known as a *covariate*. Such a model is useful for investigating changes in the location parameter due to environmental factors. For instance, shifts in climate patterns between El Niño and La Niña cause droughts and floods around the world. Thus, it is natural to investigate the dependence of extremes on ENSO, which can be done by setting $f(t)$ in (19.58) to the Niño 3.4 index, for instance. We could also generalize the model to consider more than one covariate:

$$\mu(t) = \mu_0 + \mu_1 f_1(t) + \cdots + \mu_M f_M(t). \tag{19.59}$$

We could also consider changes in the scale parameter:

$$\sigma(t) = \sigma_0 + \sigma_1 f_1(t) + \cdots + \sigma_M f_M(t), \tag{19.60}$$

and in the shape parameter

$$\xi(t) = \xi_0 + \xi_1 f_1(t) + \cdots + \xi_M f_M(t). \tag{19.61}$$

The shape parameter ξ tends to have greater uncertainty than the other parameters. As a result, changes in the shape parameter tend to be difficult to demonstrate.

The maximum likelihood method provides a general framework for quantifying changes in population parameters. In fact, the procedure is simple: replace the population parameters $\mu(t), \sigma(t), \xi(t)$ in the density function (19.47) by their functional equivalents (19.59), (19.60), and (19.61). After this substitution, the likelihood becomes a function of the parameters appearing in the functions: $\mu_0, \mu_1, \ldots$. The values of the parameters that maximize the likelihood are then identified using the same numerical techniques used to derive the original population parameters. Approximate confidence intervals for the parameters are also derived from (19.49) in the same way as the original parameters.

For our example, it turns out that ENSO has relatively little influence on mid-Atlantic US winter temperatures. On the other hand, another component of variability known as the *North Atlantic Oscillation* (NAO) does tend to influence this region. To test for a possible influence, we consider a model of the form (19.56), where the location parameter is of the form (19.58), with $f(t)$ identified with the NAO index[2]. The result is shown in Table 19.2. The maximum likelihood estimate of the slope parameter μ_1 is about -1.45, which implies that the location parameter *increases* at the rate 1.45°C per unit NAO index (recall that the sign of the data has been reversed because cold extremes are being analyzed). The standard error of this estimate is about 0.48°C, so the mean value 1.45°C is more than two standard errors from zero, indicating statistical significance (a more accurate comparison of likelihoods gives the same conclusion). The 2- and 10-year return levels from this model are plotted in Figure 19.8. Since the data is annual values, a 2-year return period corresponds to the 50th percentile (that is, the median of the extreme cold temperatures). Consistent with this, about half the dots lie below the 2-year return level (that is, 23 dots out of 46). Similarly, for a 10-year return period, about a tenth of the dots are expected to lie below the 10-year return level curve, which is indeed the case.

The return level for the GEV model is given by

$$\hat{z}_p(t) = \hat{\mu}(t) - \frac{\hat{\sigma}}{\hat{\xi}}\left(1 - r_p^{\hat{\xi}}\right), \tag{19.62}$$

which in this case depends on the NAO index through the location parameter $\hat{\mu}(t)$. The return levels for NAO index values ± 2 derived from this equation are shown

---

[2] This index was downloaded from NOAA's Climate Prediction Center website www.cpc.ncep.noaa.gov.

Table 19.2. *An example of a GEV fit to cold January extremes in the mid-Atlantic United States that includes the NAO index as a covariate for the location parameter μ, as defined in (19.57). This example follows the commands and data shown in Table 19.1. Line 10 shows the maximum likelihood estimates of μ_0, μ_1, σ, ξ, respectively; line 14 shows the associated standard errors; lines 18–21 show the covariance matrix for the parameter estimates.*

```
 1  > cold.x.tele = fevd(t.cold,location.fun=~index) # FIT GEV
 2  > cold.x.tele
 3
 4  fevd(x = t.cold, location.fun = ~index)
 5
 6  [1] "Estimation Method used: MLE"
 7  ...
 8   Estimated parameters:
 9         mu0         mu1       scale       shape
10    6.4879640  -1.4483469   3.0933949  -0.1545121
11
12   Standard Error Estimates:
13         mu0         mu1       scale       shape
14   0.5231300  0.4817077  0.3851407  0.1244832
15
16   Estimated parameter covariance matrix.
17                 mu0          mu1        scale        shape
18  mu0      0.27366503   0.03729961   0.05315883  -0.02825529
19  mu1      0.03729961   0.23204226   0.04143058  -0.02199465
20  scale    0.05315883   0.04143058   0.14833340  -0.02640878
21  shape   -0.02825529  -0.02199465  -0.02640878   0.01549606
```

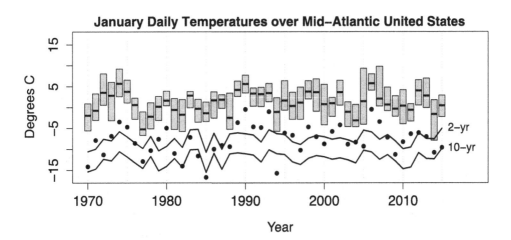

Figure 19.8 Same as Figure 19.1, except including the 2-year and 10-year return levels estimated from a GEV fit using the NAO index as a covariate on the location parameter (two solid curves).

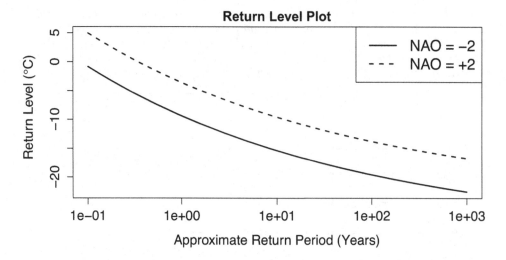

Figure 19.9 Return levels for cold extremes during January over the mid-Atlantic United States conditioned on the NAO index at values −2 (solid) and +2 (dashed), as derived from a GEV model fit using the NAO index as a covariate for the location parameter.

in Figure 19.9. For a 100-year return period, the return level is found to be about −19.6°C and −8.4°C for NAO indices −2 and +2, respectively. The goodness of fit can be assessed with a QQ-plot, which is a standard output of most software packages for extreme value analysis.

19.5 Further Topics

For further reading, we highly recommend the lucid text by Coles (2001), which covers all of the major topics of extreme value analysis. Also, Gilleland and Katz (2016) describes the valuable R package extRemes for extreme value analysis.

In many climate studies, time series are available continuously. Thus, analyzing only the maximum within a block and ignoring other extremes that may occur within a block is not an efficient use of data. A complementary approach that uses more information in the data is based on *threshold models.* Just as the GEV distribution can be justified for block maxima, another distribution known as the *Generalized Pareto Distribution (GPD)* can be justified for the probability of exceedances above a threshold. The choice of threshold is an important parameter in estimating a GPD: too large a threshold would lead to a small sample size and correspondingly large uncertainties, while too small a threshold might be too far from the distribution tail for the asymptotic approximation to hold. The choice of threshold is roughly analogous to the choice of block size in estimating the GEV.

Estimation of GPD parameters parallels that of the GEV. There also exist Bayesian estimation techniques.

The multivariate extension of extreme value theory is much more complicated than the univariate framework. This extension includes the topic of compound extremes and hazards, which is important in weather and climate studies (Davison et al., 2019). There exists an entirely separate mathematical field known as *point processes* that can be used to frame extreme value analysis. The difference in approach can be enlightening in some applications.

19.6 Conceptual Questions

1 What is a return level? What is a return period?
2 What is a return level plot? What is useful about it?
3 Why not assume a Gaussian distribution for quantifying the probability of extremes?
4 Winter season is about 90 days. When quantifying extreme cold days, does it help to increase the sample size by adding more frequent observations within the season, or adding observations in months during spring and fall?
5 What is the interpretation of the location, scale, and shape parameters?
6 What is the extremal types theorem? How does it help in analyzing extremes?
7 What is the difference between the Gumbel, Weibull, and Fréchet distributions?
8 What is the maximum likelihood method? Why is it useful?

20

Data Assimilation

> Objective analysis is a process by which meteorological observations
> distributed in space and time, and from different observing systems, are
> combined with other information – predictions from previous analyses,
> or perhaps climatology – to form a numerical representation of the state
> of the atmosphere.[1]
>
> *McPherson (1986)*

Data assimilation is a procedure for combining observations and forecasts of a
system into a single, improved description of the state of a system. Because observations and forecasts are uncertain, they are each best described by probability distributions. The problem of combining these two distributions into a new, updated
distribution that summarizes all our knowledge is solved by *Bayes theorem*. If the
distributions are Gaussian, then the parameters of the updated distribution can be
written as an explicit function of the parameters of the observation and forecast
distributions. The assumption of Gaussian distributions is tantamount to assuming
linear models for observations and state dynamics. The purpose of this chapter is
to provide an introduction to the essence of data assimilation. Accordingly, this
chapter discusses the data assimilation problem for Gaussian distributions, in which
the solution from Bayes theorem can be derived analytically. Practical data assimilation usually requires modifications of this assimilation procedure, a special case
of which is discussed in the next chapter.

---

[1] Reprinted by permission from Springer Nature: American Meteorological Society, *Mesoscale Meteorology
and Forecasting*, chapter on "Operational Objective Analysis Techniques and Potential Applications for
Mesoscale Meteorology" by Ronald D. McPherson, 1986.

20.1 The Problem

Suppose we observe a random process Y_t at certain times *in the past* and want to estimate the *current* value of the process. Let the past observations be denoted $\{o_{t-1}, o_{t-2}, \ldots\}$. Then, the most complete description of Y_t based on all *past* observations is the conditional distribution $p(y_t | o_{t-1}, o_{t-2}, \ldots)$. This distribution is the *background distribution* (also known as the *forecast* or *prior* distribution).

Suppose now that an observation of the process at time t becomes available. At this moment, we have *two* pieces of information about Y_t: the background distribution derived from all past observations $p(y_t | o_{t-1}, o_{t-2}, \ldots)$, and the newly available observation o_t. The question of how to *combine* these two pieces of information in a single distribution that incorporates everything we know about Y_t is addressed by *data assimilation*. The answer is obtained through *Bayes Theorem*, which yields the *analysis* distribution $p(y_t | o_t, o_{t-1}, \ldots)$ (also known as the *updated* or *posterior* distribution).

After observational and background distributions have been combined, the resulting analysis distribution serves as the *initial condition for the next prediction*. In this way, the states $Y_t, Y_{t+1}, \ldots$ are estimated *sequentially* as new observations become available. This process is illustrated schematically in Figure 20.1. If the distributions are Gaussian, then the resulting procedure is known as the *Kalman Filter*. We first discuss this procedure in detail for a univariate, Gaussian, Markov process, then generalize it to multivariate processes.

20.2 A Univariate Example

As a simple example, suppose Y_t is governed by the AR(1) model

$$Y_t = \phi_1 Y_{t-1} + W_t, \tag{20.1}$$

where W_t is Gaussian white noise with zero mean and variance σ_W^2, and $|\phi_1| < 1$. The behavior of this stochastic process has been discussed in Chapter 5. Suppose the state of the system at $t = 0$ is known exactly, without any uncertainty.[2] At the next time step, (20.1) tells us $Y_1 = \phi_1 y_0 + W_1$. This equation shows that even though the state of the system at $t = 0$ is known to be exactly y_0, the state at $t = 1$ will be uncertain owing to the random process W_1. Since the state at $t = 1$ is a constant plus a Gaussian, the distribution of Y_1 must be Gaussian. It is sufficient, therefore, to determine the mean and variance of Y_1 conditioned on the (perfect) observation y_0. The conditional mean is

$$\mathbb{E}[Y_1 | y_o] = \mathbb{E}[\phi_1 Y_0 + W_t | y_0] = \phi_1 y_0 + \mathbb{E}[W_1] = \phi_1 y_0, \tag{20.2}$$

---

[2] In this section, a scalar random variable is indicated by capital letters, and a specific realization is indicated by lower case letters. For instance, y is a specific realization, whereas Y is a random variable.

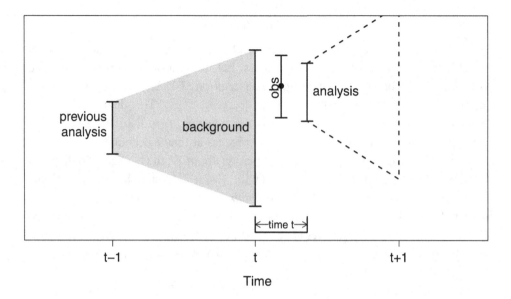

Figure 20.1 Schematic illustrating the data assimilation process. At time $t - 1$, an initial estimate of a variable ("previous analysis") is available whose confidence interval is indicated as the error bar at $t - 1$. This initial uncertainty is propagated forward to time t using the dynamical laws of the system, at which point the confidence interval has expanded to the interval labeled "background." At time t, an observation becomes available, with a confidence interval indicated in the figure. Information from the observation and background are combined into a new estimate, the "analysis." This analysis at t then becomes the initial estimate used to derive the next background at $t + 1$. The time dimension is stretched at time t to avoid putting all this information on the same vertical line.

Since y_0 is a constant and W_t has zero mean. The conditional variance is

$$\text{var}[Y_1|y_0] = \text{var}[\phi_1 Y_0 + W_t|y_0] = \text{var}[W_t|y_0] = \text{var}[W_t] = \sigma_W^2. \qquad (20.3)$$

It follows that the conditional distribution is

$$p(y_1|y_0) \sim \mathcal{N}\left(\phi_1 y_0, \sigma_W^2\right). \qquad (20.4)$$

This distribution describes everything we know about Y_1 given its value at the earlier time $t = 0$ and knowledge of the population parameters ϕ_1 and σ_W^2. Since it is the distribution of the state conditional on past observations, it is known as the background distribution.

Now, suppose that at time $t = 1$ we observe the system and obtain observation o_1. This observation contains new information about Y_1 that is not contained in the background distribution $p(y_1|y_0)$. We would like to *combine* this information with the background distribution to derive a new distribution that incorporates all available information at $t = 1$. This more comprehensive distribution is a conditional distribution denoted $p(y_1|y_0, o_1)$. To derive this distribution, recall that $p(A, B) = p(A|B)p(B)$ for any A and B; therefore

$$p(y_1|y_0, o_1) = \frac{p(y_1, y_0, o_1)}{p(y_0, o_1)}. \tag{20.5}$$

At first, this new equation may not seem helpful since the right-hand side involves two distributions that we have not specified. However, we can invoke the relation $p(A, B) = p(A|B)p(B)$ recursively to write the numerator as

$$p(y_1, y_0, o_1) = p(o_1|y_1, y_0)p(y_1|y_0)p(y_0). \tag{20.6}$$

Here, the background distribution $p(y_1|y_0)$, which is known from (20.4), appears on the right-hand side. The only other term involving y_1 is $p(o_1|y_1, y_0)$. To specify this distribution, the relation between observation O_t and true state Y_t needs to be specified. In practice, this can be difficult. For example, a thermometer measures temperature, but the measurement is imperfect and the precise relation between observation and truth is complex. For illustration purposes, suppose observation O_t and "truth" Y_t are related as

$$O_t = Y_t + R_t, \tag{20.7}$$

where R_t is a Gaussian random variable with zero mean and variance σ_R^2. R_t represents *instrumental measurement error* and is assumed to be independent of the state Y_t. In essence, observations are *independent perturbations* about the true state. The smaller the perturbations R_t, the more closely observations match the truth. This observation model implies

$$O_0 = Y_0 + R_0 \tag{20.8}$$
$$O_1 = Y_1 + R_1. \tag{20.9}$$

Although the equation for O_1 does not depend explicitly on Y_0, we cannot conclude from this that O_1 and Y_0 are independent. After all, O_1 depends on Y_1, and Y_1 depends on Y_0 through the AR(1) model, hence O_1 depends on Y_0. However, we can see from (20.9) that *if Y_1 were known*, then the only remaining part of O_1 that would be unknown is the random noise R_1, which *is* independent of Y_1 and Y_0. Therefore, if Y_1 were known, then adding Y_0 to the conditional would have no effect on the distribution of O_1. In this case, O_1 is said to be *conditionally independent* of Y_0, and this condition is expressed as

$$p(o_1|y_1, y_0) = p(o_1|y_1). \tag{20.10}$$

Observation model (20.7) immediately implies that

$$p(o_1|y_1) = \mathcal{N}(y_1, \sigma_R^2), \tag{20.11}$$

which fully specifies the distribution $p(o_1|y_1, y_0)$.

One might be concerned about the argument "if Y_1 were known" when, in practice, Y_1 never is known. No contradiction exists. To clarify this point, recall the fundamental identity $p(A, B) = p(A|B)p(B)$. Although B is assumed known in $p(A|B)$, randomness in B is still taken into account through $p(B)$. Thus, this identity does not imply B actually is known, but rather that the joint distribution can be decomposed into a part in which B is assumed to be known, and a separate part that accounts for randomness in B. This illustrates a fundamental idea: any joint distribution can be split into a part in which a variable is assumed to be known, and another part that assumes the variable is random, provided that all parts are combined at the end in accordance with probability theory. This is a common strategy for quantifying uncertainties in complicated systems.

At this point, the conditional distribution of Y_1 has been specified from (20.5) to

$$p(y_1|y_0, o_1) = \frac{p(o_1|y_1)p(y_1|y_0)p(y_0)}{p(o_1, y_0)}. \tag{20.12}$$

This equation is a special case of *Bayes' Theorem* and is in fact the general solution for *arbitrary* distributions when Y_0 and O_1 are conditionally independent given Y_1. The terms in (20.12) that depend on Y_1 have special names:

$p(o_1|y_1)$ *likelihood* of the observations
$p(y_1|y_0)$ *background* distribution; also called the *forecast* or *prior* distribution
$p(y_1|y_0, o_1)$ *analysis* distribution; also called the *posterior* distribution.

The other terms, $p(o_1, y_0)$ and $p(y_0)$, do not depend on Y_1 and therefore can be ignored since they are effectively constants. That is, the analysis distribution can be obtained by multiplying the likelihood and the background distribution, and then inferring any remaining multiplicative factors by imposing the constraint that the integral should equal one.

For notational simplicity, let us express the background distribution (20.4) as

$$p(y_1|y_0) \sim \mathcal{N}\left(\mu_B, \sigma_B^2\right). \tag{20.13}$$

To be even more explicit, the likelihood and prior distributions are

$$p(o_1 \mid y_1) = \frac{1}{\sqrt{2\pi}} \frac{1}{\sigma_R} \exp\left[-\frac{(o_1 - y_1)^2}{2\sigma_R^2}\right] \tag{20.14}$$

$$p(y_1 \mid y_0) = \frac{1}{\sqrt{2\pi}} \frac{1}{\sigma_B} \exp\left[-\frac{(y_1 - y_0)^2}{2\sigma_B^2}\right]. \tag{20.15}$$

According to (20.12), we simply multiply these distributions to derive the distribution of Y_1. Since Y_1 appears only in the argument of the exponentials, and $\exp[A]\exp[B] = \exp[A + B]$, the Y_1 terms in the exponential are

$$-\frac{(o_1 - y_1)^2}{2\sigma_R^2} - \frac{(y_1 - y_0)^2}{2\sigma_B^2}$$

$$= -y_1^2 \left(\frac{1}{2\sigma_R^2} + \frac{1}{2\sigma_B^2} \right) + y_1 \left(\frac{o_1}{\sigma_R^2} + \frac{\mu_B}{\sigma_B^2} \right) - \frac{o_1^2}{2\sigma_R^2} - \frac{\mu_B^2}{2\sigma_B^2}$$

$$= -\frac{(y_1 - \mu_A)^2}{2\sigma_A^2} + \text{terms independent of } y_1, \tag{20.16}$$

where

$$\mu_A = \mu_{Y_1 | y_0, o_1} = \left(\frac{1}{\sigma_R^2} + \frac{1}{\sigma_B^2} \right)^{-1} \left(\frac{1}{\sigma_R^2} o_1 + \frac{1}{\sigma_B^2} \mu_B \right) \tag{20.17}$$

and variance

$$\sigma_A^2 = \sigma_{Y_1 | y_0, o_1}^2 = \left(\frac{1}{\sigma_R^2} + \frac{1}{\sigma_B^2} \right)^{-1}. \tag{20.18}$$

Expression (20.16) was derived by "completing the square": if $(x-c)^2 = x^2 - 2xc + c^2$, then c can be inferred by writing the quadratic in this form and then taking half of the coefficient on x. Terms that are independent of Y_1 can be ignored since they are constants that can be inferred from the constraint that the distribution integrates to one. Finally, taking the exponential of (20.16) reveals that the distribution is Gaussian, hence it must be

$$p(y_1 \mid y_0, o_1) = \frac{1}{\sqrt{2\pi}\,\sigma_A} \exp\left[-\frac{(Y_1 - \mu_A)^2}{2\sigma_A^2} \right]. \tag{20.19}$$

This distribution is known as the *analysis distribution* (or *posterior distribution*). The analysis distribution describes everything we know about the state Y_1 based on (1) all observations up to and including time $t = 1$, (2) the relation between observations and state (20.7), and (3) the relation of Y_t to itself at different times, as expressed in (20.1). Equations (20.17) and (20.18) are scalar versions of the *Kalman Filter Update Equations*. An example of their application is shown in Figure 20.2. In this example, the observation has less uncertainty than the background, as indicated by the smaller spread and the larger peak. The analysis distribution "combines" these two distributions to produce an estimate of the state that has less overall uncertainty and lies between the background and observation.

20.3 Some Important Properties and Interpretations

To gain further insight into the analysis distribution, we consider a few simple cases. First, note that the mean analysis (20.17) can be written equivalently as

$$\mu_A = \left(\frac{\sigma_B^2}{\sigma_B^2 + \sigma_R^2} \right) o_1 + \left(\frac{\sigma_R^2}{\sigma_B^2 + \sigma_R^2} \right) \mu_B. \tag{20.20}$$

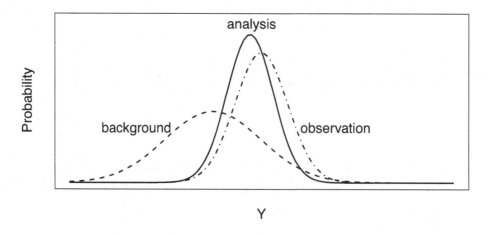

Figure 20.2 An example of an analysis distribution derived from background and observation distributions.

Since the weights add to one, this form reveals that the mean of the analysis is simply a weighted average of the background mean and the observations. The weights are such that the quantity with lesser variance has more weight.

Consider the case in which the background and observation have equal uncertainties, that is, $\sigma_R^2 = \sigma_B^2$. In this case, the analysis parameters become

$$\mu_A = \frac{o_1 + \mu_B}{2} \quad \text{and} \quad \sigma_A^2 = \frac{\sigma_B^2}{2} = \frac{\sigma_R^2}{2}. \tag{20.21}$$

This result shows that the mean of the analysis is simply the average of the observation and background mean. This seems intuitive given that the two quantities have comparable uncertainties. Furthermore, the variance of the estimate is half the variance of either estimate individually. Again, it seems sensible that the uncertainty of the analysis is less than the uncertainty in either the background state or the observation.

Consider the extreme case in which background uncertainty is much greater than observational uncertainty, that is, $\sigma_B^2 \gg \sigma_R^2$. In this case, the analysis parameters become

$$\mu_A \approx o_1 \quad \text{and} \quad \sigma_A^2 \approx \sigma_R^2. \tag{20.22}$$

This result shows that if the background uncertainty is much larger than observational uncertainty, then the background information adds little information beyond the observations – the background information can be essentially ignored. Conversely, if the observational uncertainty is much greater than the forecast distribution, that is, $\sigma_B^2 \ll \sigma_R^2$, then the analysis distribution reduces to the background

distribution, implying that the observation does not significantly change the state
distribution relative to the background distribution.

Other equivalent forms of the filtering equations (20.17) and (20.18) are

$$\mu_A = \mu_B + \frac{\sigma_B^2}{\sigma_B^2 + \sigma_R^2} (o_1 - \mu_B) \tag{20.23}$$

$$\sigma_A^2 = \sigma_B^2 - \frac{\sigma_B^4}{\sigma_B^2 + \sigma_R^2}. \tag{20.24}$$

This form shows that the analysis can be interpreted as a "predictor-corrector" procedure. Loosely speaking, the analysis starts with a "predictor" of the state, as
specified by the background distribution, and then "corrects" this distribution based
on observations. This interpretation mimics the process of "learning from observations," as the state of the system is continually updated to reflect all available
knowledge of the system.

Another property of the analysis that can be inferred from (20.24) is that the
analysis variance is always less than σ_B^2. By symmetry, the analysis variance is
also always less than σ_R^2. It follows that the analysis always has less variance, and
hence less uncertainty, than either the forecast or observation individually. This fact
is consistent with the intuition that if you combine two pieces of information, the
combination should have less uncertainty than either of the pieces individually.

20.4 Multivariate Gaussian Data Assimilation

The previous discussion focused on a single variable and a single observation. In
this section, we generalize the results to the multivariate case.

Accordingly, let $\mathbf{y}_t$ be an M-dimensional vector specifying the state of the system
at time t, and let $\mathbf{o}_t$ be a K-dimensional vector specifying observations at time t. In
practice, K is the number of observations and M is the grid size. These dimensions
differ because observations are not on the same grid as $\mathbf{y}_t$. In general, observations
are available not only at time t but also at previous times. Let $\mathbf{O}_t$ represent the set of
all observations up to and including time t. Then, we seek the distribution of $\mathbf{y}_t$ given
all available observations, $p(\mathbf{y}_t|\mathbf{O}_t)$. This distribution can be written equivalently as

$$p(\mathbf{y}_t|\mathbf{O}_t) = p(\mathbf{y}_t|\mathbf{o}_t\mathbf{O}_{t-1}), \tag{20.25}$$

because, by definition, $\mathbf{O}_t = \{\mathbf{O}_{t-1} \ \mathbf{o}_t\}$[3]. Following the same logic as in the univariate case, this distribution can be defined in terms of other distribution as

$$p(\mathbf{y}_t|\mathbf{o}_t\mathbf{O}_{t-1}) = \frac{p(\mathbf{o}_t|\mathbf{y}_t\mathbf{O}_{t-1})p(\mathbf{y}_t|\mathbf{O}_{t-1})p(\mathbf{O}_{t-1})}{p(\mathbf{o}_t\mathbf{O}_{t-1})}, \tag{20.26}$$

---

[3] To simplify notation, commas between conditional variables are suppressed in this section and the next.

which is the multivariate generalization of (20.12). Terms that do not depend explicitly on y_t (e.g., $p(o_t O_{t-1})$ and $p(O_{t-1})$) are effectively constant, and so can be inferred at the end from the fact that a distribution must integrate to unity. Therefore, we have the relation

$$p(\mathbf{y}_t|\mathbf{o}_t\mathbf{O}_{t-1}) \quad \propto \quad p(\mathbf{o}_t|\mathbf{y}_t\mathbf{O}_{t-1}) \quad p(\mathbf{y}_t|\mathbf{O}_{t-1})$$

$$\text{analysis} \qquad\qquad \text{likelihood} \qquad \text{background} \tag{20.27}$$

In the multivariate case, a common assumption about the observations is that they are a *known* linear combination of state variables, plus some random error:

$$\mathbf{o}_t = \mathbf{H}_t\mathbf{y}_t + \mathbf{r}_t. \tag{20.28}$$

In this equation, $\mathbf{H}_t$ is a $K \times M$ matrix known as the *generalized interpolation operator* and $\mathbf{r}_t$ is an independent Gaussian random variable with zero mean and covariance matrix $\mathbf{R}_t$. In general, the interpolation operator $\mathbf{H}_t$ and *statistics* of the observation errors $\mathbf{r}_t$ can depend on time. Model (20.28) immediately implies

$$p(\mathbf{o}_t|\mathbf{y}_t) = \frac{1}{\sqrt{(2\pi)^K|\mathbf{R}_t|}} \exp\left[-\frac{1}{2}(\mathbf{o}_t - \mathbf{H}_t\mathbf{y}_t)^T \mathbf{R}_t^{-1}(\mathbf{o}_t - \mathbf{H}_t\mathbf{y}_t)\right]. \tag{20.29}$$

In the univariate case, (20.7) implies a certain conditional independence. Similarly, (20.28) implies that $\mathbf{o}_t$ is *conditionally independent of previous observations*, given $\mathbf{y}_t$:

$$p(\mathbf{o}_t|\mathbf{y}_t\mathbf{O}_{t-1}) = p(\mathbf{o}_t|\mathbf{y}_t). \tag{20.30}$$

This distribution is defined in (20.29). The background distribution is assumed to be Gaussian, so it has the general form

$$p(\mathbf{y}_t|\mathbf{O}_{t-1}) = \frac{1}{\sqrt{(2\pi)^M|\mathbf{\Sigma}_t^B|}} \exp\left[-\frac{1}{2}(\mathbf{y}_t - \boldsymbol{\mu}_t^B)^T (\mathbf{\Sigma}_t^B)^{-1}(\mathbf{y}_t - \boldsymbol{\mu}_t^B)\right]. \tag{20.31}$$

The analysis distribution can be derived based on this information. Specifically, according to (20.27), the analysis distribution is simply the product of (20.31) and (20.29). This product can be manipulated into standard form by "completing the square" in a manner analogous to the univariate case. While conceptually straightforward, these algebraic manipulations require sophisticated matrix identities and give little insight into data assimilation, hence we skip the details. The result is that the distribution $p(\mathbf{y}_t|\mathbf{o}_t\mathbf{O}_{t-1})$ is Gaussian with mean and covariance matrix

$$\boldsymbol{\mu}_t^A = \boldsymbol{\mu}_t^B + \mathbf{\Sigma}_t^B\mathbf{H}_t^T \left(\mathbf{H}_t\mathbf{\Sigma}_t^B\mathbf{H}_t^T + \mathbf{R}_t\right)^{-1}(\mathbf{o}_t - \mathbf{H}_t\boldsymbol{\mu}_t^B) \tag{20.32}$$

$$\mathbf{\Sigma}_t^A = \mathbf{\Sigma}_t^B - \mathbf{\Sigma}_t^B\mathbf{H}_t^T \left(\mathbf{H}_t\mathbf{\Sigma}_t^B\mathbf{H}_t^T + \mathbf{R}_t\right)^{-1}\mathbf{H}_t\mathbf{\Sigma}_t^B. \tag{20.33}$$

These equations are the *Kalman Filter Update Equations*, and are the multivariate generalizations of (20.23) and (20.24). These equations use the following notation:

$$\mu_t^B = E[\mathbf{y}_t|\mathbf{O}_{t-1}] \qquad\qquad \Sigma_t^B = \mathrm{cov}[\mathbf{y}_t|\mathbf{O}_{t-1}] \qquad (20.34)$$

$$\mu_t^A = E[\mathbf{y}_t|\mathbf{O}_t] \qquad\qquad \Sigma_t^A = \mathrm{cov}[\mathbf{y}_t|\mathbf{O}_t]. \qquad (20.35)$$

Both equations involve the matrix

$$\mathbf{K}_t = \Sigma_t^B \mathbf{H}_t^T \left(\mathbf{H}_t \Sigma_t^B \mathbf{H}_t^T + \mathbf{R}_t\right)^{-1}, \qquad (20.36)$$

which is known as the *Kalman Gain Matrix*.

20.5 Sequential Processing of Observations

In atmospheric and oceanic data assimilation, the dimension of the observation $\mathbf{o}_t$ can exceed 1,000,000 and therefore the Kalman Filter Update Equations (20.32) and (20.33) would require inverting a million-by-million matrix. However, if the observational errors are uncorrelated, then Kalman Filter Update Equations can be solved *sequentially* without any matrix inversion! To explain how this can be done, note that the only term in (20.27) that depends on the current observation $\mathbf{o}_t$ is the likelihood function $p(\mathbf{o}_t|\mathbf{y}_t)$. If the observations are partitioned as $\mathbf{o}_t = \{\mathbf{o}_t^{(1)}, \mathbf{o}_t^{(2)}\}$, then the two parts can be modeled separately as

$$\mathbf{o}_t^{(1)} = \mathbf{H}_t^{(1)}\mathbf{y}_t + \mathbf{r}_t^{(1)} \qquad (20.37)$$

$$\mathbf{o}_t^{(2)} = \mathbf{H}_t^{(2)}\mathbf{y}_t + \mathbf{r}_t^{(2)}. \qquad (20.38)$$

Now assume that $\mathbf{r}_t^{(1)}$ and $\mathbf{r}_t^{(2)}$ are independent, which is equivalent to asserting that $\mathbf{o}_t^{(1)}$ and $\mathbf{o}_t^{(2)}$ are conditionally independent given $\mathbf{y}_t$. Under this assumption, the likelihood function can be factored in the form

$$p(\mathbf{o}_t|\mathbf{y}_t) = p(\mathbf{o}_t^{(1)}, \mathbf{o}_t^{(2)}|\mathbf{y}_t) = p(\mathbf{o}_t^{(2)}|\mathbf{y}_t)p(\mathbf{o}_t^{(1)}|\mathbf{y}_t). \qquad (20.39)$$

Therefore, the analysis equation (20.27) in the case of independent observations becomes

$$p(\mathbf{y}_t|\mathbf{o}_t\mathbf{O}_{t-1}) \propto p(\mathbf{o}_t^{(2)}|\mathbf{y}_t)p(\mathbf{o}_t^{(1)}|\mathbf{y}_t)p(\mathbf{y}_t|\mathbf{O}_{t-1}). \qquad (20.40)$$

However, the product of the last two terms is merely the analysis based on $\mathbf{o}_t^{(1)}$:

$$p(\mathbf{o}_t^{(1)}|\mathbf{y}_t)p(\mathbf{y}_t|\mathbf{O}_{t-1}) \propto p(\mathbf{y}_t|\mathbf{o}_t^{(1)}\mathbf{O}_{t-1}). \qquad (20.41)$$

This implies that the analysis based on all observations can be written as the likelihood of $\mathbf{o}_t^{(2)}$ times *the analysis based on* $\mathbf{o}_t^{(1)}$:

$$p(\mathbf{x}_t | \mathbf{o}_t^{(1)} \mathbf{o}_t^{(2)} \mathbf{O}_{t-1}) \quad \propto \quad p\left(\mathbf{o}_t^{(2)} | \mathbf{x}_t\right) \quad p(\mathbf{x}_t | \mathbf{o}_t^{(1)} \mathbf{O}_{t-1}).$$

analysis using $\mathbf{o}_t^{(1)}, \mathbf{o}_t^{(2)}$ likelihood of $\mathbf{o}_t^{(2)}$ analysis using $\mathbf{o}_t^{(1)}$ (20.42)

Thus, the assimilation of two conditionally independent observations is equivalent to the following two-step procedure: (1) first assimilate $\mathbf{o}_t^{(1)}$, which gives the *intermediate* analysis $p(\mathbf{x}_t | \mathbf{o}_t^{(1)} \mathbf{O}_{t-1})$, and then (2) assimilate $\mathbf{o}_t^{(2)}$ *using the intermediate analysis as "background"*. This argument can be applied recursively to derive the analysis *sequentially for any number of observations*. In the k'th stage of the sequential filter, the analysis based on $k - 1$ observations plays the role of the background distribution for the next assimilation. If each observation is conditionally independent of the others, then the observations can be assimilated *one-at-a-time*, and the terms $\mathbf{H}_t \hat{\mathbf{\Sigma}}^B \mathbf{H}_t$ and $\mathbf{R}_t$ reduce to scalars, in which case the Kalman filter update equations (20.32) and (20.33) do not require matrix inversions.

   Incidentally, if observational errors are correlated, they can be made uncorrelated by suitable transformation. This approach is beneficial only if $\mathbf{R}_t$ has a simple structure that makes this transformation easier to compute than the Kalman Filter Update Equations.

20.6 Multivariate Example

To illustrate some new properties of data assimilation that arise from generalizing from univariate to multivariate, suppose $\mathbf{y}_t$ is governed by the vector autoregressive model

$$\underset{M \times 1}{\mathbf{y}_t} = \underset{M \times M}{\mathbf{A}_t} \underset{M \times 1}{\mathbf{y}_{t-1}} + \underset{M \times 1}{\mathbf{w}_t}$$

(20.43)

where $\mathbf{A}_t$ is a time-dependent matrix known as the *dynamical operator* or *state transition model*, and $\mathbf{w}_t$ is Gaussian white noise with zero mean and time-dependent covariance matrix $\mathbf{Q}_t$. This model describes how $\mathbf{y}_t$ evolves in time. Initially, the state is described by the analysis distribution $p(\mathbf{y}_{t-1} | \mathbf{O}_{t-1})$. Because a linear combination of multivariate normal variables remains normal, only the mean and covariance matrix need be considered. Evaluating the mean of both sides of (20.43) *conditional* on past observations $\mathbf{O}_{t-1}$ yields

$$\mathbb{E}[\mathbf{y}_t | \mathbf{O}_{t-1}] = \mathbf{A}_t \mathbb{E}[\mathbf{y}_{t-1} | \mathbf{O}_{t-1}],$$

(20.44)

where we have used the fact that $\mathbf{w}_t$ has zero mean regardless of conditioning on $\mathbf{O}_{t-1}$. Similarly, evaluating the conditional covariance on both sides yields

$$\begin{aligned}
\text{cov}[\mathbf{y}_t | \mathbf{O}_{t-1}] &= \text{cov}[\mathbf{A}_t \mathbf{y}_{t-1} + \mathbf{w}_t | \mathbf{O}_{t-1}] \\
&= \text{cov}[\mathbf{A}_t \mathbf{y}_{t-1} | \mathbf{O}_{t-1}] + \text{cov}[\mathbf{w}_t | \mathbf{O}_{t-1}] \\
&= \mathbf{A}_t \, \text{cov}[\mathbf{y}_{t-1} | \mathbf{O}_{t-1}] \mathbf{A}_t^T + \text{cov}[\mathbf{w}_t]
\end{aligned}$$

(20.45)

where we have used the fact that $\mathbf{y}_{t-1}$ and $\mathbf{w}_t$ are independent. Using the notation (20.34) and (20.35), the background mean (20.44) and covariance matrix (20.45) become

$$\boldsymbol{\mu}_t^B = \mathbf{A}_t \boldsymbol{\mu}_{t-1}^A \quad \text{and} \quad \boldsymbol{\Sigma}_t^B = \mathbf{A}_t \boldsymbol{\Sigma}_{t-1}^A \mathbf{A}_t^T + \mathbf{Q}_t. \tag{20.46}$$

These equations show how the analysis distribution at $t-1$ is propagated to time t. When an observation $\mathbf{o}_t$ becomes available, the Kalman Filter Update Equations are solved to derive the analysis distribution for time t, which serves as the initial condition for the next assimilation cycle. This sequential procedure is the *Kalman Filter*.

To illustrate the complete forecast and data assimilation cycle, consider the choice

$$\mathbf{A}_t = \begin{pmatrix} 0.9 & 0.5 & 0.3 \\ 0.0 & 0.5 & 2.0 \\ 0.0 & 0.0 & 0.4 \end{pmatrix}, \quad \mathbf{Q}_t = \begin{pmatrix} 0 & 0 & 0 \\ 0 & 0 & 0 \\ 0 & 0 & 1 \end{pmatrix}, \quad \text{and} \quad \mathbf{y}_0 = \begin{pmatrix} 0 \\ 0 \\ 0 \end{pmatrix}. \tag{20.47}$$

This model has three variables. The specific choice for $\mathbf{Q}_t$ implies that random forcing excites only the third variable. Nevertheless, the third variable is coupled to the other two through $\mathbf{A}_t$, hence all three variables fluctuate as the model is stepped forward in time. The dynamical properties of the system are described by the eigenvectors and eigenvalues of $\mathbf{A}_t$. It can be shown that each eigenvector satisfies a *univariate* AR(1) model with AR-parameter given by the corresponding eigenvalue. Because $\mathbf{A}_t$ is triangular, the eigenvalues are simply the diagonal elements 0.9, 0.5, 0.4. The most persistent eigenvector of $\mathbf{A}_t$ is the one associated with AR-parameter $\phi_1 = 0.9$, which, according to (5.54), has time scale

$$N_1 = \frac{1 + \phi_1}{1 - \phi_1} = \frac{1 + 0.9}{1 - 0.9} = 19. \tag{20.48}$$

Other eigenvectors have shorter time scales. Thus, any autocorrelation function decays nearly to zero after 19 time steps, which defines the time scale of the dynamics.

Finally, we assume "observations" are generated using (20.28) with

$$\mathbf{H} = \begin{pmatrix} 1 & 0 & 0 \end{pmatrix} \quad \text{and} \quad R = 49. \tag{20.49}$$

We solve model (20.43) for a particular realization of the noise $\mathbf{w}_1, \mathbf{w}_2, \ldots$, which generates the realization $\mathbf{y}_1, \mathbf{y}_2, \ldots$, which may be identified with "the truth." However, we pretend we do not know the specific realization of the noise $\mathbf{w}_1, \mathbf{w}_2, \ldots$, instead we know only the statistics of the noise (that is, we know only $\mathbf{H}_t$ and $\mathbf{R}_t$). Our goal is to use data assimilation to estimate the truth $\mathbf{y}_1, \mathbf{y}_2, \ldots$ as closely as possible. The first element of the state vector, $y_1(t)$, is shown as the black curve

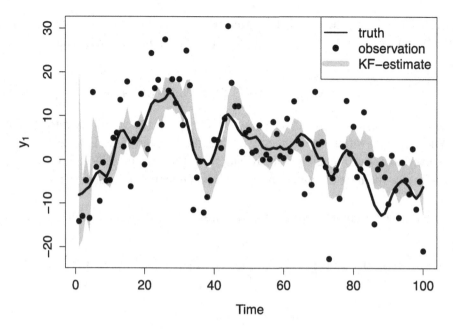

Figure 20.3 A particular realization of the first element $y_1(t)$ generated by the vector autoregressive model (20.43) (labeled "truth," solid curve) and corresponding observations (dots). Analysis values within one standard deviation of the mean analysis are indicated in grey shading.

in Figure 20.3. Note that the curve is relatively smooth and hence $y_1(t)$ is autocorrelated in time. This autocorrelation implies a dependence between $\mathbf{y}_t$ and $\mathbf{y}_{t-1}$, which is the basis for deriving a prior distribution based on past observations. This dependence is defined by the dynamical operator $\mathbf{A}_t$.

Next, we simulate "observations" by evaluating (20.28), which amounts to adding random perturbations to the true state by an amount consistent with $\mathbf{R}_t$. The choice (20.49) implies that only the first element of $\mathbf{y}_t$ is observed (with error). The particular realization of observations $\mathbf{o}_1, \mathbf{o}_2, \dots$ generated for this example are shown as dots in Figure 20.3.

Given observations $\mathbf{o}_1, \mathbf{o}_2, \dots$ and model (20.43), the only remaining information needed to perform data assimilation is an initial distribution. In this example, the choice of initial condition is not very important because its influence becomes negligible after a few dozen time steps, owing to the damping properties of $\mathbf{A}$. Here, we choose the climatological mean and covariance matrix of (20.43), which can be estimated by generating a long time series from the model and computing sample estimates of the mean and covariance matrix. The resulting sample mean defines $\boldsymbol{\mu}_0^A$ and the resulting covariance matrix defines $\boldsymbol{\Sigma}_0^A$. Then, the background distribution at $t = 1$ is obtained by evaluating (20.46),

$$\mu_1^B = \mathbf{A}\mu_0^A \quad \text{and} \quad \Sigma_1^B = \mathbf{A}\Sigma_0^A\mathbf{A}^T + \mathbf{Q}. \tag{20.50}$$

Next, the background distribution with this mean and covariance matrix is combined with the observation at $t = 1$ to compute an updated distribution from (20.32):

$$\mu_1^A = \mu_1^B + \Sigma_1^B\mathbf{H}^T\left(\mathbf{H}\Sigma_1^B\mathbf{H}^T + \mathbf{R}\right)^{-1}\left(\mathbf{o}_1 - \mathbf{H}\mu_1^B\right) \tag{20.51}$$

$$\Sigma_1^A = \Sigma_1^B - \Sigma_1^B\mathbf{H}^T\left(\mathbf{H}\Sigma_1^B\mathbf{H}^T + \mathbf{R}\right)^{-1}\mathbf{H}\Sigma_1^B. \tag{20.52}$$

The resulting analysis distribution at $t = 1$ then becomes the initial condition for the next background distribution evaluated from (20.46) for $t = 2$, and the process repeats.

The grey shading in Figure 20.3 shows the range of values within one standard deviation of the mean of the analysis distribution. This range is computed as

$$\left((\mu_t^A)_1 - \sqrt{(\Sigma_t^A)_{11}}, (\mu_t^A)_1 + \sqrt{(\Sigma_t^A)_{11}}\right). \tag{20.53}$$

Initially, the analysis uncertainty is relatively large, reflecting the fact that at the very first time step only the climatological distribution of the dynamical model is known. The spread of the analysis distribution contracts to a steady level after a dozen or so time steps. For the most part, the grey shading brackets the true state. One exception occurs around $t = 90$, but near this time the observations coincidentally happened to all fall repeatedly on one side of the truth. Such coincidences happen randomly and occur only as frequently as expected. Importantly, the analysis has much less uncertainty than the observations, as evidenced by the fact that the shading is generally closer to the solid curve than the dots.

It is interesting to consider the *second* element $y_2(t)$, shown in Figure 20.4. The analysis interval is much wider relative to the total variability than was found for the first variable. This fact implies that the second variable is more uncertain than the first, which is not surprising since the second variable is *not observed* (recall that only $y_1(t)$ + noise is observed). How can the filter estimate a quantity that is not even observed? In the present case, the filter can estimate an unobserved variable because *it is related to the first variable*, which *is* observed. Dynamically, the first element is *coupled* to the second element in the dynamical operator; this coupling is indicated by the nonzero value of $(\mathbf{A})_{12}$ in the dynamical operator $\mathbf{A}$. This coupling produces a background covariance that implies a correlation between the first and second element, which the filter exploits to estimate the second element.

20.7 Further Topics

The mathematical basis for data assimilation was laid down by Norbert Wiener and Andrey Kolmogorov in the 1940s as part of their work on *signal processing*. Wiener developed his approach during World War II to predict the position of bombers

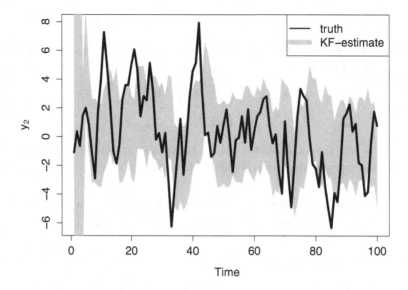

Figure 20.4 Same as Figure 20.3, except for the *second* element $y_2(t)$. No observations are shown because this variable is not observed.

from radar reflections. Unfortunately, this general framework was mathematically formidable – when Wiener's work was declassified after the war (Wiener, 1950), it was published in a yellow-bound report that engineers called the "yellow peril." Despite being motivated by engineering problems, Wiener's filter could be applied to very few practical problems, it was restricted to stationary processes, and it was difficult to generalize to multivariate systems. Later, Bode and Shannon (1950) showed that if some of the assumptions in Wiener's framework were relaxed, the framework could be generalized in other directions that were important for applications. In the 1960s, these ideas were extended considerably by Kalman (1960) and others, stimulated in part by navigation problems associated with satellites. In fact, the Kalman Filter was used to solve on-board navigation problems for the Apollo program.[4]

The Kalman Filter is a special case of a more general framework that includes the Kalman Smoother and the Kalman Predictor, which are defined as follows:

Filter	refers to estimation of $p(\mathbf{y}_t \mid \mathbf{O}_t)$
Smoother	refers to estimation of $p(\mathbf{y}_t \mid \mathbf{O}_{t+\tau})$
Predictor	refers to estimation of $p(\mathbf{y}_{t+\tau} \mid \mathbf{O}_t)$.

[4] To be fair, the history of the Kalman Filter is more complicated than this: Lauritzen (1981) notes that Thiele published a solution in 1880 that is identical to the Kalman filter in a simple case, and Simon (2006) notes that various versions of the filter were published independently of Kalman (1960), including one by Stratonovich in which the Kalman filter equations arise as the linear limit of a more general nonlinear filter. These historical details reflect a variation of Murphy's Law called Stigler's law of eponymy, which states that no scientific or mathematical discovery is named after its original discoverer.

These algorithms may be viewed as special cases of a still more general framework of optimal estimation for nonlinear or non-Gaussian systems. Furthermore, although the term "Kalman Filter" has been used in this chapter, this chapter presented only a tiny piece of the Kalman Filter. Specifically, this chapter presented only the estimation step of the Kalman Filter, which is nothing more than Bayes' theorem. The term Kalman Filter refers to a much more extensive framework that also accounts for the role of system dynamics. A dynamical system is typified by a system of ordinary differential equations, but under suitable assumptions (see Kalman, 1960) it can be reduced to the form

$$\mathbf{y}_t = \mathbf{A}_t \mathbf{y}_{t-1} + \mathbf{B}_t \mathbf{u}_t \tag{20.54}$$

$$\mathbf{o}_t = \mathbf{H}_t \mathbf{y}_t + \mathbf{r}_t. \tag{20.55}$$

where $\mathbf{u}_t$ is a K-dimensional vector and $\mathbf{B}_t$ is an $M \times K$ matrix. Because the Kalman Filter is applied sequentially, the estimation and dynamical steps may interact to produce instabilities. Thus, a basic question is whether the Kalman Filter is *stable*; that is, does the analysis remain bounded in some sense. Another question is whether a given state is *observable*. That is, can $\mathbf{y}_t$ be inferred *exactly* even in the ideal case of perfect observations (that is, $\mathbf{r}_t = \mathbf{0}$) and perfect knowledge of $\mathbf{u}_t$. Another question is whether the system is *controllable*. A system is controllable if there exists a sequence $\mathbf{u}_1, \ldots, \mathbf{u}_t$ that brings the state from $\mathbf{y}_0$ to $\mathbf{y}_t$ for any choice of $(\mathbf{y}_0, \mathbf{y}_t)$. These questions are related. Specifically, observability and controllability satisfy a *duality* principle, whereby observability in the system (20.54) and (20.55) implies controllability in the *dual* of the system, and vice versa. Also, observability and controllability imply stability. Importantly, a filter may be stable even if the dynamics are unstable. That is, the system state may grow without bound while estimation uncertainty remains bounded. A comprehensive review of these concepts and the Kalman Filter itself can be found in Maybeck (1982), Jazwinski (1970), and Simon (2006).

Stability of the Kalman Filter has been proven only for normal distributions and infinite precision. Unfortunately, numerical errors as a result of finite precision can lead to filter instability. Interestingly, the Kalman Filter can be written in different forms that are mathematically equivalent but have different numerical stabilities. One such form is the *Joseph form*. To derive the Joseph form, note that the mean analysis (20.32) can be written as

$$\boldsymbol{\mu}_t^A = \boldsymbol{\mu}_t^B + \mathbf{K}_t \left(\mathbf{o}_t - \mathbf{H}_t \boldsymbol{\mu}_t^B \right). \tag{20.56}$$

Let us assume, then, that a random variable from the analysis distribution $\mathbf{y}_A$ can be derived as a similar linear combination of observation $\mathbf{o}$ and background variable $\mathbf{y}_B$:

$$\mathbf{y}_A = \mathbf{y}_B + \mathbf{K}\,(\mathbf{o} - \mathbf{H}\mathbf{y}_B)\,, \tag{20.57}$$

where we suppress the time index t. The expectation of $\mathbf{y}_A$ with respect to the background distribution recovers (20.56). The corresponding covariance matrix is

$$\begin{aligned}
\text{cov}\left[\mathbf{y}_A - \boldsymbol{\mu}_A\right] &= \text{cov}\left[\mathbf{y}_B - \boldsymbol{\mu}_B - \mathbf{K}\,(\mathbf{o} - \mathbf{H}\boldsymbol{\mu}_B)\right] \\
&= \text{cov}\left[\mathbf{y}_B - \boldsymbol{\mu}_B - \mathbf{K}\,(\mathbf{H}\mathbf{y}_B + \mathbf{r} - \mathbf{H}\boldsymbol{\mu}_B)\right] \\
&= \text{cov}\left[(\mathbf{I} - \mathbf{K}\mathbf{H})\,(\mathbf{y}_B - \boldsymbol{\mu}_B) - \mathbf{K}\mathbf{r}\right] \\
\boldsymbol{\Sigma}_A &= (\mathbf{I} - \mathbf{K}\mathbf{H})\,\boldsymbol{\Sigma}_B\,(\mathbf{I} - \mathbf{K}\mathbf{H})^T + \mathbf{K}\mathbf{R}\mathbf{K}^T.
\end{aligned} \tag{20.58}$$

Equation (20.58) is the Joseph form of the analysis covariance matrix. This form is the sum of two symmetric, positive semi-definite matrices. As a result, the sum is guaranteed to be positive semi-definite, whereas the update equation (20.33) has no such guarantee *numerically*. In addition, the Joseph form has better matrix conditioning properties and is less sensitive to small errors in the Kalman gain matrix than (20.33) (Maybeck, 1982). Although the Joseph form can be derived from (20.33) using standard matrix identities, the derivation of (20.58) holds for *any* $\mathbf{K}$. Hence, the Joseph form provides a basis for *sub-optimal* data assimilation, where the matrix $\mathbf{K}$ can differ from the Kalman Gain.

The Kalman Filter Update Equations give optimal estimates when the system dynamics are linear and Gaussian. Unfortunately, atmospheric and oceanic data assimilation involves nonlinear dynamics. If the nonlinearity is not too strong and the dimension of the system is relatively small, then an approach that may work is the *extended Kalman Filter* (EKF). In this approach, the background covariances are derived from a *tangent linear model*, which is a linearization of the dynamical model about a nonlinear solution. This approximation is reasonable when the tangent linear model is integrated over sufficiently short times. Nevertheless, it is not practical for atmospheric and oceanic data assimilation.

The Kalman Filter Update Equations involve matrix inversions and therefore are not computationally solvable for problems characterized by a million or more variables, which is typical in atmospheric and oceanic data assimilation. In practice, the equations are simplified in certain ways and then solved using *variational* methods. To understand the variational approach, note that the Gaussian distribution has the property that the mean, median, and mode are all identical to each other. This fact implies that the mean of the analysis distribution (20.27) is also the maximum of the distribution, which can be found by taking the derivative of (20.27) with respect to $\mathbf{y}_t$ and setting the result to zero. In this approach, it is more convenient to consider -2 times the log of the distribution, which changes maxima into minima, but does not affect the location of the extremum because $-2 \log z$ is a monotonic function of z. Accordingly, we have

$$-2\log p(\mathbf{y}_t|\mathbf{o}_t\mathbf{O}_{t-1}) \propto \log{(2\pi)}^K |\mathbf{R}| + \log{(2\pi)}^M |\boldsymbol{\Sigma}_B|$$
$$+ (\mathbf{o} - \mathbf{Hy})^T \mathbf{R}^{-1} (\mathbf{o} - \mathbf{Hy}) + (\mathbf{y} - \boldsymbol{\mu}_B)^T \boldsymbol{\Sigma}_B^{-1} (\mathbf{y} - \boldsymbol{\mu}_B), \quad (20.59)$$

where the notation has been simplified by suppressing the t subscript and indicating background quantities with a B subscript. Taking the derivative of this function with respect to $\mathbf{y}$ and setting the result to zero gives

$$-2\mathbf{H}^T\mathbf{R}^{-1} (\mathbf{o} - \mathbf{Hy}) + 2\boldsymbol{\Sigma}_B^{-1} (\mathbf{y} - \boldsymbol{\mu}_B) = 0, \quad (20.60)$$

which can be solved for $\mathbf{y}$ as

$$\mathbf{y}^* = \left(\mathbf{H}^T\mathbf{R}^{-1}\mathbf{H} + \boldsymbol{\Sigma}_B^{-1}\right)^{-1} \left(\mathbf{H}^T\mathbf{R}^{-1}\mathbf{o} + \boldsymbol{\Sigma}_B^{-1}\boldsymbol{\mu}_B\right). \quad (20.61)$$

This equation is the multivariate generalization of (20.17) and can be manipulated to give the equivalent expression (20.32) (e.g., by using the Sherman–Morrison–Woodbury formula (A.10)). In addition, the second derivative of the function is

$$-2\frac{\partial^2 \log p(\mathbf{y}_t|\mathbf{o}_t\mathbf{O}_{t-1})}{\partial \mathbf{y}_t \partial \mathbf{y}_t} = \mathbf{H}^T\mathbf{R}^{-1}\mathbf{H} + \boldsymbol{\Sigma}_B^{-1}. \quad (20.62)$$

This expression can be shown to equal the *inverse* of the analysis covariance matrix (20.33) (again, by using the Sherman–Morrison–Woodbury formula (A.10)). This second partial derivative is known as the *Hessian*.

These considerations demonstrate that maximizing (20.59) would yield the same mean analysis as the Kalman Filter update equations. Although the two approaches are mathematically equivalent, they are not *computationally* equivalent. In particular, the minimum of (20.59) can be found by numerical minimization methods. Importantly, the minimization algorithm needs only the *product* of the inverse matrices with certain vectors; it does not need an explicit matrix inverse. There are numerical tricks for computing such products that avoid explicit computation of matrix inverses. This procedure of determining the analysis by minimizing a cost function is known as *variational data assimilation*.

Variational data assimilation has an interesting interpretation in terms of the method of least squares. Since the covariances $\mathbf{R}$ and $\boldsymbol{\Sigma}_B$ are fixed at each analysis update, minimizing (20.59) is equivalent to minimizing the so-called cost function

$$\text{cost function} = (\mathbf{o} - \mathbf{Hy})^T \mathbf{R}^{-1} (\mathbf{o} - \mathbf{Hy}) + (\mathbf{y} - \boldsymbol{\mu}_B)^T \boldsymbol{\Sigma}_B^{-1} (\mathbf{y} - \boldsymbol{\mu}_B). \quad (20.63)$$

The first term in this cost function measures the distance between the interpolated state $\mathbf{Hy}$ and observation $\mathbf{o}$ (e.g., the expression is never negative, because $\mathbf{R}$ is positive definite). The matrix $\mathbf{R}$ accounts for correlations between observations and differences in accuracy. In contrast, the second term in the cost function measures the distance between the state $\mathbf{y}$ and the mean of the prior distribution $\boldsymbol{\mu}_B$. This term can be interpreted as a "penalty function" that increases the cost function as the state moves away from μ_B. Thus, the state that minimizes the sum of these two functions, that is, the mean analysis μ_A, represents a balance between fitting the observations while not deviating too far from the prior distribution. This interpretation provides a framework for introducing additional constraints in data assimilation. For instance, certain implementations of data assimilation add a term to the cost function to constrain the analysis to satisfy geostrophic and hydrostatic balance.

Most surface observations, such as those by thermometers and wind vanes, have uncorrelated errors and hence are characterized by a diagonal $\mathbf{R}$. Other types of observations, such as from weather balloons, have errors that are correlated vertically but not horizontally and thus are characterized by block diagonal $\mathbf{R}$. Satellite observations tend to have locally correlated errors. In all these cases, the matrix $\mathbf{R}$ is sparse and thus the first term in (20.63) involving the inverse of $\mathbf{R}$ can be evaluated relatively easily. On the other hand, the prior covariance matrix Σ_B can be quite complex. Often, reasonable results can be obtained by assuming that the background covariance is constant and flow-independent. Typically, this covariance matrix is parameterized in terms of a small number of coefficients. One approach is to assume the forecast errors are invariant to spatial translation (that is, homogeneous) and rotation (that is, isotropic), in which case the covariances depend only on the separation between the points. Such covariance matrices can be transformed to diagonal form using spectral methods, in which case the inverse is trivial to compute. This approach is the basis of some types of *three dimensional variational assimilation* (3DVAR).

Another generalization is *four-dimensional variational assimilation* (4DVAR). In this generalization, a term is added to the cost function (20.63) that constrains the analysis such that initializing a tangent linear model with this analysis produces a forecast that is close to observations at a certain number of time steps.

As can be seen from this description, various data assimilation schemes have been developed or utilized in the meteorological and oceanographic literature, including

Optimal Interpolation   Ensemble Kalman Filter
3 - Dimensional Variational Assimilation   Ensemble Square Root Filter
4 - Dimensional Variational Assimilation   Ensemble Transform Kalman Filter
Ensemble Adjustment Kalman Filter   Extended Kalman Filter.

In fact, each of these methods assumes Gaussian distributions and linear dynamics and therefore only one optimal estimate exists. The reason different assimilation schemes exist is that they each exploit different approximations or numerical methods for the solution.

The Kalman Filter Update Equations can also be expressed equivalently in terms of the *inverse* of the covariance matrix. This form is known as the *information* filter because of its relation to the Fisher information matrix. The information filter is particularly attractive for *initializing* the assimilation because virtually no background information exists at the initial time. In such cases, the uncertainty is "large," which can be modeled by a background covariance matrix with very large eigenvalues. If $\mathbf{U}$ and $\boldsymbol{\Lambda}$ are the eigenvector and eigenvalue matrices of the background covariance matrix, then

$$\boldsymbol{\Sigma}^B = \mathbf{U}\boldsymbol{\Lambda}\mathbf{U}^T \quad \Rightarrow \quad \left(\boldsymbol{\Sigma}^B\right)^{-1} = \mathbf{U}\boldsymbol{\Lambda}^{-1}\mathbf{U}^T. \tag{20.64}$$

Thus, as an eigenvalue approaches infinity, the inverse eigenvalue approaches zero. As a result, the inverse covariance matrix remains bounded in the limit of infinite uncertainty.

An interesting application of the Kalman filter update equations is to estimate not only the state but also parameters in the dynamical model. This methodology for linear state space models is described in Harvey (1989). In certain cases, this approach is equivalent to simply augmenting the state vector $\mathbf{y}_t$ by the vector of model parameters $\boldsymbol{\beta}_t$ (DelSole, 2011). This approach requires specifying an evolution model for the parameters. An empirical approach that seems to work in practice is to use the so-called persistence model

$$\boldsymbol{\beta}_t^B = \boldsymbol{\beta}_{t-1}^A. \tag{20.65}$$

This persistence model appears to work well for model parameters that enter additively to the tendency equations, but can be problematic if the parameters are multiplied by state variables in the tendency equation. The reason for this difficulty is that multiplicative parameters change the stability properties of the dynamical system, potentially causing the model to become unstable. This instability often can be avoided by using a temporally smoothed version of the persistence model

$$\boldsymbol{\beta}_t^B = (1 - \alpha)\boldsymbol{\beta}_{t-1}^A + \alpha\boldsymbol{\beta}_{t-1}^B, \tag{20.66}$$

where α is a smoothing parameter (Yang and DelSole, 2009). A related idea is to use maximum likelihood methods to estimate parameters in the covariance matrix (Dee, 1995) or parameters for correcting biases in the analysis (Dee and da Silva, 1998).

20.8 Conceptual Questions

1 What do the following terms refer to: background distribution, prior distribution, analysis distribution, likelihood function?
2 What can you say about the relative sizes of the background error, observational error, and analysis error?
3 What is the generalized interpolation operator? What physically does it represent?
4 What is conditional independence? Why is it important in data assimilation?

5 What is variational data assimilation? What is the difference between 3DVAR and 4DVAR?

6 The measurement error R_t was assumed to be independent of Y_t (that is, flow independent). How realistic is this assumption?

7 When is it easy to assimilate observations sequentially? Why is this fact important?

8 How can data assimilation estimate a quantity that is not observed? Will data assimilation always estimate a quantity that is not observed?

9 Explain why the covariances (20.33) don't depend on the observations.

21

Ensemble Square Root Filters

... the Kalman filter could in theory be used for the optimal estimation of atmospheric states. In practice, however, it suffers from two serious drawbacks...[1]

Houtekamer and Mitchell (2005)

The previous chapter discussed *data assimilation*, which is a procedure for combining forecasts and noisy observations to derive an improved estimate of the state of a system. The procedure assumes the variables have known Gaussian distributions, but in atmospheric and oceanic data assimilation the distributions are neither Gaussian nor known. In fact, the governing dynamical equations are nonlinear and thereby generate non-Gaussian distributions even if they start with Gaussian distributions. Also, the number of state variables in the dynamical model (e.g., velocity and temperature at every grid point) often exceeds 10^5, and dealing with the corresponding $10^5 \times 10^5$ error covariance matrix is computationally challenging. Various methods for performing data assimilation with high-dimensional nonlinear models have been proposed in the literature. This chapter discusses one class of algorithms known as *Ensemble Square Root Filters*. The basic idea is to use a collection of forecasts (called an *ensemble*) to estimate the statistics of the background distribution. In addition, observational information is incorporated by *adjusting individual ensemble members* (that is, forecasts) rather than computing an entire distribution. The ensemble members may be adjusted in a variety of ways to give the same collective statistics. Despite their statistical equivalence, the filters are not computationally equivalent: certain calculations are easier to do with one filter than with another. This chapter discusses three standard filters: the Ensemble Transform Kalman Filter (ETKF), the Ensemble Square Root Filter (EnSRF), and the Ensemble Adjustment Kalman Filter (EAKF). However, when applied to chaotic dynamical models, ensemble filters often experience *filter divergence*, in which the analysis no longer tracks the truth. Filter divergence can be identified through inconsistencies between the assimilated data and their expected statistical properties. A variety of modifications have been proposed to avoid filter divergence. This chapter discusses standard modifcations known as *covariance inflation* and *covariance localization*.

---

[1] Reprinted by permission from John Wiley & Sons, *Quarterly Journal of the Royal Meteorological Society*, Ensemble Kalman Filtering, P. L. Houtekamer and H. L. Mitchell, 2005.

21.1 The Problem

As discussed in the previous chapter, data assimilation is a method for combining observations $\mathbf{o}_t$ at time t with a prior estimate of the state $\mathbf{y}_t$ at time t to derive an improved estimate of $\mathbf{y}_t$. In atmospheric and oceanic data assimilation, the dimension of these vectors, M, often exceeds 10^5, which is too large to handle by matrix computations. This chapter discusses a strategy for performing data assimilation with high-dimensional vectors.

To review data assimilation, let $\mathbf{O}_t$ represent observations available at time t and earlier: that is, let $\mathbf{O}_t = \{\mathbf{o}_t, \mathbf{o}_{t-1}, \ldots\}$. Then, the distribution of $\mathbf{y}_t$ based on observations *before* time t is described by a *background distribution* $p(\mathbf{y}_t|\mathbf{O}_{t-1})$. The background distribution is derived from a dynamical model that propagates the initial distribution from time $t-1$ to time t. The dynamical model is assumed to be perfect, in the sense that it correctly describes the evolution of initial uncertainties. For Gaussian distributions, the background distribution is specified completely by its mean and covariance matrix

$$\mu_t^B = E[\mathbf{y}_t|\mathbf{O}_{t-1}] \quad \text{and} \quad \Sigma_t^B = \text{cov}[\mathbf{y}_t|\mathbf{O}_{t-1}]. \tag{21.1}$$

At time t, new observations $\mathbf{o}_t$ become available and the question arises as to how to combine this new information with the background estimate. To answer this question, the precise relation between observations and the true state must be specified. As discussed in Chapter 20, the Kalman Filter assumes this relation is of the form

$$\mathbf{o}_t = \mathbf{H}_t \mathbf{y}_t + \mathbf{r}_t, \tag{21.2}$$

where $\mathbf{r}_t$ is Gaussian white noise with zero mean and covariance matrix $\mathbf{R}_t$, and $\mathbf{H}_t$ is the *generalized interpolation operator*. We then desire the updated distribution, known as the *analysis distribution*, characterized by the mean and covariance matrix

$$\mu_t^A = E[\mathbf{y}_t|\mathbf{O}_t] \quad \text{and} \quad \Sigma_t^A = \text{cov}[\mathbf{y}_t|\mathbf{O}_t]. \tag{21.3}$$

The solution is given by the *Kalman Filter Update Equations*:

$$\mu_t^A = \mu_t^B + \Sigma_t^B \mathbf{H}_t^T \left(\mathbf{H}_t \Sigma_t^B \mathbf{H}_t^T + \mathbf{R}_t \right)^{-1} \left(\mathbf{o}_t - \mathbf{H}_t \mu_t^B \right) \tag{21.4}$$

$$\Sigma_t^A = \Sigma_t^B - \Sigma_t^B \mathbf{H}_t^T \left(\mathbf{H}_t \Sigma_t^B \mathbf{H}_t^T + \mathbf{R}_t \right)^{-1} \mathbf{H}_t \Sigma_t^B \tag{21.5}$$

Among the assumptions listed earlier, a particularly unrealistic assumption is that the background covariance matrix Σ_t^B is *known*. In atmospheric and oceanic data assimilation, the dimension M of the state vector $\mathbf{y}_t$ is large because it contains quantities such as temperature and velocity at every grid point on the globe, leading to covariance matrices with dimensions of 10^5 or more. Dealing with such large-dimensional matrices is computationally challenging. *Ensemble filtering* is a strategy for performing data assimilation in a high-dimensional space based on two key ideas: (1) approximate the background distribution with a finite number of forecasts from the dynamical model, and (2) linearly transform these forecasts into a sample from the analysis distribution.

Note that each term in the Kalman filter update equations (21.4) and (21.5) has a subscript t. The focus in this chapter is to solve these equations for a single time step t. Therefore, the time step is fixed and the subscript is redundant. Accordingly, we drop the subscript t with the understanding that all terms will be evaluated at time t.

To explain ensemble filtering, consider a sample from the initial condition distribution. From each initial condition, the dynamical model can generate a forecast of the state. The collection of forecasts is a *forecast ensemble* and constitutes a sample from the background distribution. The forecast ensemble can then be used to estimate the mean and covariance matrix of the background distribution. Let E be the ensemble size, and let $\mathbf{e}_1, \mathbf{e}_2, \ldots, \mathbf{e}_E$ be M-dimensional vectors representing the forecast ensemble. In atmospheric and oceanic data assimilation, M is the number of prognostic variables (e.g., temperature and velocity at every grid point) and often exceeds 10^5. The sample mean, known as the *ensemble mean*, is

$$\hat{\mu} = \frac{1}{E} \sum_{i=1}^{E} \mathbf{e}_i, \tag{21.6}$$

and an unbiased estimate of the sample covariance matrix is

$$\hat{\Sigma} = \frac{1}{E-1} \sum_{i=1}^{E} \left(\mathbf{e}_i - \hat{\mu} \right) \left(\mathbf{e}_i - \hat{\mu} \right)^T, \tag{21.7}$$

where $\mathbf{e}_i - \hat{\mu}$ is called an *anomaly*. A more compact notation is to define the $M \times E$ matrix

$$\mathbf{X} = \frac{1}{\sqrt{E-1}} \left[\mathbf{e}_1 - \hat{\mu} \quad \mathbf{e}_2 - \hat{\mu} \quad \cdots \quad \mathbf{e}_E - \hat{\mu} \right], \tag{21.8}$$

in which case

$$\hat{\Sigma} = \mathbf{X}\mathbf{X}^T. \tag{21.9}$$

The matrix $\mathbf{X}$ is known as a *matrix square root* of $\hat{\boldsymbol{\Sigma}}$. We use $\mathbf{X}^B$ and $\mathbf{X}^A$ to denote ensembles drawn from the background and analysis distribution, respectively. Similarly, the respective ensemble means are $\hat{\boldsymbol{\mu}}^B$ and $\hat{\boldsymbol{\mu}}^A$, and respective covariance matrices are $\hat{\boldsymbol{\Sigma}}^B$ and $\hat{\boldsymbol{\Sigma}}^A$. The caret symbol $\hat{()}$ denotes a sample quantity, to distinguish it from population quantities.

A natural idea is to simply substitute the sample mean $\hat{\boldsymbol{\mu}}^B$ and sample covariance matrix $\hat{\boldsymbol{\Sigma}}^B$ into the Kalman Filter Update Equations (21.4) and (21.5). After this substitution, the mean $\hat{\boldsymbol{\mu}}^A$ and covariance matrix $\hat{\boldsymbol{\Sigma}}^A$ of the analysis distribution is obtained. Since the distribution is Gaussian, the analysis distribution is given by

$$p_A(\mathbf{x}) = \mathcal{N}_M(\hat{\boldsymbol{\mu}}^A, \hat{\boldsymbol{\Sigma}}^A). \tag{21.10}$$

In principle, one could generate a new analysis ensemble by drawing random samples from $p_A(\mathbf{x})$. However, the typical ensemble size E is 10–50. Therefore, the rank of $\hat{\boldsymbol{\Sigma}}^B$ is about this size. It turns out that the analysis covariance matrix derived from (21.5) *also has the same rank and the same column space* as $\hat{\boldsymbol{\Sigma}}^B$. This follows from the fact that the analysis covariance matrix (21.5) can be factored in the form

$$\hat{\boldsymbol{\Sigma}}^A = \hat{\boldsymbol{\Sigma}}^B \mathbf{W}, \tag{21.11}$$

where $\mathbf{W}$ is a suitable matrix. A standard theorem in linear algebra states that the column space of $\hat{\boldsymbol{\Sigma}}^B \mathbf{W}$ is contained in the column space of $\hat{\boldsymbol{\Sigma}}^B$ (Harville, 1997, corollary 4.2.3). Therefore, $\hat{\boldsymbol{\Sigma}}^A$ and $\hat{\boldsymbol{\Sigma}}^B$ have the same rank and the same column space. Any matrix of rank E can be represented *exactly* by a set of E vectors. When $E \ll M$, it is computationally advantageous to represent the matrix using these E vectors rather than by the $M \times M$ covariance matrix. Furthermore, the E vectors lie in the same column space as $\mathbf{X}^B$, hence the analysis ensemble must be a linear combination of $\mathbf{X}^B$. Therefore,

$$\mathbf{X}^A = \mathbf{X}^B \mathbf{C}, \tag{21.12}$$

for some matrix $\mathbf{C}$. This equation is tantamount to creating an analysis ensemble by *linearly combining* the forecast ensemble. Using somewhat different arguments, it can be shown that the analysis ensemble could be obtained as a *linear combination* of variables,

$$\mathbf{X}^A = \mathbf{C}' \mathbf{X}^B, \tag{21.13}$$

where $\mathbf{C}'$ is a matrix of rank E. In either case, the analysis is obtained by *adjusting* the forecast ensemble. This procedure is illustrated in Figure 21.1. Once $\mathbf{X}^A$ is defined through (21.12) or (21.13), the analysis covariance matrix is computed as

$$\hat{\boldsymbol{\Sigma}}^A = \mathbf{X}^A \mathbf{X}^{AT}. \tag{21.14}$$

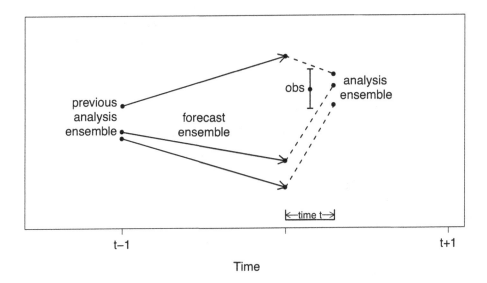

Figure 21.1 Schematic illustrating the data assimilation process based on ensemble filtering (ensemble size equals three). At time t, three distributions become available the background distribution, the observations, and the corresponding analysis distribution derived from them. The time dimension is stretched at time t to avoid putting all this information on the same vertical line.

The goal of ensemble filtering is to find the matrix $\mathbf{C}$ or $\mathbf{C}'$ such that this analysis covariance matrix (21.14) coincides *exactly* with that by the Kalman Filter Update Equations.

The matrix $\mathbf{X}^A$ is the *matrix square root* of $\hat{\mathbf{\Sigma}}^A$. Importantly, matrix square roots are not unique. For instance, in one dimension, the square root of x is $+\sqrt{x}$ and $-\sqrt{x}$. For matrices, if $\mathbf{X}^A$ is a square root of the analysis covariance matrix, then so is

$$\mathbf{X}'^A = \mathbf{X}^A \mathbf{V}^T, \tag{21.15}$$

where $\mathbf{V}$ is *any matrix with orthogonal column vectors* (that is, any matrix that satisfies $\mathbf{V}^T\mathbf{V} = \mathbf{I}$). This fact can be verified by direct substitution. Note that in one-dimension, $\mathbf{V}$ reduces to ± 1, recovering the case of real numbers.

Note that $\mathbf{X}^A$ contains only the anomaly with respect to the ensemble mean. A dynamical model must be initialized by the *full* field, which requires adding the ensemble mean back into $\mathbf{X}^A$. For future reference, the initial conditions are derived as

$$\mathbf{X}^A_{\text{initial condition}} = \sqrt{E-1}\,\mathbf{X}^A + \hat{\mu}^A \mathbf{j}^T_E, \tag{21.16}$$

where $\mathbf{j}_E$ is an E-dimensional vector of all ones.

In the following sections, we review three specific forms of square root filters.

21.1.1 Ensemble Transform Kalman Filter (ETKF)

A direct approach to deriving a square root filter is to simply substitute the appropriate square root decomposition of covariance matrices in (21.5) and manipulate:

$$\hat{\Sigma}^A = \hat{\Sigma}^B - \hat{\Sigma}^B \mathbf{H}^T \left(\mathbf{H}\hat{\Sigma}^B \mathbf{H}^T + \mathbf{R} \right)^{-1} \mathbf{H}\hat{\Sigma}^B$$

$$= \mathbf{X}^B \mathbf{X}^{BT} - \mathbf{X}^B \mathbf{X}^{BT} \mathbf{H}^T \left(\mathbf{H}\mathbf{X}^B \mathbf{X}^{BT} \mathbf{H}^T + \mathbf{R} \right)^{-1} \mathbf{H}\mathbf{X}^B \mathbf{X}^{BT}$$

$$= \mathbf{X}^B \left(\mathbf{I} - \mathbf{X}^{BT} \mathbf{H}^T \left(\mathbf{H}\mathbf{X}^B \mathbf{X}^{BT} \mathbf{H}^T + \mathbf{R} \right)^{-1} \mathbf{H}\mathbf{X}^B \right) \mathbf{X}^{BT}$$

$$\mathbf{X}^A \mathbf{X}^{AT} = \mathbf{X}^B \mathbf{D}\mathbf{X}^{BT} \tag{21.17}$$

where

$$\mathbf{D} = \mathbf{I} - \mathbf{X}^{BT} \mathbf{H}^T \left(\mathbf{H}\mathbf{X}^B \mathbf{X}^{BT} \mathbf{H}^T + \mathbf{R} \right)^{-1} \mathbf{H}\mathbf{X}^B. \tag{21.18}$$

If a square root decomposition of $\mathbf{D}$ is denoted

$$\mathbf{D} = \mathbf{C}\mathbf{C}^T, \tag{21.19}$$

then substituting this in (21.17) would yield a linear transformation of the form (21.12). However, not every square root matrix $\mathbf{C}$ is admissiable. Recall that the columns of $\mathbf{X}^B$ and $\mathbf{X}^A$ are ensemble members *centered about their respective ensemble means*. Therefore, the sum of each row should vanish. That is,

$$\mathbf{X}^B \mathbf{j}_E = \mathbf{0}. \tag{21.20}$$

Similarly, the sum of each row of $\mathbf{X}^A$ should vanish:

$$\mathbf{X}^A \mathbf{j}_E = \mathbf{X}^B \mathbf{C} \mathbf{j}_E = \mathbf{0}. \tag{21.21}$$

It is easy to see that if

$$\mathbf{C}\mathbf{j}_E = \alpha \mathbf{j}_E, \tag{21.22}$$

where α is a scalar, then (21.21) would follow from (21.20). Thus, among all possible square root matrices $\mathbf{C}$, we want to choose a particular class that satisfies (21.22), which are known as *mean preserving* (Sakov and Oke, 2008).

The matrix $\mathbf{D}$ involves the inverse of a $K \times K$ matrix, where K is the number of observations and is often very large. However, the Sherman–Morrison–Woodbury formula allows this inverse to be computed in an alternative way, namely as

$$\mathbf{D} = \left(\mathbf{I} + \mathbf{X}^{BT} \mathbf{H}^T \mathbf{R}^{-1} \mathbf{H}\mathbf{X}^B \right)^{-1}. \tag{21.23}$$

A similar trick yields the following equation for the mean analysis:

$$\mu^A = \mu^B + \mathbf{X}^B \mathbf{D}\mathbf{X}^{BT} \mathbf{H}^T \mathbf{R}^{-1} \left(\mathbf{0}_t - \mathbf{H}_t \mu_t^B \right). \tag{21.24}$$

Although $\mathbf{R}$ is also $K \times K$, its inverse can often be computed cheaply because in practice it is sparse or has some simple structure. Because the matrix $\mathbf{D}$ is an $E \times E$ matrix, it is small enough to allow computation of the eigenvector decomposition

$$\mathbf{D} = \mathbf{U}\boldsymbol{\Lambda}\mathbf{U}^T, \tag{21.25}$$

where $\mathbf{U}$ is an orthogonal matrix and $\boldsymbol{\Lambda}$ is a diagonal matrix is positive diagonal elements. Then the transformation matrix $\mathbf{C}$ that yields the matrix square root of $\mathbf{D}$ is

$$\mathbf{C} = \mathbf{U}\boldsymbol{\Lambda}^{1/2}\mathbf{V}^T, \tag{21.26}$$

where $\mathbf{V}$ is a matrix of suitable dimension such that $\mathbf{V}^T\mathbf{V} = \mathbf{I}$. The choice $\mathbf{V} = \mathbf{U}$ produces a *symmetric square root*. Happily, symmetric square roots are mean preserving. To see this, note that (21.18) and (21.20) imply $\mathbf{Dj} = \mathbf{j}$, and substituting the eigenvector decomposition (21.25) leads to the identity

$$\boldsymbol{\Lambda}\mathbf{j}' = \mathbf{j}', \tag{21.27}$$

where $\mathbf{j}' = \mathbf{U}^T\mathbf{j}$. Because the eigenvalues generally differ from one, this identity can be satisfied only if $\mathbf{j}' = \mathbf{0}$, which implies $\mathbf{U}^T\mathbf{j} = \mathbf{0}$, hence the choice $\mathbf{V} = \mathbf{U}$ in (21.26) yields a mean-preserving square root matrix. Also, Ott et al. (2004) show that the symmetric square root also has the special property that it minimizes

$$\left(\mathbf{x}^A - \mathbf{x}^B\right)^T (\hat{\boldsymbol{\Sigma}}^A)^{-1} \left(\mathbf{x}^A - \mathbf{x}^B\right), \tag{21.28}$$

in the space spanned by the ensemble members. The fact that the symmetric square root minimizes this difference-measure means that the transformation produces the smallest possible adjustment from $\mathbf{x}^B$ to $\mathbf{x}^A$. This filter is the *Ensemble Transform Kalman Filter* (ETKF) and was introduced by Bishop et al. (2001).

21.1.2 Ensemble Square Root Filter (EnSRF)

An alternative approach to constructing a square root matrix is to define the analysis as

$$\mathbf{X}^A = \mathbf{C}'\mathbf{X}^B, \tag{21.29}$$

where $\mathbf{C}'$ is determined by the condition

$$\mathbf{X}^A\mathbf{X}^{AT} = \mathbf{C}'\hat{\boldsymbol{\Sigma}}^B\mathbf{C}'^T = \hat{\boldsymbol{\Sigma}}^A. \tag{21.30}$$

Here, $\mathbf{C}'$ multiplies the forecast ensemble on the *left*, which means that the analysis is a linear combination of variables. One advantage of this approach is that the analysis ensemble is mean-preserving for any $\mathbf{C}'$ (because $\mathbf{C}'\mathbf{X}^B\mathbf{j} = \mathbf{0}$ as a result of (21.20)).

To derive $\mathbf{C}'$, we start with the Kalman filter update equation (21.5), but substitute sample estimates for the covariances:

$$\hat{\boldsymbol{\Sigma}}^A = \mathbf{C}'\hat{\boldsymbol{\Sigma}}^B\mathbf{C}'^T = \hat{\boldsymbol{\Sigma}}^B - \hat{\boldsymbol{\Sigma}}^B\mathbf{H}^T\left(\mathbf{H}\hat{\boldsymbol{\Sigma}}^B\mathbf{H}^T + \mathbf{R}\right)^{-1}\mathbf{H}\hat{\boldsymbol{\Sigma}}^B. \tag{21.31}$$

The general solution for $\mathbf{C}'$ can be derived using sophisticated matrix methods. A more intuitive solution can be derived for the case of a single observation. Any set of observations can be transformed into an uncorrelated set, which in turn can be assimilated one at a time (see Section 20.5). In this case, the terms $\mathbf{H}\hat{\boldsymbol{\Sigma}}^B\mathbf{H}^T$ and $\mathbf{R}$ reduce to scalars, implying that we seek a matrix $\mathbf{C}'$ such that

$$\mathbf{C}'\hat{\boldsymbol{\Sigma}}^B\mathbf{C}'^T = \hat{\boldsymbol{\Sigma}}^B - \beta\hat{\boldsymbol{\Sigma}}^B\mathbf{H}^T\mathbf{H}\hat{\boldsymbol{\Sigma}}^B, \tag{21.32}$$

where β is the scalar

$$\beta = 1/(\mathbf{H}\hat{\boldsymbol{\Sigma}}^B\mathbf{H}^T + \mathbf{R}). \tag{21.33}$$

With considerable foresight, consider a solution of the form

$$\mathbf{C}' = \mathbf{I} - \gamma\beta\hat{\boldsymbol{\Sigma}}^B\mathbf{H}^T\mathbf{H}, \tag{21.34}$$

where γ is a parameter to be determined. Substituting this solution into (21.32) gives

$$\hat{\boldsymbol{\Sigma}}^B - 2\gamma\beta\hat{\boldsymbol{\Sigma}}^B\mathbf{H}^T\mathbf{H}\hat{\boldsymbol{\Sigma}}^B + \gamma^2\beta^2\hat{\boldsymbol{\Sigma}}^B\mathbf{H}^T\left(\mathbf{H}\hat{\boldsymbol{\Sigma}}^B\mathbf{H}^T\right)\mathbf{H}\hat{\boldsymbol{\Sigma}}^B = \hat{\boldsymbol{\Sigma}}^B - \beta\hat{\boldsymbol{\Sigma}}^B\mathbf{H}^T\mathbf{H}\hat{\boldsymbol{\Sigma}}^B$$

Simplifying this equation gives

$$\left(\gamma^2\beta\mathbf{H}\hat{\boldsymbol{\Sigma}}^B\mathbf{H}^T - 2\gamma + 1\right)\beta\hat{\boldsymbol{\Sigma}}^B\mathbf{H}^T\mathbf{H}\hat{\boldsymbol{\Sigma}}^B = 0, \tag{21.35}$$

which holds only if the term in parentheses vanishes. Since the term in parentheses is a quadratic equation, we can use the quadratic formula to find the roots:

$$\gamma = \frac{1 \pm \sqrt{\beta\mathbf{R}}}{\beta\mathbf{H}\hat{\boldsymbol{\Sigma}}^B\mathbf{H}^T} = \frac{1}{1 \mp \sqrt{\beta\mathbf{R}}} \tag{21.36}$$

(the last equality is derived by multiplying the top and bottom by $1 \mp \sqrt{\beta\mathbf{R}}$). Which of the two solutions should be chosen? As can be seen from (21.34), the choice $\gamma = 0$ corresponds to "no adjustment." Since we generally want the *smallest* possible adjustment, we choose the solution that is closest to zero. The quantity $\beta\mathbf{R}$ is between zero and one, hence the solution for γ that is closest to zero is

$$\gamma = \frac{1 - \sqrt{\beta\mathbf{R}}}{\beta\mathbf{H}\hat{\boldsymbol{\Sigma}}^B\mathbf{H}^T} = \left(1 + \sqrt{\frac{\mathbf{R}}{\mathbf{R} + \mathbf{H}\hat{\boldsymbol{\Sigma}}^B\mathbf{H}^T}}\right)^{-1}. \tag{21.37}$$

This approach is known as the *Ensemble Square Root Filter* (EnSRF) and was proposed by Whitaker and Hamill (2002).

21.1.3 Ensemble Adjustment Kalman Filter (EAKF)

Another approach to solving for the matrix $\mathbf{C}'$ that satisfies (21.32) is to determine the SVD of $\mathbf{X}^B$, but retain only the parts associated with non-zero singular values. Thus,

$$\mathbf{X}^B = \dot{\mathbf{U}}\dot{\mathbf{S}}\mathbf{V}^T, \tag{21.38}$$

where $\dot{\mathbf{U}}$ and $\mathbf{V}$ are the singular vector matrices and $\dot{\mathbf{S}}$ is a square, diagonal matrix containing the positive singular values on the diagonal. The dots indicate that matrices in the full SVD have been truncated to retain terms associated with nonzero singular values (this is called the *economy-sized SVD*). Then, the desired square root matrix is

$$\mathbf{C}' = \mathbf{X}^B\mathbf{D}^{1/2}\mathbf{V}\dot{\mathbf{S}}^{-1}\dot{\mathbf{U}}^T, \tag{21.39}$$

which can be readily verified to satisfy

$$\mathbf{C}'\mathbf{X}^B\mathbf{X}^{BT}\mathbf{C}'^T = \mathbf{X}^B\mathbf{D}\mathbf{X}^B = \hat{\boldsymbol{\Sigma}}^A. \tag{21.40}$$

As always, $\mathbf{D}^{1/2}$ is unique up to post-multiplication by an orthogonal matrix, and so the expression for $\mathbf{C}'$ is not unique. This form of the filter is known as the *Ensemble Adjustment Kalman Filter* and was introduced by Anderson (2001).

21.2 Filter Divergence

The governing equations for atmospheric and oceanic fluids are chaotic – that is, their solutions are sensitive to small perturbations in the initial conditions. Ensemble square root filters often fail when applied to chaotic systems. To illustrate this (often dramatic) fact, we consider a "toy" mathematical model that is nonlinear and chaotic. The particular model we use is known as the Lorenz95 (or Lorenz96) model and is governed by the equations

$$\frac{dx_i}{dt} = (x_{i+1} - x_{i-2})\, x_{i-1} - x_i + f_0, \tag{21.41}$$

where $f_0 = 8$ and $i = 1, 2, \ldots, 40$ are cyclic indices. By "cyclic indices," we mean that the solution is periodic in the sense that the 40 points lie on a circle, hence $x_{i+40} = x_i$ for all i. Solutions are obtained by integrating the equations using a fourth-order Runge–Kutta scheme with a time step of 0.01. Further discussion of the characteristics of the solutions can be found in Lorenz and Emmanuel (1998).

Assimilation Experiments with Lorenz 1995 Model

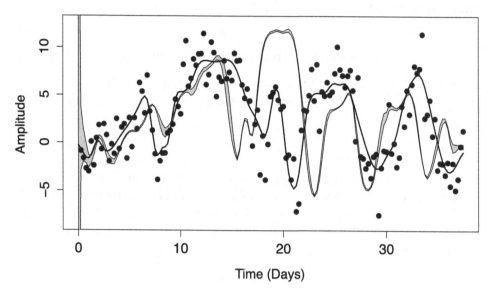

Figure 21.2 A solution of the Lorenz95 model (solid curve), which serves as "truth;" the "observations," obtained by perturbing the truth (dots); and the range of values within two standard deviations of the analysis mean (shaded region) computed from the Ensemble Square Root Filter using 20 members. Only data at the 10th grid point is shown.

Starting at a random initial condition, the model is run 100 time units, and the solutions at $i = 10$ for the last 40 time units are shown as the solid curve in Figure 21.2. We identify this solution as the "truth." Next, "observations" are obtained by adding a random perturbation from $\mathcal{N}(0, \sigma^2)$ to the truth at each grid point. This observation model is equivalent to (21.2) with the choices

$$\mathbf{R} = \sigma^2 \mathbf{I}_{40} \quad \text{and} \quad \mathbf{H} = \mathbf{I}_{40}. \tag{21.42}$$

We choose $\sigma = 4$, and, following Lorenz and Emmanuel (1998), the observations occur at every 0.05 time units. The observations at $i = 10$ are shown as the dots in Figure 21.2.

To apply the EnSRF, an initial ensemble must be defined. For this purpose, we use the climatological distribution. The climatological mean and covariance matrix are estimated by generating another 100-day solution starting from a different random initial condition. Then, a 20-member ensemble is constructed by drawing 20 random vectors from a normal distribution with this mean and covariance matrix. Starting from this ensemble, the EnSRF is applied to assimilate observations up to $t = 40$. The 95% confidence interval for the analysis at $i = 10$ is shown as grey shading in Figure 21.2 (the interval was computed using (20.53)). As can be seen, the confidence interval contains the truth only for the first few time units. After that time the analysis follows a path quite different from that of the truth. This phenomenon is known as *filter divergence*. Clearly, the analysis distribution is not representative of the truth after $t = 15$. Specifically, the confidence interval is unrealistically *small*. This phenomenon is known as *ensemble collapse*. During ensemble collapse, the analysis distribution gives the misleading impression that it is more accurate than the climatological distribution, when in fact it is not. These problems occur regardless of which filter is chosen.

Why does filter divergence and ensemble collapse occur? There is no completely satisfactory explanation (some contributing factors are discussed in Sections 21.4–21.6). Despite this, some modifications of the filter have been found to prevent filter collapse. Two of these modifications are *covariance inflation* and *covariance localization*. Before discussing these modifications, it is important to explain how filter divergence can be *diagnosed*.

21.3 Monitoring the Innovations

In the previous example, filter divergence was obvious because the analysis could be compared to the truth. In practice, the truth is not available. How can we decide if filter divergence is occurring in practical applications? A quantity that has emerged as a valuable check on the filter is the *innovation vector*

$$\mathbf{i}_t = \mathbf{o}_t - \mathbf{H}_t \boldsymbol{\mu}_t^B. \tag{21.43}$$

The innovation vector measures the discrepancy between observations and the interpolated background mean. To compute the statistics of $\mathbf{i}_t$, substitute $\mathbf{o}_t = \mathbf{H}_t \mathbf{x}_t + \mathbf{r}_t$:

$$\mathbf{i}_t = \mathbf{H}_t \left(\mathbf{x}_t - \boldsymbol{\mu}_t^B \right) + \mathbf{r}_t. \tag{21.44}$$

Taking expectations with respect to the background and observational distributions gives $\mathbb{E}[\mathbf{i}_t] = \mathbf{0}$. Similarly, the covariance matrix of the innovation vector is

$$\begin{aligned}
\mathrm{cov}[\mathbf{i}_t] &= \mathbb{E}\left[\left(\mathbf{H}_t \left(\mathbf{x}_t - \boldsymbol{\mu}_t^B \right) + \mathbf{r}_t \right) \left(\mathbf{H}_t \left(\mathbf{x}_t - \boldsymbol{\mu}_t^B \right) + \mathbf{r}_t \right)^T \right] \\
&= \mathbf{H}_t \mathbb{E}\left[\left(\mathbf{x}_t - \boldsymbol{\mu}_t^B \right) \left(\mathbf{x}_t - \boldsymbol{\mu}_t^B \right)^T \right] \mathbf{H}_t^T + \mathbb{E}\left[\mathbf{r}_t \mathbf{r}_t^T \right] \\
&= \mathbf{H}_t \boldsymbol{\Sigma}^B \mathbf{H}_t^T + \mathbf{R}_t, \tag{21.45}
\end{aligned}$$

where we have used the fact that $\mathbf{x}_t - \boldsymbol{\mu}_t^B$ and $\mathbf{r}_t$ are independent. Now, recall that if $\mathbf{x} \sim \mathcal{N}_K(\boldsymbol{\mu}, \boldsymbol{\Sigma})$, then $(\mathbf{x} - \boldsymbol{\mu})^T \boldsymbol{\Sigma}^{-1}(\mathbf{x} - \boldsymbol{\mu})$ has a chi-squared distribution with K degrees of freedom. This fact suggests examining the "Innovation Consistency Statistic" ICS:

Innovation Consistency Statistic for Lorenz 1995 Model

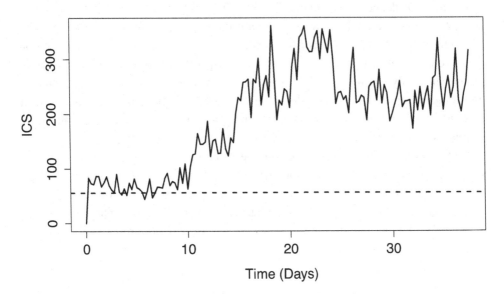

Figure 21.3 The innovation consistency statistic for the Ensemble Square Root Filter corresponding to Figure 21.2. The horizontal dashed line shows the 5% significance level.

$$\mathrm{ICS}(t) = \left(\mathbf{o}_t - \mathbf{H}_t \boldsymbol{\mu}_t^B\right)^T \left(\mathbf{H}_t \hat{\boldsymbol{\Sigma}}^B \mathbf{H}_t^T + \mathbf{R}_t\right)^{-1} \left(\mathbf{o}_t - \mathbf{H}_t \boldsymbol{\mu}_t^B\right). \qquad (21.46)$$

If the filter is working properly, then this quantity should have a chi-squared distribution with K degrees of freedom. If the background dynamics are Gaussian and Markov, it can be further proven that the innovation vector is a white Gaussian sequence (Maybeck, 1982, 229). It follows that the innovation vector should satisfy three properties: it (1) should have zero mean, (2) should be uncorrelated in time, and (3) should have covariance matrix $\mathbf{H}_t \boldsymbol{\Sigma}^B \mathbf{H}_t^T + \mathbf{R}_t$. Evidence that the innovation vector is biased, serially correlated, or has the wrong covariance matrix therefore indicates some violation of these properties, which in turn could indicate filter divergence.

A plot of ICS for the example in Figure 21.2 is shown in Figure 21.3. The figure shows that ICS exhibits strong violations of the chi-squared distribution around the same time that filter divergence is apparent in Figure 21.2. Thus, ICS appears to be a useful diagnostic as to whether the filter "is working."

21.4 Multiplicative Inflation

Let us now examine filter divergence more closely. Recall that the Kalman Filter Update Equations were derived assuming a perfectly known Gaussian background distribution. In contrast, the previous example used an *approximate* background distribution estimated from a finite ensemble. To understand the impact of sampling errors, note that the covariance update equation (21.5) is effectively a matrix *function* of the form

$$\mathbf{F}(\mathbf{B}) = \mathbf{B} - \mathbf{B}\mathbf{H}^T \left(\mathbf{H}\mathbf{B}\mathbf{H}^T + \mathbf{R}\right)^{-1} \mathbf{H}\mathbf{B}. \tag{21.47}$$

Majda and Tong (2016) show that this function is a *concave function of* **B**. A standard fact about concave functions is that they satisfy Jensen's inequality. The matrix generalization of Jensen's inequality states that if (21.47) is concave, then

$$\mathbb{E}[\mathbf{F}(\mathbf{B})] \le \mathbf{F}(\mathbb{E}[\mathbf{B}]), \tag{21.48}$$

where the matrix inequality $\mathbf{X} \le \mathbf{Y}$ means that $\mathbf{Y} - \mathbf{X}$ is positive semi-definite. This inequality implies that, on average, the variance of a linear combination of state variables in the analysis ensemble tends to be *underestimated*.

To illustrate this bias, consider an example in which there is only one state variable. Then, the analysis variance is the one-dimensional version of (21.5), which is

$$\hat{\sigma}_A^2 = \hat{\sigma}_B^2 - \frac{\hat{\sigma}_B^4}{\hat{\sigma}_B^2 + \sigma_R^2}. \tag{21.49}$$

Using this definition of $\hat{\sigma}_A^2$, Jensen's inequality implies that

$$\mathbb{E}[\hat{\sigma}_A^2] \le \sigma_A^2, \tag{21.50}$$

hence, the *sample* Kalman Filter Update Equation gives a *biased* estimate of the analysis variance. To illustrate this numerically, suppose $\hat{\sigma}_B^2$ is estimated from $E = 4$ samples from $\mathcal{N}(0, 1)$. The result of doing this 10,000 times using $\sigma_R = 1$ is shown in Figure 21.4. The mean of $\hat{\sigma}_A^2$ over the 10,000 realizations is shown as the dashed line. The population value is $\sigma_A^2 = 0.5$ (because $\sigma_B^2 = 1$ and $\sigma_R^2 = 1$) and is shown by the solid line. Clearly, the sample mean is less than the population value, thereby illustrating the bias.

This bias implies that an ensemble filter tends to underestimate the spread of the analysis ensemble. Because the filter is *sequential*, the bias is *compounded* with each analysis cycle. Therefore, the analysis variance has a tendency to get smaller with each assimilation cycle, hence $\hat{\mathbf{\Sigma}}_t^A \to 0$, and therefore $\hat{\mathbf{\Sigma}}_t^B \to 0$, which is the definition of ensemble collapse. In this limit, the update equation (21.4) implies $\hat{\boldsymbol{\mu}}_t^A \to \hat{\boldsymbol{\mu}}_t^B$, which implies that the filter merely adopts the prior estimate without modification, thereby giving filter divergence. In effect, the filter *ignores the observations*! This effect can be seen clearly in Figure 21.2– the analysis seems to ignore the observations after $t = 15$. This behavior may sound absurd, but in some sense the filter is behaving exactly as it should: given near-certainty in the background distribution, the filter *should* give little weight to the observations. After all, if you are certain, then contradictory observations do not change your mind.

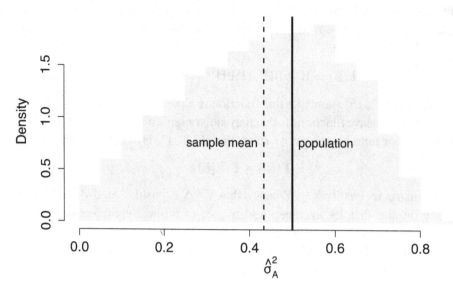

Figure 21.4 Histogram (grey shading) of the sample analysis variance $\hat{\sigma}_A^2$ when the background variance $\hat{\sigma}_B^2$ is estimated from $E = 4$ samples from $\mathcal{N}(0,1)$. The mean of the histogram samples is indicated by the vertical dashed line. The population analysis covariance σ_A^2 is indicated by the vertical solid line. 10,000 realizations were generated using $\sigma_R = 1$.

This explanation is not the full story because filter divergence does not always occur. For instance, filter divergence does not occur when the EnSRF is applied to the vector autoregressive model discussed in Section 20.6. At present, there is no general theory that accurately predicts when filter divergence will or will not occur.

Nevertheless, this reasoning suggests that the cause of filter divergence is a systematic tendency to underestimate the uncertainty. Therefore, filter divergence might be discouraged by *artificially inflating* the background variances prior to applying the Kalman filter. As an example, one could modify the forecast covariance as follows:

$$\text{replace } \mathbf{X}^B \text{ by } \alpha \mathbf{X}^B, \qquad (21.51)$$

where α, known as the *inflation factor*, is greater than unity. This method is known as *covariance inflation*. Note that $\mathbf{X}^B$ refers to deviations about the ensemble mean, rather than the state itself (see 21.8). The main question is how to select α. Experience shows that filter performance is not sensitive to α provided it is a number slightly larger than one.

To illustrate covariance inflation, the example shown in Figure 21.2 is repeated, except this time the forecast ensemble is inflated as (21.51) using the fixed

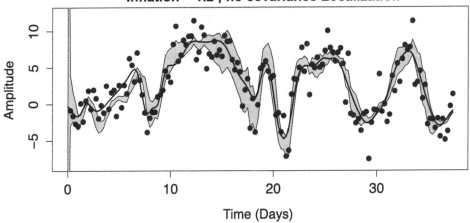

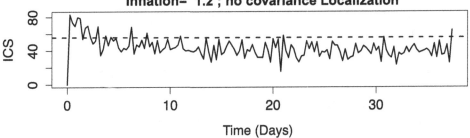

Figure 21.5 Same as Figures 21.2 and 21.3, except with covariance inflation.

value $\alpha = 1.2$. The result, shown in Figure 21.5, reveals that the analysis now tracks the truth fairly well. The Innovation Consistency Statistic is also shown in Figure 21.5 and reveals no serious inconsistencies. In this example, variance inflation has prevented ensemble collapse for 40 days, but there is no guarantee that it will prevent ensemble collapse indefinitely. In practice, one could monitor the ICS over time and increase the inflation factor whenever ICS exceeds a certain threshold for a certain period of time. An algorithm for systematically adjusting the inflation factor *adaptively* was proposed by Anderson (2007), which has become standard.

21.5 Covariance Localization

The background covariance matrix summarizes an immense amount of information. For instance, one row of the background covariance matrix is a *regression map* between the error at one spatial point and the error of other variables at other spatial points. All possible regression maps are contained in the background covariance matrix. Recall from Chapter 13 that regression maps often exhibit large spurious values by random chance. It follows that the elements of the sample covariance matrix will also exhibit large spurious values. In contrast to teleconnection maps, which involve monthly or seasonal means, regression maps of weather forecasts (e.g., six-hour forecasts) are subject to stronger physical constraints. Specifically, in atmospheric and oceanic fluids, dynamical waves communicate perturbations across space. The speed of these waves limits the extent to which initial condition errors propagate. As a result, the background covariance matrix for short lead times should be *sparse*. More precisely, matrix elements corresponding to widely spaced points should vanish. These considerations suggest that large correlations between widely spaced points are unphysical and ought to be damped toward zero as the distance between points increases. This approach is known as *covariance localization*. However, simply setting particular elements of a covariance matrix to zero often produces a matrix that is not positive semi-definite. Since all covariance matrices must be positive semi-definite, one must find a way to damp covariances to zero while also preserving positive semi-definiteness of the covariance matrix. Houtekamer and Mitchell (2001) and Hamill et al. (2001) suggested using a Schur product. This approach is based on the following definition and theorem:

Definition 21.1 (The Schur Product) *Let* $\mathbf{Y}$ *and* $\mathbf{Z}$ *be two matrices of the same dimension. The Schur product is denoted* $\mathbf{Y} \circ \mathbf{Z}$ *and defined as*

$$(\mathbf{Y} \circ \mathbf{Z})_{ij} = (\mathbf{Y})_{ij} \, (\mathbf{Z})_{ij} \, . \tag{21.52}$$

Essentially, the Schur product is an element-wise product of two matrices. The Schur product is also known as the Hadamard product.

A simple example of a Schur product is the following:

$$\begin{pmatrix} a & b \\ b & c \end{pmatrix} \circ \begin{pmatrix} x & y \\ y & z \end{pmatrix} = \begin{pmatrix} ax & by \\ by & cz \end{pmatrix} . \tag{21.53}$$

Theorem 21.1 (The Schur Product Theorem) *Suppose* $\mathbf{Y}$ *and* $\mathbf{Z}$ *are symmetric, positive semi-definite matrices of the same dimension. Then,* $\mathbf{Y} \circ \mathbf{Z}$ *is positive semi-definite.*

Importantly, if $\mathbf{Y}$ contains zero elements, then the corresponding elements of $\mathbf{Y} \circ \mathbf{Z}$ will also be zero. Thus, the Schur product theorem implies that if a positive semi-definite matrix $\mathbf{Y}$ could be found, with the property that the covariance vanishes between two points separated by a sufficiently large distance, then the Schur product of this matrix $\mathbf{Y}$ with the sample covariance matrix $\hat{\boldsymbol{\Sigma}}^B$ would also have this property, and will also be positive semi-definite. The question arises as to how to find a positive semi-definite matrix whose covariance elements vanish for sufficiently large separation distance.

Since the diagonal elements of $\mathbf{Y}$ equal one and the off-diagonal elements decay to zero, the matrix $\mathbf{Y}$ can be interpreted as a *correlation matrix*. Any row of $\mathbf{Y}$ defines a *correlation function* between the diagonal point and other points. We seek a correlation function with *local support*, which means that the correlation function is nonzero only for spatial points within a finite spatial distance of each other. Gaspari and Cohn (1999) derive a number of correlation functions with local support. In ensemble filtering, the most commonly used function is their equation (4.10), which is a fifth-order polynomial that closely matches a Gaussian function. This function is

$$
\rho(r,c) = \begin{cases} -\dfrac{r^5}{4c^5} + \dfrac{r^4}{2c^4} + \dfrac{5r^3}{8c^3} - \dfrac{5r^2}{3c^2} + 1 & 0 \le r \le c \\[2ex] \dfrac{r^5}{12c^5} - \dfrac{r^4}{2c^4} + \dfrac{5r^3}{8c^3} + \dfrac{5r^2}{3c^2} - \dfrac{5r}{c} + 4 - \dfrac{2c}{3r} & c \le r \le 2c \\[2ex] 0 & 2c \le r \end{cases} \tag{21.54}
$$

where r is the (positive) distance between points and c is a parameter that controls the length-scale over which the correlations go to zero. This length scale can be chosen based on physical considerations (e.g., propagation speed of gravity waves) or empirical considerations (e.g., estimates of correlatios based on a large forecast/observation data set).

This function defines the elements of a correlation matrix. The Schur product between this correlation matrix and the sample forecast covariance matrix will be denoted $\rho \circ \hat{\boldsymbol{\Sigma}}^B$. This product requires discretizing the polynomial. There is no guarantee that the discretized function will be positive definite, but typically the discretization is fine enough that the discretized polynomial is close to the continuous function.

It is worth emphasizing that there exist other implementations of ensemble square root filters that use nonpositive semi-definite ρ and still obtain reasonable analyses.

To implement covariance localization, we return to the original Kalman Filter Update Equations (21.4) and (21.5), substitute sample quantities, and perform the substitution

$$
\hat{\boldsymbol{\Sigma}}_B \rightarrow \rho \circ \hat{\boldsymbol{\Sigma}}_B. \tag{21.55}
$$

Thus, the equation for the mean of the analysis (21.4) becomes

$$\hat{\boldsymbol{\mu}}^{A} = \hat{\boldsymbol{\mu}}^{B} + \left[\rho \circ \hat{\boldsymbol{\Sigma}}^{B} \right] \mathbf{H}^{T} \left(\mathbf{H} \left[\rho \circ \hat{\boldsymbol{\Sigma}}^{B} \right] \mathbf{H}^{T} + \mathbf{R} \right)^{-1} \left(\mathbf{o} - \mathbf{H} \hat{\boldsymbol{\mu}}^{B} \right). \qquad (21.56)$$

If the forward interpolation operator $\mathbf{H}$ operates over short spatial scales compared to the length scale of ρ, Houtekamer and Mitchell (2001) suggest the following approximations:

$$\left[\rho \circ \hat{\boldsymbol{\Sigma}}^{B} \right] \mathbf{H}^{T} \approx \rho \circ (\hat{\boldsymbol{\Sigma}}^{B} \mathbf{H}^{T}) \qquad (21.57)$$

$$\mathbf{H} \left[\rho \circ \hat{\boldsymbol{\Sigma}}^{B} \right] \mathbf{H}^{T} \approx \rho \circ \left(\mathbf{H} \hat{\boldsymbol{\Sigma}}^{B} \mathbf{H}^{T} \right) \approx \mathbf{H} \hat{\boldsymbol{\Sigma}}^{B} \mathbf{H}^{T}, \qquad (21.58)$$

where the last approximation in (21.58) was suggested by Hamill et al. (2001). In this notation, it is understood that each element of ρ is determined by the distance between grid points of the covariance to which it is multiplied. In particular, the ρ in these equations differs from the ρ in (21.55), because the dimension of the matrices to which it is multiplied differs. With these approximations, the final equation becomes

$$\hat{\boldsymbol{\mu}}^{A} = \hat{\boldsymbol{\mu}}^{B} + \left[\rho \circ (\hat{\boldsymbol{\Sigma}}^{B} \mathbf{H}^{T}) \right] \left(\mathbf{H} \hat{\boldsymbol{\Sigma}}^{B} \mathbf{H}^{T} + \mathbf{R} \right)^{-1} \left(\mathbf{o} - \mathbf{H} \hat{\boldsymbol{\mu}}^{B} \right). \qquad (21.59)$$

This version is especially important because in atmospheric and oceanic data assimilation, the background covariance is too large to compute directly. Thus, applying localization to the background covariance matrix is difficult. Instead, as indicated earlier, the localization is applied to $\boldsymbol{\Sigma}^{B} \mathbf{H}^{T}$, which can be computed equivalently as

$$\boldsymbol{\Sigma}^{B} \mathbf{H}^{T} = \mathbf{X}^{B} \left(\mathbf{H} \mathbf{X}^{B} \right)^{T}, \qquad (21.60)$$

where $\mathbf{H} \mathbf{X}^{B}$ often has much smaller row-dimension than $\mathbf{X}^{B}$.

The next step is to implement covariance localization in the analysis perturbations. This implementation is most straightforward for the EnSRF. To the same order of approximation as discussed previously, the transformation matrix (21.34) becomes

$$\mathbf{C}' = \mathbf{I} - \gamma \beta \left[\rho \circ \hat{\boldsymbol{\Sigma}}^{B} \mathbf{H}^{T} \right] \mathbf{H}, \qquad (21.61)$$

where γ and β have the same meanings as they did in Section 21.1.2.

The impact of covariance localization on solutions of the Lorenz95 model is shown in Figure 21.6. Note that no inflation is applied in this example (that is, $\alpha = 1$). The figure shows that covariance localization dramatically improves the accuracy of the analyses and eliminates filter divergence. The innovation consistency statistic is also shown in Figure 21.6 and reveals no significant inconsistencies.

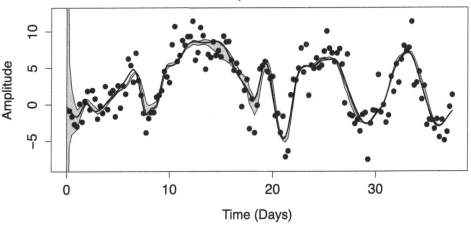

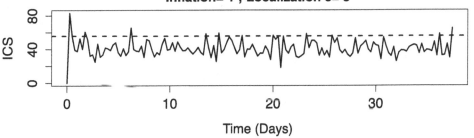

Figure 21.6 Same as Figures 21.2 and 21.3, except with covariance localization. Note that the truth (solid curve) lies within the 95% confidence interval of the analyses (grey shading).

21.6 Further Topics

Historically, square root filters were introduced to overcome numerical problems with solving (21.4) and (21.5) directly. Square root filters differ from *ensemble* filters in that square root filters solve the covariances *exactly*, whereas ensemble filters use *sample estimates*. The advantage of square root filters is that they preserve symmetry and positive definiteness of covariance matrices, and have significantly reduced storage requirements (Bierman, 1977). Square root filters were used to solve onboard navigation problems in the Apollo space program at a time when computational resources were severely limited.

The earliest proposals for ensemble filters (Evensen, 1994) were based on updating individual ensemble members using the same equation as used to update the mean, namely

$$\mathbf{X}^A = \mathbf{X}^B + \boldsymbol{\Sigma}^B \mathbf{H}^T \left(\mathbf{H}\boldsymbol{\Sigma}^B\mathbf{H}^T + \mathbf{R}\right)^{-1} \left(\mathbf{o} - \mathbf{H}\mathbf{X}^B\right). \qquad (21.62)$$

Unfortunately, this equation does not produce an analysis ensemble with the correct covariance matrix or mean. The problem is that filter treats observations $\mathbf{o}$ as random variables, whereas a naive application of Equation (21.62) would hold the observations fixed. This deficiency can be corrected by the method of *perturbed observations*, in which random noise with the appropriate distribution is added to the observations (Burgers et al., 1998; Houtekamer and Mitchell, 1998). Although adding noise to an estimation algorithm seems contrary to the goal of filtering out noise, adding such noise is necessary for the ensemble to have covariance matrices consistent with the Kalman filter update equations. Subsequently, several *deterministic* ensemble filters were derived, but these were shown to be special cases of a class of filters known as *square root filters* (Tippett et al., 2003), which have been discussed in this chapter.

Houtekamer and Zhang (2016) provide a comprehensive review of Ensemble Kalman Filters. One of the advantages of such filters is the ability to use nonlinear interpolation operators $\mathbf{H}$. Specifically, the terms $\boldsymbol{\Sigma}^B\mathbf{H}^T$ and $\mathbf{H}\boldsymbol{\Sigma}^B\mathbf{H}^T$ can be computed not only by operating $\mathbf{H}$ on the covariance matrix $\boldsymbol{\Sigma}^B$ but also by operating $\mathbf{H}$ on individual background fields, as illustrated here:

$$\boldsymbol{\Sigma}^B\mathbf{H}^T = \mathbf{X}^B \left(\mathbf{H}\mathbf{X}^B\right)^T \qquad (21.63)$$

$$\mathbf{H}\boldsymbol{\Sigma}^B\mathbf{H}^T = \left(\mathbf{H}\mathbf{X}^B\right)\left(\mathbf{H}\mathbf{X}^B\right)^T. \qquad (21.64)$$

For instance, if the observations are radiances, then $\mathbf{H}$ could be a radiative transfer model; if observations are convection rates, then $\mathbf{H}$ could be a convective parameterization scheme. However, the Kalman Filter equations are proven to be optimal only for linear interpolation operators, so the benefits of this approach may not be great for highly nonlinear operators.

In practice, the true dynamical system is not known and the available dynamical model used to generate the background ensemble contains *model error*. To account for model error, the background error covariance matrix is often inflated. This approach is admittedly crude. There is no consensus on the best way to deal with model error in data assimilation.

Recently, some operational forecast centers have been moving toward *hybrid* data assimilation systems that combine ensemble filtering with 3DVAR or 4DVAR techniques (Wang et al., 2013; Lorenc et al., 2014). The most common version is to add a fixed climatological covariance matrix to the flow-dependent background error covariance matrix with localization. The configuration giving the best analyses is a topic of considerable debate.

21.7 Conceptual Questions

1 In what ways does atmospheric and oceanic data assimilation violate the assumptions of the Kalman Filter?

2 Why not ignore inconsistencies and just evaluate the Kalman Filter equations for atmospheric and oceanic data assimilation?

3 What is a matrix square root? What is a symmetric square root?

4 Why is there more than one square root filter? How are they related?

5 What is a mean preserving transformation?

6 What is filter divergence?

7 How can you diagnose filter divergence without knowing the true state?

8 What are some ways of mitigating filter divergence?

Appendix

A.1 Useful Mathematical Relations

$$\sum_{s=0}^{N-1} \beta^s = \frac{1 - \beta^N}{1 - \beta}, \quad \beta \neq 1 \qquad \text{Geometric Series} \quad (A.1)$$

$$\sum_{s=0}^{N-1} s\beta^s = \frac{(N-1)\beta^{N+1} - N\beta^N + \beta}{(1 - \beta)^2}, \quad \beta \neq 1 \qquad (A.2)$$

$$\sum_{n=1}^{N} n = \frac{N(N+1)}{2} \qquad \text{Arithmetical Series} \quad (A.3)$$

$$\sum_{n-1}^{N} n^2 = \frac{N(N+1)(2N+1)}{6} \qquad (A.4)$$

$$\begin{pmatrix} a & b \\ c & d \end{pmatrix}^{-1} = \frac{1}{ad - bc} \begin{pmatrix} a & b \\ c & d \end{pmatrix}, \quad ad - bc \neq 0 \qquad \text{Inverse of 2 x 2 matrix} \quad (A.5)$$

$$\left(\mathbf{x}^T \mathbf{y} \right)^2 \leq \left(\mathbf{x}^T \mathbf{x} \right) \left(\mathbf{y}^T \mathbf{y} \right) \qquad \text{Cauchy–Schwartz Inequality} \quad (A.6)$$

$$\left(\mathbf{x}^T \boldsymbol{\Sigma} \mathbf{y} \right)^2 \leq \left(\mathbf{x}^T \boldsymbol{\Sigma} \mathbf{x} \right) \left(\mathbf{y}^T \boldsymbol{\Sigma} \mathbf{y} \right), \ \boldsymbol{\Sigma} \text{ pos. def.} \qquad \text{General Cauchy–Schwartz} \quad (A.7)$$

$$\left(\mathbf{x}^T \mathbf{y} \right)^2 \leq \left(\mathbf{x}^T \boldsymbol{\Sigma} \mathbf{x} \right) \left(\mathbf{y}^T \boldsymbol{\Sigma}^{-1} \mathbf{y} \right), \ \boldsymbol{\Sigma} \text{ pos. def.} \qquad \text{Extended Cauchy–Schwartz} \quad (A.8)$$

Inverse of partitioned matrices. If $\mathbf{A}$ and $\mathbf{D} - \mathbf{CA}^{-1}\mathbf{B}$ are nonsingular, then

$$\begin{pmatrix} \mathbf{A} & \mathbf{B} \\ \mathbf{C} & \mathbf{D} \end{pmatrix}^{-1} = \begin{pmatrix} \mathbf{A}^{-1} + \mathbf{A}^{-1}\mathbf{B}\left(\mathbf{D} - \mathbf{CA}^{-1}\mathbf{B}\right)^{-1}\mathbf{CA}^{-1} & -\mathbf{A}^{-1}\mathbf{B}\left(\mathbf{D} - \mathbf{CA}^{-1}\mathbf{B}\right)^{-1} \\ -\left(\mathbf{D} - \mathbf{CA}^{-1}\mathbf{B}\right)^{-1}\mathbf{CA}^{-1} & \left(\mathbf{D} - \mathbf{CA}^{-1}\mathbf{B}\right)^{-1} \end{pmatrix}$$

$$(A.9)$$

The Sherman–Morrison–Woodbury formula is

$$(\mathbf{A} + \mathbf{BDC})^{-1} = \mathbf{A}^{-1} - \mathbf{A}^{-1}\mathbf{B}\left(\mathbf{D}^{-1} + \mathbf{CA}^{-1}\mathbf{B}\right)^{-1}\mathbf{CA}^{-1}, \qquad (A.10)$$

provided that the inverses exist.

A.2 Generalized Eigenvalue Problems

Some statistical techniques lead to a *generalized eigenvalue problem* of the form

$$\mathbf{Aq} = \lambda \mathbf{Bq}, \tag{A.11}$$

where $\mathbf{A}$ and $\mathbf{B}$ are real and symmetric, $\mathbf{B}$ is positive definite, and $\mathbf{A}$ may or may not be positive definite. This equation is a generalized version of the eigenvalue problem (7.80). A nontrivial vector $\mathbf{q}$ that satisfies this equation is known as an eigenvector, and the corresponding λ is known as an eigenvalue. If M is the dimension of the matrices $\mathbf{A}$ and $\mathbf{B}$, then there are M possible solutions. For simplicity, we avoid degeneracy issues by assuming that the M eigenvalues $\lambda_1, \ldots, \lambda_M$ are distinct. For each eigenvalue λ_m, there corresponds an eigenvector $\mathbf{q}_m$. Because the matrices $\mathbf{A}$ and $\mathbf{B}$ are symmetric, the eigenvalues and eigenvectors satisfy certain properties, which are summarized in the following theorem.

Theorem A.1 (Generalized Eigenvalue Problem) *If* $\mathbf{A}$ *and* $\mathbf{B}$ *are real and symmetric matrices, and* $\mathbf{B}$ *is a positive definite matrix, then the eigenvectors and eigenvalues of* (A.11) *have the following properties:*

1 The eigenvalue λ *is real.*

2 The eigenvector $\mathbf{q}$ *is real; more precisely,* $\mathbf{q}$ *is proportional to a real vector.*

3 If $\mathbf{q}_i$ *and* $\mathbf{q}_j$ *are two eigenvectors with differing eigenvalues (that is,* $\lambda_i \neq \lambda_j$*), then*

$$\mathbf{q}_i^T \mathbf{Aq}_j = 0 \quad and \quad \mathbf{q}_i^T \mathbf{Bq}_j = 0. \tag{A.12}$$

These latter relations are known as orthogonality conditions.

A trick to proving this theorem is to transform the generalized eigenvalue problem (A.11) into a standard eigenvalue problem. To do this, let the eigenvector decomposition of $\mathbf{B}$ be $\mathbf{UDU}^T$. Define

$$\mathbf{W} = \mathbf{D}^{-1/2}\mathbf{U}^T. \tag{A.13}$$

This matrix has the property $\mathbf{WBW}^T = \mathbf{I}$. Also, the matrix is invertible and satisfies $\mathbf{W}^T\mathbf{W}^{-1T} = \mathbf{I}$. Multiplying $\mathbf{W}$ on the left of (A.11) gives

$$\mathbf{WAq} = \lambda\mathbf{WBq}$$

$$\mathbf{WA}\left(\mathbf{W}^T\mathbf{W}^{-1T}\right)\mathbf{q} = \lambda\mathbf{WB}\left(\mathbf{W}^T\mathbf{W}^{-1T}\right)\mathbf{q}$$

$$\left(\mathbf{WAW}^T\right)\left(\mathbf{W}^{-1T}\mathbf{q}\right) = \lambda\mathbf{I}\left(\mathbf{W}^{-1T}\mathbf{q}\right)$$

$$\left(\mathbf{WAW}^T\right)\tilde{\mathbf{q}} = \lambda\tilde{\mathbf{q}}, \tag{A.14}$$

where we define the transformed eigenvector $\tilde{\mathbf{q}} = \mathbf{W}^{-1T}\mathbf{q}$. Because $\mathbf{A}$ is symmetric, (A.14) is a standard eigenvalue problem *with a symmetric matrix.* Therefore, we know immediately that the eigenvalues $\lambda_1, \ldots, \lambda_M$ are real, and that the eigenvectors $\tilde{\mathbf{q}}_1, \ldots, \tilde{\mathbf{q}}_M$ are real and orthogonal. Furthermore, the transformation (A.13) consists of real elements; therefore its inverse is real and the untransformed eigenvectors $\mathbf{q}_1, \ldots, \mathbf{q}_M$ must be real. This proves the first two properties of Theorem A.1. To prove the third, note that orthogonality of the transformed vectors means

$$\tilde{\mathbf{q}}_i^T \tilde{\mathbf{q}}_j = 0 \quad \text{for any } i \neq j. \tag{A.15}$$

Substituting the definition of the transformed matrix yields

$$\tilde{\mathbf{q}}_i^T \tilde{\mathbf{q}}_j = \mathbf{q}_i^T \mathbf{W}^{-1} \mathbf{W}^{-1^T} \mathbf{q}_j = \mathbf{q}_i^T \mathbf{B} \mathbf{q}_j = 0 \quad \text{for any } i \neq j. \tag{A.16}$$

This orthogonality constraint immediately implies $\mathbf{q}_i^T \mathbf{A} \mathbf{q}_j = 0$ for $i \neq j$, as can be seen by substituting (A.11). This proves the third property of Theorem A.1.

These orthogonality properties can be summarized more concisely by introducing

$$\mathbf{Q} = \begin{bmatrix} \mathbf{q}_1 & \mathbf{q}_2 & \cdots & \mathbf{q}_M \end{bmatrix}. \tag{A.17}$$

Recall that eigenvectors are unique up to a multiplicative constant. It is convention to chose this constant such that

$$\mathbf{q}_i^T \mathbf{B} \mathbf{q}_i = 1. \tag{A.18}$$

Then, the orthogonality properties (A.12) and normalization (A.18) imply

$$\mathbf{Q}^T \mathbf{B} \mathbf{Q} = \mathbf{I}. \tag{A.19}$$

Furthermore, the eigenvalue problem (A.11) can be written as

$$\mathbf{A}\mathbf{Q} = \mathbf{B}\mathbf{Q}\mathbf{\Lambda}, \tag{A.20}$$

where $\mathbf{\Lambda}$ is a diagonal matrix whose diagonal elements equal the eigenvalues $\lambda_1, \ldots, \lambda_M$. Multiplying both sides of this equation on the left by $\mathbf{Q}^T$ and invoking (A.19) shows that

$$\mathbf{Q}^T \mathbf{A} \mathbf{Q} = \mathbf{\Lambda}. \tag{A.21}$$

A.3 Derivatives of Quadratic Forms and Traces

In some statistical analyses, the derivative of a quadratic form with respect to a vector or matrix needs to be computed. For instance, the following identity is often needed:

$$\frac{\partial \mathbf{x}^T \mathbf{A} \mathbf{x}}{\partial \mathbf{x}} = \left(\mathbf{A} + \mathbf{A}^T \right) \mathbf{x}. \tag{A.22}$$

The basic procedure for evaluating such a derivative is to write the quadratic form explicitly in index-notation form. For instance, this quadratic form can be written as

$$\mathbf{x}^T \mathbf{A} \mathbf{x} = \sum_j \sum_k x_j x_k a_{jk} \Rightarrow x_j x_k a_{jk}, \tag{A.23}$$

where the arrow means we use the implicit notation that indices that appear twice are summed. Taking the derivative with respect to x_i yields

$$\left(\frac{\partial \mathbf{x}^T \mathbf{A} \mathbf{x}}{\partial \mathbf{x}} \right)_i = \frac{\partial}{\partial x_i} \left(x_j x_k a_{jk} \right) \tag{A.24}$$

$$= \left(\delta_{ij} x_k a_{jk} + x_j \delta_{ik} a_{jk} \right) \tag{A.25}$$

$$= a_{ik} x_k + a_{ji} x_j \tag{A.26}$$

$$= \left(\mathbf{A}\mathbf{x} + \mathbf{A}^T \mathbf{x} \right)_i, \tag{A.27}$$

where we have used the identity

$$\frac{\partial v_i}{\partial v_k} = \delta_{ij} = \begin{cases} 0 & \text{if } i \neq k \\ 1 & \text{if } i = k \end{cases}. \tag{A.28}$$

A similar technique can be applied to traces, as the following example demonstrates.

Example A.1 *Show*

$$\frac{\partial \, \text{tr}[\mathbf{AB}]}{\partial \mathbf{A}} = \mathbf{B}^T. \tag{A.29}$$

As above, expand the trace in index notation and then take the derivative:

$$\left(\frac{\partial \, \text{tr}[\mathbf{AB}]}{\partial \mathbf{A}} \right)_{ij} = \frac{\partial a_{kl} b_{lk}}{\partial a_{ij}} = \delta_{ki} \delta_{lj} b_{lk} = b_{ji}. \tag{A.30}$$

The following identities are simply listed but can be proven in the same way:

$$\frac{\partial \mathbf{x}^T \mathbf{A} \mathbf{y}}{\partial \mathbf{x}} = \mathbf{A} \mathbf{y} \tag{A.31}$$

$$\frac{\partial \, \text{tr}[\mathbf{A}^T \mathbf{A}]}{\partial \mathbf{A}} = 2\mathbf{A} \tag{A.32}$$

$$\frac{\partial \, \text{tr}[\mathbf{A}^T \mathbf{B} \mathbf{A}]}{\partial \mathbf{A}} = \left(\mathbf{B} + \mathbf{B}^T \right) \mathbf{A}. \tag{A.33}$$

References

Agresti, A. 2012. *Categorical Data Analysis*. Wiley.

Akaike, H. 1973. Information Theory and an Extension of the Maximum Likelihood Principle. Pages 267–281 of: Petrov, B. N., and Czáki, F. (eds.), *2nd International Symposium on Information Theory*. Akademiai Kiadó.

Allen, M. R., and Stott, P. A. 2003. Estimating Signal Amplitudes in Optimal Fingerprinting, Part I: Theory. *Clim. Dyn.*, **21**, 477–491.

Allen, M. R., and Tett, S. F. B. 1999. Checking for Model Consistency in Optimal Fingerprinting. *Clim. Dyn.*, **15**, 419–434.

Anderson, E., Bai, Z., Bishof, C., Demmel, J., Dongarra, J., du Croz, J., Greenbaum, A., Hammarling, S., McKenny, A., Ostouchov, S., and Sorensen, D. 1992. *LAPACK User's Guide*. Philadelphia, PA: Society for Industrial and Applied Mathematics. 235pp.

Anderson, J. L. 2001. An Ensemble Adjustment Filter for Data Assimilation. *Mon. Wea. Rev.*, **129**, 2884–2903.

Anderson, J. L. 2007. An Adaptive Covariance Inflation Error Correction Algorithm for Ensemble Filters. *Tellus*, **59A**, 210–224.

Anderson, T. W. 1984. *An Introduction to Multivariate Statistical Analysis*. Wiley-Interscience.

Archer, D. 2010. *The Global Carbon Cycle*. Princeton University Press.

Baker, M. 2016. Is There a Reproducibility Crisis? *Nature*, **533**, 452–454.

Bartlett, M. S. 1937. Properties of Sufficiency and Statistical Tests. *Proceedings of the Royal Society of London. Series A – Mathematical and Physical Sciences*, **160**(901), 268–282.

Bartlett, M. S. 1946. On the Theoretical Specification and Sampling Properties of Autocorrelated Time-Series. *Supplement to the Journal of the Royal Statistical Society*, **8**(1), 27–41.

Battisti, D. S., and Sarachik, E. S. 1995. Understanding and Predicting ENSO. *Rev. Geophys.*, **33**(S2), 1367–1376.

Bender, Ralf, and Lange, Stefan. 2001. Adjusting for multiple testing: When and How? *J. Clin. Epid.*, **54**(4), 343–349.

Benjamini, Y., and Hochberg, Y. 1995. Controlling the False Discovery Rate: A Practical and Powerful Approach to Multiple Testing. *Journal of the Royal Statistical Society. Series B (Methodological)*, **57**(1), 289–300.

Berger, J. O. 1985. *Statistical Decision Theory and Bayesian Analysis*. 2nd ed. Springer.

Bernstein, D. S. 2009. *Matrix Mathematics: Theory, Facts, and Formulas*. Princeton University Press.

Bierman, G. J. 1977. *Factorization Methods for Discrete Sequential Estimation.* Mathematics in Science and Engineering. Academic Press. 241 pp.

Bishop, Craig H., Etherton, Brian, and Majumdar, Sharanya J. 2001. Adaptive Sampling with the Ensemble Transform Kalman Filter. Part I: Theoretical Aspects. *Mon. Wea. Rev.,* **129**, 420–436.

Bode, H. W., and Shannon, C. E. 1950. A Simplified Derivation of Least Squares Smoothing and Prediction Theory. *Proc. IRE,* **38**, 417–425.

Box, G. E. P. 1953. Non-Normality and Tests on Variances. *Biometrika,* **40**(3/4), 318–335.

Box, George E. P., and Watson, G. S. 1962. Robustness to Non-normality of Regression Tests. *Biometrika,* **49**, 93–106.

Box, George E. P., Jenkins, Gwilym M., and Reinsel, Gregory C. 2008. *Time Series Analysis: Forecasting and Control.* 4th ed. Wiley-Interscience.

Bretherton, Christopher S., Widmann, Martin, Dymnikov, Valentin P., Wallace, John M., and Bladé, Ileana. 1999. The Effective Number of Spatial Degrees of Freedom of a Time-Varying Field. *J. Climate,* **12**(7), 1990–2009.

Brockwell, P. J., and Davis, R. A. 1991. *Time Series: Theory and Methods.* 2nd ed. Springer Verlag.

Broomhead, D. S., and King, G. P. 1986. Extracting Qualitative Dynamics from Experimental Data. *Physica D,* **20**, 217–236.

Burgers, Gerrit, van Leeuwen, Peter Jan, and Evensen, Geir. 1998. Analysis Scheme in the Ensemble Kalman Filter. *Mon. Wea. Rev.,* **126**, 1719–1724.

Burnham, K. P., and Anderson, D. R. 2002. *Model Selection and Multimodel Inference: A Practical Information-theoretic Approach.* 2nd ed. Springer.

Cane, Mark A., Zebiak, Stephen E., and Dolan, Sean C. 1986. Experimental Forecasts of El Niño. *Nature,* **321**(6073), 827–832.

Cane, Mark A., Clement, Amy C., Murphy, Lisa N., and Bellomo, Katinka. 2017. Low-Pass Filtering, Heat Flux, and Atlantic Multidecadal Variability. *J. Climate,* **30**(18), 7529–7553.

Chin, Edwin H. 1977. Modeling Daily Precipitation Occurrence Process with Markov Chain. *Water Resources Research,* **13**(6), 949–956.

Cohen, Jonathan P. 1982. Convergence Rates for the Ultimate and Pentultimate Approximations in Extreme-Value Theory. *Advances in Applied Probability,* **14**(4), 833–854.

Coles, S. 2001. *An Introduction to Statistical Modeling of Extreme Values.* Springer.

Conover, W. J. 1980. *Practical Nonparametric Statistics.* 2nd ed. Wiley-Interscience.

Cover, T. M., and Thomas, J. A. 1991. *Elements of Information Theory.* Wiley-Interscience.

Cressie, N., and Wikle, C. K. 2011. *Statistics for Spatio-Temporal Data.* Wiley.

Cryer, J. D., and Chan, K.-S. 2010. *Time Series Analysis with Applications in R.* Springer.

Davison, A., Huser, R., and Thibaud, E. 2019. *Handbook of Environmental and Ecological Statistics.* CRC Press.

Dee, Dick P. 1995. On-line Estimation of error Covariance Parameters for Atmospheric Data Assimilation. *Mon. Wea. Rev.,* **123**, 1128–1145.

Dee, Dick P., and da Silva, Arlindo M. 1998. Data Assimilation in the Presence of Forecast Bias. *Quart. J. Roy. Meteor. Soc.,* **124**, 269–297.

DelSole, T. 2004. Predictability and Information Theory Part I: Measures of Predictability. *J. Atmos. Sci.,* **61**, 2425–2440.

DelSole, T. 2011. State and Parameter Estimation in Stochastic Dynamical Models. *ECMWF Proceedings on Representing Model Uncertainty and Error in Numerical Weather and Climate Prediction Models,* 255–262.

DelSole, T., and Chang, P. 2003. Predictable Component Analysis, Canonical Correlation Analysis, and Autoregressive Models. *J. Atmos. Sci.,* **60**, 409–416.

DelSole, T., and Shukla, J. 2009. Artificial Skill Due to Predictor Screening. *J. Climate*, **22**, 331–345.

DelSole, T., and Tippett, M. K. 2009. Average Predictability Time: Part II: Seamless Diagnosis of Predictability on Multiple Time Scales. *J. Atmos. Sci.*, **66**, 1188–1204.

DelSole, T., and Tippett, M. K. 2014. Comparing Forecast Skill. *Mon. Wea. Rev.*, **142**, 4658–4678.

DelSole, T., and Tippett, M. K. 2017. Predictability in a Changing Climate. *Clim. Dyn.*, Oct.

DelSole, T., and Tippett, M. K. 2020. Comparing Climate Time Series – Part 1: Univariate Test. *Adv. Stat. Clim. Meteorol. Oceanogr.*, **6**(2), 159–175.

DelSole, T. and Tippett, M. K. 2021a. A Mutual Information Criterion with applications to Canonical Correlation Analysis and graphical models. Stat, 10(1):e385.

DelSole, T., and Tippett, M. K. 2021b. Correcting the Corrected AIC. *Statistics & Probability Letters*, 109064.

DelSole, T., and Yang, X. 2011. Field Significance of Regression Patterns. *J. Climate*, **24**, 5094–5107.

DelSole, T., Trenary, L., Yan, X., and Tippett, Michael K. 2018. Confidence Intervals in Optimal Fingerprinting. *Clim. Dyn.*, Jul.

dePillis, J. 2002. *777 Mathematical Conversation Starters*. America Mathematical Society.

Déqué, M. 1988. 10-day Predictability of the Northern Hemisphere Winter 500-mb Height by the ECMWF Operational Model. *Tellus*, **40A**, 26–36.

Déqué, M., Royer, J. F., Stroe, R., and France, Meteo. 1994. Formulation of Gaussian Probability Forecasts Based on Model Extended-range Integrations. *Tellus A*, **46**(1), 52–65.

Dommenget, Dietmar, and Latif, Mojib. 2002. A Cautionary Note on the Interpretation of EOFs. *J. Climate*, **15**(2), 216–225.

Efron, B., and Tibshirani, R. J. 1994. *An Introduction to the Bootstrap*. Chapman & Hall.

Efthimiou, C. 2011. *Introduction to Functional Equations: Theory and Problem-solving Strategies for Mathematical Competitions and Beyond (MSRI Mathematical Circles Library)*. American Mathematical Society.

Evensen, G. 1994. Sequential Data Assimilation with a Nonlinear Quasi-geostrophic Model using Monte Carlo Methods to Forecast Error Statistics. *J. Geophys. Res.*, **99**, 1043–1062.

Fan, Yun, and van den Dool, Huug. 2008. A Global Monthly Land Surface Air Temperature Analysis for 1948–present. *Journal of Geophysical Research: Atmospheres*, **113**(D1).

Farrell, Brian F., and Ioannou, Petros J. 1996. Generalized Stability Theory. Part I: Autonomous Operators. *J. Atmos. Sci.*, **53**, 2025–2040.

Farrell, Brian F., and Ioannou, Petros J. 2001. Accurate Low-Dimensional Approximation of the Linear Dynamics of Fluid Flow. *J. Atmos. Sci.*, **58**(18), 2771–2789.

Feynman, R. P., Leighton, R. B., and Sands, M. 1977. *The Feynman Lectures on Physics, Vol. 1: Mainly Mechanics, Radiation, and Heat*. Addison-Wesley.

Fisher, R. A. 1915. Frequency Distribution of the Values of the Correlation Coefficient in Samples from an Indefinitely Large Population. *Biometrika*, **10**(4), 507–521.

Fisher, R. A. 1921. On the 'Probable Error' of a Coefficient of Correlation Deduced from a Small Sample. *Metron*, **1**, 1–32.

Fisher, R. A. 1936. Has Mendel's Work Been Rediscovered? *Annals of Science*, **1**(2), 115–137.

Fisher, R. A., and Tippett, L. H. C. 1928. Limiting Forms of the Frequency Distribution of the Largest or Smallest Member of a Sample. *Mathematical Proceedings of the Cambridge Philosophical Society*, **24**(4), 180–190.

Flury, B. N. 1985. Analysis of Linear Combinations with Extreme Ratios of Variance. *J. Amer. Stat. Assoc.*, **80**, 915–922.

Fujikoshi, Y. 1985. Selection of Variables in Discriminant Analysis and Canonical Correlation Analysis. Pages 219–236 of: Krishnaiah, P. R. (ed), *Multivariate Analysis VI: Proceedings of the Sixth International Symposium on Multivariate Analysis*. Elsevier.

Fujikoshi, Y., Ulyanov, V. V., and Shimizu, R. 2010. *Multivariate Statistics: High-dimensional and Large-Sample Approximations*. John Wiley and Sons.

Gaspari, Gregory, and Cohn, Stephen E. 1999. Construction of Correlation Functions in Two and Three Dimensions. *Quart. J. Roy. Meteor. Soc.*, **125**, 723–757.

Gelman, A., Carlin, J. B., Stern, H. S., and Rubin, D. B. 2004. *Bayesian Data Analysis*. 2nd ed. Chapman and Hall.

Ghil, M., Allen, M. R., Dettinger, M. D., Ide, K., Kondrashov, D., Mann, M. E., Robertson, A. W., Saunders, A., Tian, Y., Varadi, F., and Yiou, P. 2002. Advanced Spectral Methods for Climatic Time Series. *Rev. Geophys.*, **40**, 3.1–3.41.

Gilleland, E., and Katz, R. 2016. extRemes 2.0: An Extreme Value Analysis Package in R. *J. Stat. Software*, **72**(1), 1–39.

Glantz, S. A, and Slinker, B. K. 2016. *Primer of Applied Regression and Analysis of Variance*. 3rd ed. McGraw-Hill.

Gneiting, T., and Raftery, A. E. 2007. Strictly Proper Scoring Rules, Prediction, and Estimation. *J. Am. Stat. Assoc.*, **102**, 359–378.

Golub, Gene H., and Van Loan, Charles F. 1996. *Matrix Computations*. Third ed. Baltimore: The Johns Hopkins University Press. 694 pp.

Good, P. I. 2005. *Permutation, Parametric and Bootstrap Tests of Hypotheses*. 3rd ed. Springer.

Good, P. I. 2006. *Resampling Methods*. 3rd ed. Birkhäuser.

Hamill, Thomas M , Whitaker, Jeffrey S., and Snyder, Chris. 2001. Distance-Dependent Filtering of Background Error Covariance Estimates in an Ensemble Kalman Filter. *Mon. Wea. Rev.*, **129**, 2776–2790.

Hamill, Thomas M., Bates, Gary T., Whitaker, Jeffrey S., Murray, Donald R., Fiorino, Michael, Galarneau, Thomas J., Zhu, Yuejian, and Lapenta, William. 2013. NOAA's Second-Generation Global Medium-Range Ensemble Reforecast Dataset. *Bull. Amer. Meteor. Soc.*, **94**(10), 1553–1565.

Hannan, E. J., and Quinn, B. G. 1979. The Determination of the Order of an Autoregression. *Journal of the Royal Statistical Society. Series B (Methodological)*, **41**(2), 190–195.

Hansen, James, and Lebedeff, Sergej. 1987. Global Trends of Measured Surface Air Temperature. *J. Geophys. Research: Atmospheres*, **92**(D11), 13345–13372.

Hansen, M. H., and Yu, B. 2001. Model Selection and the Principle of Minimum Description Length. *J. Amer. Stat. Assoc.*, **96**, 746–774.

Harrell, F. 2001. *Regression Modeling Strategies: With Applications to Linear Models, Logistic Regression, and Survival Analysis*. Springer.

Harvey, A. C. 1989. *Forecasting, Structural Time Series Models and the Kalman Filter*. Cambridge University Press.

Harville, D. A. 1997. *Matrix Algebra from a Statistician's Perspective*. Springer.

Hasselmann, K. 1976. Stochastic Climate Models I. Theory. *Tellus*, **28**, 473–485.

Hastie, T., Tibshirani, R., and Friedman, J. H. 2009. *Elements of Statistical Learning*. 2nd ed. Springer.

Hays, J. D., Imbrie, John, and Shackleton, N. J. 1976. Variations in the Earth's Orbit: Pacemaker of the Ice Ages. *Science*, **194**(4270), 1121–1132.

Henderson, Douglas A., and Denison, Daniel R. 1989. Stepwise Regression in Social and Psychological Research. *Psychological Reports*, **64**(1), 251–257.

Hodges, J. L., and Lehmann, E. L. 1956. The Efficiency of Some Nonparametric Competitors of the t-Test. *The Annals of Mathematical Statistics*, **27**(2), 324–335.

Hoeting, J. A., Madigan, D., Raftery, A. E., and Volinsky, C. T. 1999. Bayesian Model Averaging: A Tutorial. *Statist. Sci.*, **14**, 382–417.

Horn, R. A., and Johnson, C. R. 1985. *Matrix Analysis*. New York: Cambridge University Press. 561 pp.

Hosking, J. R. M., Wallis, J. R., and Wood, E. F. 1985. Estimation of the Generalized Extreme-Value Distribution by the Method of Probability-Weighted Moments. *Technometrics*, **27**(3), 251–261.

Hotelling, H. 1933. Analysis of a Complex of Statistical Variables into Principal Components. *J. Educational Psychology*, **24**, 417–441.

Hotelling, Harold. 1936. Relations between Two Sets of Variates. *Biometrika*, **28**(3/4), 321–377.

Houtekamer, P. L., and Mitchell, Herschel L. 1998. Data Assimilation Using an Ensemble Kalman Filter Technique. *Mon. Wea. Rev.*, **126**, 796–811.

Houtekamer, P. L., and Mitchell, Herschel L. 2001. A Sequential Ensemble Kalman Filter for Atmospheric Data Assimilation. *Mon. Wea. Rev.*, **129**, 123–137.

Houtekamer, P. L., and Mitchell, Herschel L. 2005. Ensemble Kalman Filtering. *Quarterly J. Roy. Met. Soc.*, **131**(613), 3269–3289.

Houtekamer, P. L., and Zhang, Fuqing. 2016. Review of the Ensemble Kalman Filter for Atmospheric Data Assimilation. *Mon. Wea. Rev.*, **144**(12), 4489–4532.

Huang, Boyin, Banzon, Viva F., Freeman, Eric, Lawrimore, Jay, Liu, Wei, Peterson, Thomas C., Smith, Thomas M., Thorne, Peter W., Woodruff, Scott D., and Zhang, Huai-Min. 2015. Extended Reconstructed Sea Surface Temperature Version 4 (ERSST.v4). Part I: Upgrades and Intercomparisons. *J. Climate*, **28**(3), 911–930.

Huang, Boyin, Thorne, Peter W., Banzon, Viva F., Boyer, Tim, Chepurin, Gennady, Lawrimore, Jay H., Menne, Matthew J., Smith, Thomas M., Vose, Russell S., and Zhang, Huai-Min. 2017. Extended Reconstructed Sea Surface Temperature, Version 5 (ERSSTv5): Upgrades, Validations, and Intercomparisons. *J. Climate*, **30**(20), 8179–8205.

Hurvich, C. M., and Tsai, C.-L. 1989. Regression and Time Series Model Selection in Small Samples. *Biometrika*, **76**(2), 297–307.

Hyndman, Rob J., and Fan, Yanan. 1996. Sample Quantiles in Statistical Packages. *The American Statistician*, **50**(4), 361–365.

International Committee of Medical Journal Editors. 2019. *Recommendations for the Conduct, Reporting, Editing, and Publication of Scholarly Work in Medical Journals*.

Ioannidis, J. P. A. 2005. Why Most Published Research Findings Are False. *PLOSMed*, **2**, 696–701.

Jazwinski, A. H. 1970. *Stochastic Processes and Filtering Theory*. New York: Academic Press. 376 pp.

Jenkins, G. M, and Watts, D. G. 1968. *Spectral Analysis and Its Applications*. Holden-Day.

Johnson, R. A., and Bhattacharyya, G. K. 1992. *Statistics: Principles and Methods*. 2nd ed. John Wiley and Sons.

Johnson, R. A., and Wichern, D. W. 2002. *Applied Multivariate Statistical Analysis*. 5th ed. Prentice-Hall.

Jolliffe, I. T. 2002. *Principal Component Analysis*. Springer.

Jolliffe, Ian T. 2003. A Cautionary Note on Artificial Examples of EOFs. *J. Climate*, **16**(7), 1084–1086.

Jones, P. D., Osborn, T. J., and Briffa, K. R. 1997. Estimating Sampling Errors in Large-Scale Temperature Averages. *J. Climate*, **10**(10), 2548–2568.

Kalman, R.E. 1960. A New Approach to Linear Filtering and Prediction Problems. *Trans. ASME, Ser. D, J. Basic Eng.*, **82**, 35–45.

Katz, Richard W. 1977. Precipitation as a Chain-dependent Process. *J. Appl. Meteor.*, **16**, 671–676.

Kendall, M. G. 1946. *Contributions to the Study of Oscillatory Time-Series*.

Kirchmeier-Young, Megan C., Zwiers, Francis W., and Gillett, Nathan P. 2016. Attribution of Extreme Events in Arctic Sea Ice Extent. *J. Climate*, **30**(2), 553–571.

Kolmogorov, A. 1992. 15. On The Empirical Determination of a Distribution Law. Pages 139–146 of: Shiryayev, A. N. (ed.), *Selected Works of A. N. Kolmogorov: Volume II Probability Theory and Mathematical Statistics*. Springer Netherlands.

Kotz, S., and Nadarajah, S. 2000. *Extreme Value Distributions: Theory and Applications*. Imperial College Press.

Krüger, T. 2013. *Discovering the Ice Ages: International Reception and Consequences for a Historical Understanding of Climate*. Brill.

LaJoie, E., and DelSole, T. 2016. Changes in Internal Variability Due to Anthropogenic Forcing: A New Field Significance Test. *J. Climate*, **29**(15), 5547–5560.

Lauritzen, S. L. 1981. Time Series Analysis in 1880: A Discussion of Contributions Made by T. N. Thiele. *Int. Statist. Rev.*, **49**, 319–331.

Lawley, D. N. 1956. Tests of Significance for the Latent Roots of Covariance and Correlation Matrices. *Biometrika*, **43**, 128–136.

Lehman, E. L., and Romano, J. P. 2005. *Testing Statistical Hypotheses*. Springer.

Leith, C. E. 1978. Predictability of climate. *Nature*, **276**(5686), 352–355.

Leung, Lai-Yung, and North, Gerald R. 1990. Information Theory and Climate Prediction. *J. Climate*, **3**, 5–14.

Livezey, Robert E., and Chen, W.Y. 1983. Statistical Field Significance and Its Determination by Monte Carlo Techniques. *Mon Wea. Rev.*, **111**, 46–59.

Ljung, G. M., and Box, G. E. P. 1978. On a measure of lack of fit in time series models. *Biometrika*, **65**, 297–303.

Lorenc, Andrew C., Bowler, Neill E., Clayton, Adam M., Pring, Stephen R., and Fairbairn, David. 2014. Comparison of Hybrid-4DEnVar and Hybrid-4DVar Data Assimilation Methods for Global NWP. *Mon. Wea. Rev.*, **143**(1), 212–229.

Lorenz, E. N. 1963. Deterministic Nonperiodic Flow. *J. Atmos. Sci.*, **20**, 130–141.

Lorenz, E. N. 1969. Three Approaches to Atmospheric Predictability. *Bull. Am. Meteor. Soc.*, 345–351.

Lorenz, E. N., and Emmanuel, K. A. 1998. Optimal Sites for Supplementary Weather Observations: Simulation with a Small Model. *J. Atmos. Sci.*, **55**, 399–414.

Madden, R. A., and Julian, P. R. 1994. Observations of the 40-50-day Tropical Oscillation-A Review. *Mon. Wea. Rev.*, **122**, 814–837.

Majda, A., and Tong, X. 2016. Robustness and Accuracy of Finite Ensemble Kalman Filters in Large Dimensions. *Comm. Pure Appl. Math*, **submitted**.

Makkonen, Lasse. 2008. Bringing Closure to the Plotting Position Controversy. *Communications in Statistics: Theory and Methods*, **37**(3), 460–467.

Mallows, C. L. 1973. Some Comments on C_P. Technometrics, **15**(4), 661–675.

Mann, H. B., and Whitney, D. R. 1947. On a Test of Whether One of Two Random Variables Is Stochastically Larger than the Other. *Ann. Math. Statist.*, 50–60.

Mardia, K. V., Kent, J. T., and Bibby, J. M. 1979. *Multivariate Analysis*. Academic Press.

Maybeck, P. S. 1982. *Stochastic Models, Estimation, and Control*. Vol. 1. Academic Press.

McPherson, Ronald D. 1986. Operational Objective Analysis Techniques and Potential Applications for Mesoscale Meteorology. Pages 151–172 of: Ray, Peter S. (ed.), *Mesoscale Meteorology and Forecasting*. American Meteorological Society.

McQuarrie, A. D. R., and Tsai, C.-L. 1998. *Regression and Time Series Model Selection*. World Scientific Publishing.

Mendenhall, W., Scheaffer, R. L., and Wackerly, D. D. 1986. *Mathematical Statistics with Applications*. 3rd ed. PWS Publishers.

Miller, A. 2002. *Subset Selection in Regression*. 2nd ed. Chapman and Hall.

Mood, A. M. 1954. On the Asymptotic Efficiency of Certain Nonparametric Two-Sample Tests. *Ann. Math. Statist.*, **25**(3), 514–522.

Morice, Colin P., Kennedy, John J., Rayner, Nick A., and Jones, Phil D. 2012. Quantifying Uncertainties in Global and Regional Temperature Change Using an Ensemble of Observational Estimates: The HadCRUT4 data set. *J Geophys. Res.-Atmospheres*, **117**.

Muirhead, R. J. 2009. *Aspects of Multivariate Statistical Theory*. John Wiley & Sons.

Mundry, Roger, and Nunn, Charles L. 2009. Stepwise Model Fitting and Statistical Inference: Turning Noise into Signal Pollution. *The American Naturalist*, **173**(1), 119–123.

Noble, B., and Daniel, J. W. 1988. *Applied Linear Algebra*. 3rd ed. Prentice-Hall.

North, Gerald R., Bell, Thomas L., Cahalan, Robert F., and Moeng, Fanthune J. 1982. Sampling Errors in the Estimation of Empirical Orthogonal Functions. *Mon. Wea. Rev.*, **1982**, 699–706.

Olson, Chester L. 1974. Comparative Robustness of Six Tests in Multivariate Analysis of Variance. *J. Am. Stat. Assoc.*, **69**(348), 894–908.

Ott, Edward, Hunt, Brian R., Szunyogh, Istvan, Zimin, Aleksey V., Kostelich, Eric J., Corazza, Matteo, Kalnay, Eugenia, Patil, D. J., and Yorke, James A. 2004. A Local Ensemble Kalman Filter for Atmospheric Data Assimilation. *Tellus*, **A56**, 415–428.

Penland, C., and Sardeshmukh, P. D. 1995. The Optimal-growth of Tropical Sea-surface Temperature Anomalies. *J. Climate*, **8**, 1999–2024.

Press, W. H., Teukolsky, S. A., Vetterling, W. T., and Fannerty, B. P. 2007. *Numerical Recipes*. 3rd ed. Cambridge University Press.

Priestley, M. B. 1981. *Spectral Analysis and Time Series*. Academic Press.

Rao, C. R. 1973. *Linear Statistical Inference and Its Applications*. 2nd ed. John Wiley & Sons.

Raymo, M. E., Lisiecki, L. E., and Nisancioglu, Kerim H. 2006. Plio-Pleistocene Ice Volume, Antarctic Climate, and the Global $\delta^{18}O$ Record. *Science*, **313**(5786), 492–495.

Reeves, J., Chen, J., Wang, X. L., Lund, R., and Lu, Q. Q. 2007. A Review and Comparison of Changepoint Detection Techniques for Climate Data. *J. Appl. Meteor. Climatol.*, **46**, 900–915.

Rencher, A. C., and Pun, F. C. 1980. Inflation of R^2 in Best Subset Regression. *Technometrics*, **22**, 49–53.

Rencher, A. C., and Schaalje, G. B. 2008. *Linear Models in Statistics*. 2nd ed. Wiley-Interscience.

Renwick, James A., and Wallace, John M. 1995. Predictable Anomaly Patterns and the Forecast Skill of Northern Hemisphere Wintertime 500-mb height fields. *Mon. Wea. Rev.*, **123**, 2114–2131.

Ribes, A., Azais, J.-M., and Planton, S. 2009. Adaptation of the Optimal Fingerprint Method for Climate Change Detection Using a Well-conditioned Covariance Matrix Estimate. *Climate Dynamics*, **33**, 707–722.

Rosset, Saharon, and Tibshirani, Ryan J. 2020. From Fixed-X to Random-X Regression: Bias-Variance Decompositions, Covariance Penalties, and Prediction Error Estimation. *J. Am. Stat. Assoc.*, **115**(529), 138–151.

Saha, S., Moorthi, S., Wu, X., Wang, J., Nadiga, S., Tripp, P., Behringer, D., Hou, Y.-T., Chuang, Hui-ya, Iredell, M., ., Ek, M., Meng, J., Yang, R., Mendez, M. P., van den

Dool, H., Zhang, Q., Wang, W., Chen, M., and Becker, E. 2014. The NCEP Climate Forecast System Version 2. *J. Climate*, **27**, 2185–2208.

Sakov, P., and Oke, P. R. 2008. Implications of the Form of the Ensemble Transformation in the Ensemble Square Root Filters. *Mon. Wea. Rev.*, **136**, 1042–1053.

Scheffé, Henry. 1970. Practical Solutions of the Behrens-Fisher Problem. *J. Am. Stat. Assoc.*, **65**(332), 1501–1508.

Schneider, Tapio, and Griffies, Stephen. 1999. A Conceptual Framework for Predictability Studies. *J. Climate*, **12**, 3133–3155.

Schneider, Tapio, and Held, Isaac M. 2001. Discriminants of Twentieth-century Changes in Earth Surface Temperatures. *J. Climate*, **14**, 249–254.

Shukla, J. 1981. Dynamical Predictability of Monthly Means. *Mon. Wea. Rev.*, **38**, 2547–2572.

Shumway, Robert H., and Stoffer, David S. 2006. *Time Series Analysis and Its Applications: with R Examples*. 2nd ed. Springer.

Siegert, Stefan, Bellprat, Omar, Ménégoz, Martin, Stephenson, David B., and Doblas-Reyes, Francisco J. 2016. Detecting Improvements in Forecast Correlation Skill: Statistical Testing and Power Analysis. *Mon. Wea. Rev.*, **145**(2), 437–450.

Simon, D. 2006. *Optimal State Estimation. Kalman, H_∞, and Nonlinear Approaches*. Wiley-Interscience.

Smith, R. L. 1985. Maximum Likelihood Estimation in a Class of Nonregular Cases. *Biometrika*, **72**(1), 67–90.

Spiegel, M. R., Schiller, J., and Srinivasan, R. A. 2000. *Schaum's Outline of Theory and Problems in Probability and Statistics*. 2nd ed. McGraw-Hill.

Stigler, S. M. 1986. *The History of Statistics*. Harvard University Press.

Strang, G. 1988. *Linear Algebra and Its Applications*. Harcourt Brace Jovanovich.

Strang, G. 2016. *Introduction to Linear Algebra*. Wellesley-Cambridge Press.

Taylor, Geoffrey I. 1962. Gilbert, Thomas Walker. 1868–1958. *Biographical Memoirs of Fellows of the Royal Society*, **8**, 167–174.

Taylor, K. E., Stouffer, R. J., and Meehl, G. A. 2012. An Overview of CMIP5 and the Experimental Design. *Bull. Am. Meteor. Soc.*, **93**, 485–498.

Thompson, Bruce. 1995. Stepwise Regression and Stepwise Discriminant Analysis Need Not Apply Here: A Guidelines Editorial. *Educational and Psychological Measurement*, **55**(4), 525–534.

Tian, S., Hurvich, C. M., and Simonoff, J. S. 2020. Selection of Regression Models under Linear Restrictions for Fixed and Random Designs. *ArXiv e-prints*, 28.

Tippett, Michael K., Anderson, J. L., Bishop, Craig H., Hamill, Thomas M., and Whitaker, J. S. 2003. Ensemble Square-root Filters. *Mon. Wea. Rev.*, **131**, 1485–1490.

Torrence, C., and Webster, P. J. 1998. The Annual Cycle of Persistence in the El Niño/Southern Oscillation. *Quart. J. Roy. Meteor. Soc.*, **124**, 1985–2004.

Van Huffel, S., and Vandewalle, J. 1991. *The Total Least Squares Problem*. Society for Industrial and Applied Mathematics.

Ventura, Valérie, Paciorek, Christopher J., and Risbey, James S. 2004. Controlling the Proportion of Falsely Rejected Hypotheses When Conducting Multiple Tests with Climatological Data. *J. Climate*, **17**(22), 4343–4356.

Venzke, S., Allen, M. R., Sutton, R. T., and Rowell, D. P. 1999. The Atmospheric Response over the North Atlantic to Decadal Changes in Sea Surface Temperature. *J. Climate*, **12**, 2562–2584.

Vinod, H. D. 1976. Canonical Ridge and Econometrics of Joint Production. *Journal of Econometrics*, **4**(2), 147–166.

von Storch, Hans, and Zwiers, Francis W. 1999. *Statistical Analysis in Climate Research*. Cambridge University Press.

Walker, G. T., and Bliss, E. W. 1932. World Weather V. *Mem. Roy. Meteor. Soc.*, 53–84.

Walker, Gilbert. 1928. World Weather. *Quarterly Journal of the Royal Meteorological Society*, **54**(226), 79–87.

Wang, Xuguang, Parrish, David, Kleist, Daryl, and Whitaker, Jeffrey. 2013. GSI 3DVar-Based Ensemble–Variational Hybrid Data Assimilation for NCEP Global Forecast System: Single-Resolution Experiments. *Mon. Wea. Rev.*, **141**(11), 4098–4117.

Weare, Bryan C., and Nasstrom, John S. 1982. Examples of Extended Empirical Orthogonal Function Analyses. *Mon. Wea. Rev.*, **110**(6), 481–485.

Wheeler, M. C., and Hendon, H. 2004. An All-season Real-time Multivariate MJO Index: Development of an Index for Monitoring and Prediction. *Mon. Wea. Rev.*, **132**, 1917–1932.

Whitaker, Jeffery, and Hamill, Thomas M. 2002. Ensemble Data Assimilation without Perturbed Observations. *Mon. Wea. Rev.*, **130**, 1913–1924.

Whittingham, M. J., Stephens, P. A., Bradbury, R. B., and Freckleton, R. P. 2006. Why Do We Still Use Stepwise Modelling in Ecology and Behaviour? *J. Anim. Ecol.*, **75**(5), 1182–1189.

Wiener, N. 1950. *Extrapolation, Interpolation, and Smoothing of Stationary Time Series: With Engineering Applications*. MIT Press.

Wilcoxon, Frank. 1945. Individual Comparisons by Ranking Methods. *Biometrics Bulletin*, **1**(6), 80–83.

Wilks, D. S. 2006. On "Field Significance" and the False Discovery Rate. *J. Appl. Meteor. Climatol.*, **45**(9), 1181–1189.

Wilks, D. S. 2011. *Statistical Methods in the Atmospheric Sciences: An Introduction*. 3rd ed. Academic Press.

Wilks, D. S. 2016. "The Stippling Shows Statistically Significant Grid Points": How Research Results are Routinely Overstated and Overinterpreted, and What to Do about it. *Bulletin of the American Meteorological Society*, **97**(12), 2263–2273.

Wilks, S. S. 1932. Certain Generalizations in the Analysis of Variance. *Biometrika*, **24**(3–4), 471–494.

Wills, Robert C., Schneider, Tapio, Wallace, John M., Battisti, David S., and Hartmann, Dennis L. 2018. Disentangling Global Warming, Multidecadal Variability, and El Niño in Pacific Temperatures. *Geophys. Res. Lett.*, **45**(5), 2487–2496.

Witten, Daniela M., Tibshirani, Robert, and Hastie, Trevor. 2009. A Penalized Matrix Decomposition, with Applications to Sparse Principal Components and Canonical Correlation Analysis. *Biostatistics*, **10**(3), 515–534.

Wyrtki, Klaus. 1975. El Niño: The Dynamic Response of the Equatorial Pacific Ocean to Atmospheric Forcing. *J. Phys. Oceanogr.*, **5**(4), 572–584.

Yang, X., and DelSole, T. 2009. Using the ensemble Kalman Filter to Estimate Multiplicative Model Parameters. *Tellus*, **61**, 601–609.

Yule, G. Udny. 1927. On a Method of Investigating Periodicities in Disturbed Series, with Special Reference to Wolfer's Sunspot Numbers. *Phil. Trans. R. Soc. Lond. A, Containing Papers of a Mathematical or Physical Character*, **226**, 267–298.

Zhang, Xuebin, Alexander, Lisa, Hegerl, Gabriele C., Jones, Philip, Tank, Albert Klein, Peterson, Thomas C., Trewin, Blair, and Zwiers, Francis W. 2011. Indices for Monitoring Changes in Extremes Based on Daily Temperature and Precipitation Data. *WIREs Clim. Chan.*, **2**(6), 851–870.

Zwiers, W. F. 1996. Interannual Variability and Predictability in an Ensemble of AMIP Climate Simulations Conducted with the CCC GCM2. *Clim. Dyn.*, **12**(12), 825–847.

Index

Printed in the United States
by Baker & Taylor Publisher Services